AF361802

Nanotechnology in Animal Science

Nanotechnology in Animal Science

Editors

Antonio Gonzalez-Bulnes

Nesrein M. Hashem

MDPI • Basel • Beijing • Wuhan • Barcelona • Belgrade • Manchester • Tokyo • Cluj • Tianjin

Editors
Antonio Gonzalez-Bulnes
CEU Universities
Spain

Nesrein M. Hashem
Alexandria University
Egypt

Editorial Office
MDPI
St. Alban-Anlage 66
4052 Basel, Switzerland

This is a reprint of articles from the Topical Collection published online in the open access journal *Animals* (ISSN 2076-2615) (available at: https://www.mdpi.com/journal/animals/topical_collections/Nanotechnology_in_Animal_Science).

For citation purposes, cite each article independently as indicated on the article page online and as indicated below:

LastName, A.A.; LastName, B.B.; LastName, C.C. Article Title. *Journal Name* **Year**, *Volume Number*, Page Range.

ISBN 978-3-0365-5945-2 (Hbk)
ISBN 978-3-0365-5946-9 (PDF)

Contents

About the Editors vii

Preface to "Nanotechnology in Animal Science" ix

Nesrein M. Hashem and Antonio Gonzalez-Bulnes
Nanotechnology and Reproductive Management of Farm Animals: Challenges and Advances
Reprinted from: *Animals* **2021**, *11*, 1932, doi:10.3390/ani11071932 1

Nesrein M. Hashem, Hossam R. EL-Sherbiny, Mohamed Fathi and Elshymaa A. Abdelnaby
Nanodelivery System for Ovsynch Protocol Improves Ovarian Response, Ovarian Blood Flow
Doppler Velocities, and Hormonal Profile of Goats
Reprinted from: *Animals* **2022**, *12*, 1442, doi:10.3390/ani12111442 21

Nagwa I. El-Desoky, Nesrein M. Hashem, Ahmed G. Elkomy and Zahraa R. Abo-Elezz
Improving Rabbit Doe Metabolism and Whole Reproductive Cycle Outcomes via Fatty
Acid-Rich *Moringa oleifera* Leaf Extract Supplementation in Free and Nano-Encapsulated Forms
Reprinted from: *Animals* **2022**, *12*, 764, doi:10.3390/ani12060764 35

**Katrin Quester, Sarahí Rodríguez-González, Laura González-Dávalos, Carlos Lozano-Flores,
Adriana González-Gallardo, Santino J. Zapiain-Merino, Armando Shimada, et al.**
Chitosan Nanoparticles Containing Lipoic Acid with Antioxidant Properties as a Potential
Nutritional Supplement
Reprinted from: *Animals* **2022**, *12*, 417, doi:10.3390/ani12040417 53

**Rasha M. M. Abou Elez, Ibrahim Elsohaby, Nashwa El-Gazzar, Hala M. N. Tolba,
Eman N. Abdelfatah, Samah S. Abdellatif, Ahmed Atef Mesalam, et al.**
Antimicrobial Resistance of *Salmonella enteritidis* and *Salmonella typhimurium* Isolated from
Laying Hens, Table Eggs, and Humans with Respect to Antimicrobial Activity of
Biosynthesized Silver Nanoparticles
Reprinted from: *Animals* **2021**, *11*, 3554, doi:10.3390/ani11071932 67

**Saengrawee Thammawithan, Oranee Srichaiyapol, Pawinee Siritongsuk, Sakda Daduang,
Sompong Klaynongsruang, Nuvee Prapasarakul and Rina Patramanon**
Anisotropic Silver Nanoparticles Gel Exhibits Antibacterial Action and Reduced Scar Formation
on Wounds Contaminated with Methicillin-Resistant *Staphylococcus pseudintermedius* (MRSP) in
a Mice Model
Reprinted from: *Animals* **2021**, *11*, 3412, doi:10.3390/ani11123412 85

**Ahmed Dawod, Noha Osman, Hanim S. Heikal, Korany A. Ali, Omaima M. Kandil,
Awad A. Shehata, Hafez M. Hafez, et al.**
Impact of Nano-Bromocriptine on Egg Production Performance and Prolactin Expression in
Layers
Reprinted from: *Animals* **2021**, *11*, 2842, doi:10.3390/ani11102842 99

**Pantu-Kumar Roy, Ahmad-Yar Qamar, Bereket-Molla Tanga, Seonggyu Bang,
Gyeonghwan Seong, Xun Fang, Ghangyong Kim, et al.**
Modified *Spirulina maxima* Pectin Nanoparticles Improve the Developmental Competence of In
Vitro Matured Porcine Oocytes
Reprinted from: *Animals* **2021**, *11*, 2483, doi:10.3390/ani11092483 111

**Amr El-Nile, Mahmoud Elazab, Hani El-Zaiat, Kheir El-Din El-Azrak, Alaa Elkomy,
Sobhy Sallam and Yosra Soltan**
In Vitro and *In Vivo* Assessment of Dietary Supplementation of Both Natural or Nano-Zeolite in
Goat Diets: Effects on Ruminal Fermentation and Nutrients Digestibility
Reprinted from: *Animals* **2021**, *11*, 2215, doi:10.3390/ani11082215 . . . . **125**

**Catrenar De Silva, Norazah Mohammad Nawawi, Murni Marlina Abd Karim,
Shafinaz Abd Gani, Mas Jaffri Masarudin, Baskaran Gunasekaran
and Siti Aqlima Ahmad**
The Mechanistic Action of Biosynthesised Silver Nanoparticles and Its Application in
Aquaculture and Livestock Industries
Reprinted from: *Animals* **2021**, *11*, 2097, doi:10.3390/ani11072097 . . . . **141**

**Hidayat Mohd Yusof, Nor'Aini Abdul Rahman, Rosfarizan Mohamad,
Uswatun Hasanah Zaidan and Anjas Asmara Samsudin**
Antibacterial Potential of Biosynthesized Zinc Oxide Nanoparticles against Poultry-Associated
Foodborne Pathogens: An In Vitro Study
Reprinted from: *Animals* **2021**, *11*, 2093, doi:10.3390/ani11072093 . . . . **159**

**Sameh A. Abdelnour, Mahmoud Alagawany, Nesrein M. Hashem, Mayada R. Farag,
Etab S. Alghamdi, Faiz Ul Hassan, Rana M. Bila, et al.**
Nanominerals: Fabrication Methods, Benefits and Hazards, and Their Applications in
Ruminants with Special Reference to Selenium and Zinc Nanoparticles
Reprinted from: *Animals* **2021**, *11*, 1916, doi:10.3390/ani11071916 . . . . **181**

**Agata Lange, Agnieszka Grzenia, Mateusz Wierzbicki, Barbara Strojny-Cieslak,
Aleksandra Kalińska, Marcin Gołebiewski, Daniel Radzikowski, et al.**
Silver and Copper Nanoparticles Inhibit Biofilm Formation by Mastitis Pathogens
Reprinted from: *Animals* **2021**, *11*, 1884, doi:10.3390/ani11071884 . . . . **203**

**Ayodele Olaolu Oladejo, Yajuan Li, Xiaohu Wu, Bereket Habte Imam, Jie Yang, Xiaoyu Ma,
Zuoting Yan, et al.**
Modulation of Bovine Endometrial Cell Receptors and Signaling Pathways as a
Nanotherapeutic Exploration against Dairy Cow Postpartum Endometritis
Reprinted from: *Animals* **2021**, *11*, 1516, doi:10.3390/ani11061516 . . . . **215**

**Eman M. Hassanein, Nesrein M. Hashem, Kheir El-Din M. El-Azrak,
Antonio Gonzalez-Bulnes, Gamal A. Hassan and Mohamed H. Salem**
Efficiency of GnRH–Loaded Chitosan Nanoparticles for Inducing LH Secretion and Fertile
Ovulations in Protocols for Artificial Insemination in Rabbit Does
Reprinted from: *Animals* **2021**, *11*, 440, doi:10.3390/ani11020440 . . . . **237**

About the Editors

Nesrein M. Hashem

Prof. Dr. Nesrein M. Hashem is a staff member at the Animal and Fish Production Department, Faculty of Agriculture, Alexandria University. She started her early career as a demonstrator in 2000 and she has occupied the position of Professor of Animal Physiology at the Faculty since 2020. Professor Nesrein has professional experience in the field of research, teaching, and society service. She has published about 80 scientific articles in peer-reviewed journals (55 articles in Scopus) and international conferences and 3 book chapters on different aspects of animal biological sciences, animal production/reproduction, environment, toxicology, and nanotechnology (Google Scholar website: https://scholar.google.com/citations?user=n4cw5V0AAAAJ&hl=en). She has an H-index of 17 on Google Scholar and an H-index of 13 at Scopus. In addition, she has served as an editor/guest editor for many ISI journals (e.g., Animals, Frontiers in Veterinary Science). She has served as a reviewer for more than 30 ISI journals (https://publons.com/researcher/1178750/nesrein-m-hashem/). She has supervised nine M.Sc. and Ph.D. students to date, and all the thesis results were published in high-ranked ISI journals (Q1 journals). She has acted as a PI, Co-PI, and researcher in many research projects (6 projects) funded by STDF, ASRT, and other research support entities. She has attended about 40 national and international conferences and workshops. She also has international collaboration with peer-colleges from different universities/institutes worldwide (Spain, France, Colombia, India, Pakistan, and Iran) in the field of animal biology, veterinary, and biotechnology science. She awarded the State Encouragement Award for Women, Academy of Scientific Research & Technology (ASRT), the Ministry of Scientific Research, Egypt in 2019 and Alexandria University Award for Scientific Encouragement in 2017. Professor Nesrein has 22 years teaching experience and teaches many undergraduate and postgraduate courses for students in both ordinary and language education programs at the faculty in the field of biology, physiology, endocrinology, hematology, environment and adaptation, biotechnology, and animal reproduction. Moreover, she has many social and community service activities. She has also organized several local and international conferences, workshops, and training programs.

Preface to "Nanotechnology in Animal Science"

This topical collection on "Nanotechnology in Animal Science" was devoted to publish recent advances in the field of nanotechnology applications in animal science. The collection's scope has grown in line with the emergence and overspread of nanotechnology in many areas of life and science, including animal and biological science. In this collection, we aimed to publish unique academic work that has been conducted worldwide to show recent perspectives and prospectives regarding the application of nanotechnology in animal science, with an emphasis on domestic and food-producing animals. To date, the collection has published 15 articles and review articles on this topic. The articles published in this collection show the importance of nanotechnology applications to improve animals' productivity and welfare, and the possibility of producing different engineered nanoparticles to either replace some drugs/chemical and/or improve the existing ones' bioactivity. Moreover, the published studies refer to the hazards that nanoparticles can post to humans, animals, and the environment and possible interventions to minimize such hazards. We hope that by publishing this cutting-edge work, we can pave the way for more studies in the field to enrich our knowledge of this promising field.

Antonio Gonzalez-Bulnes and Nesrein M. Hashem

Editors

Review

Nanotechnology and Reproductive Management of Farm Animals: Challenges and Advances

Nesrein M. Hashem [1,*] **and Antonio Gonzalez-Bulnes** [2,*]

[1] Department of Animal and Fish Production, Faculty of Agriculture (El-Shatby), Alexandria University, Alexandria 21545, Egypt
[2] Departamento de Produccion y Sanidad Animal, Facultad de Veterinaria, Universidad CardenalHerrera-CEU, CEU Universities, C/Tirant lo Blanc, 7, 46115 Alfara del Patriarca, Valencia, Spain
* Correspondence: nesreen.hashem@alexu.edu.eg (N.M.H.); antonio.gonzalezbulnes@uchceu.es (A.G.-B.)

Simple Summary: The emergence of nanotechnology paves the way for innovating countless applications in the agricultural and livestock production sector. In the field of reproductive management of farm animals, nanotechnology offers unconventional and innovative solutions for the existing reproductive management challenges. The main concept of nanotechnology comes through their ability to modulate drug behavior and consequently their biological effects (e.g., male effect). In this review, the challenges of the current reproductive management in farm animals will be discussed in line with the possible solutions offered by applying nanotechnology.

Abstract: Reproductive efficiency of farm animals has central consequences on productivity and profitability of livestock farming systems. Optimal reproductive management is based on applying different strategies, including biological, hormonal, nutritional strategies, as well as reproductive disease control. These strategies should not only guarantee sufficient reproductive outcomes but should also comply with practical and ethical aspects. For example, the efficiency of the biological- and hormonal-based reproductive strategies is mainly related to several biological factors and physiological status of animals, and of nutritional strategies, additional factors, such as digestion and absorption, can contribute. In addition, the management of reproductive-related diseases is challenged by the concerns regarding the intensive use of antibiotics and the development of antimicrobial resistant strains. The emergence of nanotechnology applications in livestock farming systems may present innovative and new solutions for overcoming reproductive management challenges. Many drugs (hormones and antibiotics), biological molecules, and nutrients can acquire novel physicochemical properties using nanotechnology; the main ones are improved bioavailability, higher cellular uptake, controlled sustained release, and lower toxicity compared with ordinary forms. In this review, we illustrate advances in the most common reproductive management strategies by applying nanotechnology, considering the current challenges of each strategy.

Keywords: nano-delivery system; reproductive management; bio stimulation; nutrition; hormones; antibiotics; reproductive diseases; livestock

Citation: Hashem, N.M.; Gonzalez-Bulnes, A. Nanotechnology and Reproductive Management of Farm Animals: Challenges and Advances. *Animals* **2021**, *11*, 1932. https://doi.org/10.3390/ani11071932

Academic Editors: Luisa Bogliolo, Mirco Corazzin and Marcella Guarino

Received: 8 June 2021
Accepted: 24 June 2021
Published: 29 June 2021

Publisher's Note: MDPI stays neutral with regard to jurisdictional claims in published maps and institutional affiliations.

1. Introduction

In livestock farming systems, reproductive efficiency has central consequences on the productivity, profitability, and sustainability of farms. The reproductive performance of farm animals determines the efficiency of milk and/or meat production, either directly or through managing decisions, such as replacement and culling rates. Optimal reproductive management is based on applying precision strategies which also needs to consider costs, animals' welfare, environmental impacts, and human health. Most of the reproductive management practices are ready for their use in commercial livestock farms after selecting the strategy which meets goals of every farm [1,2]. Such strategies may include one or more bio stimulation tools (e.g., male effect), reproductive assisted techniques (mainly estrous synchronization and artificial insemination), nutritional management, and prevention/treatment of reproductive diseases [3–5].

Although these reproductive management strategies are widely and predominantly applied in different livestock production systems, their efficiency is challenged by several practical and ethical aspects. For example, hormonal-based reproductive therapies are the preferred method for reproductive management; however, their effectiveness is highly dependent on their pharmacokinetics and pharmacodynamics, which may be affected by biological factors [6,7]. The male effect is a sexual bio stimulation method that confers an opportunity to eliminate the intensive use of hormones in reproductive management and the production of hormone residues-free animal products; however, its outputs are challenged by the sexual activity of both males and females, male to female ratio, and age and the experience of the male [4,8–10]. Similarly, nutritional management practices for improving the reproductive performance of farm animals may be negatively impacted by the lack of nutrients bioavailability and insufficient delivery of required nutrients [11,12]. Lastly, the management of reproductive-related diseases is challenged by the concerns regarding the intensive use of antibiotics and the development of antimicrobial-resistant strains [13–15].

In view of these considerations, the emergence of novel technologies, such as nanotechnology, paves the way for countless applications in agricultural and livestock production. The most important and promising application of nanotechnologies in the livestock production sector is in the field of nano-drug delivery systems. Many drugs, biological molecules, and nutrients can acquire novel physicochemical properties by using nanotechnology, such as improved bioavailability, higher mobility and cellular uptake, controlled sustained release of the drug at the target site, lower toxicity compared with other compounds, improved enzymatic actions, and increased mucoadhesive properties [16]. In this review, we aim to illustrate possible advances in the most common reproductive management strategies by applying nanotechnology.

2. Biological Stimulation Management, Male Effect

2.1. Challenges of Male Effect Applications

Pheromones are volatile chemical substances secreted by specific organs and scent glands of different animals. In mammals, including livestock species, pheromones play crucial roles in the communication between animals, signaling many behavioral and physiological processes. The role of pheromones in regulating several reproductive events in farm animals is well documented and can be indicated through observing changes in animals' activity following direct (physical) and/or indirect (spatial) exposure to specific sexual pheromones [4,17]. The most known pheromones-based phenomenon that can prime and regulate many reproductive events in farm animals is known as the "Male Effect" [18]. Pheromones in the wool, wax, and urine originating from sexually active males have a significant influence on the onset of puberty, the resumption of sexual activity of seasonal anestrous females, and the reduction in postpartum anestrus in many farm animals, specifically sheep, goats, and swine [19,20]. Each of these effects is mediated by olfactory chemical signals, pheromones, released from the male and affecting the hypothalamic system of the stimulated female to generate pulses of gonadotropin-releasing hormone (GnRH) modulating sexual activity [21]. The male effect is well-recognized and applied in reproductive management [22–24]. In small ruminants, for example, many breeds exert different degrees of seasonality, which constrains the progression of the breeding program of commercial flocks, being restricted within a definite period of the year [25]. The resumption of sexual and ovulation activities of the seasonal anestrous females depends on the activation of the neuro-endocrine system by male pheromone signals and the subsequent modulation in LH pulse frequency and amplitude after sudden exposure to males. Usually, ovulation occurs within 2–3 days after male introduction to a large number of females [26]. From a practical point of view, the male effect not only has the advantage of advancing the breeding season by about 4–6 weeks or more, but it can also provide an acceptable degree of female synchrony at the time of breeding and thus the subsequent lambing/kidding time. In context, the male effect is successfully used to induce ovulation in prepubertal ewe lambs and lactational anestrous ewes. The male effect can be used with 35 to 40 days postpartum ewes to decrease the time between deliveries [19].

In pigs, the presence of boars reduces the postpartum period in lactating sows and the age of onset of puberty in gilts by about 30 days. These effects were early linked toolfactory cues because a pen previously occupied by a boar and presumably pervaded with a boar's odor is effective in inducing early puberty [17]. Apparently, priming pheromones remaining in the boar's pen after his removal are sufficient to induce early puberty [21]. The same was observed for sheep and goats, as pheromones in the wool, wax, and urine are sufficient to stimulate females to ovulate.

The male effect, being a sexual bio stimulation method, confers an opportunity to diminish, or even eliminate in the future, the intensive use of hormones in reproductive management and promote the production of hormone residues-free animal products. However, the outputs of this method, when applied at a farm scale, are challenged by several limiting factors. In brief, the main such factors are latitude, breed, sexual activity, hormonal status, age, and experience of the male [8–10], the depth of female anestrous status [4], male to female ratio, and the length of isolation period [27]. In fact, the application of such a method for induction of estrus needs the isolation of males from females for a certain period and reintroduction thereafter to stimulate females' sexual activity (estrus and ovulation) and/or to introduce new males that have not previously been kept in the herd [27]. In many cases, the male effect is applied during out of season, during which males also may undergo weak sexual activity. For this reason, many studies have referred to the importance of priming males with exogenous stimuli, such as hormonal treatments (mainly melatonin [28] and testosterone [8]) and photostimuli [29,30], to speed up the sexual activity of males during the natural anestrous season. Overall, these challenges can negatively impact the application of the male effect at a large/field scale as a biological reproductive management protocol [4].

2.2. Nanotechnology Approaches for Developing Male Effect Procedure

The male effect, as previously described, depends on the effect of olfactory signals generated by male pheromones on the hypothalamic–pituitary–gonadal axis of the females. Previous research has shown that the physical or visual contact between male and female is not necessary to affect the sexual behavior of the female. Early studies confirmed the capacity of priming pheromones collected from the wool, wax, and urine of rams to stimulate ewes to ovulate early in the breeding season [31,32]. Similarly, a jar containing the odor of the buck around the location of females can be used as an aid to detect estrus in goats due to the strong characteristic seasonal odor of bucks pheromones [21]. A recent study tested the effectiveness of a novel 3-molecule boar pheromone spray to improve the reproductive performance of sows (BOARBETTER) [33]. Meta-analysis of the results showed significant increases in total born piglets by 0.49 per litter and born alive piglets by 0.37 per litter, with the treatment being modulated by the age of the sow, sincethe treatment increased total born piglets by 0.88 ($p < 0.05$) per litter and pigs born alive by 0.73 ($p < 0.05$) pigs per litter when considering first to third parties. The overall conclusion of this study is that the management of animals' olfactory environment could be a cost effective, safe, and meaningful reproductive management tool for improving reproductive performance of sows [33].

Hence, the delivery of sexual pheromones to females seems the main mechanism mediating the male effect. Actually, the initial olfactory perception starts with the binding of the volatile molecules of the pheromones with olfactory binding proteins in olfactory-specific cells, such as the main olfactory epithelium [18]. Thereafter, pheromone signals can be transferred through the airways of nasal mucosa to olfactory centers in the brain to stimulate the hypothalamic–pituitary region to release reproductive hormones secretion. Furthermore, pheromones can be transferred through systematic circulations after passing tight junctions of olfactory epithelial cells [18,34,35]. In this context, it was suggested that volatile molecules of the pheromones, in spray form, could be more effective than the liquid form of the pheromones for delivering signals through the main olfactory epithelium receptors [33].

An aerosol nano-drug delivery system may facilitate the delivery of different drugs, including pheromones, through the olfactory pathways, nose-to-brain delivery and/or through the systematic circulation into target sites, passing the brain–blood and other biological barriers [14,34]. Despite few studies performed on applications of an aerosol nano-drug delivery system in the field of livestock production and veterinary medicine, Pamungkas et al. [35] proved the efficiency of chitosan-TPP nanoparticles for nasal human chorionic gonadotropins (hCG) delivery to induce ovulation in cows. Moreover, many vaccines have been developed using nanomaterials to be delivered through the nasal pathway [14].

In this regard, different nanomaterials can be used as a novel delivery system to implement sexual priming pheromones ready for female's treatment through nasal spray treatments. Specifically, the chemical compounds that act as pheromones are identified and can be isolated from different sites of the animal body including, wool wax, saliva, urine, and feces, of sexually active males. Fabrication of pheromones in an aerosol nano-formula may aid in overcoming some obstacles facing the male effect application at a farm scale. In this way, such therapies can be used at any season of the year regardless of the sexual activity of the males and the need for isolating or priming males' sexual activitybefore introduction into the group of females. Moreover, it may also reduce the number of males on the farm (Figure 1). Actually, the emergence of aerosol nano-drug delivery technology opens the way for developing numerous pheromones-based protocols for biological control of reproductive events, such as the male effect, estrous detection, postpartum anestrous, puberty, pregnancy diagnosis, sexual activity of males, and mother-offspring relationship [36,37]. These could be achieved through producing commercial bio stimulating pheromones and/or pheromones-based bioassay kits. As an example, urinary pheromones 4-methyl phenol (4-mp, p-Cresol), 9-octadecenoic acid (oleic acid), and luteinizing hormone (LH) are suggested to develop nanoparticle-based bioassay kits for detection of estrus or ovulation in buffaloes [37].

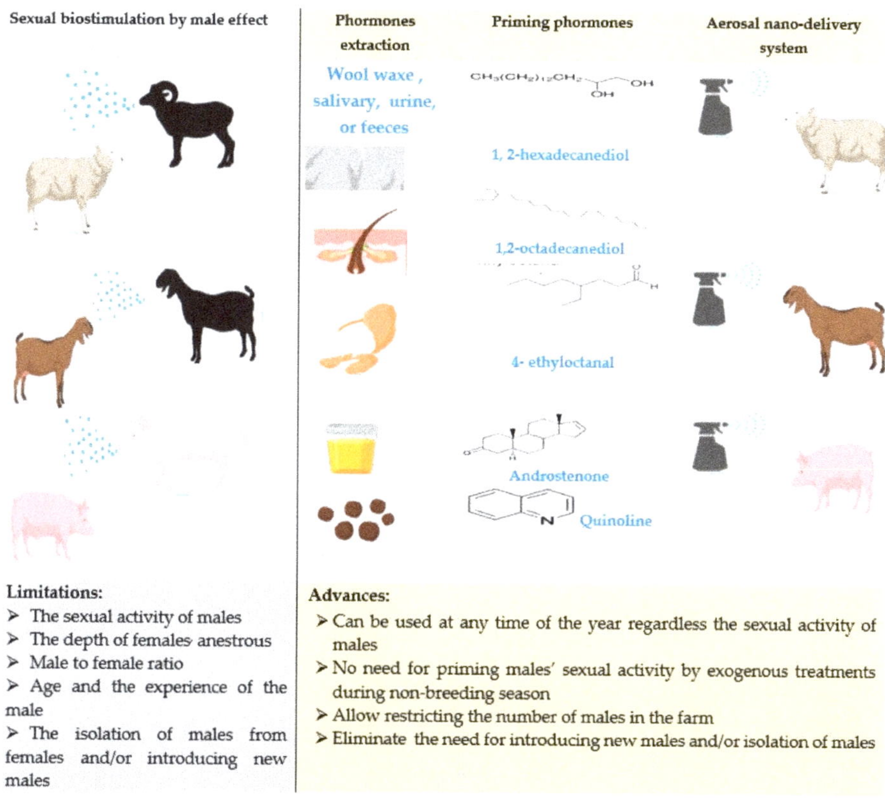

Figure 1. Possible applications of an aerosol nano-delivery system in pheromones-dependent sexual bio stimulation (male effect).

3. Hormonal Based-Treatments

3.1. Importance and Challenges of Hormonal Based-Treatments

Currently, despite the attempts to use biological strategies for managing reproduction in farm animals, the procedures, including exogenous hormonal therapies, cannot be excluded from the farming system. Hormones-based treatments are an efficient tool to improve fertility and farm business profitability. Several survey studies confirmed the importance of hormonal-based protocols for reproductive management; e.g., around 80% of practitioners from 714 inorganic-dairy farms in England confirmed the importance of hormones for controlled and efficient reproductive management [38]; similarly, 87% (103 of 153) of managers from large dairy herds (average herd size of 613 cows) in the USA confirmed the use of hormonal synchronization or timed artificial insemination in their reproductive management programs [3]. Hormones are needed, and therefore widely used, for implementing different assisted reproductive techniques (estrous synchronization/induction, timed artificial insemination, and superovulation), improving reproductive efficiency, and treating reproductive pathologies in both females and males [8,39–42].

The most important hormones used for managing fertility in farm animals are gonadotrophins, progesterone, estradiol, testosterone, melatonin, and prostaglandins. The effectiveness of these hormones mainly depends on their pharmacokinetics and pharmacodynamics [6]. Some hormones, such as gonadotropin-releasing hormone (GnRH) and prostaglandin $F_{2\alpha}$ ($PGF_{2\alpha}$), have low molecular weights and a short lifespan, which restricts the sustained delivery of the hormones to the target sites and therefore their biological activity. Other hormones, such as glycoprotein gonadotropins (follicle-stimulating hormones, FSH; luteinizing hormone, LH; human chorionic gonadotropin, hCG; and equine chorionic gonadotropin, eCG), may stimulate the immune system and formation of specific antibodies, leading to a refractoriness to repeated gonadotropic treatments in different farm animal species [43,44]. In addition, repeated treatments with these hormones are associated with low fertility and other reproductive outcomes. It has been shown that anti-eCG antibodies can hindereCGbioactivitiesvia two mechanisms. First, through preventing the interaction of eCG with its receptors; second, by a conformational change in eCG by anti-eCG antibodies, which can inhibit eCG bioactivities. It is worth noting that these modulations of eCG bioactivities by its antibodiesmainly impact the FSH bioactivity of eCG rather than the LH bioactivity, which is essential for the recruitment and development of ovarian follicles, and therefore fertility is affected after repeated treatments [45,46]. In this context, ewes subjected up to three times to eCG/FSH based-super ovulatory treatments showed a lower fertilization rate and lower number of total recovered and viable embryos at the second and third recoveries when compared to the first flushing [44]. In another study, goats developing high eCG antibody concentrations after repeated eCG treatments exhibited a lower kiddingrate (41.3%) than other females (66.7%). The decreased fertility of these goats was associated with a delay in estrous rate and the preovulatory LH-surge [47]. Similarly, in rabbit does, repeated treatments with recombinant human FSH, increased FSH antibody levels in these females at the moment of the third and fourth superovulation treatments [43].

In addition to the aforementioned challenges facing the use of hormones for managing reproduction/fertility in farm animals, animal health and welfare and environmental-related issues are also raising challenges. For example, a shortage ofeCG availability is expected due to aspects related to animal welfare, as this hormone is obtained by bleeding pregnant mares. The ongoing societal pressure against companies manufacturing the hormone may prevent further hormone production in the near future [48].

Finally, conventional hormonal delivery systems may also upset the environmental ecosystem balance due to the spread of hormonal residues and carrier materials into the environment. The most evident example is the use of progesterone-impregnated intravaginal devices, which were developed for controlling the estrous cycle in different farm animals [49]. These devices are mainly composed of silicon polymers loaded with progesterone, which need to be loaded with high progesteroneconcentrations to release enough hormones to the vaginal mucosa, which increases the risk of hormone emissions to the environment and the transmission of the hormone directly to breeders/workers or indirectly to consumers through animal products [50].

3.2. Nanotechnology Approaches for Developing Hormonal Based-Treatments

Nano-delivery of hormones has emerged as a new pharmacological approach. Several engineered nanoparticles have been proposed as novel platforms for the protection and controlled release of reproductive hormones, including gonadotropin and steroid hormones. There are, to date, several studies reporting the use of nano-hormone delivery systems in the field of livestock reproductive management. These studies showed an enhancement of the pharmacokinetics and pharmacodynamics of the hormonal treatments, specifically, those of low molecular weight and short lifespan, such as GnRH [7,51–54]. In addition, the opportunity to use biodegradable materials as a matrix for progesteronedelivery instead of silicon-based materials was shown [55–57] to sustain the release of some hormones, such as melatonin in in vitro production media [58–60], and to change the route of hormone administration by enhancing mucosal absorption even if these hormones have high molecular weight [35]. Overall, these studies show that the use of nano-hormone delivery systems provides many advantages to hormonal-based treatments, such as decreasing hormone dosage, changing the route of administration, increasing animal welfare, and decreasing the risk of exposition to different hormones by workers and technicians (Table 1).

Table 1. Characteristics and expected advances by the implementation of nanotechnologies in the use of reproductive hormones for farm animals' reproductive management.

Figure	Technique	Particle Characteristics	Expected Advances
GnRH-chitosan-TPP NPs [7]	Ionic-gelation	Size = 212 nm, PdI = 0.295, Zp = 8.0 mV, EE = 90%	➤ Optimizing route of administration
GnRH-chitosan-TPP NPs [53]		Size = 93.91 nm, PdI = 0.302, Zp = 11.6mV, EE = 91.2%	➤ Decreasing dosage
GnRH-chitosan-dextran sulfate NPs [51]	Ionic-gelation	EE = 40–50%	➤ Increasing bioavailability ➤ Increasing animal welfare
hCG-chitosan-TPP NPs [35]	Ionic-gelation	-	
P$_4$-chitosan-TPP-Tween 80 [61]	Spray-drying	Size = 1 and 7 µm, EE = 69–75%	
P$_4$-polymethyl-methacrylate-nanospheres [57]	Miniemulsion polymerization	size = 150–200 nm, EE > 69%	➤ Improving environmental and human health
P$_4$-polymethyl-methacrylate-nanocapsules [57]		size = 240–300 nm, EE > 90%	
P$_4$-polylactic acid NPs [55]	Solution blow spinning	Size = 289–441 nm	
Melatonin-loaded lipid-core Nps [58]	Interfacial deposition of polymer	size = 168 nm PdI = 0.08	➤ Enhancing quality and increasing production rates of bovine blastocysts produced by in vitro fertilization
Melatonin loaded-lipid (olive oil) NPs [60]	Hot homogenization-ultrasonication	Size = 119nm, PdI = 0.09, EE = 94%	➤ Sustained release during IVF

GnRH = gonadotropin releasing hormone, TPP = tripolyphosphate, NPs = nanoparticles, hCG = human chorionic gonadotropin, P$_4$ = progesterone, PdI = polydispersity, Zp = zeta potential, and EE = encapsulation efficiency, and IVF = in vitro fertilization.

The more interesting trend is to use nano-hormone delivery systems to change the biological behavior of the hormones. This hypothesis is specifically raised in response to the direction towards the banning of eCG production and the lack of other products with similar activity. Among the different gonadotropins used for ovarian stimulation, eCG has a unique feature as it displays pronounced FSH-like activity in addition to its LH-like activity when used in species other than the equine [62,63]. Thus, the shortage/disappearance of this hormone will negatively impact many assisted reproductive techniques, specifically those depending on the stimulation of follicular growth. In these terms, Santos-Jimenez et al. [64] showed the possibility of the use GnRH as an eCG alternative if GnRH release behavior is slowed and sustained for a longer time, avoiding the occurrence of early LH-surge. In this study, diluting GnRH in propylene-glycol was effective in achieving this purpose and has been used as an eCG alternative in a CIDR-based estrous synchronization protocol tailored for sheep. Accordingly, nano-hormone delivery systems may aid in changing the biological behavior of GnRH. In these terms, a preliminary study by Hashem and co-authors (unpublished data), aimed at GnRH-chitosan-TPP nanoparticles as a potential alternative to eCG for stimulating ovarian follicles growth before artificial insemination in rabbits has been carried out. The results of this study showed the ability of GnRH-chitosan-TPP nanoparticles to induce greater follicular growth and formation of ovulation points than eCG following mating when it was subcutaneously administered at 2 µg/doe (Figure 2).

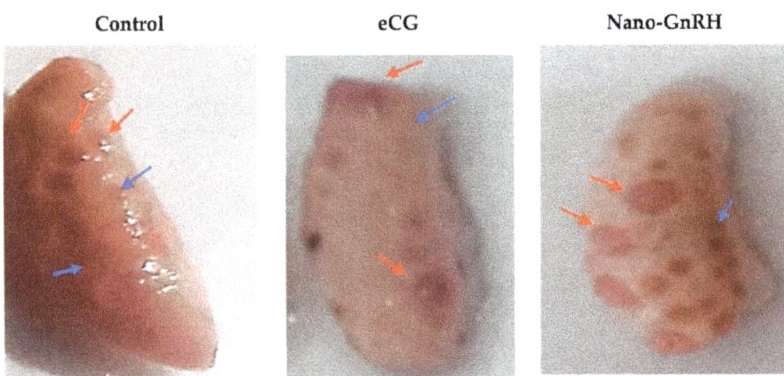

Figure 2. Two-days ovulation points (red arrows) and growing follicles (blue arrows) in ovaries of control rabbits or those treated with 25 IU equine chorionic gonadotropin (eCG) or 0.2µ gonadotropin-releasing hormone (GnRH)-chitosan- tripolyphosphate (TPP) nanoparticles. Higher numbers of growing follicles and ovulation points and larger diameters of ovulation points could be observed in GnRH-chitosan-TPP nanoparticles (unpublished data).

4. Nutritional Management

4.1. Importance and Challenges of Nutritional Management

Nutritional management is one of the most important strategies that can be used to control reproduction in different farm animals. Nutrition can affect the reproductive efficiency of farm animals at different reproductive windows, including the development of the reproductive organs during fetal life, puberty, and active breeding life [65]. Accurate synchronization between the periods of feeding and specific reproductive events enables achieving maximum benefits of nutrition. For example, feeding small ruminants with high energy-containing diets (concentrate and/or high energy feed additives, e.g., protected fats/glycerol) before breeding/mating season for a long term (about two months, [20,66]) or short term (around 7–10 days before mating, [12]) has been found to improve reproductive performance of both males (spermatogenesis, semen quality, and libido; [24]) and females (folliculogenesis, ovulation rate, and estrous rate; [12]), leading to improved fecundity. In dairy animals, specifically in high milk-producing animals, proper nutritional management of the transition period (late pregnancy to early lactation) aids in decreasing the risks of negative energy balance and its consequences on animal health and postpartum reproductive performance (resumption of ovarian cycles and estrus and embryo survival) [67,68].

It is worth mentioning that the relationship between nutrition and reproduction is far more than meeting energy and protein requirements or maintaining body weight and body condition score [4]. The role of specific nutrients in controlling specific reproductive events is also described in enormous studies and has been known as "functional effects" [68–70]. In this regard, amino acids have been shown as one of the most functional nutrients required for the competence of several reproductive events [71]. Amino acids are the main nutritional elements for the oviduct and uterine histotroph and are an essential component in the amniotic and allantoic fluids, supporting the important role of amino acids for normal embryonic and fetal development. In cows, the concentration of methionine, histidine, and lysine in the uterine lumen has been found to increase more than 10-fold during embryo elongation (days 14 to 18). An inadequate supply of these amino acids can hinder the rapid growth of the embryo between days 14 and 19 in the pregnant cow and, afterward, the subsequent growth of embryonic, fetal, and placental tissues [72]. Feeding rumen-protected methionine is able to improve the embryonic size and pregnancy maintenance in multiparous cows [73]. In context, some fatty acids, specifically long-chain polyunsaturated fatty acids originated from oily seeds (linseed, soybean, rapeseed, and sunflower seed), fish oil, and fat-based feed additives (tallow and protected fats) can affect metabolism, hormonal balance, gonads functions, quality of gametes, embryo survival, and establishment of pregnancy in different farm animals [12,74–76]. Cows supplemented with linseed oil (source of linolenic, C18:3) tended (p = 0.07) to show higher plasma progesterone levels, and a higher conception rate on the first artificial insemination when compared with soybean supplemented cows and control cows [76]. Furthermore, some minerals, such as selenium, zinc, calcium, and phosphorus, have specific effects on reproduction in both males [77] and females [75]. As an example, feeding organic seleniumimprovesneutrophilfunction around parturition, immune responsivenessin multiparous cows, uterine health, and increasessecond-service pregnancy per artificial insemination [75].

Imbalanced nutrition causes body weight loss, poor body condition, delayed puberty, long days open, ovarian dysfunction, hormonal imbalances, and, thus, causes infertility. The adjustment of nutritional requirements of farm animals for energy and different nutrients (protein, fat, vitamins, and micro- and/or macro-minerals) is crucially required to achieve optimal reproductive performance [65]. Actually, meeting animals' requirements during different reproductive windows is not always an easy matter and is greatly challenged by many limiting factors. In many cases, animals cannot obtain their optimal nutritional requirements and show symptoms of nutrient deficiency; however, the deficiency may not be due to the lack of offered feed and/or nutrients. In fact, nutrient utilization efficiency depends on many factors, such as feed intake, digestion and absorption efficiency, and the physiological and metabolic status of animals [65,66]. In most farm animals, the periods around mating, late pregnancy, and early lactation are associated with behavioral, metabolic, and physiological changes that may affect both feed intake and nutrients utilization [76]. For example, late pregnant animals, specifically ruminants, cannot obtain enough energy intake due to the limited capacity of rumen/stomach with pregnancy advancement, leading to a symptom of negative energy balance [78]. In such a state, providing high corn/grain-based diets results in several health problems and reproductive disorders, mainly aci0dosis. The symptoms of negative energy balance have been solved by increasing diet energy density by including fats, oils, glycerol, and propylene glycol [79]. However, improper use of such supplementations may drive negative effects on rumen microorganism action and rumen fermentation, impairing the overall performance of the animal [80]. For instance, glycerol is rapidly fermented in the rumen and negatively affects microbial fermentation [80]; furthermore, long-chain polyunsaturated fatty acids have toxic effects on rumen microorganisms too [81]. Therefore, the amounts of these supplementations should be restricted in the diet.Similarly, feeding diets high in crude protein or inadequate in fermentable carbohydratescan result in inefficient protein utilization and excess absorption of rumen ammonia [82]. Increased circulating ammonia levels can change the uterine environment and have toxic effects on an embryo, increasing the risks of embryonic mortality [83]. On the other hand, the biological role of some specific nutrients, such as unsaturated fatty acids and amino acids, could be impaired by the action of enzymatic hydrolysis of microorganisms. Rumen bacteria biohydrogenation toxic dietary unsaturated fatty acids to a saturated fatty acid to protect their cellular construction [81]. This mechanism results in a higher outflow of saturated fatty acids to the small intestine for digestion and absorption [84], decreasing the availability of these functional fatty acids to reproductive organs.

4.2. Nanotechnology Approaches for Improving Nutritional Management Outputs

Specific engineered nanomaterials can be implemented to provide various solutions to nutritional manipulation challenges raised during feeding producing animals (Table 2). In the field of animal nutrition, nano-encapsulation of active feed ingredients and processing of nano-minerals, in particular trace minerals of poor bioavailability, are rapidly growing technologies that are used to protect targeted feed ingredients from degradation (during processing, storage, and in the rumen) and to ensure effective and sustained delivery of valuable feed ingredients to target sites [85]. In addition, feed ingredients in nano-forms show novel physicochemical properties (mainly small size and high surface area) that improve its absorption, bioavailability, and utilization by animal tissues. However, up to date, studies showing the effect of nano-feed ingredients on animal reproduction are limited. Few studies have shown the positive effects of nano-minerals on the reproductive performance of farm animals [86–89].

Table 2. Characteristics and expected advances by the implementation of nanotechnologies in nutritional management for farm animals' reproductive management.

Formula	Technique	Particle Characteristics	Expected Advances
Zinc oxide NPs [89]	Commercial product	Size = 30.92 nm Zp = 32.16 mV	➢ Increase bioavailability and reduce the negative effects of toxic concentrations in in vitro reproductive assisted techniques
Selenium oxide NPs [89]	Commercial product	Size = 78.47 nm Zp = −20.36 mV	
Selenium oxide NPs [86]	Chemical reduction method using ascorbic acid and acacia gum	Size = 45.00 nm	➢ Increase absorption and bioavailability and reduce the negative effects of toxic concentrations during late pregnancy
Fish oil or soy oil -in-water NPs Soy oil-fish oil or rapeseed-fish oil-in-water NPs [81]	Nanoemulsion	-	➢ Improve post-ruminal supply of PUFA ➢ Decreased transformation rate of PUFA to SFA in the bio-hydrogenation environment
Solid lipid-lysine NPs [90]	Ultrasonic processor	Size = 200–500 nm Zp = < −30 mV EE = 40–90%	➢ Improve post-ruminal supply of lysine amino acid
Alginate-chitosan-glycerol NPs [85]	Ionic-gelation	Size = 3 mm EE = 78.1%	➢ Encapsulating glycerol to bypass rumen fermentation

Nps = nanoparticles, Zp = zeta potential, EE = encapsulation efficiency, PUFA = poly unsaturated fatty acids, and SFA = saturated fatty acids.

In a study by El-Sherbiny et al. [81], nano-emulsification of soybean and fish oil, aimed at post-rumen delivery of long-chain polyunsaturated fatty acids, significantly increased the proportions of long-chain polyunsaturated fatty acids (oleic, linoleic, and linolenic acids) in the rumen fluid without negative effects on rumen fermentation efficiency. Similarly, Albuquerque et al. [90] used solid lipid nanoparticles (composed of arachidonic or stearic acids) to protect lysine from enzymatic hydrolysis by ruminal microbiota and to increase rumen-bypass of lysine to the small intestine. In this study, the lipid–lysine nano-formula showed stability against microbial hydrolysis in the rumen for up to 24h. In addition, Gawad and Fellner [85] applied encapsulation techniques to protect glycerol from microbial fermentation using alginate or alginate–chitosan; results showed the efficiency of alginate–chitosan mixture to minimize encapsulated glycerol release into the rumen culture and to increase amounts of glycerol delivered to the lower digestive tract; however, the particle size was not at nano-scale [85]. Having in mind these findings, nanotechnology can improve nutrient utilization and the bioavailability of specific nutrients needed during specific periods of the reproductive cycle, by solving the imbalance between elevated needs of energy and active feed ingredients (amino acids, fatty acids, and minerals), and other biological processes, such as rumen fermentation.

In addition, the opportunity to use lowerquantitiesof nano-feed ingredients to get the same effects as ordinary forms may decrease feeding costs and the release of some materials, mainly minerals, to the environment decreasing pollution.

5. Management of Reproductive-Related Diseases

5.1. Importance and Challenges of Antibiotic Applications

Animal production, and specifically reproduction, is usually associated with the episode of reproductive diseases. Postpartum diseases, specifically endometritis caused by different bacterial species (mainly *Escherichia coli*, *Staphylococcus aureus*, *Bacillus cereus*, *Pseudomonas aeruginosa*, *Prevotella melaninogenica*, and *Arcanobacterium pyogenes*), are accompanied by impaired reproductive performance reflected in reduced conception rate and increased risk of reproductive culling [91]. The hazard of pregnancy, days elapsed from calving until pregnancy, reached 0.60, 0.31, and 0.24 in cows with metritis, clinical end metritis, and subclinical end metritis, respectively, when compared to healthy herd mates [92]. Pregnancy-associated diseases, such as toxoplasmosis (*Toxoplasma gondii*) and neosporosis *(Neospora caninum)*, are protozoan diseases that lead to significant economic loss in farm ruminants. Worldwide, toxoplasmosis infection is a zoonosis that mainly causes reproductive failure in small ruminants, whereas neosporosis infection is a common zoonosis that causes abortion in cattle [93]. In dairy farms, bovine mastitis, mainly caused by *Staphylococcus aureus* causes significant economic losses due to severe declines in milk yield (about 380 tons of milk are lost every year in the world), dumped milk, reproductive disorders, and expenses paid to the replacement of infected animals, increased costs of pharmacologic costs, and replacing tainted milk [15,94]. Additionally, the contamination of raw milk with *Staphylococcus aureus* raises public health problems throughout the food chain.

Overall, these diseases have negative impacts on the animals' reproductive efficiency and welfare, public health, and the final profit of the production process. The clinical symptoms of most of these bacterial and/or zoonotic diseases are mediated directly by microbial products (endotoxin) and tissue damage, or indirectly by inflammatory (cytokines and eicosanoids) and/or oxidative stress (nitric oxide) mediators [93]. These changes have a negative impact on sperm function and quality (sperm motility and sperm phagocytosis), ovarian function (follicular steroidogenesis and growth, ovulation, and ova competence development to blastocyst), uterine competence (implantation failure), and embryonic development (retarded development into blastocyst) [15,92].

Currently, antibiotic-based therapy is the commonly recommended therapy for tackling different microbial/protozoan diseases, including reproductive-related diseases. The effectiveness of antibiotics-based therapies is controlled by the pharmacokinetics of the drug. The delivery of antibiotics into targeted infected sites depends on the rate of absorption and distribution of the drug, which can be limited by different biological factors as the stability of the antibiotics against degradation by gastrointestinal enzymes (oral administration), blood hydrolytic enzymes (parenteral administration), drug solubility, and thus cellular uptake and bioavailability. Furthermore, some diseases cause fibrous damage in infected tissues limiting the penetration of the antibiotics into infected sites when local treatment is applied, such as in the direct infusion of the drug into the uterus in endometritis cases or through teats in mastitis cases [15,95].

Despite the limiting biological and therapeutic efficiency of the current antibiotics-based therapies, emerging concerns regarding the development of antimicrobial-resistant species add other limitations on the applications of antibiotic-based therapies, specifically in food-producing farm animals. The fear of developing more wild pathogenic microbial species, creating infectious and cross-transited microbial species, transferring of antibiotic residues into animal products (meat and milk), and the release of antibiotics into the environment are all aspects that should be taken into account [96]. Antimicrobial resistance leads not only to an encumbrance on public health but also extends to the risk of therapy failure and repeated infection, along with subsequent economic impacts. Actually, these factors make the treatment of reproductive-related diseases (mastitis [15], toxoplasmosis, and neosporosis [93]) by antibiotics a controversial strategy. Accordingly, novel, safe, and effective antibiotics-based therapeutic approaches are needed, particularly when treatments are directed to food-producing farm animals [93].

5.2. Nanotechnology Approaches

Many studies have shown the opportunity of using many engineered nanomaterials (e.g., liposomes, polymeric nanoparticles, solid lipid nanoparticles, nanogels, and inorganic nanoparticles), which are synthesized with specific physicochemical properties to overcome the therapeutic limitations of antibiotics-based therapies [94,97–99].

The use of nano-formula for antibiotics-based therapies may offer additional advantages over conventional antibiotics formula, such as (1) reducing the dose of the antibiotic, (2) allowing efficient delivery of the antibiotic to the infected sites, (3) shortening the therapeutic timing and side effects, and (4) preventing burst release and degradation of the antibiotic [94]. Nanomaterials may be protective against the rapid degradation of the antibiotic and may improve its delivery to the infected site, but, moreover, nanomaterials themselves could be engineered to show cytotoxic and destructive properties against microorganisms. Moreover, some nanoparticles have destructive effects on the bacterial cell membrane, enzymes, and functional and structural cell proteins mainly through evoking cellular oxidative pathways, in addition to their ability to inhibit the formation of bacterial biofilm, to induce changes in the gene expression, and to stimulate innate and adaptive immunity [97]. Additionally, nanoparticles could be engineered to hinder the bacterial adhesion, colonization, and biofilm development of bacteria [15]. Furthermore, nanomaterials have the ability to incorporate one or more drugs without any effect on the structure of the compound but increasing its pharmacological action [98].

Specifically, in swine, enrofloxacin antibiotic is used to treat several bacterial infections, such as *Pasteurella*, *Mycoplasma*, *Escherichia coli*, or *Salmonella*, with an intramuscularly recommended dose average between 2.5 to 5 mg enrofloxacin/kg Bw/day for 3 to 5 days. Paudel et al. [99] showed that enrofloxacin-loaded poly(lactic-co-glycolic acid) nanoparticles may be delivered orally in a suspension in drinking water, and the minimum inhibitory concentration against *Escherichia coli* was reduced by 23% compared to free enrofloxacin alone. Such finding, combined with increased bioavailability, maybe an interesting first step to reduce the dose of enrofloxacin and, therefore, its side effects (including the propagation of antibiotics resistance). El-Zawawy et al. [100] reported that incorporating triclosan into the lipid bilayer of liposomes allowed its usein lower doses, which in turn reduced its biochemical adverse effects. In another study, sodium dodecyl sulfate-coated atovaquone nanosuspensionsconsiderably increased the therapeutic efficiency against experimentally acquired and reactivated toxoplasmosis by improving the passage of gastrointestinal and blood–brain barriers [101]. Similarly, tilmicosin (a semi-synthetic macrolide antibiotic)-loaded hydrogenated castor oil with lower dosage showed better therapeutic efficacy than free tilmicosin for *Staphylococcus aureus* mastitis infectiondue to the enhanced bioavailability and sustained-release performance [102]. Recently, nano drugs have also been used as astrategy to solve the multi-drug resistance and intracellular persistence of *Staphylococcus aureus*, which is associated with the subclinical and relapsing infection of bovine mastitis [94]. Yang et al. [13] showed the possibility of prolonging post-antibiotic effects and thus dosing intervals when amoxicillinnanoparticles are used for treating bovine mastitis. This would decrease the rate of antibiotic use and the costs of medication.

Recently, the combination of the advantages of nano-drug delivery technology and alternative medicine, which depends on the usage of natural products with antimicrobial activity, opens the way to innovative natural and safe antibiotic alternatives. Numerous studies have shown probiotic species, microbial extracts, and plant secondary metabolites (essential oils and polyphenols) as potential antimicrobial agents [103,104]. A nano-formulacontaining 0.4–10% of oregano oil was developed for treating dairy cow endometritis. The uterine infusion of this nano-formula for 2–5 days showed a remarkable curative effect, being able to diminish inflammation, sterilizing uterus, draining pus by contracting the uterus, and promoting the recovery of the uterine function (Patent: CN104288222A, china https://patents.google.com/patent/CN104288222A/en; accessed on 12 February 2021). In another study, poly(lactic-coglycolic) acid(PLGA)-epigallocatechin gallate-deoxycyclin nanoparticles have been successfully used as an assisted-endometritis therapy [105]. In context, chitosan-TPP nanoparticles have been used for treating mastitis [106].

Additionally, many metal nanoparticles, such as silver oxide (Ag_2O), gold (Au), zinc oxide (ZnO), titanium dioxide (TiO_2), and copper oxide (CuO), have shown effective antimicrobial activity against a broad spectrum of microorganisms [19]. These approaches may give an opportunity to completely substitute antibiotics-based treatments with more safe therapies. The emergence of biological biosynthesis procedures of nanometal (silver) using microorganisms (*Escherichia coli, Acinetobacter species, Staphylococcus aureus, Pseudomonas aeruginosa*, and *Klebsiella pneumoniae*) and/or natural reducing agents (polyphenols, flavonoids, and phenolic biomolecules of Camellia, green tea, and black tea leaf extracts)has encouraged the use of nano metals as an alternative to antibiotics, meeting both therapeutical and environmental aspects [103,107]. In this context, apigenin (a polyphenolic compound) was successfully used to synthesis silver nanoparticles with a size of 10 nm.These nanoparticles showed antimicrobial activity against pathogenic bacteria *Prevotella melaninogenica* and *Arcanobacterium pyogenes* isolated from endometritis infected uterine discharges by inhibiting cell viability and biofilm formation in a dose-and time-dependent manner [91]. Similarly, Yuan et al. [108] confirmed the antibacterial activity of biologically synthesized silver nanoparticles against two multiple drug-resistant strains of *Pseudomona saeruginosa* and *Staphylococcus aureus* isolated from mastitis-infected goats milk samples.

Regarding the toxicity of such nanomedicines to animal tissues, Radzikowski et al. [109] confirmed the potential of commercially available silver nanoparticles, copper nanoparticles, and their combination in decreasing the viability of mastitis-borne pathogens without showing toxic effects on mammary gland tissues. Furthermore, Paudel et al. [99] confirmed the lower toxicity of enrofloxacin entrapped nanoparticles to mammalian cells relative to a free drug as the incorporation of the drug into the PLGA matrix minimized the production of reactive oxygen speciesevoked by the antibiotic.

A summary of studies on the nano drugs developed to treat reproductive-related diseases is shown in Table 3.

Table 3. Summary of studies on the nano drugs developed to treat reproductive-related diseases.

Type of Drug	Formula	Technique	Particle Characteristics	Drug Activity	Usage
Antibiotic [99]	Enrofloxacin-poly lactic-co-glycolic acid NPs	-	Size = 102 nm PdI = 0.095 Zp = −32 mV	Antimicrobial agent against *Staphylococcus aureus, Escherichia coli*	Endometritis and mastitis treatment
Antibiotic [102]	Tilmicosin-loaded hydrogenated castor oil NPs	Hot homogenization and ultrasonication	Size = 343 nm PdI = 0.33 Zp = 7.9 mV EE = 60.4%	Antimicrobial agent against *Staphylococcus aureus*	Mastitis treatment
Antibiotic [100]	Triclosan-loaded liposome NPs	Dehydration-rehydration	Size = 53.3 nm EE = 90%	Antimicrobial agent against *Toxoplasma gondii*	Toxoplasmosis treatment
Antibiotic [101]	Atovaquone-poloxamer 188 - sodium dodecyl sulfate	-	-	Antimicrobial agent against *Toxoplasma gondii*	Toxoplasmosis treatment
Nitric oxide (NO) [110]	NO-alginate-chitosan NO-chitosan-TPP	-	Size= 270–375 nm PdI=0.27–0.31 Zp = 16−17 mV	Antimicrobial agent against *Staphylococcus aureus, Escherichia coli*	Mastitis treatment
Metal [91]	Silver NPs	Biosynthesis by apigenin	Size = 10 nm	Antimicrobial agent against *Prevotella melaninogenica* and *Arcanobacterium pyogenes*	Antibiotic alternative for endometritis treatment
Metal [108]	Silver NPs	Biosynthesis by quercetin	Size = 20 nm Zp= 37.7mV	Antimicrobial agent against *Staphylococcus aureus* and *Pseudomonas aeruginosa*	Antibiotic alternative for mastitis treatment
Chitosan [106]	Chitosan-TPP Nps	Ionotropic gelation	Size = 19.1 nm PdI = 0.41 Zp = 49.9 mV Yield particle = 92.8%	Antimicrobial agent against *Pseudomona* sp.	Antibiotic alternative for mastitis treatment
Antibiotic + polyphenol [105]	Poly(lactic-co-glycolic) acid-epigallocatechin gallate-doxycycline Nps Singh et al., 2015	Modified double emulsion solvent evaporation/extraction technique	Size = 176 to 211 nm PdI = 0.124 to 0.466 EE= 78.5 to 86.3%	Anti-inflammatory agent	Assisted-endometritis therapy
Essential oil [1]	Oregano oil Nps	-	-	Antimicrobial agent against *Staphylococcus aureus, Escherichia coli, Streptococcus* spp.	Antibiotic alternative for endometritis treatment

Nps = nanoparticles, Zp = zeta potential, PdI = Polydispersityindex, and EE = encapsulation efficiency.[1] https://patents.google.com/patent/CN104288222A/en; accessed on 12 February 2021.

6. Conclusions

In this review, we illustrated the existing challenges and limitations of the most commonly applied reproductive management strategies in farm animals. As illustrated, the efficiency of the reproductive management strategies is restricted by the ease of application in the field scale, physiological status and behavior of animals, drug availability and uptake, in addition to environmental aspects, such as the release of antibiotic/hormone residues. Nanotechnology presents innovative and alternative solutions, such as that posed for solving challenges of the male effect by using an aerosol nano-drug delivery system of pheromones. Applying nanotechnology may also solve the problem of the lack of some hormones in the near future by changing the pharmacokinetics and pharmacodynamicsbehavior of hormones allowing the use of the hormones in other unconventional biological applications. Furthermore, challenges related to nutritional management can also be solved mainly through improving nutrients bioavailability during sensitive reproductive windows, which may interfere with nutrients availability and utilization. Lastly, nanotechnology can improve the efficiency of the antibiotic and/or creating natural antibiotic alternatives that may restrict the development of antibiotics resistant microbial species.However, this review presents ambitious solutions for the existing reproductive management challenges in the light of available literature; more studies are required to test the efficiency of such strategies.

Author Contributions: Conceptualization, N.M.H.; resources, N.M.H.; writing—original draft preparation, N.M.H.; writing—review and editing, N.M.H. and A.G.-B.; visualization, N.M.H. and A.G.-B. All authors have read and agreed to the published version of the manuscript.

Funding: No fund was received for this review article.

Institutional Review Board Statement: Not applicable.

Informed Consent Statement: Not applicable.

Conflicts of Interest: The authors declare no conflict of interest.

References

1. Olynk, N.; Wolf, C. Economic analysis of reproductive management strategies on US commercial dairy farms. *J. Dairy Sci.* **2008**, *91*, 4082–4091. [CrossRef]
2. Smith, M.F.; Geisert, R.D.; Parrish, J.J. Reproduction in domestic ruminants during the past 50 yr: Discovery to application. *J. Anim. Sci.* **2018**, *96*, 2952–2970. [CrossRef]
3. Caraviello, D.; Weigel, K.; Fricke, P.; Wiltbank, M.; Florent, M.; Cook, N.; Nordlund, K.; Zwald, N.; Rawson, C. Survey of management practices on reproductive performance of dairy cattle on large US commercial farms. *J. Dairy Sci.* **2006**, *89*, 4723–4735. [CrossRef]
4. Delgadillo, J.A.; Martin, G.B. Alternative methods for control of reproduction in small ruminants: A focus on the needs of grazing industries. *Anim. Front.* **2015**, *5*, 57–65. [CrossRef]
5. Hashem, N.M.; Hassanein, E.M.; Simal-Gandara, J. Improving reproductive performance and health of mammals using honeybee products. *Antioxidants* **2021**, *10*, 336. [CrossRef] [PubMed]
6. Hashem, N.M.; Gonzalez-Bulnes, A. State-of-the-art and prospective of nanotechnologies for smart reproductive management of farm animals. *Animals* **2020**, *10*, 840. [CrossRef]
7. Hassanein, E.; Hashem, N.; El-Azrak, K.; Gonzalez-Bulnes, A.; Hassan, G.; Salem, M. Efficiency of GnRH–Loaded Chitosan Nanoparticles for Inducing LH Secretion and Fertile Ovulations in Protocols for Artificial Insemination in Rabbit Does. *Animals* **2021**, *11*, 440. [CrossRef] [PubMed]
8. Tejada, L.M.; Meza-Herrera, C.A.; Rivas-Munoz, R.; Rodriguez-Martinez, R.; Carrillo, E.; Mellado, M.; Veliz-Deras, F.G. Appetitive and consummatory sexual behaviors of rams treated with exogenous testosterone and exposed to anestrus dorper ewes: Efficacy of the male effect. *Arch. Sex Behav.* **2017**, *46*, 835–842. [CrossRef] [PubMed]
9. Ungerfeld, R. Socio-sexual signalling and gonadal function: Opportunities for reproductive management in domestic ruminants. *Soc. Reprod. Fertil. Suppl.* **2007**, *64*, 207. [CrossRef]
10. Chasles, M.; Chesneau, D.; Moussu, C.; Delgadillo, J.A.; Chemineau, P.; Keller, M. Sexually active bucks are efficient to stimulate female ovulatory activity during the anestrous season also undertemperate latitudes. *Anim. Reprod. Sci.* **2016**, *168*, 86–91. [CrossRef] [PubMed]
11. Izquierdo, A.C.; Gutiérrez, J.P.; Hernández, W.M.; Mancera, A.V.; Crispín, R.H. Obtención, evaluación y manipulación del semen de verraco en una unidad de producción mexicana. *Rev. Vet.* **2016**, *26*, 69–74. [CrossRef]

12. Hashem, N.; El-Zarkouny, S. Effect of short-term supplementation with rumen-protected fat during the late luteal phase on reproduction and metabolism of ewes. *J. Anim. Physiol. Anim. Nutr.* **2014**, *98*, 65–71. [CrossRef]
13. Yang, X.; Ouyang, W.; Sun, J.; Li, X. Post-antibiotic effect of Amoxicillin nanoparticles against main pathogenic bacteria of Bovine mastitis in vitro. *J. Northwest A F Univ. Nat. Sci. Ed.* **2009**, *37*, 1–6.
14. Cerbu, C.; Kah, M.; White, J.C.; Astete, C.E.; Sabliov, C.M. Fate of biodegradable engineered nanoparticles used in veterinary medicine as delivery systems from a one health perspective. *Molecules* **2021**, *26*, 523. [CrossRef] [PubMed]
15. Algharib, S.A.; Dawood, A.; Xie, S. Nanoparticles for treatment of bovine Staphylococcus aureus mastitis. *Drug Deliv.* **2020**, *27*, 292–308. [CrossRef]
16. Osama, E.; El-Sheikh, S.M.; Khairy, M.H.; Galal, A.A. Nanoparticles and their potential applications in veterinary medicine. *J. Adv. Vet. Res.* **2020**, *10*, 268–273.
17. Signoret, J.; Lindsay, D. The male effect in domestic mammals: Effect on LH secretion and ovulation—importance of olfactory cues. *Olfaction Endocr. Regul.* **1982**, 63–72.
18. Murata, K.; Tamogami, S.; Itou, M.; Ohkubo, Y.; Wakabayashi, Y.; Watanabe, H.; Okamura, H.; Takeuchi, Y.; Mori, Y. Identification of an olfactory signal molecule that activates the central regulator of reproduction in goats. *Curr. Biol.* **2014**, *24*, 681–686. [CrossRef]
19. Ferreira-Silva, J.C.; Burnett, T.A.; Souto, P.F.M.P.; Filho, P.C.B.G.; Pereira, L.C.; Araujo, M.V.; Moura, M.T.; Oliveira, M.A.L. Progesterone (P4), luteinizing hormone (LH) levels and ovarian activity in postpartum Santa Inês ewes subject to a male effect. *Arch. Anim. Breed.* **2017**, *60*, 95–100. [CrossRef]
20. Meza-Herrera, C.A.; Cano-Villegas, O.; Flores-Hernandez, A.; Veliz-Deras, F.G.; Calderon-Leyva, G.; Guillen-Munoz, J.M.; de la Pena, C.G.; Rosales-Nieto, C.A.; Macias-Cruz, U.; Avendano-Reyes, L. Reproductive outcomes of anestrous goats supplemented with spineless Opuntia megacantha Salm-Dyck protein-enriched cladodes and exposed to the male effect. *Trop. Anim. Health Prod.* **2017**, *49*, 1511–1516. [CrossRef] [PubMed]
21. Rekwot, P.; Ogwu, D.; Oyedipe, E.; Sekoni, V. The role of pheromones and biostimulation in animal reproduction. *Anim. Reprod. Sci.* **2001**, *65*, 157–170. [CrossRef]
22. De Santiago-Miramontes, M.; Rivas-Muñoz, R.; Muñoz-Gutiérrez, M.; Malpaux, B.; Scaramuzzi, R.; Delgadillo, J. The ovulation rate in anoestrous female goats managed under grazing conditions and exposed to the male effect is increased by nutritional supplementation. *Anim. Reprod. Sci.* **2008**, *105*, 409–416. [CrossRef]
23. Delgadillo, J.A.; Gelez, H.; Ungerfeld, R.; Hawken, P.A.; Martin, G.B. The 'male effect' in sheep and goats—Revisiting the dogmas. *Behav. Brain Res.* **2009**, *200*, 304–314. [CrossRef] [PubMed]
24. Martin, G.; Milton, J.; Davidson, R.; Hunzicker, G.B.; Lindsay, D.; Blache, D. Natural methods for increasing reproductive efficiency in small ruminants. *Anim. Reprod. Sci.* **2004**, *82*, 231–245. [CrossRef]
25. Hashem, N.; El-Zarkouny, S.; Taha, T.; Abo-Elezz, Z. Effect of season, month of parturition and lactation on estrus behavior and ovarian activity in Barki x Rahmani crossbred ewes under subtropical conditions. *Theriogenology* **2011**, *75*, 1327–1335. [CrossRef]
26. Rosa, H.; Bryant, M. The 'ram effect' as a way of modifying the reproductive activity in the ewe. *Small Rumin. Res.* **2002**, *45*, 1–16. [CrossRef]
27. Zarazaga, L.; Gatica, M.; Hernández, H.; Gallego-Calvo, L.; Delgadillo, J.; Guzmán, J. The isolation of females from males to promote a later male effect is unnecessary if the bucks used are sexually active. *Theriogenology* **2017**, *95*, 42–47. [CrossRef]
28. Zarazaga, L.; Gatica, M.; Hernández, H.; Chemineau, P.; Delgadillo, J.; Guzmán, J. Photoperiod-treated bucks are equal to melatonin-treated bucks for inducing reproductive behaviour and physiological functions via the "male effect" in Mediterranean goats. *Anim. Reprod. Sci.* **2019**, *202*, 58–64. [CrossRef] [PubMed]
29. Zarazaga, L.A.; Gatica, M.C.; Hernandez, H.; Keller, M.; Chemineau, P.; Delgadillo, J.A.; Guzman, J.L. The reproductive response to the male effect of 7- or 10-month-old female goats is improved when photostimulated males are used. *Animal* **2019**, *13*, 1658–1665. [CrossRef] [PubMed]
30. Ponce, J.; Velázquez, H.; Duarte, G.; Bedos, M.; Hernández, H.; Keller, M.; Chemineau, P.; Delgadillo, J. Reducing exposure to long days from 75 to 30 days of extra-light treatment does not decrease the capacity of male goats to stimulate ovulatory activity in seasonally anovulatory females. *Domest. Anim. Endocrinol.* **2014**, *48*, 119–125. [CrossRef]
31. Knight, T.; Lynch, P. Source of ram pheromones that stimulate ovulation in the ewe. *Anim. Reprod. Sci.* **1980**, *3*, 133–136. [CrossRef]
32. Izard, M. Pheromones and reproduction in domestic animals. *Pheromones Reprod. Mamm.* **1983**, *253*.
33. McGlone, J.J.; Devaraj, S.; Garcia, A. A novel boar pheromone mixture induces sow estrus behaviors and reproductive success. *Appl. Anim. Behav. Sci.* **2019**, *219*, 104832. [CrossRef]
34. Dugasa, H.L.; Williams, R.O., III. Nanotechnology for pulmonary and nasal drug delivery. *Nanotechnol. Drug Deliv. Vol. Two Nano Eng. Strateg. Nanomed. Against Sev. Dis.* **2016**, *102*.
35. Pamungkas, F.A.; Sianturi, R.S.G.; Wina, E.; Kusumaningrum, D.A. Chitosan nanoparticle of hCG (Human Chorionic Gonadotrophin) hormone in increasing induction of dairy cattle ovulation. *Indones. J. Anim. Vet. Sci.* **2016**, *21*, 34–40.
36. Kekan, P.M.; Ingole, S.D.; Sirsat, S.D.; Bharucha, S.V.; Kharde, S.D.; Nagvekar, A.S. The role of pheromones in farm animals—A review. *Agric. Rev.* **2017**, *38*. [CrossRef]
37. Archunan, G. Reproductive enhancement in buffalo: Looking at urinary pheromones and hormones. *Iran. J. Vet. Res.* **2020**, *21*, 163.

38. Higgins, H.M.; Ferguson, E.; Smith, R.F.; Green, M.J. Using hormones to manage dairy cow fertility: The clinical and ethical beliefs of veterinary practitioners. *PLoS ONE* **2013**, *8*, e62993. [CrossRef] [PubMed]

39. Hashem, N.; El-Azrak, K.; El-Din, A.N.; Taha, T.; Salem, M. Effect of GnRH treatment on ovarian activity and reproductive performance of low-prolific Rahmani ewes. *Theriogenology* **2015**, *83*, 192–198. [CrossRef] [PubMed]

40. Hashem, N.; El-Zarkouny, S.; Taha, T.; Abo-Elezz, Z. Oestrous response and characterization of the ovulatory wave following oestrous synchronization using PGF2α alone or combined with GnRH in ewes. *Small Rumin. Res.* **2015**, *129*, 84–87. [CrossRef]

41. Hashem, N.; Aboul-Ezz, Z. Effects of a single administration of different gonadotropins on day 7 post-insemination on pregnancy outcomes of rabbit does. *Theriogenology* **2018**, *105*, 1–6. [CrossRef] [PubMed]

42. Sun, T.-C.; Li, H.-Y.; Li, X.-Y.; Yu, K.; Deng, S.-L.; Tian, L. Protective effects of melatonin on male fertility preservation and reproductive system. *Cryobiology* **2020**, *95*, 1–8. [CrossRef] [PubMed]

43. De Castro, M.V.; Cortell, C.; Mocé, E.; Marco-Jiménez, F.; Joly, T.; Vicente, J. Effect of recombinant gonadotropins on embryo quality in superovulated rabbit does and immune response after repeated treatments. *Theriogenology* **2009**, *72*, 655–662. [CrossRef] [PubMed]

44. Forcada, F.; Ait Amer-Meziane, M.; Abecia, J.A.; Maurel, M.C.; Cebrian-Perez, J.A.; Muino-Blanco, T.; Asenjo, B.; Vazquez, M.I.; Casao, A. Repeated superovulation using a simplified FSH/eCG treatment for in vivo embryo production in sheep. *Theriogenology* **2011**, *75*, 769–776. [CrossRef] [PubMed]

45. Herve, V.; Roy, F.; Bertin, J.; Guillou, F.; Maurel, M.C. Antiequine chorionic gonadotropin (eCG) antibodies generated in goats treated with eCG for the induction of ovulation modulate the luteinizing hormone and follicle-stimulating hormone bioactivities of eCG differently. *Endocrinology* **2004**, *145*, 294–303. [CrossRef] [PubMed]

46. Kara, E.; Dupuy, L.; Bouillon, C.; Casteret, S.; Maurel, M.-C. Modulation of gonadotropins activity by antibodies. *Front. Endocrinol.* **2019**, *10*. [CrossRef] [PubMed]

47. Roy, F.; Maurel, M.-C.; Combes, B.; Vaiman, D.; Cribiu, E.P.; Lantier, I.; Pobel, T.; Delétang, F.; Combarnous, Y.; Guillou, F. The negative effect of repeated equine chorionic gonadotropin treatment on subsequent fertility in Alpine goats is due to a humoral immune response involving the major histocompatibility complex. *Biol. Reprod.* **1999**, *60*, 805–813. [CrossRef]

48. Manteca Vilanova, X.; De Briyne, N.; Beaver, B.; Turner, P.V. Horse welfare during equine Chorionic Gonadotropin (eCG) production. *Animals* **2019**, *9*, 1053. [CrossRef]

49. Rathbone, M.J.; Burke, C.R. Controlled release intravaginal veterinary drug delivery. In *Long Acting Animal Health Drug Products*; Springer: New York, NY, USA, 2013; pp. 247–270.

50. De Graaff, W.; Grimard, B. Progesterone-releasing devices for cattle estrus induction and synchronization: Device optimization to anticipate shorter treatment durations and new device developments. *Theriogenology* **2018**, *112*, 34–43. [CrossRef]

51. Fernández-Serrano, P.; Casares-Crespo, L.; Viudes-de-Castro, M.P. Chitosan–dextran sulphate nanoparticles for Gn RH release in rabbit insemination extenders. *Reprod. Domest. Anim.* **2017**, *52*, 72–74. [CrossRef]

52. Casares-Crespo, L.; Fernández-Serrano, P.; Viudes-De-Castro, M.P. Protection of GnRH analogue by chitosan-dextran sulfate nanoparticles for intravaginal application in rabbit artificial insemination. *Theriogenology* **2018**, *116*, 49–52. [CrossRef] [PubMed]

53. Hashem, N.; Sallam, S. Reproductive performance of goats treated with free gonadorelin or nanoconjugated gonadorelin at estrus. *Domest. Anim. Endocrinol.* **2020**, *71*, 106390. [CrossRef] [PubMed]

54. Rather, M.A.; Sharma, R.; Gupta, S.; Ferosekhan, S.; Ramya, V.; Jadhao, S.B. Chitosan-nanoconjugated hormone nanoparticles for sustained surge of gonadotropins and enhanced reproductive output in female fish. *PLoS ONE* **2013**, *8*, e57094. [CrossRef] [PubMed]

55. Oliveira, J.E.; Medeiros, E.S.; Cardozo, L.; Voll, F.; Madureira, E.H.; Mattoso, L.H.C.; Assis, O.B.G. Development of poly (lactic acid) nanostructured membranes for the controlled delivery of progesterone to livestock animals. *Mater. Sci. Eng. C* **2013**, *33*, 844–849. [CrossRef]

56. Helbling, I.M.; Ibarra, J.C.; Luna, J.A. The optimization of an intravaginal ring releasing progesterone using a mathematical model. *Pharm. Res.* **2014**, *31*, 795–808. [CrossRef]

57. Fogolari, O.; Felippe, A.C.; Leimann, F.V.; Gonçalves, O.H.; Sayer, C.; Araújo, P.H.H.D. Method validation for progesterone determination in poly (Methyl methacrylate) nanoparticles synthesized via miniemulsion polymerization. *Int. J. Polym. Sci.* **2017**, *2017*. [CrossRef]

58. Remião, M.H.; Lucas, C.G.; Domingues, W.B.; Silveira, T.; Barther, N.N.; Komninou, E.R.; Basso, A.C.; Jornada, D.S.; Beck, R.C.R.; Pohlmann, A.R. Melatonin delivery by nanocapsules during in vitro bovine oocyte maturation decreased the reactive oxygen species of oocytes and embryos. *Reprod. Toxicol.* **2016**, *63*, 70–81. [CrossRef]

59. Li, Y.; Zhao, X.; Wang, L.; Liu, Y.; Wu, W.; Zhong, C.; Zhang, Q.; Yang, J. Preparation, characterization and in vitro evaluation of melatonin-loaded porous starch for enhanced bioavailability. *Carbohydr. Polym.* **2018**, *202*, 125–133. [CrossRef]

60. Siahdasht, F.N.; Farhadian, N.; Karimi, M.; Hafizi, L. Enhanced delivery of melatonin loaded nanostructured lipid carriers during in vitro fertilization: NLC formulation, optimization and IVF efficacy. *RSC Adv.* **2020**, *10*, 9462–9475. [CrossRef]

61. Helbling, I.M.; Busatto, C.A.; Fioramonti, S.A.; Pesoa, J.I.; Santiago, L.; Estenoz, D.A.; Luna, J.A. Preparation of TPP-crosslinked chitosan microparticles by spray drying for the controlled delivery of progesterone intended for estrus synchronization in cattle. *Pharm. Res.* **2018**, *35*, 1–15. [CrossRef]

62. Combarnous, Y.; Salesse, R.; Garnier, J. Physico-chemical properties of pregnant mare serum gonadotropin. *Biochim. et Biophys. Acta BBA Protein Struct.* **1981**, *667*, 267–276. [CrossRef]

63. STEWART, F.; Allen, W.; Moor, R.M. Pregnant mare serum gonadotrophin: Ratio of follicle-stimulating hormone and luteinizing hormone activities measured by radioreceptor assay. *J. Endocrinol.* **1976**, *71*, 371–382. [CrossRef]

64. Santos-Jimenez, Z.; Guillen-Gargallo, S.; Encinas, T.; Berlinguer, F.; Veliz-Deras, F.G.; Martinez-Ros, P.; Gonzalez-Bulnes, A. Use of propylene-glycol as a cosolvent for gnrh in synchronization of estrus and ovulation in sheep. *Animals* **2020**, *10*, 897. [CrossRef] [PubMed]

65. Cordova-Izquierdo, A. Best Practices in Animal Reproduction: Impact of Nutrition on Reproductive Performance Livestock. *J. Adv. Dairy Res.* **2016**, *4*, 152. [CrossRef]

66. Cannas, A.; Tedeschi, L.O.; Atzori, A.S.; Lunesu, M.F. How can nutrition models increase the production efficiency of sheep and goat operations? *Anim. Front.* **2019**, *9*, 33–44. [CrossRef] [PubMed]

67. Zhao, C.; Bai, Y.; Fu, S.; Wu, L.; Xu, C.; Xia, C. Follicular fluid proteomic profiling of dairy cows with anestrus caused by negative energy balance. *Ital. J. Anim. Sci.* **2021**, *20*, 650–663. [CrossRef]

68. Scaramuzzi, R.; Martin, G. The importance of interactions among nutrition, seasonality and socio-sexual factors in the development of hormone-free methods for controlling fertility. *Reprod. Dom. Anim.* **2008**, *43*, 129–136. [CrossRef] [PubMed]

69. Wu, G. Functional amino acids in growth, reproduction, and health. *Adva. Nutr.* **2010**, *1*, 31–37. [CrossRef] [PubMed]

70. Peñagaricano, F.; Souza, A.H.; Carvalho, P.D.; Driver, A.M.; Gambra, R.; Kropp, J.; Hackbart, K.S.; Luchini, D.; Shaver, R.D.; Wiltbank, M.C. Effect of maternal methionine supplementation on the transcriptome of bovine preimplantation embryos. *PLoS ONE* **2013**, *8*, e72302. [CrossRef]

71. Abdelnour, S.A.; Al-Gabri, N.A.; Hashem, N.M.; Gonzalez-Bulnes, A. Supplementation with proline improves haemato-biochemical and reproductive indicators in male rabbits affected by environmental heat-stress. *Animals* **2021**, *11*, 373. [CrossRef]

72. Wiltbank, M.; Shaver, R.; Toledo, M.; Carvalho, P.; Baez, G.; Follendorf, T.; Lobos, N.; Luchini, D.; Souza, A. Potential benefits of feeding methionine on reproductive efficiency of lactating dairy cows. *Four State Dairy Nutr. Manag.* **2014**, *4*, 19–26.

73. Toledo, M.Z.; Baez, G.M.; Garcia-Guerra, A.; Lobos, N.E.; Guenther, J.N.; Trevisol, E.; Luchini, D.; Shaver, R.D.; Wiltbank, M.C. Effect of feeding rumen-protected methionine on productive and reproductive performance of dairy cows. *PLoS ONE* **2017**, *12*, e0189117. [CrossRef]

74. Thatcher, W.; Santos, J.E.; Staples, C.R. Dietary manipulations to improve embryonic survival in cattle. *Theriogenology* **2011**, *76*, 1619–1631. [CrossRef]

75. Sturmey, R.; Reis, A.; Leese, H.; McEvoy, T. Role of fatty acids in energy provision during oocyte maturation and early embryo development. *Reprod. Dom. Anim.* **2009**, *44*, 50–58. [CrossRef]

76. Castro, T.; Martinez, D.; Isabel, B.; Cabezas, A.; Jimeno, V. Vegetable oils rich in polyunsaturated fatty acids supplementation of dairy cows' diets: Effects on productive and reproductive performance. *Animals* **2019**, *9*, 205. [CrossRef] [PubMed]

77. Hosny, N.S.; Hashem, N.M.; Morsy, A.S.; Abo-Elezz, Z.R. Effects of Organic Selenium on the Physiological Response, Blood Metabolites, Redox Status, Semen Quality, and Fertility of Rabbit Bucks Kept Under Natural Heat Stress Conditions. *Front. Vet. Sci.* **2020**, *7*. [CrossRef]

78. Zereu, G. Factors affecting feed intake and its regulation mechanisms in ruminants a review. *Int. J. Livest. Res.* **2016**, *6*, 19–40.

79. Hippen, A.R.; DeFrain, J.M.; Linke, P.L. Glycerol and other energy sources for metabolism and production of transition dairy cows. In Proceedings of the Florida Ruminant Nutrition Symposium, Gainesville, FL, USA, 29–30 January 2008.

80. Shin, J.; Wang, D.; Kim, S.; Adesogan, A.; Staples, C. Effects of feeding crude glycerin on performance and ruminal kinetics of lactating Holstein cows fed corn silage-or cottonseed hull-based, low-fiber diets. *J. Dairy Sci.* **2012**, *95*, 4006–4016. [CrossRef] [PubMed]

81. El-Sherbiny, M.; Cieslak, A.; Pers-Kamczyc, E.; Szczechowiak, J.; Kowalczyk, D.; Szumacher-Strabel, M. A nanoemulsified form of oil blends positively affects the fatty acid proportion in ruminal batch cultures. *J. Dairy Sci.* **2016**, *99*, 399–407. [CrossRef]

82. Hackmann, T.J.; Firkins, J.L. Maximizing efficiency of rumen microbial protein production. *Front. Microbiol.* **2015**, *6*, 465. [CrossRef]

83. Hammon, D.; Holyoak, G.; Dhiman, T. Association between blood plasma urea nitrogen levels and reproductive fluid urea nitrogen and ammonia concentrations in early lactation dairy cows. *Anim. Reprod. Sci.* **2005**, *86*, 195–204. [CrossRef]

84. Boerman, J.; Lock, A. Effect of unsaturated fatty acids and triglycerides from soybeans on milk fat synthesis and biohydrogenation intermediates in dairy cattle. *J. Dairy Sci.* **2014**, *97*, 7031–7042. [CrossRef]

85. Gawad, R.; Fellner, V. Evaluation of glycerol encapsulated with alginate and alginate-chitosan polymers in gut environment and its resistance to rumen microbial degradation. *Asian Australas. J. Anim. Sci.* **2019**, *32*, 72. [CrossRef]

86. Kachuee, R.; Abdi-Benemar, H.; Mansoori, Y.; Sánchez-Aparicio, P.; Seifdavati, J.; Elghandour, M.M.; Guillén, R.J.; Salem, A.Z. Effects of sodium selenite, L-selenomethionine, and selenium nanoparticles during late pregnancy on selenium, zinc, copper, and iron concentrations in Khalkhali Goats and their kids. *Biol. Trace Elem. Res.* **2019**, *191*, 389–402. [CrossRef]

87. Jahanbin, R.; Yazdanshenas, P.; Amin Afshar, M.; Mohammadi Sangcheshmeh, A.; Varnaseri, H.; Chamani, M.; Nazaran, M.H.; Bakhtiyarizadeh, M.R. Effect of zinc nano-complex on bull semen quality after freeze-thawing process. *Anim. Prod.* **2015**, *17*, 371–380.

88. Khalil, W.A.; El-Harairy, M.A.; Zeidan, A.E.; Hassan, M.A. Impact of selenium nano-particles in semen extender on bull sperm quality after cryopreservation. *Theriogenology* **2019**, *126*, 121–127. [CrossRef]

89. Shahin, M.A.; Khalil, W.A.; Saadeldin, I.M.; Swelum, A.A.-A.; El-Harairy, M.A. Comparison between the effects of adding vitamins, trace elements, and nanoparticles to shotor extender on the cryopreservation of dromedary camel epididymal spermatozoa. *Animals* **2020**, *10*, 78. [CrossRef] [PubMed]

90. Albuquerque, J.; Casal, S.; de Jorge Páscoa, R.N.M.; Van Dorpe, I.; Fonseca, A.J.M.; Cabrita, A.R.J.; Neves, A.R.; Reis, S. Applying nanotechnology to increase the rumen protection of amino acids in dairy cows. *Sci. Rep.* **2020**, *10*, 1–12. [CrossRef] [PubMed]

91. Gurunathan, S.; Choi, Y.-J.; Kim, J.-H. Antibacterial efficacy of silver nanoparticles on endometritis caused by Prevotella melaninogenica and Arcanobacterum pyogenes in dairy cattle. *Int. J. Mol. Sci.* **2018**, *19*, 1210. [CrossRef]

92. Vallejo-Timaran, D.A.; Arango-Sabogal, J.C.; Reyes-Velez, J.; Maldonado-Estrada, J.G. Postpartum uterine diseases negatively impact the time to pregnancy in grazing dairy cows from high-altitude tropical herds. *Prev. Vet. Med.* **2020**, *185*, 105202. [CrossRef] [PubMed]

93. Sánchez-Sánchez, R.; Vázquez, P.; Ferre, I.; Ortega-Mora, L.M. Treatment of toxoplasmosis and neosporosis in farm ruminants: State of knowledge and future trends. *Curr. Top. Med. Chem.* **2018**, *18*, 1304–1323. [CrossRef]

94. Zhou, K.; Li, C.; Chen, D.; Pan, Y.; Tao, Y.; Qu, W.; Liu, Z.; Wang, X.; Xie, S. A review on nanosystems as an effective approach against infections of Staphylococcus aureus. *Int. J. Nanomed.* **2018**, *13*, 7333–7347. [CrossRef]

95. Olsen, J.E.; Christensen, H.; Aarestrup, F.M. Diversity and evolution of blaZ from Staphylococcus aureus and coagulase-negative staphylococci. *J. Antimicrob. Chemother.* **2006**, *57*, 450–460. [CrossRef] [PubMed]

96. Piotr, S.; Marta, S.; Aneta, F.; Barbara, K.; Magdalena, Z. Antibiotic resistance in Staphylococcus aureus strains isolated from cows with mastitis in the eastern Poland and analysis of susceptibility of resistant strains to alternative non-antibiotic agents: Lysostaphin, nisin and polymyxin B. *J. Vet. Med Sci.* **2013**. [CrossRef]

97. Wang, L.; Hu, C.; Shao, L. The antimicrobial activity of nanoparticles: Present situation and prospects for the future. *Int. J. Nanomed.* **2017**, *12*, 1227. [CrossRef] [PubMed]

98. Gholipourmalekabadi, M.; Mobaraki, M.; Ghaffari, M.; Zarebkohan, A.; Omrani, V.F.; Urbanska, A.M.; Seifalian, A. Targeted drug delivery based on gold nanoparticle derivatives. *Cur. Pharmac. Des.* **2017**, *23*, 2918–2929. [CrossRef] [PubMed]

99. Paudel, S.; Cerbu, C.; Astete, C.E.; Louie, S.M.; Sabliov, C.; Rodrigues, D.F. Enrofloxacin-impregnated PLGA nanocarriers for efficient therapeutics and diminished generation of reactive oxygen species. *ACS Appl. Nano Mat.* **2019**, *2*, 5035–5043. [CrossRef]

100. El-Zawawy, L.A.; El-Said, D.; Mossallam, S.F.; Ramadan, H.S.; Younis, S.S. Triclosan and triclosan-loaded liposomal nanoparticles in the treatment of acute experimental toxoplasmosis. *Experime. Parasitol.* **2015**, *149*, 54–64. [CrossRef] [PubMed]

101. Shubar, H.M.; Lachenmaier, S.; Heimesaat, M.M.; Lohman, U.; Mauludin, R.; Mueller, R.H.; Fitzner, R.; Borner, K.; Liesenfeld, O. SDS-coated atovaquone nanosuspensions show improved therapeutic efficacy against experimental acquired and reactivated toxoplasmosis by improving passage of gastrointestinal and blood–brain barriers. *J. Drug Target.* **2011**, *19*, 114–124. [CrossRef]

102. Wang, X.; Zhang, S.; Zhu, L.; Xie, S.; Dong, Z.; Wang, Y.; Zhou, W. Enhancement of antibacterial activity of tilmicosin against Staphylococcus aureus by solid lipid nanoparticles in vitro and in vivo. *Vet. J.* **2012**, *191*, 115–120. [CrossRef]

103. Mozafari, M.; Torkaman, S.; Karamouzian, F.M.; Rasti, B.; Baral, B. Antimicrobial applications of nanoliposome encapsulated silver nanoparticles: A potential strategy to overcome bacterial resistance. *Curr. Nanosci.* **2021**, *17*, 26–40. [CrossRef]

104. Garzon, S.; Laganà, A.S.; Barra, F.; Casarin, J.; Cromi, A.; Raffaelli, R.; Uccella, S.; Franchi, M.; Ghezzi, F.; Ferrero, S. Novel drug delivery methods for improving efficacy of endometriosis treatments. *Expert Opin. Drug Deliv.* **2020**, *18*, 1–13. [CrossRef] [PubMed]

105. Singh, A.K.; Chakravarty, B.; Chaudhury, K. Nanoparticle-Assisted Combinatorial Therapy for Effective Treatment of Endometriosis. *J. Biomed. Nanotechnol.* **2015**, *11*, 789–804. [CrossRef]

106. Rivera Aguayo, P.; Bruna Larenas, T.; Alarcon Godoy, C.; Cayupe Rivas, B.; Gonzalez-Casanova, J.; Rojas-Gomez, D.; Caro Fuentes, N. Antimicrobial and antibiofilm capacity of chitosan nanoparticles against wild type strain of Pseudomonas sp. isolated from milk of cows diagnosed with bovine mastitis. *Antibiotics* **2020**, *9*, 551. [CrossRef] [PubMed]

107. Das, C.A.; Kumar, V.G.; Dhas, T.S.; Karthick, V.; Govindaraju, K.; Joselin, J.M.; Baalamurugan, J. Antibacterial activity of silver nanoparticles (biosynthesis): A short review on recent advances. *Biocatal. Agric. Biotechnol.* **2020**, 101593. [CrossRef]

108. Yuan, Y.G.; Peng, Q.L.; Gurunathan, S. Effects of Silver Nanoparticles on Multiple Drug-Resistant Strains of Staphylococcus aureus and Pseudomonas aeruginosa from Mastitis-Infected Goats: An Alternative Approach for Antimicrobial Therapy. *Int. J. Mol. Sci.* **2017**, *18*, 569. [CrossRef]

109. Radzikowski, D.; Kalińska, A.; Ostaszewska, U.; Gołębiewski, M. Alternative solutions to antibiotics in mastitis treatment for dairy cows-a review. *Anim. Sci. Pap. Rep.* **2020**, *38*, 117–133.

110. Cardozo, V.F.; Lancheros, C.A.; Narciso, A.M.; Valereto, E.C.; Kobayashi, R.K.; Seabra, A.B.; Nakazato, G. Evaluation of antibacterial activity of nitric oxide-releasing polymeric particles against Staphylococcus aureus and Escherichia coli from bovine mastitis. *Int. J. Pharm.* **2014**, *473*, 20–29. [CrossRef]

Article

Nanodelivery System for Ovsynch Protocol Improves Ovarian Response, Ovarian Blood Flow Doppler Velocities, and Hormonal Profile of Goats

Nesrein M. Hashem [1],*, Hossam R. EL-Sherbiny [2], Mohamed Fathi [2] and Elshymaa A. Abdelnaby [2]

1   Department of Animal and Fish Production, Faculty of Agriculture (El-Shatby), Alexandria University, Alexandria 21545, Egypt
2   Theriogenology Department, Faculty of Veterinary Medicine, Cairo University, Giza 12211, Egypt; hossamelsherbiny7353575@cu.edu.eg (H.R.E.-S.); mido2022@yahoo.com (M.F.); elshymaa.ahmed@cu.edu.eg (E.A.A.)
*   Correspondence: nesreen.hashem@alexu.edu.eg

Simple Summary: Drug delivery systems depending on nanotechnology have been recently created to improve biological activity of drugs including hormones. Nanoformulated drugs have different pharmacokinetic properties in biological systems compared to their conventional forms due to the acquired new physicochemical properties such as lesser size, high size-to-weight ratio, and different surface charges and shapes. These properties enable them to efficiently deliver into target sites coping with existing biological barriers. Ovsynch protocol, GPG, is one of the most important estrous synchronization protocols due to the possibility of applying timed insemination with this protocol. This protocol is commonly used in many farm animals, as it aids in herd management and eliminates the need for the detection of estrus. However, there are still some shortcomings that restrict the outcomes of this protocol, such as scattering ovulation time, short luteal phase and inadequate luteal function, and low conception rates. Therefore, this study aims to evaluate ovarian response, blood flow of the ovarian and luteal arteries, and hormonal profile of goats receiving either a standard Ovsynch protocol or an Ovsynch protocol delivered via a nanodelivery system with different dosages of hormones. The present study is the first one that poses the idea that the nanodelivery system for the Ovsynch protocol enables lower hormone dose administration and improves the results of the protocol by inducing tighter synchrony of ovulation and better luteal function of synchronized goats.

Abstract: Fifteen cyclic, multiparous goats were equally stratified and received the common Ovsynch protocol (GPG: intramuscular, IM, injection of 50 mg gonadorelin, followed by an IM injection of 125 µg cloprostenol 7 days later, and a further IM injection of 50 mg gonadorelin 2 days later) or the Ovsynch protocol using nanofabricated hormones with the same dosages (NGPG) or half dosages (HNGPG) of each hormone. The ovarian structures and ovarian and luteal artery hemodynamic indices after each injection of the Ovsynch protocol using B-mode, color, and spectral Doppler scanning were monitored. Levels of blood serum progesterone (P_4), estradiol (E_2), and nitric oxide (NO) were determined. After the first gonadotrophin-releasing hormone (GnRH) injection, the number of large follicles decreased ($p = 0.02$) in NGPG and HNGPG, compared with GPG. HNGPG resulted in larger corpus luteum (CL) diameters ($p = 0.001$), and improved ovarian and luteal blood flow, compared with GPG and NGPG. Both NGPG and HNGPG significantly increased E_2 and NO levels compared with GPG. HNGPG increased ($p < 0.001$) P_4 levels compared with GPG, whereas NGPG resulted in an intermediate value. After prostaglandin $F_{2\alpha}$ ($PGF_{2\alpha}$) injection, HNGPG had the largest diameter of CLs ($p = 0.001$) and significantly improved ovarian blood flow compared with GPG and NGPG. Both NGPG and HNGPG increased ($p = 0.007$) NO levels, compared with GPG. E_2 level was increased ($p = 0.028$) in HNGPG, compared with GPG, whereas NGPG resulted in an intermediate value. During the follicular phase, HNGPG increased ($p = 0.043$) the number of medium follicles, shortened ($p = 0.04$) the interval to ovulation, and increased ($p < 0.001$) ovarian artery blood flow and levels ($p < 0.001$) of blood serum P_4, E_2, and NO, compared with GPG and NGPG. During the luteal phase, the numbers of CLs were similar among different experimental groups, whereas

---

Citation: Hashem, N.M.; EL-Sherbiny, H.R.; Fathi, M.; Abdelnaby, E.A. Nanodelivery System for Ovsynch Protocol Improves Ovarian Response, Ovarian Blood Flow Doppler Velocities, and Hormonal Profile of Goats. *Animals* **2022**, *12*, 1442. https://doi.org/10.3390/ani12111442

Academic Editor: Giuseppina Basini

Received: 8 April 2022
Accepted: 1 June 2022
Published: 3 June 2022

Publisher's Note: MDPI stays neutral with regard to jurisdictional claims in published maps and institutional affiliations.

---

the diameter of CLs, luteal blood flow, and levels of blood serum P_4 and NO increased ($p < 0.001$) in HNGPG, compared with GPG and NGPG. Conclusively, the nanodelivery system for the Ovsynch protocol could be recommended as a new strategy for improving estrous synchronization outcomes of goats while enabling lower hormone dose administration.

Keywords: nanodelivery system; GPG; estrous synchronization; blood flow; ovary

1. Introduction

One of the most widely used assisted reproductive techniques in farm animals is estrous cycle synchronization. Several estrous synchronization protocols based on the use of P_4 or $PGF_{2\alpha}$ analogs as key regulatory hormones have been developed [1]. Nevertheless, these protocols result in sufficient synchrony of the estrous cycle, and they do not eliminate the need for estrus detection. With the widespread of the artificial insemination process and the increased intensive livestock production system, the need for more practicable estrous synchronization protocols has become crucial [2]. The inclusion of gonadotropins, specifically GnRH, in estrous synchronization procedures, allows for more regulation of ovarian function by modulating follicular waves and the corpus luteum life span [3,4]. The full control of the ovarian activity helps set the ovulation time and thus apply fixed time-artificial insemination, eliminating the need for estrous detection as one of the most laborious farm works. To date, the popular estrous synchronization protocol that confers these advantages is "Ovsynch" or "GPG," which includes a GnRH-$PGF_{2\alpha}$-GnRH treatment sequence [5,6]. This protocol has gained increasing attention, and it is applied to synchronize estrous/ovarian cycles in different farm animal species, including goats. In goats, this procedure aids in herd management, particularly large herds, by concentrating on kidding and related farm practices in a short time [4]. Nevertheless, promising synchronization results have been obtained when applying such a protocol in various farm animals, including dairy cows, sheep, and goats [3,6,7]. There are still some defects that restrict the outcomes of this protocol, such as scattering ovulation time, short luteal phase and inadequate luteal function, and low conception rates [3,8]. Part of estrous synchronization protocols efficiency is ascribed to the pharmacokinetics of hormones and their bioavailability [1]. Because both GnRH and $PGF_{2\alpha}$ have a small molecular weight and short half-life, improving the pharmacokinetics and bioavailability of these hormones may enhance their biological action and estrous synchronization outcomes [9]. Recently, nanotechnology and drug delivery systems have been used to improve drugs, hormones, and biological activity. Nanodelivered drugs have varying pharmacokinetic properties in biological systems different from their conventional forms [10]. Nanodrugs have lesser size, high size to weight ratio, and different surface charge and shapes, compared with the original drugs [11]. These criteria can be controlled by applying different carrying materials and fabrication conditions, aiming for better delivery of drugs into target sites coping with different biological barriers (e.g., lytic enzymes and blood barriers). Few studies have referred to the improved biological activity of nanofabricated GnRH compared with the conventional form. For example, nano-GnRH improved artificial insemination outcomes of rabbits [12] and luteal function of pregnant goats [9] even when GnRH dosage was reduced. Hence, this study aimed to evaluate ovarian response, Doppler sonographic analysis of ovarian and luteal arteries, and hormonal profile of goats synchronized for estrus with conventional Ovsynch protocol or nanodelivery Ovsynch protocol.

2. Materials and Methods

2.1. Fabrication and Characterization of Hormone-Conjugated Nanoparticles

Chitosan (Cat No. AL1234 00100; Alpha Chemica, Maharashtra, India) and sodium tripolyphosphate (TPP; Thermo Fisher GmbH, Kandel, Germany) were used to create a nanocarrier polymer using the ionic gelation method described by Hashem et al. [9]. Briefly, chitosan (0.1%, *w/v*) was vigorously stirred in an aqueous acidic solution (1%, *v/v*) to obtain chitosan cation nanoparticles. Furthermore, we prepared an aqueous solution of TPP (0.1 %, *w/v*). The chitosan-TPP nanoparticles complex were prepared by slowly dropping the TPP solution into the chitosan solution (2 chitosan: 1 TPP) under constant magnetic stirring (1200 rpm) for 2 h. Either gonadorelin solution (Ovurelin, 100 μg gonadorelin (as acetate), /mL, Bayer New Zealand Limited, Manukau, Auckland, New Zealand) or prostaglandin $F_{2\alpha}$ solution (PGF$_{2\alpha}$, Estrumate, 250 μg cloprostenol/mL, equivalent to 263 μg cloprostenol sodium/mL, Intervet/Merck Animal Health) was constantly dropped under magnetic stirring (1200 rpm) to the chitosan-TPP nanoparticle solution (1 mL hormone: 3 mL chitosan-TPP nanoparticles complex) for 2 h. A dynamic light scattering leaser and Doppler electrophoresis (Zetasizer Nano ZS, Malvern Instruments Ltd., Worcestershire, UK) were used to identify physicochemical characteristics (the size, polydispersity (PdI), and zeta potential) of chitosan-TPP nanoparticles conjugated or not with GnRH or PGF$_{2\alpha}$.

2.2. Animals and Experimental Design

This study was conducted at the Experimental Farm of the Faculty of Veterinary Medicine, Cairo University, Egypt. All experimental protocols and procedures were performed according to the regulations of the Veterinary Animal Care Committee of the Faculty of Veterinary Medicine, Cairo University (Ethical approval number: Vet CU 8/03/2022/423). Fifteen cyclic, multiparous Baladi goats of age 4.5 ± 0.5 years and with average body weight and body condition scores of 35.5 ± 1.5 kg and 3.42 ± 0.57, respectively, were used in this study. All goats were healthy with no reproductive difficulties. They were managed as one flock under the same optimal management conditions at the Faculty of Veterinary Medicine's research farm. The study was conducted during the natural breeding season. Goats received their nutritional requirements according to NRC recommendations [13].

During the experiment, goats were separated into three equal experimental groups ($n = 5$/group). The first group (GPG) received common Ovsynch protocol (an IM injection of 50 mg gonadorelin, followed by an IM injection of 125 mg cloprostenol 7 days later, and a further IM injection of 50 mg gonadorelin 2 days later), whereas the second and third groups received the Ovsynch protocol using nanofabricated hormones with the same dosage (NGPG) or half dosage (HNGPG) of the common protocol.

2.3. Ultrasonography Evaluation of Ovarian Structures and Hemodynamic Pattern

Ovarian structures and ovarian artery and luteal blood flow were evaluated using a pulsed-wave Doppler ultrasound scanner equipped with a transrectal 5–7.5 MHz linear-array transrectal transducer (EXAGO, Echo Control Medical, Angoulême, France) in color and spectral modes. The standard velocity and Doppler filter of the Doppler ultrasound were set at 25 cm/s and 100 Hz, respectively. The gate cursor was set to 0.5 mm in width. The angle of intonation was 46° [14,15]. During the experimental period, one operator was responsible for ultrasonography examination to avoid variations in collected data. Goats were subjected to intensive ultrasonography examinations after each Ovsynch protocol injection. The response to the first GnRH injection was assessed by scanning ovaries 5 days after the first GnRH injection. The response to the PGF$_{2\alpha}$ injection was assessed by scanning ovaries 2 days after PGF$_{2\alpha}$ injection. Furthermore, ovaries were scanned again at 0, 12, and 24 h after the second GnRH injection (presenting follicular phase) and on days 5, 10, and 15 of the synchronized estrous cycle (presenting luteal phase).

In each scanning session, we recorded the number and size category of each visible ovarian follicles (small follicles, ≥ 2 to ≤ 3 mm in diameter; medium follicles, >3 to <5 mm in diameter; and large follicles, ≥ 5 mm in diameter). The number and diameters of corpora lutea (CLs) were also recorded. The time of the disappearance of the dominant follicle at the follicular phase was used to estimate the time of ovulation onset by calculating the time interval between the first observation of the dominant follicle at 0 time and the midway point of the last observation of the dominant follicle (at 12 or 24 h).

For hemodynamic pattern evaluation, the blood flow indices of both ovarian and luteal arteries were determined using the spectral mode at the same time as ovarian structure evaluation. The electronic caliper of the ultrasound recorded several diameters of follicles and CLs. The vascularization of the ovarian follicle and CL was determined by the color flow mode with the presence of a pulsed-wave spectral graph showing Doppler parameters [16]. Blood flow indices of both ovarian and luteal arteries was determined using the spectral mode. Pulsatility index (PI), resistance index (RI), and peak systolic velocity (PSV) were used as blood flow index parameters [14].

2.4. Determination of Ovarian Steroids and Nitric Oxide

Blood samples were collected via jugular vein puncture corresponding to the times of ultrasound examination. Serum was harvested and stored at $-20\,°C$ for hormonal and NO analyses. P_4 (EIA-1561) and E_2 (EIA-2693) were analyzed using competitive enzyme-linked immunosorbent assay kits (DRG Diagnostics GmbH, Marburg, HE, Germany). The sensitivity of the method was 0.055 ng/mL for P_4 and 9.9 pg/mL for E_2. The intra- and inter-assay correlation coefficients were, respectively, 6.86% and 5.59% for P_4 and 2.71% and 9.39% for E_2. For NO assay, each serum sample was combined with an identical quantity of freshly organized Griess reagent and incubated for 10 min at room temperature, and absorbance was measured at 560 nm using a microtiter plate reader with the assay sensitivity of 0.225 mmol/L [17].

2.5. Statistical Analysis

The data collected after the initial GnRH and $PGF_{2\alpha}$ injections, including ovarian response, ovarian and luteal blood flow indicators, and hormonal profile, were analyzed using the generalized linear model procedure of SAS (version 9 edition. Cary, NC: SAS Inst, Inc; 2004). The used model was $y_{ij} = \mu + T_i + e_{ij}$in, where y_{ij} is the observed value of the dependent variable, μ is the overall mean, T_i is the fixed effect of the ith treatment (i = 1:4), and e_{ij} is the residual error. Least-squares procedures using a mixed model, considering the time of data collection as repeated measurements, were used to assess the effect of estrous synchronization protocols on ovarian response, ovarian and luteal blood flow indices, and hormonal profile during the follicular and luteal phases of the synchronized estrous cycle. The treatment (T, estrous synchronization protocol: GPG, NGPG, and HNGPG), the day of blood sampling and/or ovarian examination (S), and their interaction (T × S) were introduced as fixed effects and individual goats as random effects. The used model was $y_{ijk} = \mu + T_i + S_j + (T \times S)_{ij} + e_{ijk}$, where y_{ijk} is the observed value of the dependent variable determined from a sample taken from each animal, μ is the overall mean, T_i is the fixed effect of the ith treatment (i = 1:3), S_j is the fixed effect of the jth sampling/examination time (j = 1:3), $(T \times S)_{ij}$ is the interaction between treatment and day of the estrous cycle, and e_{ijk} is the residual error. All results were shown as means ($\pm$SEM), and differences between the means of different experimental groups were detected using Duncan's new multiple range test with a *p*-value of less than 0.05.

3. Results

3.1. Physicochemical Characteristics of Ovsynch Hormone Nanoparticles

Table 1 shows physicochemical characteristics of chitosan-TPP nanoparticles and hormone-loaded chitosan-TPP nanoparticles. Results showed that chitosan-TPP nanoparticles and hormone-loaded chitosan-TPP nanoparticles were in nanosize ($\approx$100 to 200 nm). The addition of hormones to the chitosan-TPP nanoparticles increased their size, indicating the binding of both molecules together. The range of PdI values of chitosan-TPP nanoparticles hormone loaded-chitosan-TPP nanoparticles was between 0.288 and 0.545. All fabricated nanoparticles (free or hormone-loaded) had positive zeta potential values.

Table 1. Physicochemical properties (particle size, polydispersity index (PdI), and zeta potential) of chitosan-TPP nanoparticles, gonadotropin-releasing hormone (GnRH)-chitosan-TPP nanoparticles, and prostaglandin $F_{2\alpha}$ ($PGF_{2\alpha}$)-chitosan-TPP nanoparticles.

Nanoparticles	Particle Size (nm)	PdI	Zeta Potential (mV)
Chitosan-TPP	110.25 ± 9.24	0.288 ± 0.05	17.36 ± 0.245
GnRH-chitosan-TPP	200.5 ± 4.25	0.442 ± 0.09	22.5 ± 0.24
$PGF_{2\alpha}$-chitosan-TPP	198.8 ± 15.20	0.545 ± 0.12	20.1 ± 0.31

3.2. Effect of Common or Nanodelivered Ovsynch Protocols on Ovarian Structures

In response to the first GnRH injection, the numbers of the small, medium, total follicles, and numbers of CLs were similar among the experimental groups, whereas the numbers of large follicles decreased ($p = 0.024$) in NGPG and HNGPG, compared with GPG. HNGPG had greater diameter of CLs ($p = 0.001$) than GPG and NGPG (Table 2).

Table 2. Ovarian response to Ovsynch protocol using conventional hormones (GPG: 50 mg gonadorelin and 125 mg cloprostenol) or nanodelivered hormones with different doses (NGPG: 50 mg gonadorelin and 125 mg cloprostenol and HNGPG: 25 mg gonadorelin and 62.5 mg cloprostenol) (means $\pm$ standard error of the mean, SEM).

Variable	Treatment (T)			SEM	*p*-Value
	GPG	NGPG	HNGPG		T
Response to the first GnRH injection					
Number of small follicles (≥ 2 to 3 mm)	6.00	5.80	6.40	0.181	0.585
Number of medium follicles (≥ 2 to ≤ 3 mm)	6.40	6.00	6.60	0.148	0.315
Number of large follicles (≥ 5 mm)	3.00 [a]	2.00 [b]	1.80 [b]	0.256	0.024
Number of total follicles	15.40	13.80	14.80	0.535	0.293
Number of corpora lutea	2.00	2.00	2.60	0.145	0.092
Diameter of corpus luteum (mm)	4.51 [b]	4.27 [b]	5.84 [a]	0.125	0.001
Response to $PGF_{2\alpha}$ injection					
Number of small follicles (≥ 2 to ≤ 3 mm)	5.80	5.60	5.40	0.106	0.493
Number of medium follicles (>3 to <5 mm)	6.00	5.20	5.80	0.151	0.198
Number of large follicles (≥ 5 mm)	3.40	3.00	3.20	0.117	0.564
Number of total follicles	15.20 [a]	13.80 [b]	14.40 [ab]	0.191	0.038
Number of corpora lutea	2.00	2.00	2.60	0.117	0.148
Diameter of corpus luteum (mm)	4.48 [b]	4.22 [b]	5.78 [a]	0.152	0.001
Follicular phase					
Number of small follicles (≥ 2 to ≤ 3 mm)	4.40	4.20	4.47	0.014	0.436
Number of medium follicles (>3 to <5 mm)	5.67 [b]	5.33 [b]	6.13 [a]	0.203	0.043
Number of large follicles (≥ 5 mm)	3.20	3.47	3.33	0.186	0.607
Number of total follicles	11.92	11.56	12.39	0.303	0.174
Dominant follicle diameter (mm)	6.07	5.96	6.21	0.077	0.414
Diameter of secondary dominant follicle (mm)	5.38	5.42	5.82	0.098	0.106
Diameter of tertiary dominant follicle (mm)	5.08	5.12	5.42	0.081	0.105
Ovulation time (h)	15.60 [a]	15.60 [a]	10.82 [b]	2.74	0.040
Luteal phase					
Number of corpora lutea	1.80	2.07	2.00	0.253	0.455
Diameter of corpus luteum (mm)	7.62 [b]	7.81 [ab]	8.03 [a]	0.310	0.001

GnRH: gonadotropin-releasing hormone (gonadorelin), $PGF_{2\alpha}$: prostaglandin $F_{2\alpha}$ (cloprostenol), follicular phase: ovaries were scanned at 0, 12, and 24 h after the second GnRH injection, and luteal phase: ovaries were scanned on days 5, 10, and 15 of the synchronized estrous cycle. Within a row, means with different superscript letters (a and b) are significantly different ($p < 0.05$).

After $PGF_{2\alpha}$ injection, the total number of follicles decreased ($p = 0.038$) in NGPG compared with GPG, whereas HNGPG recorded an intermediate value. The diameter of CLs was greater ($p = 0.001$) in HNGPG than in GPG and NGPG (Table 2).

During the follicular phase, the numbers of small and large follicles, numbers of total follicles, and diameter of dominant follicles were similar among the experimental groups

(Table 2). HNGPG increased ($p = 0.043$) the number of medium follicles compared with GPG and NGPG. HNGPG shortened ($p = 0.04$) the time interval between the second GnRH and onset of ovulation compared with GPG and NGPG (Table 2). During the luteal phase, numbers of CLs were similar among different experimental groups, whereas the diameter of CLs increased ($p < 0.001$) in HNGPG, compared with GPG and NGPG (Table 2).

3.3. Effect of Common or Nanodelivered Ovsynch Protocols on Hemodynamic Indices

In response to the first GnRH injection, HNGPG decreased ($p = 0.001$) the ovarian artery pulsatility index (PI) and resistance index (RI), compared with GPG and NGPG. HNGPG decreased ($p = 0.044$) the luteal artery RI and increased ($p = 0.041$) PSV, in comparison with GPG and NGPG (Table 3).

In response to $PGF_{2\alpha}$ injection, HNGPG decreased ovarian artery PI ($p = 0.003$) and RI ($p = 0.002$), whereas increased ($p = 0.003$) luteal artery RI compared to GPG and NGPG (Table 3).

During the follicular phase, the least ($p = 0.003$) Doppler indices (PI and RI) of ovarian artery associated with greatest ($p < 0.001$) PSV were observed in HNGPG, whereas opposite patterns of Doppler indices were observed for the luteal artery (Table 3 and Figure 1). During the luteal phase, PSV of the ovarian artery declined ($p < 0.001$) and was associated with a significant elevation of both Doppler indices ($p < 0.001$) in HNGPG, whereas opposite patterns of Doppler indices were observed for the luteal artery (Table 3 and Figure 1).

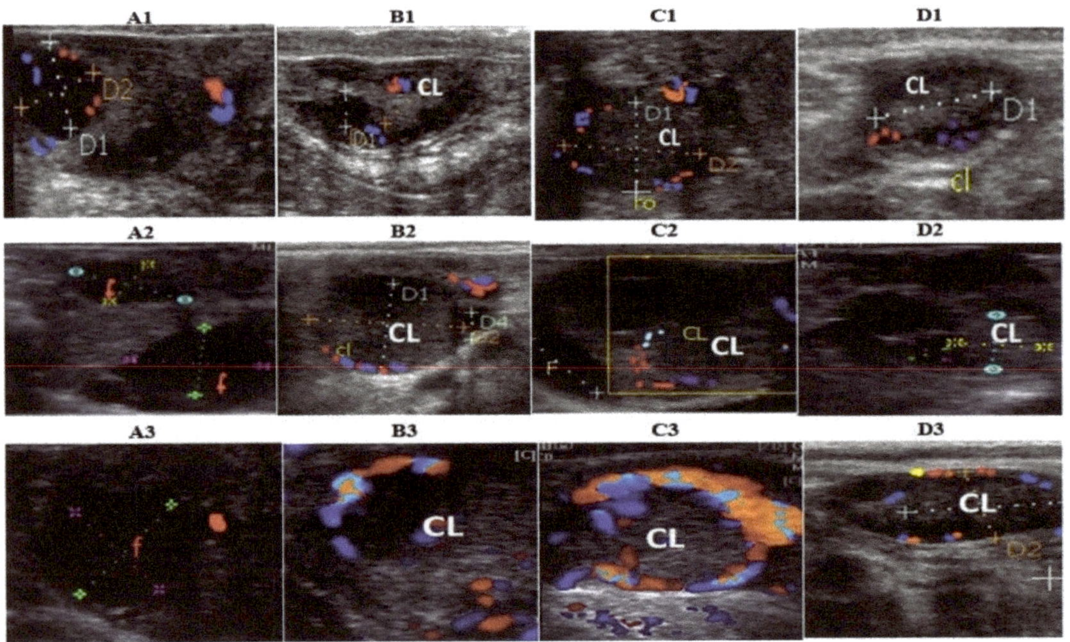

Figure 1. B-mode and colored ultrasonograms showing the preovulatory dominant follicle (Follicle: f; **A1**, **A2**, and **A3**) at the follicular phase and corpus luteum (CL) on days 5 (**B1**, **B2**, and **B3**), 10 (**C1**, **C2**, and **C3**), 15 (**D1**, **D2**, and **D3**) at the luteal phase in GPG, NGPG, and HNGPG protocols, respectively. Blue colored areas refer to the blood flow away from the probe (blood supply towards the organ/tissue) and red colored areas refer to the blood flow towards the probe (blood return from the organ/tissue towards blood circulation).

Table 3. Ovarian artery and luteal artery hemodynamic patterns in response to Ovsynch protocol using conventional hormones (GPG: 50 mg gonadorelin and 125 mg cloprostenol) or nanodelivered hormones with different doses (NGPG: 50 mg gonadorelin and 125 mg cloprostenol and HNGPG: 25 mg gonadorelin and 62.5 mg cloprostenol) (means ± standard error of the mean, SEM).

Variable	Treatment (T)			SEM	*p*-Value
	GPG	NGPG	HNGPG		T
Response to the first GnRH injection					
Ovarian artery					
PI	1.55 [a]	1.53 [a]	1.43 [b]	0.013	0.001
RI	0.53 [a]	0.53 [a]	0.41 [b]	0.014	0.001
PSV cm/s	16.03	14.48	15.01	0.203	0.134
Luteal artery					
PI	1.25	1.20	1.28	0.015	0.152
RI	0.57 [a]	0.56 [a]	0.44 [b]	0.024	0.044
PSV cm/s	12.25 [b]	12.27 [b]	13.52 [a]	0.187	0.041
Response to PGF$_{2\alpha}$ injection					
Ovarian artery					
PI	1.55 [a]	1.54 [a]	1.45 [b]	0.012	0.003
RI	0.63 [a]	0.62 [a]	0.55 [b]	0.014	0.002
PSV cm/s	12.94 [b]	12.95 [b]	14.65 [a]	0.219	0.003
Luteal artery					
PI	1.25	1.23	1.28	0.013	0.364
RI	0.57 [b]	0.57 [b]	0.66 [a]	0.014	0.047
PSV cm/s	12.75	12.42	13.08	0.214	0.438
Follicular phase					
Ovarian artery					
PI	0.91 [a]	0.88 [a]	0.81 [b]	0.025	0.003
RI	0.85 [a]	0.81 [a]	0.71 [b]	0.006	<0.001
PSV cm/s	9.76 [b]	9.77 [b]	10.62 [a]	0.111	<0.001
Luteal artery					
PI	1.24 [b]	1.25 [b]	1.32 [a]	0.024	0.020
RI	0.61 [b]	0.64 [a]	0.66 [a]	0.022	0.019
PSV cm/s	12.52 [a]	12.57 [a]	12.01 [b]	0.320	0.005
Luteal phase					
Ovarian artery					
PI	1.32 [b]	1.38 [b]	1.51 [a]	0.025	<0.001
RI	0.59 [b]	0.60 [b]	0.64 [a]	0.016	<0.001
PSV cm/s	12.25 [a]	12.17 [a]	11.25 [b]	0.164	<0.001
Luteal artery					
PI	1.04 [a]	1.02 [b]	1.02 [b]	0.017	<0.001
RI	0.48 [a]	0.45 [b]	0.41 [c]	0.007	<0.001
PSV cm/s	13.59 [b]	13.83 [b]	14.53 [a]	0.071	<0.001

GnRH: Gonadotropin-releasing hormone (gonadorelin), PI: pulsatility index, RI: resistance index, PSV: peak systolic velocity, PGF$_{2\alpha}$: prostaglandin F$_{2\alpha}$ (cloprostenol), follicular phase: ovaries were scanned at 0, 12, and 24 h after the second GnRH injection, and luteal phase: ovaries were scanned on days 5, 10, and 15 of the synchronized estrous cycle. Within a row, means with different superscript letters (a and b) are significantly different ($p < 0.05$).

3.4. Effect of Common or Nanodelivered Ovsynch Protocols on Hormonal Profile and Nitric Oxide Levels

In response to the first GnRH injection, both NGPG and HNGPG increased ($p < 0.001$ and $p < 0.001$; respectively) blood serum E$_2$ and NO levels, compared with GPG. HNGPG increased ($p < 0.001$) levels of blood serum P$_4$, compared with GPG, whereas NGPG resulted in an intermediate value (Table 4).

Table 4. Blood serum levels of ovarian steroids and nitric oxide in response to Ovsynch protocol using conventional hormones (GPG: 50 mg gonadorelin and 125 mg cloprostenol) or nanodelivered hormones with different doses (NGPG: 50 mg gonadorelin and 125 mg cloprostenol and HNGPG: 25 mg gonadorelin and 62.5 mg cloprostenol) (means $\pm$ standard error of the mean, SEM).

Variable	Treatment (T)			SEM	p-Value
	GPG	NGPG	HNGPG		T
Response to first GnRH injection					
Progesterone (ng/mL)	5.11 [b]	5.18 [ab]	5.27 [a]	0.397	<0.001
Estradiol (pg/mL)	24.11 [b]	29.93 [a]	31.08 [a]	1.39	<0.001
Nitric oxide (µmol/L)	24.91 [c]	31.25 [b]	44.29 [a]	3.52	<0.001
Response to $PGF_{2\alpha}$ injection					
Progesterone (ng/mL)	0.96	0.50	0.63	0.056	0.081
Estradiol (pg/mL)	31.07 [b]	36.40 [ab]	46.21 [a]	1.41	0.028
Nitric oxide (µmol/L)	53.69 [b]	62.90 [a]	64.61 [a]	1.38	0.007
Follicular phase					
Progesterone (ng/mL)	0.53 [a]	0.45 [b]	0.56 [a]	0.100	<0.001
Estradiol (pg/mL)	36.64 [b]	40.53 [b]	46.66 [a]	1.59	<0.001
Nitric oxide (µmol/L)	29.96 [b]	32.42 [b]	44.76 [a]	0.475	<0.001
Luteal phase					
Progesterone (ng/mL)	4.36 [b]	4.41 [b]	4.92 [a]	0.064	<0.001
Estradiol (pg/mL)	22.14	24.77	23.71	1.69	0.609
Nitric oxide (µmol/L)	35.38 [b]	42.16 [b]	54.27 [a]	0.642	0.020

GnRH: gonadotropin-releasing hormone (gonadorelin), $PGF_{2\alpha}$: prostaglandin $F_{2\alpha}$ (cloprostenol), follicular phase: ovaries were scanned at 0, 12, and 24 h after the second GnRH injection, and luteal phase: ovaries were scanned on days 5, 10, and 15 of the synchronized estrous cycle. Within a row, means with different superscript letters (a, b, and c) are significantly different ($p < 0.05$).

In response to $PGF_{2\alpha}$ injection, both NGPG and HNGPG increased ($p = 0.007$) NO levels, compared with GPG (Table 4). Blood serum E_2 levels increased ($p = 0.028$) in HNGPG, compared with GPG, whereas NGPG resulted in an intermediate value (Table 4).

During the follicular phase, the least blood serum P_4 level was observed in NGPG. HNGPG increased ($p < 0.001$) blood serum E_2, P_4, and NO levels, compared with GPG and NGPG. During the luteal phase, blood serum P_4 and NO levels increased ($p < 0.001$ and $p = 0.020$; respectively) in HNGPG, compared with GPG and NGPG, whereas no changes in blood serum E_2 levels were observed among the experimental groups (Table 4).

4. Discussion

In this study, we aimed to evaluate the efficiency of the Ovsynch estrous synchronization protocol as one of the most crucial estrous synchronization protocols in goats when GnRH and $PGF_{2\alpha}$ are delivered using a nano-based drug delivery system. Several studies have reported the advantages of engineered nanodrugs, including hormones, and their efficient biological activities due to the changes in physicochemical properties of materials in nanoforms [18,19]. Engineered nanohormones have a longer half-life time and sustained release, greater cellular uptake, and more efficient passage across epithelial or endothelial barriers [1]. These advantages can improve the delivery of the hormone to the target sites and thus the final action of the hormone. Most nanoparticles used in delivery systems are 50–250 nm in size. This size allows particles to move through various barriers and cell pores with ease, enhancing cellular uptake [20]. In this context, the success of any estrous synchronization protocol primarily depends on the pharmacokinetics and pharmacodynamics hormones. The Ovsynch estrous synchronization protocol depends on the use of GnRH and $PGF_{2\alpha}$. Given the fact that both hormones have a short lifespan and low molecular weight, they are susceptible to degradation easily by different lytic enzymes of the systemic circulation, restricting the sustained delivery of the hormones to the target sites and therefore their biological activity.

In this study, the conjugation of GnRH and $PGF_{2\alpha}$ to chitosan-TPP nanoparticles resulted in particle size of ≤ 200 nm, acceptable PdI (<0.6), and positively charged nanoparticles with acceptable zeta potential (>15 mV) for both hormones [9,21]. These properties can facilitate the delivery and uptake of both hormones by their target organs. For GnRH, where the brain is the main target organ, such physiochemical properties are suitable for efficient drug delivery to the brain. Nanoparticles with sizes ranging from 50 to 200 nm, a PdI ≤ 0.4, and a positive surface charge (up to 15 mV) are efficient for drug delivery to the brain [22]. These results can be confirmed in our study and several previous studies, as nano-GnRH even if used with a low dosage improved several reproductive events in farm animals, such as ovulation [12] and CL luteinization [9].

The main target organ for $PGF_{2\alpha}$ is the CL, and the main barrier to $PGF_{2\alpha}$ activity is its rapid degradation through the circulatory system, particularly the pulmonary system. However, in this study, the reduction of $PGF_{2\alpha}$ dosage to half did not adversely affect its luteolytic activity. To the best of our knowledge, there are no available studies in the field of estrous synchronization of farm animals that discussed the biological efficiency of nanoengineered $PGF_{2\alpha}$. However, some studies referred to the importance of transforming prostaglandin analogs to ensure the accumulation and sustained release of these hormones a long time [23,24].

Results of this study show the relevance of nanoengineered GnRH and $PGF_{2\alpha}$ in improving the response of the ovary, ovarian hemodynamic patterns, and hormonal profile to the Ovsynch protocol after each injection, resulting in better ovulatory wave characteristics and subsequent luteal functions of the synchronized cycle. In Ovsynch protocol, the supposed role of the first GnRH injection is to trigger ovulation of the LH-responsive follicle and/or to luteinize growing follicles. In this study, the first GnRH in both NGPG and HNGPG decreased the number of large follicles, compared with GPG. This is one of the intended effects of the first GnRH injection, as the termination of the existing follicular wave enables the development of a new follicular wave.

In this study, the first GnRH injection of HNGPG increased the diameter of CLs, as well as P_4 levels on day five after the first GnRH injection. These effects may be related to the positive effects of the first GnRH of HNGPG on ovarian hemodynamic patterns (increased PSV and decreased PI and/or RI). The improvements in ovarian hemodynamic patterns can be ascribed to the increased levels of blood serum E_2 and NO. Both E_2 and NO have vasodilatation effects, resulting in improved vascular blood flow, supporting the development of ovarian structures and the hormone synthesis and secretion [14]. The level of P_4 during the emergence of the new follicular wave is a crucial factor in the development of a new follicular wave, as greater P_4 levels stimulate follicular turnover and boost the growth of a new follicular wave [25,26].

In this study, two main findings should be highlighted in response to the $PGF_{2\alpha}$ injection. First, the decrease in blood serum P_4 levels was not associated with a decrease in the number and/or diameters of CLs 2 days after $PGF_{2\alpha}$ injection, compared with those recorded before the $PGF_{2\alpha}$ injection, in all experimental groups. These results are in agreement with those previously reported on the mechanism of $PGF_{2\alpha}$-induced luteolysis.

The luteolysis process is divided into two stages: a rapid drop in the functionality of the CL, resulting in lower P_4 levels, and a phase of structural regression, resulting in luteal tissue shrinkage [27]. Second, both NGPG and HNGPG improved blood flow of the ovarian artery and increased blood serum E_2 and NO levels, when compared with GPG. Classically, $PGF_{2\alpha}$ induces luteolysis by decreasing luteal blood flow, resulting in hypoxia; however, before the initial steps of the luteolytic cascade, $PGF_{2\alpha}$ increases luteal blood flow and CL vascularization [28,29]. Additionally, the increased blood serum E_2 and NO levels in NGPG and HNGPG may contribute to this effect through their vasodilation effects [30,31]. These findings support the relevance of the $PGF_{2\alpha}$ in nanoform for the expected role of this injection even when the dose is reduced to half of the conventional dose. Therefore, it can be said that administering $PGF_{2\alpha}$ using a nanodelivery system may improve the biological function of $PGF_{2\alpha}$.

One of the parameters used for judging the effectiveness of the Ovsynch protocol is the ability of the protocol to induce tighter synchrony of ovulation and to shorten the time required from the second GnRH injection to ovulation. Tighter and shorter ovulation periods allow for easier applications of timed artificial insemination, as fertility variations due to scattering ovulations relative to insemination time are minimized, avoiding ova aging after ovulation [3,8]. In our study, HNGPG significantly reduced the time interval between the second GnRH and ovulation, resulting in one of the most crucial goals of the Ovsynch protocol. HNGPG-treated goats had better ovulatory wave characteristics, larger diameters of dominant and subdominant follicles, increased number of medium follicles, and higher blood serum E_2 levels compared to GPG and NGPG. These effects can be associated with various positive effects of nanodelivered hormones on ovulatory follicle characteristics, ovarian blood flow, and the hormonal milieu during ovulatory wave emergence. HNGPG increased blood serum P_4 levels during the emergence of the ovulatory wave, the period after the first GnRH injection. Elevated levels of blood serum P_4 for the short term (5–7 days) are critical for stimulating ovarian follicles turnover and emergence of new ovulatory waves [25]. Additionally, these positive effects may be associated with increased levels of blood serum E_2 and NO and the subsequent improvements in ovarian artery blood flow [32].

Interestingly, the positive effects of HNGPG extended to the luteal phase of the synchronized estrous cycle, as goats of this group showed the greatest diameter of CLs and blood serum P_4 levels. One reason restricting the efficiency of an estrous synchronization protocol is the inadequacy of CL functioning and/or formation of short-lived CL, inhibiting subsequent fertility and pregnancy maintenance. In species like goats, the CL is the main source of P_4, which is required during the entire pregnancy for the completion of a successful pregnancy [6]. The improved structure and functionality of CL during the luteal phase of the synchronized estrous cycle in HNGPG can be ascribed to the improved luteal blood supply, as the vascularization pattern of CL blood vessels and endothelial cells play a crucial role in CL function [33]. In this context, NO has been identified as a major mediator of increased CL blood flow in cattle [34]. Moreover, several studies have reported the ability of GnRH administered around ovulation time and/or during the early luteal phase to improve the functioning of newly formed CL by triggering the release of pituitary luteinizing hormone, LH, and subsequent luteinization of granulosa and theca cells [35]. This effect depends on the ability of GnRH to induce sufficient gonadotropin release [23], mainly LH, from the pituitary gland; consequently, long-lasting release of GnRH, as expected, following transforming GnRH to nanoform, may improve its biological efficiency by facilitating sustained surge of gonadotropins.

Finally, the question that should be addressed in this study is why NGPG did not improve the characteristics of the ovulatory follicle, CL structure and functionality, and HNGPG. In fact, there is no elucidation of these findings. However, it can be speculated that providing GnRH and/or $PGF_{2\alpha}$ by a nanodelivery system with the same doses used in conventional Ovsynch protocol could result in improper responses. Long-lasting release of the hormones may lead to receptors' refractions and a lack of response to the hormone [36]. Nevertheless, to confirm or reject this assumption, more studies are required to investigate the dosage effect, release pattern of the nanofabricated hormones in the biological systems, and the interplay between these factors and animal response. Moreover, future studies have to show more information on the effect of such protocol on the reproductive performance of goats after insemination and subsequent fertility and pregnancy outcomes.

5. Conclusions

Conclusively, the results of this study confirmed the ability of HNGPG treatment to enhance the ovarian and luteal blood flow at the follicular and luteal phases, characteristics of the ovulatory wave, E_2 synthesis at the follicular phase, and corpus luteum function and P_4 synthesis at the luteal phase. Accordingly, the nanodelivery system for hormones of

Ovsynch protocol can be recommended as a new promising reproduction management practice for improving Ovsynch protocol estrous synchronization outcomes of goats.

Author Contributions: Conceptualization, N.M.H.; methodology, N.M.H., E.A.A., H.R.E.-S. and M.F.; software, N.M.H. and E.A.A.; data curation, N.M.H. and E.A.A.; writing—original draft preparation, N.M.H., E.A.A., H.R.E.-S. and M.F.; writing—review and editing, N.M.H. All authors have read and agreed to the published version of the manuscript.

Funding: This research received no external funding.

Institutional Review Board Statement: This study was conducted at the Experimental Farm of the Faculty of Veterinary Medicine, Cairo University, Egypt. All experimental protocols and procedures were performed according to the regulations of the Veterinary Animal Care Committee of the Faculty of Veterinary Medicine (Ethical approval number: Vet CU 8/03/2022/423), Cairo University. The nano- hormones were fabricated and characterized at the Lab of Animal Physiology of the Faculty of Agriculture, and the Central Lab of the Faculty of Pharmacy, Alexandria Uinversity.

Informed Consent Statement: Not applicable.

Data Availability Statement: The data presented in this study are available on request from the corresponding author. The data are not publicly available because of privacy.

Acknowledgments: The authors thank Fayez Salib, Farm Superintendent, in the Faculty of Veterinary Medicine, Cairo University, for providing the facilities to perform this study and the availability of goats anytime during the study.

Conflicts of Interest: The authors declare no conflict of interest.

References

1. Hashem, N.M.; Gonzalez-Bulnes, A. Nanotechnology and reproductive management of farm animals: Challenges and advances. *Animals* **2021**, *11*, 1932. [CrossRef] [PubMed]
2. Gupta, P.P.; Nikhil Kumar Tej, J.; Johnson, P.; Nandi, S.; Mondal, S.; Kaushik, K.; Krishna, K. Efficiency of different synchronization protocols on oestrous response and rhythmic changes in 17β-oestradiol and progesterone hormone concentration in Salem Black goats. *Biol. Rhythm Res.* **2021**, *52*, 11–21. [CrossRef]
3. Hashem, N.; El-Zarkouny, S.; Taha, T.; Abo-Elezz, Z. Oestrous response and characterization of the ovulatory wave following oestrous synchronization using PGF$_{2\alpha}$ alone or combined with GnRH in ewes. *Small Rumin. Res.* **2015**, *129*, 84–87. [CrossRef]
4. Holtz, W.; Sohnrey, B.; Gerland, M.; Driancourt, M.-A. Ovsynch synchronization and fixed-time insemination in goats. *Theriogenology* **2008**, *69*, 785–792. [CrossRef]
5. Nur, Z.; Nak, Y.; Nak, D.; ÜSTÜNER, B.; Tuna, B.; Şİmşek, G.; Sağirkaya, H. The use of progesterone-supplemented Co-synch and Ovsynch for estrus synchronization and fixed-time insemination in nulliparous Saanen goat. *Turk. J. Vet. Anim. Sci.* **2013**, *37*, 183–188.
6. Panjaitan, B.; Dewi, A.; Nasution, F.F.; Adam, M.; Siregar, T.N.; Thasmi, C.N.; Syafruddin, S. Comparison of estrous performance and progesterone level of kacang goats induced by PGF$_{2\alpha}$ versus ovsynch protocol. In Proceedings of the E3S Web of Conferences, Banda Aceh, Indonesia, 15–17 October 2019; p. 01045.
7. Borchardt, S.; Pohl, A.; Carvalho, P.; Fricke, P.; Heuwieser, W. Effect of adding a second prostaglandin F2α injection during the Ovsynch protocol on luteal regression and fertility in lactating dairy cows: A meta-analysis. *J. Dairy Sci.* **2018**, *101*, 8566–8571. [CrossRef]
8. Ali, A.; Hayder, M.; Saifelnaser, E. Ultrasonographic and endocrine evaluation of three regimes for oestrus and ovulation synchronization for sheep in the subtropics. *Reprod. Domest. Anim.* **2009**, *44*, 873–878. [CrossRef]
9. Hashem, N.; Sallam, S. Reproductive performance of goats treated with free gonadorelin or nanoconjugated gonadorelin at estrus. *Domest. Anim. Endocrinol.* **2020**, *71*, 106390. [CrossRef]
10. Rather, M.; Bhat, I.; Gireesh-Babu, P.; Chaudhari, A.; Sundaray, J.; Sharma, R. Molecular characterization of kisspeptin gene and effect of nano–encapsulted kisspeptin-10 on reproductive maturation in Catla catla. *Domest. Anim. Endocrinol.* **2016**, *56*, 36–47. [CrossRef]
11. Hashem, N.M.; Gonzalez-Bulnes, A. State-of-the-art and prospective of nanotechnologies for smart reproductive management of farm animals. *Animals* **2020**, *10*, 840. [CrossRef]
12. Hassanein, E.M.; Hashem, N.M.; El-Azrak, K.E.-D.M.; Gonzalez-Bulnes, A.; Hassan, G.A.; Salem, M.H. Efficiency of GnRH–loaded chitosan nanoparticles for inducing LH secretion and fertile ovulations in protocols for artificial insemination in rabbit does. *Animals* **2021**, *11*, 440. [CrossRef]

13. National Research Council (US), Committee on Nutrient Requirements of Small Ruminants; National Research Council, Committee on the Nutrient Requirements of Small Ruminants, Board on Agriculture, Division on Earth and Life Studies. *Nutrient Requirements of Small Ruminants: Sheep, Goats, Cervids, and New World Camelids*; China Legal Publishing House: Beijing, China, 2007.

14. Abdelnaby, E.A.; El-Maaty, A.M.A.; Ragab, R.; Seida, A. A ovsynch produced larger follicles and corpora lutea of lower blood flow associated lower ovarian and uterine blood flows, estradiol and nitric oxide in cows: Ovsynch ovarian response and blood flow of cows. *J. Adv. Vet. Res.* **2020**, *10*, 165–176.

15. Abouelela, Y.S.; Yasin, N.A.; Khattab, M.A.; El-Shahat, K.; Abdelnaby, E.A. Ovarian, uterine and luteal hemodynamic variations between pregnant and non-pregnant pluriparous Egyptian buffalos with special reference to their anatomical and histological features. *Theriogenology* **2021**, *173*, 173–182. [CrossRef] [PubMed]

16. Abdelnaby, E.; El-Maaty, A.A. Luteal blood flow and growth in correlation to circulating angiogenic hormones after spontaneous ovulation in mares. *Bulg. J. Vet. Med.* **2017**, *20*, 97–109. [CrossRef]

17. Abdelnaby, E.A.; El-Maaty, A.M.A.; Ragab, R.S.; Seida, A.A. Assessment of Uterine vascular perfusion during the estrous cycle of mares in connection to circulating leptin and nitric oxide concentrations. *J. Equine Vet. Sci.* **2016**, *39*, 25–32. [CrossRef]

18. Saha, R.; Bhat, I.A.; Charan, R.; Purayil, S.B.P.; Krishna, G.; Kumar, A.P.; Sharma, R. Ameliorative effect of chitosan-conjugated 17α-methyltestosterone on testicular development in Clarias batrachus. *Anim. Reprod. Sci.* **2018**, *193*, 245–254. [CrossRef] [PubMed]

19. Munari, C.; Ponzio, P.; Alkhawagah, A.; Schiavone, A.; Mugnai, C. Effects of an intravaginal GnRH analogue administration on rabbit reproductive parameters and welfare. *Theriogenology* **2019**, *125*, 122–128. [CrossRef]

20. Danaei, M.; Dehghankhold, M.; Ataei, S.; Hasanzadeh Davarani, F.; Javanmard, R.; Dokhani, A.; Khorasani, S.; Mozafari, M. Impact of particle size and polydispersity index on the clinical applications of lipidic nanocarrier systems. *Pharmaceutics* **2018**, *10*, 57. [CrossRef]

21. Barbari, G.R.; Dorkoosh, F.A.; Amini, M.; Sharifzadeh, M.; Atyabi, F.; Balalaie, S.; Tehrani, N.R.; Tehrani, M.R. A novel nanoemulsion-based method to produce ultrasmall, water-dispersible nanoparticles from chitosan, surface modified with cell-penetrating peptide for oral delivery of proteins and peptides. *Int. J. Nanomed.* **2017**, *12*, 3471. [CrossRef]

22. Saraiva, C.; Praça, C.; Ferreira, R.; Santos, T.; Ferreira, L.; Bernardino, L. Nanoparticle-mediated brain drug delivery: Overcoming blood–brain barrier to treat neurodegenerative diseases. *J. Control. Release* **2016**, *235*, 34–47. [CrossRef]

23. Ishihara, T.; Yamashita, Y.; Takasaki, N.; Yamamoto, S.; Hayashi, E.; Tahara, K.; Takenaga, M.; Yamakawa, N.; Ishihara, T.; Kasahara, T. Prostaglandin E1-containing nanoparticles improve walking activity in an experimental rat model of intermittent claudication. *J. Pharm. Pharmacol.* **2013**, *65*, 1187–1194. [CrossRef] [PubMed]

24. Kanaya, T.; Miyagawa, S.; Kawamura, T.; Sakai, Y.; Masada, K.; Nawa, N.; Ishida, H.; Narita, J.; Toda, K.; Kuratani, T. Innovative therapeutic strategy using prostaglandin I 2 agonist (ONO1301) combined with nano drug delivery system for pulmonary arterial hypertension. *Sci. Rep.* **2021**, *11*, 7292. [CrossRef] [PubMed]

25. Savio, J.; Thatcher, W.; Morris, G.; Entwistle, K.; Drost, M.; Mattiacci, M. Effects of induction of low plasma progesterone concentrations with a progesterone-releasing intravaginal device on follicular turnover and fertility in cattle. *Reproduction* **1993**, *98*, 77–84. [CrossRef] [PubMed]

26. Macmillan, K.L. Recent advances in the synchronization of estrus and ovulation in dairy cows. *J. Reprod. Dev.* **2010**, *56*, S42–S47. [CrossRef] [PubMed]

27. Herzog, K.; Brockhan-Lüdemann, M.; Kaske, M.; Beindorff, N.; Paul, V.; Niemann, H.; Bollwein, H. Luteal blood flow is a more appropriate indicator for luteal function during the bovine estrous cycle than luteal size. *Theriogenology* **2010**, *73*, 691–697. [CrossRef] [PubMed]

28. Alves, N.; Torres, C.; Guimarães, J.; Moraes, E.; Costa, P.; Silva, D. Ovarian follicular dynamic and plasma progesterone concentration in Alpine goats on the breeding season. *Arq. Bras. De Med. Veterinária E Zootec.* **2018**, *70*, 2017–2022. [CrossRef]

29. Jonczyk, A.W.; Piotrowska-Tomala, K.K.; Skarzynski, D.J. Comparison of intra-CL injection and peripheral application of prostaglandin F2α analog on luteal blood flow and secretory function of the bovine corpus luteum. *Front. Vet. Sci.* **2021**, *8*, 811809. [CrossRef]

30. Zoller, D.; Lüttgenau, J.; Steffen, S.; Bollwein, H. The effect of isosorbide dinitrate on uterine and ovarian blood flow in cycling and early pregnant mares: A pilot study. *Theriogenology* **2016**, *85*, 1562–1567. [CrossRef]

31. Musicki, B.; Liu, T.; Lagoda, G.A.; Bivalacqua, T.J.; Strong, T.D.; Burnett, A.L. Endothelial nitric oxide synthase regulation in female genital tract structures. *J. Sex. Med.* **2009**, *6*, 247–253. [CrossRef]

32. Mattioli, M.; Barboni, B.; Turriani, M.; Galeati, G.; Zannoni, A.; Castellani, G.; Berardinelli, P.; Scapolo, P.A. Follicle activation involves vascular endothelial growth factor production and increased blood vessel extension. *Biol. Reprod.* **2001**, *65*, 1014–1019. [CrossRef]

33. Miyamoto, A.; Shirasuna, K. Luteolysis in the cow: A novel concept of vasoactive molecules. *Anim. Reprod. (AR)* **2018**, *6*, 47–59.

34. Shirasuna, K.; Watanabe, S.; Asahi, T.; Wijayagunawardane, M.P.; Sasahara, K.; Jiang, C.; Matsui, M.; Sasaki, M.; Shimizu, T.; Davis, J.S. Prostaglandin F 2α increases endothelial nitric oxide synthase in the periphery of the bovine corpus luteum: The possible regulation of blood flow at an early stage of luteolysis. *Reproduction* **2008**, *135*, 527–539. [CrossRef] [PubMed]

35. Hashem, N.; El-Azrak, K.; El-Din, A.N.; Taha, T.; Salem, M. Effect of GnRH treatment on ovarian activity and reproductive performance of low-prolific Rahmani ewes. *Theriogenology* **2015**, *83*, 192–198. [CrossRef] [PubMed]
36. Junaidi, A.; Williamson, P.; Cummins, J.; Martin, G.; Blackberry, M.; Trigg, T. Use of a new drug delivery formulation of the gonadotrophin-releasing hormone analogue Deslorelin for reversible long-term contraception in male dogs. *Reprod. Fertil. Dev.* **2003**, *15*, 317–322. [CrossRef] [PubMed]

Article

Improving Rabbit Doe Metabolism and Whole Reproductive Cycle Outcomes via Fatty Acid-Rich *Moringa oleifera* Leaf Extract Supplementation in Free and Nano-Encapsulated Forms

Nagwa I. El-Desoky, Nesrein M. Hashem *, Ahmed G. Elkomy and Zahraa R. Abo-Elezz

Department of Animal and Fish Production, Faculty of Agriculture (El-Shatby), Alexandria University, Alexandria 21545, Egypt; enagwa278@gmail.com (N.I.E.-D.); ahmed.elkoumi@alexu.edu.eg (A.G.E.); mkosba@hotmail.com (Z.R.A.-E.)
* Correspondence: nesreen.hashem@alexu.edu.eg

Citation: El-Desoky, N.I.; Hashem, N.M.; Elkomy, A.G.; Abo-Elezz, Z.R. Improving Rabbit Doe Metabolism and Whole Reproductive Cycle Outcomes via Fatty Acid-Rich *Moringa oleifera* Leaf Extract Supplementation in Free and Nano-Encapsulated Forms. *Animals* **2022**, *12*, 764. https://doi.org/10.3390/ani12060764

Academic Editor: Angela Trocino

Received: 10 February 2022
Accepted: 15 March 2022
Published: 18 March 2022

Publisher's Note: MDPI stays neutral with regard to jurisdictional claims in published maps and institutional affiliations.

Simple Summary: Under intensive rabbit production systems, due to the increased energy requirements of reproductive events, specifically pregnancy and lactation, rabbit does may confront several metabolic disorders as a result of energy imbalance. *Moringa oleifera* leaf ethanolic extract (ME) is one of the phytogenic extracts that has an impressive range of phytochemicals, specifically fatty acids (FAs). These phytochemicals may be biologically effective to support metabolism and reproductive functions of rabbit does during different reproductive cycle events. However, the high FAs content of ME makes them highly susceptible to lipid oxidation, diminishing their nutritional value and biological effects. In this study, we aimed to test the effects of FAs of ME either in a free-from or in a nano-encapsulated form on metabolism, immunity, milk production, milk composition, and reproductive performance of rabbit does during different physiological status (premating, mating, pregnancy, and lactation). The results showed that ME improved health, metabolism, immune functions, milk production and composition, and reproductive performance of rabbit does. These effects remained obvious even when a lower dose of ME was used in a nano-encapsulated form.

Abstract: The effects of free and nano-encapsulated ME supplementations on the metabolism, immunity, milk production and composition, and reproductive performance of rabbit does during premating, mating, pregnancy, and lactation were investigated. Multiparous rabbit does (n = 26 per group) received 50 mg of free ME (FME) daily, 25 mg of nano-encapsulated ME (HNME), or 10 mg of nano-encapsulated ME (LNME) per kilogram of body weight or were not supplemented (C) during a whole reproductive cycle. The ME contained 30 fatty acids with 54.27% total unsaturated fatty acids (USFAs). The fatty acid encapsulation efficiency of alginate nanoparticles was 70.46%. Compared with the C group, rabbits in all ME treatments had significantly increased body weight, feed intake, and glucose concentration and significantly decreased non-esterified free fatty acids and β-hydroxybutyrate concentrations. Rabbits supplemented with ME also had significantly increased white blood cell counts, phagocytic activity, lysozyme activity, and immunoglobulin G and decreased interleukin-1β concentrations. Moreover, ME supplementation significantly increased the concentrations of colostrum immunoglobulins, milk yield and energy content, and milk USFAs (omega-3 and 6). Rabbit does in the ME treatments had significantly higher conception and parturition rates and better litter characteristics than the C rabbit does. These results demonstrate the positive role of ME fatty acids on the health status and productive and reproductive performance of rabbit does at different physiological stages. Compared with the FME treatment, these parameters were further improved in rabbits that received nano-encapsulated ME at lower doses, illustrating how nano-encapsulation technology improves the bioavailability of ME.

Keywords: Moringa; encapsulation; reproduction; milk; fatty acid; immunoglobulins

---

1. Introduction

On rabbit farms, rabbit does are considered biological units used for producing new individuals; thus, their reproductive efficiency governs production and subsequently, economic returns. Rabbits are typically bred in intensive production systems, which results in the simultaneous occurrence of more than one reproductive event, mainly pregnancy and lactation. This intensive production system represents a great metabolic burden on rabbit does, as they must balance their nutritional requirements with pregnancy and lactation requirements to sustain milk production and/or the growth of the fetuses [1,2].

Imbalances in the nutritional/energy requirements of rabbit does can lead to serious metabolic disorders and negative energy balance, which decreases the reproductive performance of rabbit does [3]. The body condition and energy balance of rabbit does have been shown to affect their short-term and long-term reproductive efficiency [4]. If rabbit does lack adequate energy sources and nutritional requirements during their reproductive cycle, they may develop metabolic disorders, such as glycemia (shortage of circulating glucose levels), and exhibit increased levels of undesirable metabolites, such as non-esterified free fatty acids (NEFAs) and β-hydroxybutyrate (β-HB), due to body fat lipolysis [5,6]. Additionally, negative energy balances are associated with disrupted immunity and elevated inflammatory reactions [7].

Many attempts have been made to improve the body condition and energy status of rabbit does during sensitive reproductive windows using energy-rich diets. However, feed intake does not meet the energy deficit, leading to the mobilization of body reserves, lost body energy, and inadequate reproductive performance [5,8]. Functional nutrients/molecules, such as fatty acids (FAs), amino acids, and vitamins that can modulate metabolism and/or support physiological functions may represent an effective intervention [9,10]. Among these functional nutrients/molecules, FAs have many vital biological roles in mammals. The hepatocytes use FAs as precursors for synthesizing essential energy-yielding metabolites, such as triglycerides [11], and are used as an energy substrate by oocytes and embryos [12]. Moreover, FAs are necessary for gamete/embryo differentiation, growth, and immune system function [12] via several modes of action. Many FAs act as functional molecules for many reproductive events; numerous studies have reported the positive effects of FAs on ovulation, fertilization, pregnancy outcomes, and milk production in rabbits [9,13–15]. For example, the ovulation rate and the number of normal embryos are positively correlated with monounsaturated FAs (MUFAs) concentrations at the time of mating [12,15]. In the pre- and post-implantation periods of rabbit does, supplementation with docosahexaenoic acid methyl ester (DHA) and eicosapentaenoic acid methyl ester (EPA) increases plasma progesterone levels and improves pregnancy outcomes [16,17].

Moringa oleifera is a tropical plant rich in biologically active substances, including FAs, amino acids, vitamins, minerals, and phytochemicals. Several studies confirmed the enrichment of this plant with biologically active FAs, specifically polyunsaturated FAs (PUFAs) [18–20]. However, PUFAs are highly susceptible to lipid oxidation, which diminishes their nutritional value and biological effects. To overcome this problem, encapsulation processes can be used to protect products rich in PUFAs from the oxidation reactions that occur when PUFAs are exposed to oxygen, metal ions, high temperatures, and light [21].

In this study, the effects of FAs in either free-from or nano-encapsulated form of ME on metabolism, immunity, milk production, milk composition, and reproductive performance in rabbit does during different physiological stages (premating, mating, pregnancy, and lactation) were investigated.

2. Materials and Methods

This study was conducted at the Laboratory of Rabbit Physiology Research, Agricultural Experimental Station, Faculty of Agriculture, Alexandria University, Egypt. The experimental procedures were revised and approved by the Alexandria University-Institutional Animal Care and Use Committee (AU-IACUC) with approval number AU 08 21 07 26 2 82.

2.1. Moringa oleifera Leaf Extraction and Nanofabrication

Moringa (*Moringa oleifera*) leaves were extracted using 70% ethanol solution, filtrated, and evaporated to complete dryness. The dried ME was used to fabricate a sodium alginate nano-complex using calcium chloride ($CaCl_2$) as a cross-linking agent by adopting the ionic-gelation method [22].

2.2. Nano-Encapsulated ME Physicochemical Properties and Fatty Acid Profile

In our previous study [22], we determined the physicochemical properties of alginate–$CaCl_2$ nano-encapsulated ME assayed using a dynamic light scattering nanoparticle analyzer. The mean size of the fabricated nanoparticles was 93.69 nm, the zeta potential was 8.95 mV, and the polydispersity (PdI) was 0.442. The encapsulation efficiency of alginate–$CaCl_2$ nanoparticles for ME phenolic compounds was 57.43% [22].

In this study, further analyses were conducted to identify the FA profile of ME and the FA encapsulation efficiency (EE) of the alginate–$CaCl_2$ nanoparticles. For this purpose, 10 mL of each ME raw and supernatant (free ME, nonencapsulated fractions) sample was mixed with 50 mL of a methanol-chloroform mixture (2:1) in a separation funnel. The mixture was vigorously shaken for 5 min and left for separation. The mixture was centrifuged for 10 min at 2500 rpm to achieve layer separation. The chloroform layer was removed, filtered through a coarse filter paper, and evaporated until completely dry [23]. For fat methylation, a weight of 50 mg of lipids of both raw extracted ME and supernatant (free ME, nonencapsulated fractions obtained by centrifugation of the ME after the nano-encapsulation process) was incubated with 10 mL of a sulfuric acid (1%) and methanol mixture (1:100) in a water bath at 90 °C for 90 min. After cooling, 8 mL of water and 5 mL of petroleum ether were added, and the mixture was shaken vigorously. The ether layer was withdrawn and evaporated to dryness. FA methyl esters were quantified using a Thermo Scientific gas chromatograph GC Trace 1300 coupled with an EI Mass spectrometer ISQ 7000 model (Thermo Fisher Scientific, 168 Third Avenue Waltham, MA, USA) equipped with Thermo TR-50 MS capillary column (30 m in length $\times$ 250 μm in diameter $\times$ 0.25 μm in thickness of film). Spectroscopic detection by GC–MS involved an electron ionization system that utilized high energy electrons (70 eV), MS transfer line temperature 300 °C and ion source temperature 300 °C. Pure helium gas (99.995%) was used as the carrier gas with a flow rate of 1 mL/min. The initial temperature was set at 60 °C for 2 min, then increased to 100 °C at a rate of 10 °C/min kept for 5 min, then with 10 °C/min to 150 °C and kept for 5 min, then with 10 °C/min to 200 °C and kept for 5 min, then with 10 °C/min to 250 °C and kept for 20 min. One microliter of the prepared extracts was injected in a splitless mode. The concentrations of individual FAs in raw and supernatant of ME were calculated by dividing the relative area for each detected individual FA by the total area for Fas and expressed as g/100 g FA methyl esters [24]. The EE (%) of ME Fas by sodium alginate–$CaCl_2$ nanoparticles were estimated by determining the FA concentration of each FA in the raw ME (before encapsulation, C raw) and the resultant supernatant following the collection of the nano-complex particles (C supernatant), using the following equation: EE (%) = C raw–C supernatant/C raw $\times$ 100.

2.3. Animal Management and Experimental Design

One hundred and four, multiparous, V-line rabbit does (a maternal synthetic line selected based on litter size at weaning [25]), weighing 2.75 $\pm$ 0.18 kg, were managed under similar housing (The average of ambient temperature, relative humidity, and daylength was 30.04 $\pm$ 2.05 °C, 78.39 $\pm$ 3.01%, and 13.20 $\pm$ 0.73 h, respectively) and hygiene conditions. Each doe was kept in an individual standard galvanized battery cage (60 cm L $\times$ 55 cm W $\times$ 40 cm H) equipped with feeders and automatic drinkers. Rabbit does were fed a pellet diet (18.32% crude protein and 10.76 MJ/kg digestible energy) containing 18% barley, 25% wheat bran, 6% yellow corn, 18% soybean, 27% alfalfa hay, 3% molasses, 1% di-calcium phosphate, and 2% NaCl and premix, meeting their daily nutritional requirements as recommended by the National Research Council [26]. Rabbit does were randomly divided into four experimental groups (n = 26 per group): 0 mg/kg body weight (BW) free ME ©, 50 mg/kg BW free ME (FME), 25 mg/kg BW nano-encapsulated ME (HNME), and 10 mg/kg BW nano-encapsulated ME (LNME). The treatments were orally administered to rabbit does daily for a complete reproductive cycle (around 75 days), including premating (10 days pre-insemination), mating (insemination day), pregnancy (30 days), and lactation (30 days). The dose of ME for each rabbit doe was individually added to 50 mL of water, which is less than the daily consumption rate to ensure full consumption of each dose. Then clean tap water was offered to each doe *ad libitum*. This process was daily repeated during the entire experimental period. Estrus synchronization was accomplished by administering 25 IU of equine chorionic gonadotropin (Gonaser®, Hipra, Spain) via intramuscular injection (IM) to each doe, which was followed 48 h later by the IM administration of 0.8 µg of gonadotropin-releasing hormone (0.8 µg buserelin; Receptal, Boxmeer, Holland) to induce ovulation. Does were immediately artificially inseminated with 0.2 mL (15×10^6 sperm/insemination) of fresh diluted (1:5) pooled semen collected from previously proven-fertile rabbit bucks [27].

2.4. Physiological Variables

Each rabbit doe was weighed weekly in the morning before offering feed. The feed intake (g/day) was calculated daily by subtracting the unconsumed feed from the total amount of offered feed [28].

2.5. Blood Sampling and Analysis

Blood samples were collected from the marginal ear veins of eight randomly selected does using heparinized tubes (blood collection vacuum tubes, REF: G40111, NEW VAC, China) during premating (10 days after the beginning of the treatments), mating (insemination day), pregnancy (days 10 and 20 of pregnancy), and lactation (day 7). Each sample was divided into whole blood and separated blood plasma ($2000 \times g$ for 20 min at $-4\,^\circ$C to obtain plasma) for immune variable and biochemical attribute analyses.

2.5.1. Blood Biochemical Attributes

Blood plasma glucose concentrations were colorimetrically measured using commercial test kits (SPINREACT, Girona, Spain). Blood plasma β-HB and NEFA concentrations were determined via the kinetic enzymatic method using commercial kits (DiaSys Diagnostic Systems, Holzheim, Germany).

2.5.2. Immune Variables

Whole blood samples were used to assess immune variables, namely white blood cells (WBCs), WBC differential count, and phagocytic activity. To determine the phagocytic activity, a mixture (1:1) of whole blood and *Staphylococcus albus* (1.0×10^5 cells/mL) in phosphate-buffered solution (pH = 7.2) was incubated for 30 min at 37 °C. A smear of the mixture was prepared, dried, and fixed with methanol for 30 min. The smear was processed via Levowitz–Weber staining for 2 min and washed three times with distilled water. Phagocytic cells with engulfed bacteria were counted using a light microscope at $100 \times$ magnification, and phagocytic activity was calculated as the percentage of phagocytic cells containing bacterial cells. The plasma lysozyme activity was determined by mixing a 50-µL plasma sample with 3 mL of *Micrococcus lysodekticus* bacterial suspension. The absorbance of the mixture was measured at 570 nm directly after plasma addition (A_1) and again after incubation for 30 min (A_2) at 37 °C. The plasma lysozyme activity was calculated using the following formula: lysozyme activity = (A_1–A_2)/A_2 [29].

Interleukin-1β (IL-1β) in the blood plasma samples was determined using a commercial kit (Cat. No. MBS262525, MyBioSource, Inc., San Diego, CA, USA). The sensitivity of the method was 5 pg/mL, and the intra- and inter-assay precisions were $\geq 8\%$ and $\geq 12\%$, respectively. Immunoglobulin G (IgG) and immunoglobulin M (IgM) were assessed by the colorimetric method using commercial kits (IBL America Immuno-Biological Laboratories, Inc., Spring Lake Park MN, USA); the sensitivity and specificity of the assays exceeded 96%.

2.6. Colostrum and Milk Analysis

2.6.1. Colostrum Collection and Analysis

Colostrum samples were collected from eight randomly selected does within 8 h post-parturition. For this purpose, nests were checked three times a day, and colostrum samples were collected once the parturition process was completed. Colostrum samples were collected manually by gently massaging the mammary gland of the doe. The concentrations of colostrum immunoglobulin (Ig) fractions (IgM, IgA, IgG, IgE, and IgD) were determined using commercial kits (IBL America Immuno-Biological Laboratories, Inc., Spring Lake Park, MN, USA).

2.6.2. Milk Collection and Analysis

Milk samples were collected on days 7, 14, and 21 using an 'air vacuum pump' from all nipples of the mammary gland and milk yield was calculated by the weight-suckle-weight method [1]. Briefly, kits were separated from their dams for 24 h to prevent free suckling. Then, does were treated with oxytocin to stimulate milk ejection, and a 10 mL sample was obtained. Next, does were allowed to nurse their kids. The kits were weighed before suckling and again after suckling. The sum of the difference between the weight of kits before and after suckling and the weight of the collected milk samples represents the milk yield of each doe.

Milk samples were diluted with deionized distilled water at a proportion of 1:2 to facilitate the analysis. Approximately 10 g of the diluted sample was weighed into a silica crucible and placed in an oven at 70 °C until dry. The oven temperature was then increased to 105 °C for 3 h until a constant weight was reached to obtain the total solids. The total moisture was calculated as the difference between the fresh sample weight and the total solids weight. The contents of protein, fat, and total solids in milk samples were determined according to [30] and the energy concentration in milk was estimated [31]. Milk FAs were extracted and identified as described in Section 2.2.

2.7. Productive and Reproductive Performance

Fertility and pregnancy output variables including conception rate, parturition rate, and litter size and litter weight of kits at birth (day 0 of the age of kits), and at weaning (day 30 of the age of kits) were recorded. The conception rate ([number of does diagnosed positive on day 10/number of inseminated does] × 100), parturition rate ([number of delivered does/number of inseminated does] × 100), litter size at birth and weaning (total rabbits born and weaned per each doe), and litter weight at birth and weaning were recorded [27].

2.8. Statistical Analysis

The Statistical Analysis Software package (Version 8. Cary, NC, USA; 2001) was used to analyze all results. Variables assessed more than once (i.e., the reproductive cycle stages) were analyzed using the MIXED procedure for repeated measurement with a model considering fixed and random effects. The fixed effects of treatment (C, FME, HNME, and LNME), status (premating, mating, pregnancy, and lactation), and the treatment/status interaction on physiological, immunological, and biochemical variables were assessed. The rabbit does effect was introduced as a random factor. One-way ANOVA was used to assess the treatment effects on colostrum, milk, and litter characteristics (number and weight). The chi-square test was used to assess the effects of treatments on conception and parturition rates. Duncan's multiple range test was used to detect differences among treatment means. Results are presented as the least square mean (± pooled standard error of the mean [SEM]). The significance level of the statistical analysis tests was set at $p < 0.05$.

3. Results

3.1. Fatty Acid Profile of ME and EE

The ME FA profile analysis identified 30 FAs (Table 1). The major detected FAs were palmitic acid methyl ester C16:0, oleic acid methyl ester C18:1n-9, lignoceric acid methyl ester C24:0, gamma-linolenic acid methyl ester C18:3n-6, caprylic acid methyl ester C8:0, behenic acid methyl ester C22:0, and docosahexaenoic acid methyl ester C22:6n-3. The concentration of total UFAs was higher than the total SFAs. Odd-chain FAs were also detected (Table 1).

Table 1. Fatty acids (FAs) profile and encapsulation efficiency of each individual identified FA of nano-encapsulated *Moringa* leaves ethanolic extract (ME).

FAs	FA, g/100 g FA Methyl Esters
Caprylic acid methyl ester, C8:0	5.30
Capric acid methyl ester, C10:0	0.61
Undecanoic acid methyl ester, C11:0	1.01
Lauric acid methyl ester, C12:0	0.63
Tridecanoic acid methyl ester, C13:0	0.64
Myristic acid methyl ester, C14:0	1.18
Pentadecanoic acid methyl ester, C15:0	0.81
Palmitic acid methyl ester, C16:0	11.45
Heptadecanoic acid methyl ester, C17:0	1.45
Stearic acid methyl ester, C18:0	3.72
Arachidic acid methyl ester, C20:0	1.97
Heneicosanoic acid methyl ester, C21:0	2.55
Behenoic acid methyl ester, C22:0	5.14
Tricosanoic acid methyl ester, C23:0	3.26
Lignoceric acid methyl ester, C24:0	6.02
Myristoleic acid methyl ester, C14:1n-9	1.53
Pentadecenoic acid methyl ester, C15:1n-5	1.71
Palmitoleic acid methyl ester, C16:1n-7	2.12
Heptadecenoic acid methyl ester, C17:1n-7	2.00
Oleic acid methyl ester, C18:1n-9	9.04
Elaidic acid methyl ester, C18:1n-9t	3.82
Eicosenoic acid methyl ester, C20:1n-9	2.62
Erucic acid methyl ester, C22:1n-9	2.57
Nervonic acid methyl ester, C24:1n-9	3.93
Linolenic acid methyl ester (LA), C18:2n-6	3.68
Gama-Linolenic acid methyl ester (GLA), C18:3n-6	5.98
Dihomo-gamma-linolenic acid (DGLA), C20:4n-6	3.89
Eicosatrienoic acid methyl ester (ETE), C20:3n-3	3.53
Eicosapentaenoic acid methyl ester (EPA), C20:5n-3	3.40
Docosahexaenoic acid methyl ester (DHA), C22:6n-3	4.45
Saturated fatty acid	45.73
Unsaturated fatty acid	54.27
Mono-unsaturated fatty acid	29.34
Poly-unsaturated fatty acid	24.93
Poly-unsaturated fatty acid/Saturated fatty acid	0.54
Total odd FAs	13.43
Omega-3 FAs	11.38
Omega-6 FAs	13.55
Omega-9 FAs	21.98
Omega-6 FAs/Omega-3 FAs	1.19

The EE of alginate–CaCl$_2$ for Me FAs reached 100% for 14 FAs and ranged from 0 to 57% for the remaining FAs. The range of EE of alginate–CaCl$_2$ for Me FAs was between 49.1% to 80.45% (Table 2).

Table 2. Encapsulation efficiency of alginate-$CaCL_2$ for moringa extract fatty acids (FAs).

FAs Category	Encapsulation Efficiency [1], %
Saturated FAs	71.03
Unsaturated FAs	69.62
Mono-unsaturated FAs	77.64
Poly-unsaturated FAs	60.26
Total odd FAs	80.45
Omega-3 FAs	71.4
Omega-6 FAs	49.1
Omega-9 FAs	68.70

[1] Encapsulation efficiency (EE, %) = FA concentration in raw MLEE-FA concentration in MLEE supernatant/ FA concentration in raw MLEE × 100.

3.2. BW, Feed Intake, and Energy-Related Metabolites

The effects of different ME treatments on the BW, feed intake, blood plasma glucose, β-HB, and NEFAs of rabbit does during the experimental period are shown in Figure 1. Compared with the C group, all ME treatments significantly increased BW ($p < 0.001$, Figure 1a) and feed intake ($p < 0.001$, Figure 1b), and the highest values were observed in the HNME treatment. The ME treatments significantly increased blood plasma glucose concentrations ($p < 0.001$, Figure 1c) compared with the C group, and the highest values were observed in the LNME and HNME treatments. This effect began at mating and continued through pregnancy and lactation. The ME treatments significantly decreased the concentrations of blood plasma NEFAs ($p < 0.001$, Figure 1d) and β-HB ($p < 0.005$, Figure 1e) compared with the C group, and the lowest values were observed in the LNME treatment. These effects started at mating and continued during pregnancy and lactation periods, except the effect on NEFAs, which started earlier at premating.

3.3. Blood Immune Variables

The effects of ME treatments on the blood immune system variables of rabbit does during the experimental period are shown in Figure 2a–e. Compared with the C group, all ME treatments increased numbers ($p < 0.001$) of WBCs and percentages ($p < 0.001$) of lymphocytes, monocytes, eosinophils, phagocytic activity, and lysozyme activity ($p < 0.04$), whereas the percentage of neutrophils was decreased ($p < 0.001$). The largest changes were detected in HNME and LNME, followed by FME. These changes were observed at different physiological stages, starting at premating. Compared with the C group, the ME treatments decreased ($p < 0.001$) blood plasma interleukin concentrations at all physiological stages and the lowest value was observed in the LNME treatment. The ME treatments significantly increased ($p < 0.001$) blood plasma immunoglobulin G and M compared with the C group; LNME treatments resulted in the highest value. These effects started at mating and continued through pregnancy and lactation.

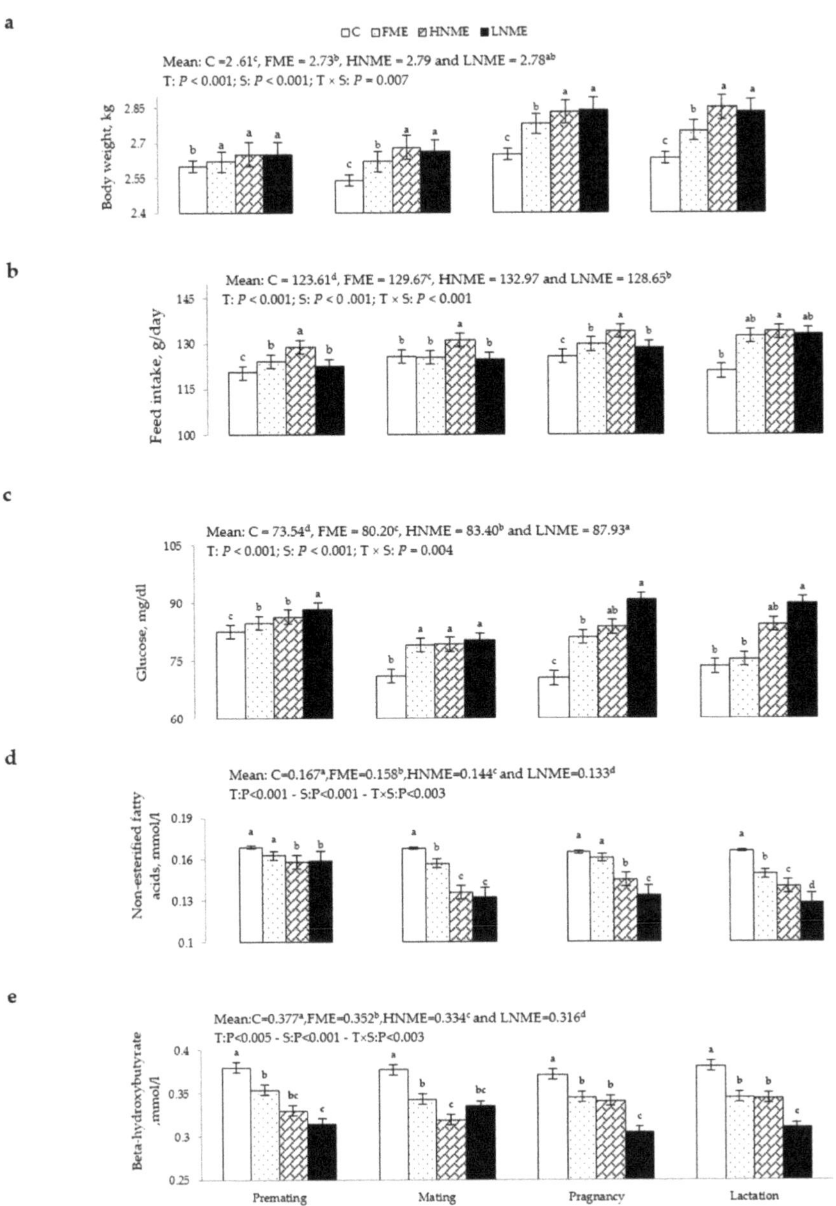

Figure 1. Means (± SEM) treatment by physiological status effects on body weight (**a**), feed intake (**b**), blood plasma glucose (**c**), non-esterified free fatty acids (**d**), and β-hydroxybutyra© (**e**) of multiparous rabbit does supplemented with 0 mg/kg BW (C), 50 mg/kg BW free ME (FME), 25 mg/kg BW nano-encapsulated ME (HNME), and 10 mg/kg BW nano-encapsulated ME (LNME). The means of different treatments within the same physiological status with different lowercase letter superscripts differ significantly (p < 0.05).

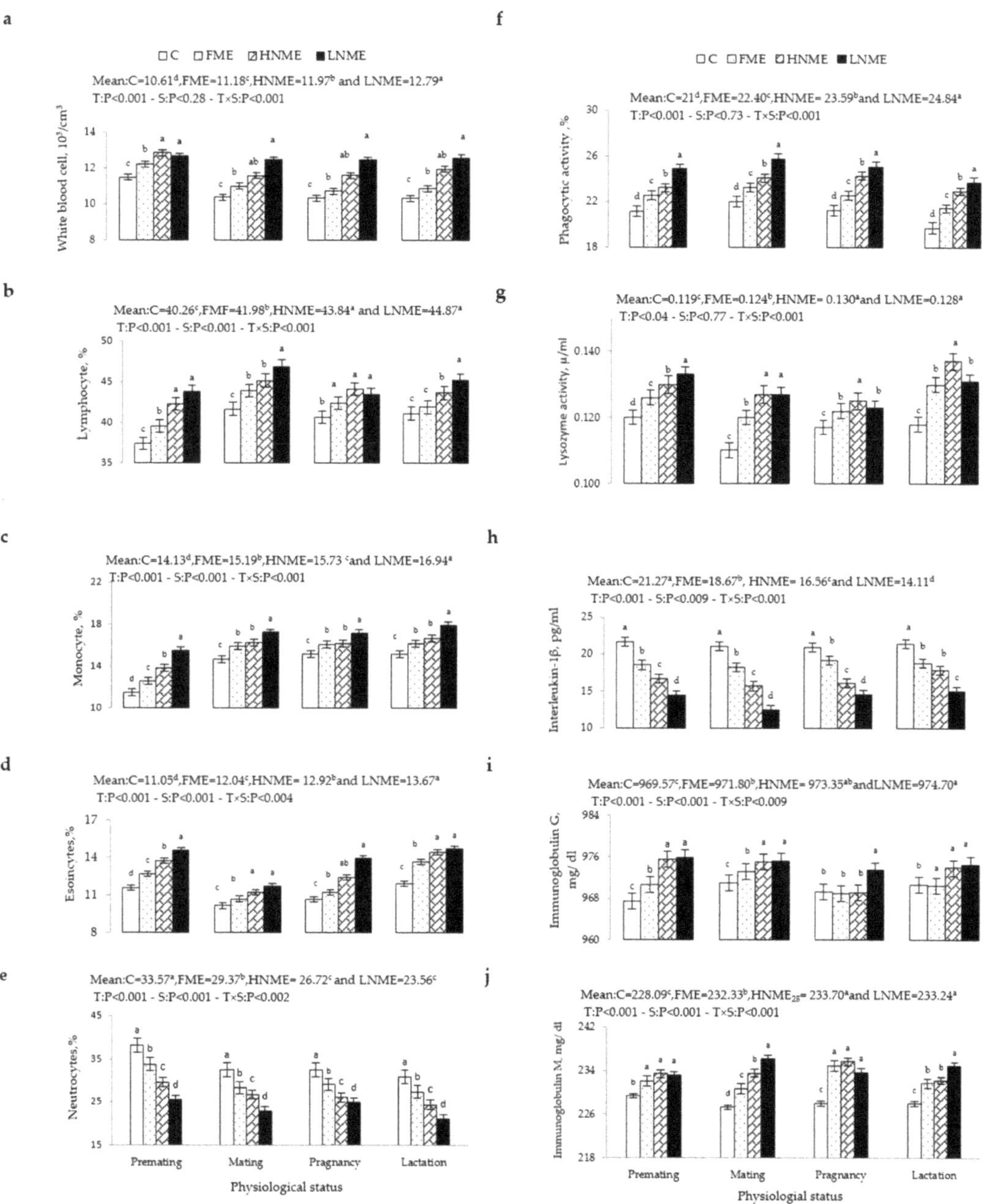

Figure 2. Means (± SEM) treatment by physiological status interaction effects on immunity variables; white blood cell (**a**), lymphocytes (**b**), monocytes (**c**), eosinocytes (**d**), neutro©es (**e**), phagocytic activity (**f**), lysozyme activity (**g**), interleukin-1β (**h**), immunoglobulin G (**i**) and immunoglobulin M (**j**) of multiparous rabbit does supplemented with 0 mg/kg BW (C), 50 mg/kg BW free ME (FME), 25 mg/kg BW nano-encapsulated ME (HNME), and 10 mg/kg BW nano-encapsulated ME (LNME). The means of different treatments within the same physiological status with different lowercase letter superscripts differ significantly ($p < 0.05$).

3.4. Colostrum Immunoglobulin, Milk Yield, and Milk Composition

The effects of ME treatments on colostrum immunoglobulin of rabbit does during the experimental period are shown in Table 3. Compared with the C group, the ME treatments significantly increased ($p = 0.001$) the concentrations of colostrum immunoglobulin (IgG, IgA, IgM, and IgD) but did not affect ($p = 0.160$) IgE concentrations. The highest values were observed in LNME and HNME treatments, followed by FME. The ME treatments significantly increased milk yield ($p = 0.001$) and milk composition (protein, $p = 0.001$; total solids, $p = 0.001$; and energy milk contents, $p = 0.005$) and tended to increase fat content ($p = 0.07$) compared to the C group. The highest values were observed in the LNME treatment.

Table 3. Effects of free and nano-encapsulated *Moringa oleifera* leaf ethanolic extract (ME) supplementations on colostrum, milk yield and milk composition of multiparous rabbit does during the experimental period (mean ± SEM).

Variable	Treatment [1]				SEM	*p* Value
	C	FME	HNME	LNME		
Colostrum immunoglobulin, mg/dL						
Immunoglobulin M	225.32 [c]	230.85 [b]	237.15 [a]	232.14 [b]	2.27	0.001
Immunoglobulin A	73.22 [b]	75.23 [a]	76.89 [a]	75.56 [a]	0.91	0.001
Immunoglobulin G	964.20 [c]	968.21 [b]	972.54 [a]	968.23 [b]	2.35	0.001
Immunoglobulin E	12.99	13.22	13.52	13.39	0.07	0.160
Immunoglobulin D	28.86 [c]	31.83 [b]	35.52 [a]	35.53 [a]	0.82	0.001
Milk yield and composition						
Milk yield, g/day	117.50 [c]	159.44 [b]	161.52 [ab]	169.86 [a]	2.46	0.001
Milk Composition, %						
Protein	11.79 [d]	12.33 [c]	12.68 [b]	13.39 [a]	0.02	0.001
Fat	13.38	13.50	14.12	14.69	0.33	0.07
Total solids	27.91 [c]	29.54 [bc]	31.77 [ab]	32.51 [a]	2.23	0.001
Energy, MJ/kg	8.50 [c]	8.56 [bc]	9.04 [ab]	9.41 [a]	0.06	0.005

[1] Multiparous rabbit supplemented with 0 kg BW (C), 50 mg/kg BW free ME (FME), 25 mg/kg BW nano-encapsulated ME (HNME), and 10 mg/kg BW nano-encapsulated ME (LNME). Means within the raw having different superscripts (a, b, c) differ significantly ($p < 0.05$).

3.5. Milk Fatty Acid Profile and Fatty Acid Health Indices

From the 20 FAs studied, 13 were affected by the nano-encapsulated ME treatments (Table 4). For SFAs, all ME treatments significantly increased the concentrations of C16:0 to C20:0 and significantly decreased C15:0 compared with the C group. However, the ME-treatments significantly decreased the concentration of C15:0 in milk compared with the C group. All ME treatments significantly increased milk USFAs ($p = 0.007$), MUFAs (C16:1n-7 and C18:1n-9; $p = 0.001$), PUFAs, omega-6 (C18:2n-6, C20:4n-6; $p = 0.014$) compared with the C group. Higher concentrations of omega-3 FAs (C18:2n-3, C18:3n-3, C20:5n-3, and C22:6n-3) were recorded in the LNME and HNME treatments compared with the FME and C treatments (Table 4).

3.6. Reproductive Performance

The effects of ME treatments on the reproductive performance of rabbit does during the experimental period are shown in Table 5. Compared with the C group, the ME treatments significantly increased conception ($p = 0.004$) and parturition ($p = 0.003$) rates, the total litter sizes at birth and weaning ($p = 0.003$ and $p < 0.001$, respectively), and the litter weights at birth and weaning ($p = 0.005$ and $p < 0.001$, respectively); the highest values were observed in LNME and HNME treatments, followed by the FME treatment (Table 5).

Table 4. Effects of free and nano-encapsulated *Moringa oleifera* leaf ethanolic extract (ME) supplementations on milk fatty acids (FAs) profile of multiparous rabbit does during the experimental period (mean ± SEM).

FAs, g/100 g FA Methyl Esters	Treatment [1]				SEM	*p*-Value
	C	FME	HNME	LNME		
Butyric acid, C4:0	0.101	0.112	0.110	0.108	0.002	0.616
Caproic acid methyl ester, C6:0	0.410	0.410	0.411	0.403	0.001	0.378
Caprylic acid methyl ester, C8:0	25.59	25.97	26.27	26.13	0.11	0.167
Capric acid methyl ester, C10:0	21.76	21.59	21.74	22.18	0.12	0.476
Lauric acid methyl ester, C12:0	2.66	2.67	2.77	2.79	0.08	0.938
Myristic acid methyl ester, C14:0	1.55	1.54	1.57	1.57	0.01	0.729
Pentadecanoic cid methyl ester, C15:0	0.827 [a]	0.757 [b]	0.816 [a]	0.773 [ab]	0.01	0.039
Palmitic acid methyl ester, C16:0	12.94 [ab]	13.30 [a]	12.56 [b]	12.79 [b]	0.01	0.030
Heptadecanoic acid methyl ester, C17:0	0.713 [c]	0.755 [b]	0.744 [bc]	0.789 [a]	0.009	0.006
Stearic acid methyl ester, C18:0	2.66 [b]	3.00 [a]	2.78 [b]	2.80 [b]	0.04	0.005
Arachidic acid methyl ester, C20:0	0.125 [b]	0.182 [ab]	0.179 [ab]	0.221 [a]	0.01	0.013
Myristoleic acid methyl ester, C14:1 n-9	0.125 [b]	0.159 [a]	0.164 [a]	0.161 [a]	0.01	0.021
Palmitoleic acid methyl ester, C16:1 n-7	1.52 [c]	1.70 [a]	1.66 [ab]	1.63 [b]	0.02	0.001
Oleic acid methyl ester, C18:1 n-9	11.39 [b]	11.39 [b]	11.42 [ab]	11.43 [a]	0.007	0.004
Conjugated Linoleic Acid (CLA), C18:2 n-3	0.074 [c]	0.080 [b]	0.085 [a]	0.086 [a]	0.002	0.001
Linolenic acid methyl ester (LA), C18:2 n-6	12.21 [b]	12.49 [a]	12.60 [a]	12.56 [a]	0.05	0.030
α-Linolenic acid methyl ester (ALA), C18:3 n-3	2.47 [c]	2.50 [b]	2.51 [b]	2.55 [a]	0.01	0.007
Arachidonic acid methyl ester(ARA), C20:4 n-6	0.537 [ab]	0.523 [b]	0.534 [b]	0.548 [a]	0.003	0.015
Eicosapentaenoic acid methyl ester(EPA), C20:5n-3	0.049 [c]	0.068 [b]	0.074 [a]	0.073 [a]	0.004	0.001
Docosahexaenoic acid(DHA), C22:6n-3	0.061 [b]	0.065 [b]	0.070 [a]	0.076 [a]	0.004	0.001
Degree of FAs saturation, g/100 g FA methyl ester						
Saturated FAs	72.27	71.72	71.62	71.70	0.22	0.282
Unsaturated FAs	27.73 [b]	28.28 [a]	28.38 [a]	28.30 [a]	0.08	0.007
Monounsaturated FAs	13.04 [c]	13.28 [a]	13.26 [ab]	13.19 [b]	0.03	0.001
Polyunsaturated FAs	14.69 [b]	15.00 [a]	15.12 [a]	15.11 [a]	0.06	0.014
Unsaturated FAs / Saturated FAs	0.400	0.402	0.405	0.401	0.001	0.552

[1] Multiparous rabbit supplemented with 0 mg/kg BW (C), 50 mg/kg BW free ME (FME), 25 mg/kg BW nano-encapsulated ME (HNME), and 10 mg/kg BW nano-encapsulated ME (LNME). Means within the raw having different superscripts (a, b, c) differ significantly (*p* < 0.05).

Table 5. Effects of free and nano-encapsulated *Moringa oleifera* leaf ethanolic extract (ME) supplementations on the reproductive performance of multiparous rabbit does during the experimental period (mean ± SEM).

Variable	Treatment [1]				SEM	*p*-Value
	C	FME	HNME	LNME		
Conception rate, %	76.9 [c] (20/26)	84.61 [bc] (22/26)	92.3 [b] (24/26)	96.15 [a] (25/26)	-	0.004
Parturition rate, %	69.23 [c] (18/26)	84.61 [b] (22/26)	88.46 [ab] (23/26)	92.3 [a] (24/26)	-	0.003
Litter size at birth	6.33 [bc]	5.95 [c]	7.17 [b]	7.86 [a]	2.4	0.004
No. live litter sizes	5.16 [c]	5.90 [ab]	6.65 [b]	7.34 [a]	3.7	0.003
No. dead litter sizes	1.17 [a]	0.05 [b]	0.52 [ab]	0.52 [ab]	2.2	0.130
Litter weight at birth, g	298.36 [c]	324.17 [b]	340.65 [b]	409.30 [a]	50.08	0.005
Litter size at weaning	5.27 [c]	5.92 [bc]	6.64 [b]	7.21 [a]	0.63	< 0.001
Litter weight at weaning, g	1699.1 [c]	2376.2 [b]	2796.8 [ab]	3144.6 [a]	174.46	< 0.001

[1] Multiparous rabbit supplemented with 0 mg/kg BW (C), 50 mg/kg BW free ME (FME), 25 mg/kg BW nano-encapsulated ME (HNME), and 10 mg/kg BW nano-encapsulated ME (LNME). Means within the raw having different superscripts (a, b, c) differ significantly (*p* < 0.05).

4. Discussion

Recent research has focused on the development of phytogenic-based feed additives due to their impressive biological activities and safety for humans, animals, and the environment. However, the effectiveness of such feed additives is challenged by the sensitivity of these compounds to different environmental and industrial factors, causing reductions in nutritional value, bioavailability, product quality, and stability. In this context, encapsulation technology presents an innovative solution for maintaining the biological activities of sensitive active components [21,32].

In this study, ME was selected as a source of FAs because several studies have reported the richness of ME with biologically active FAs, specifically PUFAs [19,33]. PUFAs contain many double bonds, which make them highly susceptible to lipid oxidation, which diminishes their nutritional value and biological effects. Encapsulation can protect high n3/n6-containing FA products from oxidation reactions upon exposure to oxidation-inducing factors, such as oxygen, metal ions, high temperatures, and light [21]. This effect was confirmed in this study. Alginate proved to be an excellent encapsulating material that encapsulated 71.3% of SFAs, 69.62% of UFAs, 77.64% of MUFAs, and 60.26% of PUFAs in ME. Among the biopolymer materials used for encapsulation processes, alginate has the potential to form a film or coating due to its colloidal properties. Alginate has been found to effectively coat materials with different active components, including lipid sources, such as essential oil, fish oil, and plant oil (seed, leaves, flower) [32,34,35]. In this study, although all ME treatments improved the metabolism and productive and reproductive traits of rabbit does, the encapsulated ME resulted in better responses and productivity (milk production and reproductive performance), even at the lower ME concentration. These findings support the benefits of encapsulation for maintaining the biological activities of natural phytogenic compounds intended for use as feed additives. Similar results were previously obtained by [22,36].

In this study, we compared the effects of ME in free and encapsulated forms on the metabolic status and immunity of rabbit does during a whole reproductive cycle (from premating to weaning). We hypothesized that ME FAs might change colostrum/milk yield and composition to improve the viability and growth performance of litters. Rabbits are typically bred in intensive production systems, meaning they are pregnant, suckling, or both for most of their lifetime. These reproductive events are very costly in terms of energy [8].

In this study, we observed positive effects of ME supplementation on BW and metabolism. These effects may be mediated by different mechanisms. The first mechanism may be related to improved BW and metabolism due to changes in the feed intake of rabbit does, specifically during pregnancy and lactation. In this study, rabbit does supplemented with ME showed higher feed intake during pregnancy and lactation periods. Normally, doe feed intake decreases at the end of pregnancy due to the limited space available in the gastrointestinal tract and the lack of available carbohydrates [16], resulting in a negative energy balance due to the transfer of the body fat mass into the fetuses. During lactation, the feed intake of females increases very rapidly after kindling (60–75%), but this increase is insufficient to cover the requirements due to maintenance and milk production. Thus, a very negative energy balance and considerable mobilization of body fat are often observed during the first lactation. About 80% of the energy for reproduction comes from feed intake and about 20% from fat mobilization. Moreover, the energy deficit increases when females are concurrently pregnant and lactating [1]. The improved feed intake in ME-supplemented groups can be attributed to the effect of some FAs on energy-regulating hormones, such as leptin. Leptin suppresses food intake, inducing weight loss [15,16]. The inclusion of EPA and DHA in pregnant and lactating rabbit females is associated with decreased leptin levels (higher leptinemia) [37], which may partially explain the increased feed intake of ME-supplemented rabbit does. The increased feed intake could also be related to improved feed digestibility and increased flow rate of feed bolus as reported previously by [38], allowing does to consume more feed.

The second mechanism related to the improved metabolism of ME-treated does may be the increased availability of energy-yielding metabolites. All ME-supplemented rabbit does had better energy status, as indicated by higher body weights and blood plasma glucose and lower lipid metabolites (NEFAs and β-HB) during reproductive events, specifically pregnancy and lactation. In rabbits, NEFAs, glucose, and long-chain FAs play a role in the relationship between energy balance and reproductive efficiency [39]. During the post-partum period, mammary glands and fetoplacental units use substrates, such as glucose, long-chain FAs, and free FAs as energy sources [2]. Thus, the lack of these substrates drives does to use body fat to meet the increased energy requirements of growing fetuses and/or milk lactation. In this study, ME supplementation improved the concentration of blood plasma glucose and provided FAs as energy sources. These findings are consistent with many previous studies reporting that ME is a good source of energy-yielding nutrients that can be easily used by animals for biological events, such as growth [40,41] and reproduction [22,33]. The increase in glucose concentrations in ME-treated rabbit does might be due to improved gluconeogenesis. Interestingly, glucose cannot be synthesized from FAs because they are cleaved by β-oxidation into acetyl coenzyme A (CoA), which subsequently enters the citric acid cycle and is oxidized to CO_2. However, the last three carbon atoms of odd-chain FAs generate propionyl CoA during β-oxidation and are thus partly gluconeogenic. Because ME contains 13.43% of odd-FAs, this pathway provides a conceivable reason for increased glucose concentrations in this study [42].

In this study, ME supplementation not only improved the energy status of rabbit does but also improved their immune status. ME-supplemented rabbit does exhibited better innate (phagocytes and their phagocytic activity and lysozyme activity) and humor (immunoglobulin G & M) immunity indicators than non-supplemented rabbit does. These findings are in accordance with those reported in several previous studies [38,43,44]. In context, Isitua and Ibeh [45] reported that rabbits fed *Moringa* leaves showed a significant increase in CD4 (T-helper) cells, which evoke cell-mediated immunity and help B-cells to produce antibodies. The positive role of ME on immune system function can be ascribed to the presence of considerable quantities of various FAs in ME. Immune cells contain FAs and, thus, their reactivity and functioning can be modulated by the profile of dietary fats. Supplementary dietary FAs may influence the immune status via several mechanisms, such as the inhibition of the arachidonic acid (AA) metabolic process, production of anti-inflammatory mediators, modification of intracellular lipids, and activation of nuclear receptors. ME contains high concentrations of omega-3 FAs (11.38%), EPA, and DHA, which stimulate phagocytic activity by macrophages in addition to their anti-inflammatory, anti-proliferative and anti-atherosclerotic activities [10]. In this study, ME contained high concentrations of omega-6 FAs (13.55%), which are linked to inflammation, mainly because AA is the precursor of pro-inflammatory lipid mediators. However, the concentrations of IL-1B (an inflammatory factor) were lower in ME-treated rabbit does. This finding can be explained by the fact that ME naturally has the recommended ratio between omega-6 and omega-3 PUFAs, which should be from 3:1 to 1:1 to provide positive effects on immunity and health [46].

The positive energy and immune status of ME-supplemented rabbit does were reflected in their productive and reproductive performance. ME-supplemented rabbit does had better colostrum and milk yield and composition as well as enhanced fertility traits. Colostrum is a special type of milk formed during the last days of pregnancy and the first few days after birth. The main components of colostrum in rabbits are proteins and fats. Thus, the nutritional requirements for these components are elevated, specifically toward the end of pregnancy and the early stages of lactation. For example, the coefficients of variation in total protein concentration reach their maximum during the early stages of lactation (5–26%), and immunoglobulins account for most of the total protein in colostrum [47]. As reported in previous studies [38,47], we found that ME improved colostrum protein and fat contents due to its high amino acid and FA concentrations. The improved protein content and different immunoglobulin (IgG, IgA, and IgM) in colostrum reflects an increased availability of amino acids for immunoglobulin synthesis [48] and improves the passive transfer of immunoglobulin to the mammary gland. Moreover, some FAs identified in ME have been shown to improve immunoglobulin contents in colostrum;

for example, colostrum IgG concentrations were increased in animals fed an n-3 PUFA-rich diet [9].

Supplementation of ME also improved both milk yield and milk composition, specifically protein and fat content. Given that the ME-treated rabbit does had increased BWs and blood metabolites (higher glucose and lower fat metabolites) than the control rabbit does during the energy-consuming lactation period, it can be said that ME played a role in supporting energy requirements and modulating metabolism in favor of milk production without negative effects on energy balance status. In fact, feeding either Moringa leaf or ME has been shown to increase milk production in many farm animals, including rabbits [49,50]. Moringa contains plenty of nutrients, such as amino acids, FAs, and vitamins, which are precursors for milk synthesis [51]. Moreover, it contains phytosterols, phytoestrogens that can boost both lactogenesis and mammogenesis [18]. Moringa is rich in phytosterols, such as sitosterol, stigmasterol, and campesterol, which are precursors for hormones. These phytochemicals, along with high amino acid contents, can enhance estrogen excretion, stimulating mammogenesis and milk production [20].

In the present study, all ME treatments significantly increased the concentrations of milk USFAs, MUFAs (C16:1n-7 and C18:1n-9), omega-6 PUFAs (conjugated linoleic acid [CLA] and AA), omega-3 PUFAs (linoleic acid [LA], alpha-linolenic acid, EPA, and DHA) compared to the C group. As identified from the ME FA profile, ME has considerable concentrations of omega-3 PUFAs (EPA and DHA) and omega-6 PUFAs (LA, gamma-linolenic acid, and dihomo-gamma-linolenic acid), which are implicated in PUFA long-chain biosynthetic pathways. LA is the main precursor for omega-6 FA pathways; it can be used as a precursor for longer omega-6 PUFAs via the effects of elongase and desaturase enzymes [46]. This pathway seems to be enhanced in this study, as ME-treated rabbit does had higher concentrations of longer omega-6 PUFAs (CLA and AA) in their milk. The same trend was observed for omega-3 PUFAs, and omega-3 EPA can be used as a precursor for omega-3 DHA biosynthesis [16,17]. Overall, ME supplementation to rabbit does during lactation improved FA bioavailability to the mammary gland and activated FA biosynthetic pathways, leading to different FA profiles in the milk of ME-supplemented rabbit does. These positive effects might also be related to the improved energy status and the availability of energy-yielding nutrients, such as glucose, during lactation. It has been confirmed that the FA elongation process can be stopped if glucose, the acetyl-CoA source, is not adequately available [2,4].

Results of the present study confirmed the superior reproductive performance of the ME-supplemented rabbit does, which had better fertility and pregnancy outcomes than the non-supplemented (C) rabbit does. The positive effect of ME on the energy status of rabbit does is needed to achieve better reproductive performance. Several studies reported that higher BWs of rabbit does around mating time and early pregnancy are associated with better artificial insemination outputs [52,53]. In addition to the positive role of ME on body weight and metabolism, the unique FA composition of ME might have other positive effects. As reported in other studies, FA-rich diets can improve fertility and pregnancy outcomes in rabbits [15,46]. In fact, FAs are an important element for many reproductive events through various modes of action. Numerous FAs can positively influence reproduction by altering the ovarian follicles and corpus luteum function via improved energy status and increasing precursor levels for the synthesis of reproductive steroids and prostaglandins [54]. They also improve the competence of oocytes and boost embryo development [55] and pregnancy and fetal development [56]. For example, PUFAs, such as EPA and DHA can improve the quality of embryos by decreasing apoptosis rates and improving cell membrane integrity [15,57]. Moreover, omega-3 PUFAs are associated with low PGF2α production in uterine and placental tissues, decreasing the susceptibility to abortion and/or preterm parturition [58].

In this study, the ME-treated rabbit does had better pregnancy outcomes (litter sizes and weights at birth and weaning) than the control rabbit does. The developing fetuses and the offspring in the first days of lactation are totally dependent on their dams to provide the nutrition required for growth and development. Thus, such positive effects of ME on pregnancy outcomes might begin early and be related to improved oocyte quality, embryo development, and implantation during the early stages of pregnancy by providing specific FAs and/or energy sources. Additionally, during the later stages of pregnancy, most fetal lipids are delivered from maternal circulation through the placenta and are obtained from the diet or fat lipolysis [59], constituting the fat deposits on which the survival of the newborn depends. In this study, no decreases in dams body weight were observed in the ME-treated rabbit does. This finding leads us to speculate that ME was effective to support metabolism during pregnancy, providing the lipid/FA requirements for fetuses without depleting the dams' fat stores. Because rabbit kits are altricial, the improved pregnancy outcomes may also be related to the improved growth rate and viability of litters after birth [18]. ME improved both colostrum immunoglobulin contents and milk production and composition and increased concentrations of some FAs that are important for boosting litter growth and health. Focusing on the important FAs for rabbit kit growth and development, higher EPA and DHA contents in milk were associated with the development of the nervous system because these FAs are used for membrane phospholipid synthesis [15,18]. Other FAs, such as AA, have been shown to boost the immune system and health status of kits [14].

5. Conclusions

In the present study, ME FAs served as a good source of energy and provided functional FAs for the ideal completion of specific physiological events related to milk production and reproduction. These effects were reflected in the improved metabolism, immune status, milk production, and reproductive performance of rabbit does at different reproductive stages. These effects were observed in rabbits supplemented with free and encapsulated forms of ME. However, the nano-encapsulated form allowed for an 80% reduction (10 mg/kg BW) in the optimal dose (50 mg/kg BW) without affecting the treatment efficiency, highlighting the importance of nano-encapsulation for improving FA bioavailability.

Author Contributions: Conceptualization, N.I.E.-D., N.M.H. and A.G.E.; methodology, N.I.E.-D. and N.M.H.; validation, N.I.E.-D. and N.M.H.; formal analysis, N.M.H.; investigation, N.I.E.-D. and N.M.H.; data curation, N.I.E.-D. and N.M.H.; writing—original draft preparation, N.I.E.-D. and N.M.H.; writing—review and editing, N.M.H. and supervision, Z.R.A.-E., A.G.E. and N.M.H. and All authors have read and agreed to the published version of the manuscript.

Funding: This research received no external funding.

Institutional Review Board Statement: The experimental procedures were previously assessed and approved by Alexandria University-Institutional Animal Care and Use Committee (AU-IACUC, AU 08 21 07 26 2 82).

Informed Consent Statement: Not applicable.

Data Availability Statement: The data presented in this study are available on request from the corresponding author. The data are not publicly available because of privacy.

Conflicts of Interest: The authors declare no conflict of interest.

References

1. Maertens, L.; Lebas, F.; Szendro, Z. Rabbit milk: A review of quantity, quality and non-dietary affecting factors. *World Rabbit. Sci.* **2006**, *14*, 205–230. [CrossRef]
2. Castellini, C.; Dal Bosco, A.; Arias-Álvarez, M.; Lorenzo, P.L.; Cardinali, R.; Rebollar, P.G. The main factors affecting the reproductive performance of rabbit does: A review. *Anim. Reprod. Sci.* **2010**, *122*, 174–182. [CrossRef] [PubMed]
3. Menchetti, L.; Barbato, O.; Sforna, M.; Vigo, D.; Mattioli, S.; Curone, G.; Tecilla, M.; Riva, F.; Brecchia, G. Effects of diets enriched in linseed and fish oil on the expression pattern of toll-like receptors 4 and proinflammatory cytokines on gonadal axis and reproductive organs in rabbit buck. *Oxidative Med. Cell. Longev.* **2020**, *2020*, 4327470. [CrossRef] [PubMed]
4. Castellini, C. Reproductive activity and welfare of rabbit does. *Ital. J. Anim. Sci.* **2007**, *6* (Suppl. 1), 743–747. [CrossRef]

5. Martínez-Paredes, E.; Ródenas, L.; Martínez-Vallespín, B.; Cervera, C.; Blas, E.; Brecchia, G.; Boiti, C.; Pascual, J. Effects of feeding programme on the performance and energy balance of nulliparous rabbit does. *Animal* **2012**, *6*, 1086–1095. [CrossRef] [PubMed]

6. Marchiani, S.; Vignozzi, L.; Filippi, S.; Gurrieri, B.; Comeglio, P.; Morelli, A.; Danza, G.; Bartolucci, G.; Maggi, M.; Baldi, E. Metabolic syndrome-associated sperm alterations in an experimental rabbit model: Relation with metabolic profile, testis and epididymis gene expression and effect of tamoxifen treatment. *Mol. Cell. Endocrinol.* **2015**, *401*, 12–24. [CrossRef] [PubMed]

7. Brown, A.; Vallenari, A.; Prusti, T.; de Bruijne, J.; Babusiaux, C.; Biermann, M.; Creevey, O.; Evans, D.; Eyer, L.; Hutton, A. Gaia Early Data Release 3-Summary of the contents and survey properties (Corrigendum). *Astron. Astrophys.* **2021**, *650*, C3.

8. Fortun-Lamothe, L. Energy balance and reproductive performance in rabbit does. *Anim. Reprod. Sci.* **2006**, *93*, 1–15. [CrossRef]

9. Mateo, R.; Carroll, J.; Hyun, Y.; Smith, S.; Kim, S. Effect of dietary supplementation of n-3 fatty acids and elevated concentrations of dietary protein on the performance of sows. *J. Anim. Sci.* **2009**, *87*, 948–959. [CrossRef]

10. Al-Khalaifah, H. Modulatory effect of dietary polyunsaturated fatty acids on immunity, represented by phagocytic activity. *Front. Vet.* **2020**, *7*, 672. [CrossRef] [PubMed]

11. Nielsen, S.; Karpe, F. Determinants of VLDL-triglycerides production. *Curr. Opin. Lipidol.* **2012**, *23*, 321–326. [CrossRef] [PubMed]

12. Hadjadj, I.; Hankele, A.-K.; Armero, E.; Argente, M.-J.; de la Luz García, M. Fatty Acid Profile of Blood Plasma at Mating and Early Gestation in Rabbit. *Animals* **2021**, *11*, 3200. [CrossRef] [PubMed]

13. Rooke, J.; Sinclair, A.; Edwards, S. Feeding tuna oil to the sow at different times during pregnancy has different effects on piglet long-chain polyunsaturated fatty acid composition at birth and subsequent growth. *Br. J. Nutr.* **2001**, *86*, 21–30. [CrossRef] [PubMed]

14. Betancourt López, C.A.; Bernal Santos, M.G.; Vázquez Landaverde, P.A.; Bauman, D.E.; Harvatine, K.J.; Srigley, C.T.; Becerra Becerra, J. Both Dietary Fatty Acids and Those Present in the Cecotrophs Contribute to the Distinctive Chemical Characteristics of New Zealand Rabbit Milk Fat. *Lipids* **2018**, *53*, 1085–1096. [CrossRef] [PubMed]

15. Rebollar, P.; García-García, R.; Arias-Álvarez, M.; Millán, P.; Rey, A.I.; Rodríguez, M.; Formoso-Rafferty, N.; De la Riva, S.; Masdeu, M.; Lorenzo, P. Reproductive long-term effects, endocrine response and fatty acid profile of rabbit does fed diets supplemented with n-3 fatty acids. *Anim. Reprod. Sci.* **2014**, *146*, 202–209. [CrossRef]

16. Rodríguez, M.; García-García, R.; Arias-Álvarez, M.; Formoso-Rafferty, N.; Millán, P.; López-Tello, J.; Lorenzo, P.; González-Bulnes, A.; Rebollar, P. A diet supplemented with n-3 polyunsaturated fatty acids influences the metabomscic and endocrine response of rabbit does and their offspring. *J. Anim. Sci.* **2017**, *95*, 2690–2700. [PubMed]

17. Rodríguez, M.; García-García, R.; Arias-Álvarez, M.; Millán, P.; Febrel, N.; Formoso-Rafferty, N.; López-Tello, J.; Lorenzo, P.; Rebollar, P. Improvements in the conception rate, milk composition and embryo quality of rabbit does after dietary enrichment with n-3 polyunsaturated fatty acids. *Animal* **2018**, *12*, 2080–2088. [CrossRef] [PubMed]

18. Sri, W.; Sri, W.; Hendrawan, S. The effects of adding Moringa oleifera leaves extract on rabbit does' milk production and mammary gland histology. *Russ. J. Agric. Socio-Econ. Sci.* **2019**, *8*, 296–304.

19. Saini, R.; Shetty, N.; Prakash, M.; Giridhar, P. Effect of dehydration methods on retention of carotenoids, tocopherols, ascorbic acid and antioxidant activity in Moringa oleifera leaves and preparation of a RTE product. *J. Food Sci. Technol.* **2014**, *51*, 2176–2182. [CrossRef]

20. Gopalakrishnan, L.; Doriya, K.; Kumar, D.S. Moringa oleifera: A review on nutritive importance and its medicinal application. *Food Sci. Hum. Wellness* **2016**, *5*, 49–56. [CrossRef]

21. Ruiz, P.A.; Morón, B.; Becker, H.M.; Lang, S.; Atrott, K.; Spalinger, M.R.; Scharl, M.; Wojtal, K.A.; Fischbeck-Terhalle, A.; Frey-Wagner, I. Titanium dioxide nanoparticles exacerbate DSS-induced colitis: Role of the NLRP3 inflammasome. *Gut* **2017**, *66*, 1216–1224. [CrossRef] [PubMed]

22. El-Desoky, N.I.; Hashem, N.M.; Gonzalez-Bulnes, A.; Elkomy, A.G.; Abo-Elezz, Z.R. Effects of a Nanoencapsulated Moringa Leaf Ethanolic Extract on the Physiology, Metabolism and Reproductive Performance of Rabbit Does during Summer. *Antioxidants* **2021**, *10*, 1326. [CrossRef]

23. Egan, W.F. *Frequency Synthesis by Phase Lock*; Wiley: New York, NY, USA, 1981.

24. Tsiplakou, E.; Mountzouris, K.; Zervas, G. Concentration of conjugated linoleic acid in grazing sheep and goat milk fat. *Livest. Sci.* **2006**, *103*, 74–84. [CrossRef]

25. Estany, J.; Baselga, M.; Blasco, A.; Camacho, J. Mixed model methodology for the estimation of genetic response to selection in litter size of rabbits. *Livest. Prod. Sci.* **1989**, *21*, 67–75. [CrossRef]

26. Carmichael, W.; Gorham, P.; Biggs, D. Two laboratory case studies on the oral toxicity to calves of the freshwater cyanophyte (blue-green alga) Anabaena flos-aquae NRC-44-1. *Can. Vet. J.* **1977**, *18*, 71. [PubMed]

27. Hashem, N.; Aboul-Ezz, Z. Effects of a single administration of different gonadotropins on day 7 post-insemination on pregnancy outcomes of rabbit does. *Theriogenology* **2018**, *105*, 1–6. [CrossRef] [PubMed]

28. Hosny, N.S.; Hashem, N.M.; Morsy, A.S.; Abo-Elezz, Z.R. Effects of organic selenium on the physiological response, blood metabolites, redox status, semen quality, and fertility of rabbit bucks kept under natural heat stress conditions. *Front. Vet. Sci.* **2020**, *7*, 290. [CrossRef] [PubMed]

29. Hashem, N.M.; Shehata, M.G. Antioxidant and Antimicrobial Activity of Cleomedroserifolia (Forssk.) Del. and Its Biological Effects on Redox Status, Immunity, and Gut Microflora. *Animals* **2021**, *11*, 1929. [CrossRef] [PubMed]

30. Jimidar, M.; Hartmann, C.; Cousement, N.; Massart, D. Determination of nitrate and nitrite in vegetables by capillary electrophoresis with indirect detection. *J. Chromatogr. A* **1995**, *706*, 479–492. [CrossRef]

31. Xiccato, G.; Parigi-Bini, R.; Dalle Zotte, A.; Carazzolo, A.; Cossu, M. Effect of dietary energy level, addition of fat and physiological state on performance and energy balance of lactating and pregnant rabbit does. *Anim. Sci.* **1995**, *61*, 387–398. [CrossRef]

32. Rijo, P.; Matias, D.; Fernandes, A.S.; Simões, M.F.; Nicolai, M.; Reis, C.P. Antimicrobial plant extracts encapsulated into polymeric beads for potential application on the skin. *Polymers* **2014**, *6*, 479–490. [CrossRef]

33. El-Desoky, N.; Hashem, N.; Elkomy, A.; Abo-Elezz, Z. Physiological response and semen quality of rabbit bucks supplemented with Moringa leaves ethanolic extract during summer season. *Animal* **2017**, *11*, 1549–1557. [CrossRef] [PubMed]

34. Ji, M.; Sun, X.; Guo, X.; Zhu, W.; Wu, J.; Chen, L.; Wang, J.; Chen, M.; Cheng, C.; Zhang, Q. Green synthesis, characterization and in vitro release of cinnamaldehyde/sodium alginate/chitosan nanoparticles. *Food Hydrocoll.* **2019**, *90*, 515–522. [CrossRef]

35. Fernando, I.P.S.; Lee, W.; Han, E.J.; Ahn, G. Alginate-based nanomaterials: Fabrication techniques, properties, and applications. *Chem. Eng. J.* **2020**, *391*, 123823. [CrossRef]

36. Hashem, N.M.; Hosny, N.S.; El-Desoky, N.I.; Shehata, M.G. Effect of Nanoencapsulated Alginate-Synbiotic on Gut Microflora Balance, Immunity, and Growth Performance of Growing Rabbits. *Polymers* **2021**, *13*, 4191. [CrossRef] [PubMed]

37. Chankuang, P.; Linlawan, A.; Junda, K.; Kuditthalerd, C.; Suwanprateep, T.; Kovitvadhi, A.; Chundang, P.; Sanyathitiseree, P.; Yinharnmingmongkol, C. Comparison of Rabbit, Kitten and Mammal Milk Replacer Efficiencies in Early Weaning Rabbits. *Animals* **2020**, *10*, 1087. [CrossRef] [PubMed]

38. Hashem, N.; Soltan, Y.; El-Desoky, N.; Morsy, A.; Sallam, S. Effects of Moringa oleifera extracts and monensin on performance of growing rabbits. *Livest. Sci.* **2019**, *228*, 136–143. [CrossRef]

39. Filly, A.; Fernandez, X.; Minuti, M.; Visinoni, F.; Cravotto, G.; Chemat, F. Solvent-free microwave extraction of essential oil from aromatic herbs: From laboratory to pilot and industrial scale. *Food Chem.* **2014**, *150*, 193–198. [CrossRef] [PubMed]

40. Khan, I.U.; Sajid, S.; Javed, A.; Sajid, S.; Shah, S.U. Comparative diagnosis of typhoid fever by polymerase chain reaction and widal test in Southern Districts (Bannu, Lakki Marwat and DI Khan) of Khyber Pakhtunkhwa, Pakistan. *Acta Sci. Malays.* **2017**, *1*, 12–15. [CrossRef]

41. Khalid, A.R.; Yasoob, T.B.; Zhang, Z.; Yu, D.; Feng, J.; Zhu, X.; Hang, S. Supplementation of Moringa oleifera leaf powder orally improved productive performance by enhancing the intestinal health in rabbits under chronic heat stress. *J. Therm. Biol.* **2020**, *93*, 102680. [CrossRef] [PubMed]

42. Lennarz, W.J.; Lane, M.D. *Encyclopedia of Biological Chemistry*; Academic Press: Cambridge, MA, USA, 2013.

43. El-Gazzar, H.I.; Gad, S.S.; Mubarak, M.H.; El-Deep, G.S.; GabAlla, A.A. Effect of Nitrogen fertilizer Levels and Six Canola Genotypes on Physicochemical Properties of (*Brassica napus* L.) Cultivated under North Sinai Conditions. *Sinai J. Appl. Sci.* **2019**, *8*, 35–50. [CrossRef]

44. Mousa, S. *Boosting Refugee Outcomes: Evidence from Policy, Academia, and social Innovation*; Academia, and Social Innovation (2 October 2018); Stanford Immigration Policy Lab: Stanford, CA, USA, 2018.

45. Isitua, C.; Ibeh, I. Toxicological assessment of aqueous extract of Moringa oleifera and Caulis bambusae leaves in rabbits. *J. Clin. Toxicol. S* **2013**, *12*, 4.

46. Rodríguez, M.; Rebollar, P.G.; Mattioli, S.; Castellini, C. n-3 PUFA sources (precursor/products): A review of current knowledge on rabbit. *Animals* **2019**, *9*, 806. [CrossRef] [PubMed]

47. Sun, J.; Li, Y.; Pan, X.; Nguyen, D.N.; Brunse, A.; Bojesen, A.M.; Rudloff, S.; Mortensen, M.S.; Burrin, D.G.; Sangild, P.T. Human Milk Fortification with Bovine Colostrum Is Superior to Formula-Based Fortifiers to Prevent Gut Dysfunction, Necrotizing Enterocolitis, and Systemic Infection in Preterm Pigs. *J. Parenter. Enter. Nutr.* **2019**, *43*, 252–262. [CrossRef] [PubMed]

48. Declerck, I.; Dewulf, J.; Piepers, S.; Decaluwé, R.; Maes, D. Sow and litter factors influencing colostrum yield and nutritional composition. *J. Anim. Sci.* **2015**, *93*, 1309–1317. [CrossRef] [PubMed]

49. Alameda-Pineda, X.; Horaud, R. A geometric approach to sound source localization from time-delay estimates. *IEEE/ACM Trans. Audio Speech Lang. Process.* **2014**, *22*, 1082–1095. [CrossRef]

50. Adesina, O.; Oladapo, O.; Adeniran, O.; Obehioye, O.; Iyabode, O. Blood transfusion in obstetrics: Attitude and perceptions of pregnant women. *Trop. J. Obstet. Gynaecol.* **2014**, *31*, 38–44.

51. Hewitson, J.P.; Grainger, J.R.; Maizels, R.M. Helminth immunoregulation: The role of parasite secreted proteins in modulating host immunity. *Mol. Biochem. Parasitol.* **2009**, *167*, 1–11. [CrossRef] [PubMed]

52. Benayad, A.; Taghouti, M.; Benali, A.; Aboussaleh, Y.; Benbrahim, N. Nutritional and technological assessment of durum wheat-faba bean enriched flours, and sensory quality of developed composite bread. *Saudi J. Biol. Sci.* **2021**, *28*, 635–642. [CrossRef] [PubMed]

53. Balthazar, C.F.; Guimarães, J.T.; Rocha, R.S.; Pimentel, T.C.; Neto, R.P.; Tavares, M.I.B.; Graça, J.S.; Alves Filho, E.G.; Freitas, M.Q.; Esmerino, E.A. Nuclear magnetic resonance as an analytical tool for monitoring the quality and authenticity of dairy foods. *Trends Food Sci. Technol.* **2021**, *108*, 84–91. [CrossRef]

54. De Mattos, A.M.; Olyaei, A.J.; Bennett, W.M. Nephrotoxicity of immunosuppressive drugs: Long-term consequences and challenges for the future. *Am. J. Kidney Dis.* **2000**, *35*, 333–346. [CrossRef]

55. Cerri, R.L.; Rutigliano, H.M.; Chebel, R.C.; Santos, J.E. Period of dominance of the ovulatory follicle influences embryo quality in lactating dairy cows. *Reproduction* **2009**, *137*, 813–823. [CrossRef] [PubMed]

56. Volpato, S.; Cavalieri, M.; Guerra, G.; Sioulis, F.; Ranzini, M.; Maraldi, C.; Fellin, R.; Guralnik, J.M. Performance-based functional assessment in older hospitalized patients: Feasibility and clinical correlates. *J. Gerontol. Ser. A Biol. Sci. Med. Sci.* **2008**, *63*, 1393–1398. [CrossRef] [PubMed]

57. Escamilla, A.; Bautista, M.; Zafra, R.; Pacheco, I.; Ruiz, M.T.; Martínez-Cruz, S.; Méndez, A.; Martínez-Moreno, A.; Molina-Hernández, V.; Pérez, J. Fasciola hepatica induces eosinophil apoptosis in the migratory and biliary stages of infection in sheep. *Vet. Parasitol.* **2016**, *216*, 84–88. [CrossRef] [PubMed]
58. Jones, M.L.; Mark, P.J.; Mori, T.A.; Keelan, J.A.; Waddell, B.J. Maternal dietary omega-3 fatty acid supplementation reduces placental oxidative stress and increases fetal and placental growth in the rat. *Biol. Reprod.* **2013**, *88*, 1–8. [CrossRef] [PubMed]
59. Herrera, E.; Ortega-Senovilla, H. Lipid metabolism during pregnancy and its implications for fetal growth. *Curr. Pharm. Biotechnol.* **2014**, *15*, 24–31. [CrossRef] [PubMed]

Article

Chitosan Nanoparticles Containing Lipoic Acid with Antioxidant Properties as a Potential Nutritional Supplement

Katrin Quester [1], Sarahí Rodríguez-González [2], Laura González-Dávalos [2], Carlos Lozano-Flores [2], Adriana González-Gallardo [3], Santino J. Zapiain-Merino [1], Armando Shimada [2], Ofelia Mora [2] and Rafael Vazquez-Duhalt [1,*]

[1] Centro de Nanociencias y Nanotecnología, Universidad Nacional Autónoma de México, Km 107 Carretera Tijuana-Ensenada, Ensenada 22860, Mexico; quester@cnyn.unam.mx (K.Q.); santino.zapiain@gmail.com (S.J.Z.-M.)

[2] Laboratorio de Rumiología y Metabolismo Nutricional (RuMeN), FES-C, Universidad Nacional Autónoma de México, Blvd. Juriquilla 3001, Querétaro 76230, Mexico; sarahi.r.g.200@gmail.com (S.R.-G.); lauragdavalos@yahoo.com.mx (L.G.-D.); lozancarlos2020@gmail.com (C.L.-F.); shimada@unam.mx (A.S.); ofemora66@unam.mx (O.M.)

[3] Instituto de Neurobiología, Universidad Nacional Autónoma de México, Blvd. Juriquilla 3001, Querétaro 76230, Mexico; gallardog@unam.mx

[*] Correspondence: rvd@ens.cnyn.unam.mx

Citation: Quester, K.; Rodríguez-González, S.; González-Dávalos, L.; Lozano-Flores, C.; González-Gallardo, A.; Zapiain-Merino, S.J.; Shimada, A.; Mora, O.; Vazquez-Duhalt, R. Chitosan Nanoparticles Containing Lipoic Acid with Antioxidant Properties as a Potential Nutritional Supplement. *Animals* **2022**, *12*, 417. https://doi.org/10.3390/ani12040417

Academic Editors: Antonio Gonzalez-Bulnes and Nesreen Hashem

Received: 11 January 2022
Accepted: 27 January 2022
Published: 10 February 2022

Simple Summary: Alfa-lipoic acid (ALA) is an important antioxidant that could be added to animal feed as a nutritional supplement. To improve its stability in the digestive system, ALA was encapsulated in chitosan nanoparticles. The nanoparticles containing ALA were stable in stomach-like conditions and were able to cross the intestinal barrier. Chitosan-based nanoparticles seem to be an attractive administration method for antioxidants, or other sensible additives, in food.

Abstract: The addition of the antioxidant α-lipoic acid (ALA) to a balanced diet might be crucial for the prevention of comorbidities such as cardiovascular diseases, diabetes, and obesity. Due to its low half-life and instability under stomach-like conditions, α-lipoic acid was encapsulated into chitosan nanoparticles (Ch-NPs). The resulting chitosan nanoparticles containing 20% w/w ALA (Ch-ALA-NPs) with an average diameter of 44 nm demonstrated antioxidant activity and stability under stomach-like conditions for up to 3 h. Furthermore, fluorescent Ch-ALA-NPs were effectively internalized into 3T3-L1 fibroblasts and were able to cross the intestinal barrier, as evidenced by everted intestine in vitro experiments. Thus, chitosan-based nanoparticles seem to be an attractive administration method for antioxidants, or other sensible additives, in food.

Keywords: antioxidant; cell internalization; chitosan nanoparticles; everted intestine; lipoic acid

1. Introduction

Reactive oxygen species (ROS) are highly reactive molecules generated as by-products of oxygen metabolism in aerobic organisms [1]. Several diseases, such as hypertension and diabetes, amongst others, are associated with ROS generation [2,3]. Aging is one of the known interrupters of the balance between ROS generation and antioxidant defenses [4], and an age-related increase in oxidative damage to DNA, proteins and lipids was shown [5], particularly in cardiac and vascular tissues [6]. To slow down this oxidation-related damage, the antioxidant vitamins A, C, and E are popular nutritional supplements. However, clinical studies have failed to confirm their benefits in the prevention of cardiovascular diseases [7–9].

α-Lipoic acid (ALA), also known as thioctic acid or 1,2-dithiolane-3-pentanoic acid, is a natural antioxidant found in plants as well as in animal cells, where it is commonly found in mitochondria and acting as co-factor for a number of enzyme complexes that are involved in energy generation. Both existing forms of ALA—the oxidized (disulfide) and the reduced (dithiol; dihydrolipoic acid or DHLA) forms—showed antioxidant properties, and the ALA/DHLA couple is often referred to as the "universal antioxidant" due to its ability to not only improve but also restore intrinsic antioxidant systems after quenching free radicals [10–13]. Besides its great antioxidant potential, ALA was shown to efficiently remove heavy metals from the blood stream [11,13,14]. ALA has gained considerable attention as a dietary supplement because of its antioxidant activity and considerable anti-aging, anti-inflammatory, detoxifying, cognitive, cardiovascular, anti-cancer, and neuroprotective properties [15]. Some studies indicate that ALA might play an important role in the treatment of severe diseases such as diabetes mellitus [16–18], Alzheimer's disease [19–25], obesity [26–30], multiple sclerosis [31–33], and schizophrenia [34–36].

Our research group previously demonstrated that broilers fed a diet supplemented with 40 mg/kg ALA for 7 weeks had significantly decreased liver levels of thiobarbituric acid reactive substances and hydroxyl radicals, whereas their total glutathione pools were increased compared to birds fed the same diet but without ALA [37]. We also demonstrated that the mRNA expression levels of the genes encoding four enzymes involved in energy metabolism (glutathione S-transferase theta 1 (GSTT1), oxoglutarate (alpha-ketoglutarate) dehydrogenase (OGDH), pyruvate dehydrogenase (lipoamide) alpha 1 (PDHA1), and dihydrolipoamide S-succinyl transferase (DLST)), as well as those encoding sirtuins 1 and 3, showed significant decreases in ALA-treated broilers compared to untreated controls ($p < 0.01$). These results confirm that the responses of broilers to ALA were associated with the down-regulation of certain liver enzymes, especially those involved in glucose metabolism and the tricarboxylic acid (TCA) cycle [38].

Nevertheless, the amount of this antioxidant synthesized by the organism does not meet bodily needs [39] unless supplementarily administered through diet. In addition, ALA presents a low half-life and bioavailability due to hepatic degradation, reduced solubility, and instability in the stomach environment [40]. Therefore, improving ALA administration is the main challenge, and the encapsulation of ALA into biocompatible nanoparticles seems to be a promising alternative.

Chitosan is a polymer produced after deacetylation of the natural polymer chitin, which is the main component of the shells of crabs and other crustaceans. The United States Food and Drug Administration (FDA) and the European Commission, among others, have approved chitosan as safe for use in food and drugs. Chitosan has been widely used for the encapsulation of proteins and therapeutical enzymes [41–44] or as biocatalytic nanoparticles [45]. Importantly, among the biopolymers used as nanocarriers, it has attracted significant attention since it is positively charged [46]. Therefore, this polysaccharide can be customized for a wide variety of chemical and ionic side-chain reactions [44]. Chitosan nanoparticles are known to easily achieve a sustained slow release of the cargo as well as increase bioavailability and therapeutic efficiency, while being biodegradable and non-toxic [47]. The production process for nanoparticles such as chitosan NPs could be improved to gain the advantages of ease of production, high yield at low cost, and larger loading capacity [48–52].

In this study, chitosan-α-lipoic acid nanoparticles (Ch-ALA-NPs) with antioxidant activity were synthesized, and their stability under stomach-like conditions was assayed. Furthermore, the cell internalization of chitosan nanoparticles in intestinal cells and fibroblasts and the ability to cross the intestinal mucosal barrier in vitro by using the everted intestine technique were demonstrated.

2. Materials and Methods

2.1. Synthesis of Chitosan-α-Lipoic Acid Nanoparticles (Ch-ALA-NPs)

The chitosan nanoparticles were prepared based on the protocol described by [53] with the following modifications: Chitosan obtained from the deacetylation of shrimp shell chitin was purchased from Future Foods (Tlalnepantla de Baz, Mexico). The preparation showed 90.4% deacetylation, a viscosity of 54 mPa·s in 1% acetic acid, and an average molecular weight of 290 kDa. The chitosan solution was prepared by dissolving 2.5 mg mL^{-1} of chitosan in 2% acetic acid while stirring at approximately 200–300 rpm overnight at room temperature, and then centrifuging for 15 min at 8000× g at 4 °C. The resulting supernatant was again centrifuged for 20 min at 11,000× g at 4 °C. Then, the supernatant was filtered through a 1.2 µm membrane, and the pH was adjusted to 4.5.

Nanoparticle synthesis was achieved using an automatic pipet system designed by our research group and equipped with an insulin syringe to control the addition of reagents at a constant rate. The reagent addition was performed at room temperature and constant agitation of approximately 800 rpm. To 5 mL of the chitosan solution, 200 µL of α-lipoic acid solution (6 mg mL^{-1} in 20% ethanol) were added at 0.09 mm/s. After 15 min, 1 mL of tripolyphosphate pentasodium (TPP; Sigma-Aldrich, St. Louis, MO, USA) solution (0.25 mg mL^{-1}) was added at 0.09 mm/s and left to react for 1 h. Then, 100 µL of 2.5% glutaraldehyde (Sigma-Aldrich, USA) were added at 0.09 mm/s and again left for 1 h. Afterwards, the solution was centrifuged for 30 min at 2000× g at room temperature (Sorvall Legend RT centrifuge) to remove chitosan aggregates. The resulting supernatant was then ultracentrifuged for 2 h at 60,000× g at 4 °C (Beckman Coulter Optima XPN-100 ultracentrifuge). Subsequently, the pellet, containing the chitosan-α-lipoic acid nanoparticles (Ch-ALA-NPs), was resuspended in 100 mM phosphate butter (pH 6), filtered through a 0.2 µm membrane, and stored at 4 °C until further modification.

2.2. Synthesis of Chitosan-Green Fluorescent-Nanoparticles Ch-GFP-NPs

Chitosan nanoparticles labeled by encapsulating carboxylated green fluorescent protein (GFP) were synthetized as previously reported [44]. In brief, GFP was reacted with malonic acid in the presence of carbodiimide (EDC) (N-(3-dimethyl-aminopropyl)-N'-ethylcarbodiimide hydrochloride) and N-hydroxysuccinimide (NHS). The amount of malonic acid was twice the amount of free carboxylic groups on the protein surface on a molar basis, and the amounts of EDC and NHS were ten and five times the amount of malonic acid, respectively. The mixtures were allowed to react for 3 h in rotation at room temperature and then dialyzed against MES (50 mM, pH 6). The encapsulation of carboxylated GFP was performed by ionic gelation with TPP as described above.

2.3. Morphological Analyses of the Nanoparticles

The hydrodynamic diameter and Zeta potential of all preparations were determined by the DLS technique on a Zetasizer NanoZS (Malvern, UK). To confirm the size and shape of the nanoparticles, high-resolution transmission electron microscopy (HR-TEM) was performed using a JEOL JEM-2010. Five µL of a 1000 times diluted sample of the nanoparticles were placed onto a carbon-coated 300 mesh copper grid (Ted Pella, Inc.; Redding, CA, USA) and then stained with phosphotungstic acid (PTA).

2.4. Cargo Capacity and Antioxidant Activity of the Ch-ALA-NPs

To estimate the amount of encapsulated α-lipoic acid, HPLC analyses were performed using the supernatant after ultracentrifugation. The resulting concentration of free α-lipoic acid was subtracted from the initial α-lipoic acid concentration used for nanoparticle synthesis.

Antioxidant activity of free and Ch-ALA-NPs was determined by using the inhibition of pyrogallol autooxidation method reported by [54], with some modifications. First, 50 mM Tris-HCl buffer (pH 8.2, containing 1 mM EDTA) was oxygenized with air. Pyrogallol (2 mM) was dissolved in this oxygenized buffer. The capacity of the free ALA and Ch-ALA-NPs to inhibit the spontaneous pyrogallol oxidation in the presence of atmospheric oxygen was determined using equivalent concentrations of the nanoparticles and free α-lipoic acid and monitored in a UV-vis spectrophotometer (Lambda 25, Perkin Elmer, Waltham, MA, USA) at 420 nm.

2.5. Release of α-Lipoic Acid from the Ch-ALA-NPs

The release of α-lipoic acid from the chitosan nanoparticles was determined in HCl (pH 2) at 37 °C for 3 h, to simulate the stomach environment. A suspension of Ch-ALA-NPs (10 mL) was added to a dialysis membrane with a molecular weight cut-off (MWCO) of 14 kDa. The dialysis membrane containing the nanoparticle suspension was transferred into 20 mL of HCl (pH 2). Under constant agitation at room temperature, 1 mL samples of HCl were measured at 330 nm at different time points to determine the corresponding amounts of released α-lipoic acid. The released ALA was quantified by using a standard curve.

2.6. Fluorophore Labeled Nanoparticles

In order to monitor and quantify the nanoparticles containing α-lipoic acid in the everted intestine experiments, the nanoparticles were conjugated with a fluorophore (Fluorescein isothiocyanate, FITC), and produced as follows: One mL of ALA solution in isopropanol (100 mg/mL) was added to 35 mL solution of chitosan (2%) in acetic acid, pH 5.5. The ALA solution was added at a rate of 0.09 mm/s using the automatic pipet system designed by our research group and described above. After 15 min, 250 μL of a solution of FITC in methanol (10 mg/mL) were added at a rate of 0.03 mm/s. Then, 7 mL of 0.25 % TPP were added at a rate of 0.09 mm/s, and the mixture was shaken for 1 h to produce Ch-ALA-FITC-NPs by ionic gelation. Finally, 1 mL of glutaraldehyde (2.5%) was added with the same system and shaken for 1 h. The nanoparticle suspension was centrifuged at 3500 rpm for 30 min to eliminate the bigger particles, and the supernatant was then centrifuged at 60,000× *g* for 2 h. The pellet containing the Ch-ALA-FITC-NPs was resuspended in Milli Q water.

2.7. Internalization of Ch-GFP-NPs into Fibroblasts 3T3-L1

Fibroblast 3T3-L1 (CL-173, ATCC) cells (1×10^5) were cultivated in DMEM containing 10% fetal bovine serum, 1% penicillium and 1% streptomycin in 6-well plates (Corning, NY, USA) containing microscope glass slides and incubated at 37 °C and 5% CO_2. Once they reached a confluence of 80–85%, Ch-GFP-NPs were added in different concentrations (5, 10, 20 and 50 μL mL^{-1}) to the medium and incubated for 24 and 48 h. Afterward, the glass slides, on which cells adhered, were washed with PBS, stained for 15 min with MitoTracker Red CMXRos (Invitrogen, Waltham, MA, USA), and fixed with 4% PFA. Cell nuclei were stained with 4′,6-diamidino-2-phenylindole (DAPI) (Invitrogen, Carlsbad, CA, USA). Cells were then analyzed with a confocal laser scanning microscope (LSM 780, Carl Zeiss Inc., Oberkochen, Germany) using a DPSS laser with emission at 561 nm and a HeNe laser with emission at 633 nm. Obtained images were analyzed using the software ZEN Blue Edition (Zeiss, Jena, Germany) and ImageJ (NIH, Bethesda, MD, USA).

The 3D reconstruction was performed with Amira software (Thermo Fisher Scientific, Waltham, MA, USA) by visualizing, processing, and analyzing data obtained by 780 LSM confocal microscope (Zeiss, Jena, Germany). Nuclei were seen in blue and mitochondria in red, while nanoparticles were in green; the observed signal was processed by the isosurface and volume rendering module. This visualization was carried out using immersive reality within the Cave Automatic Virtual Environment (CAVE) of the LAVIS-UNAM laboratory.

2.8. Intestine Barrier Crossing of Ch-ALA-FITC-NPs by In Vitro Everted Intestine

Intestinal tissues were taken from Wistar rats and chickens (*Gallus gallus*) (donated from a local farm), to compare NP translocation through the intestine barrier in two species by everted intestine experiments. Wistar rats were housed and used at Universidad Nacional Autónoma de México (UNAM) according to regulations of the Mexican government regarding the use of laboratory animals for research purposes (CICUAL-UNAM, NOM-062-ZOO-1999 4.2.2). Animals were sacrificed with CO_2 and by cervical dislocation by trained personnel. This protocol was approved by the Institute's Research Ethics Committee (Comité de Ética en la Investigación, INB-UNAM) with register #065. Intestinal tissue was obtained from 20–21 day old male rats, and 6-month-old chickens. Twelve cm of intestine from each rat or 10 cm from each chicken were dissected and placed in a Petri dish containing Krebs-Ringer buffer at 37 °C. Immediately, the eversion of the intestine was carefully realized using a Pasteur pipette. The intestinal content was removed, and the tissue was washed with the same buffer. The lower end of the intestine was closed using a nylon thread, 600 µL Krebs-Ringer buffer were added, and the other end of the intestine was closed as well. The sealed intestine was then transferred into a 15 mL Falcon tube, previously gasified with CO_2, containing 50 µL/mL Ch-ALA-FITC-NPs in Krebs-Ringer buffer, and incubated for 60 min at 37 °C under agitation. Subsequently, the intestinal content was centrifuged for 1 min, and fluorescence of the supernatant was spectrofluorimetrically measured at an excitation of 495 nm and an emission of 519 nm. Control experiments were carried out in the same conditions but without adding nanoparticles. The percentage of nanoparticle internalization was estimated by comparing the fluorescence of the samples to the fluorescence of the initial nanoparticle solution. All experiments were performed in triplicate.

2.9. Statistical Analysis

Significant differences in all determinations were analyzed by one-way ANOVA test, followed by post hoc Tukey rank test. Statistical significance was set at $p < 0.05$ and expressed as compact letter display. Statistical analysis was performed on five independent replicates using the software STATISTICA 8.0™ (StatSoft Inc., Tulsa, OK, USA).

3. Results

Chitosan nanoparticles were produced by ionic gelation in the presence of ALA. Preliminary experiments to determine the maximal ALA loading in the chitosan nanoparticles were performed. A fixed amount of chitosan (12.5 mg) and different concentrations of ALA from 0.5 to 5 mg were used to produce chitosan ALA-loaded nanoparticles (Ch-ALA-NPs). The ALA content was then determined spectrophotometrically. Higher ALA concentrations than 5 mg induced precipitation and no nanoparticle formation.

The aim of this work was to produce chitosan nanoparticles as carriers for ALA. Four different preparations were synthetized: control nanoparticles without ALA (Ch-NPs), chitosan nanoparticles containing green fluorescent protein (Ch-GFP-NPs) to monitor cell internalization of nanoparticles by confocal microscopy, FITC-modified and -containing ALA particles (Ch-ALA-FITC-NPs) to evaluate intestinal barrier crossing by the everted intestine technique by fluorescence, and the ALA-containing chitosan nanoparticles (Ch-ALA-NPs) (Table 1). The nanoparticles formed with a Ch:ALA ratio of 2.5:1 showed an ALA content of 20% *w/w*, and this ratio was selected for further experiments.

Table 1. Hydrodynamic diameter and Zeta potential of different preparations of chitosan-based nanoparticles.

Nanoparticle Preparation	Hydrodynamic Diameter (nm)	Zeta Potential (mV)
Ch-NPs	88.4 (±28.2) [a]	49 (±1.9) [a]
Ch-GFP-NPs	96.7 (±35.2) [a]	45 (±2.2) [b]
Ch-ALA-NPs	44.1 (±20.8) [b]	32 (±0.8) [c]
Ch-ALA-FITC-NPs	84.6 (±28.2) [a]	28 (±0.2) [d]

[a,b,c,d] Significant differences from four independent determinations were analyzed by one-way ANOVA test, followed by post hoc Tukey rank test. Statistical significance was set at $p < 0.05$.

Dynamic light scattering (DLS) analyses (Table 1) revealed a hydrodynamic diameter of approximately 44.1 (±20.8) nm, which was confirmed by high-resolution transmission electron microscopy (HR-TEM) (Figure 1). The nanoparticle size was half that of the nanoparticles without ALA (Ch-NPs) and could be due to a strong interaction between the carboxylic group of ALA and the amino group of glucosamine monomers of chitosan. This interaction was also supported by the reduction of zeta potential from 49 mV to 32 mV when the ALA was loaded into nanoparticles (Table 1).

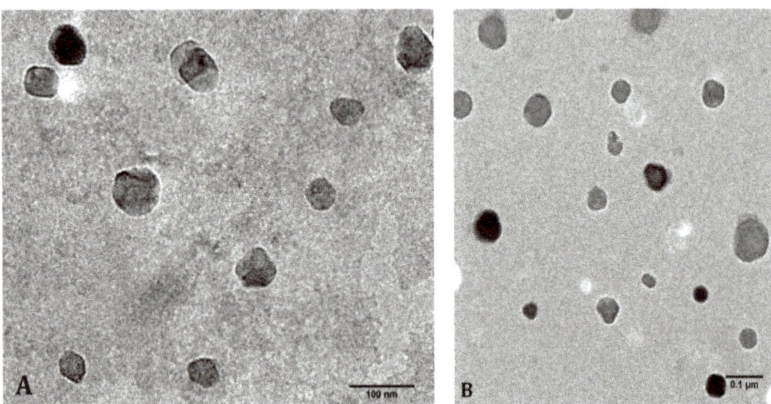

Figure 1. HR-TEM images of synthesized Ch-ALA-NPs reveal an average diameter of 44 nm. Scale bars are 100 nm in (**A**), and 0.1 μm in (**B**).

The antioxidant activity of both free and nano-encapsulated ALA was also determined (Figure 2). The inhibition of the spontaneous oxidation of pyrogallol in the presence of atmospheric oxygen was used as the antioxidation assay. The antioxidant activity was not significantly affected by the encapsulation, Ch-ALA-NPs being slightly more efficient than free ALA, and both preparations were able to completely inhibit this reaction at the same ALA equivalents. Chitosan-only nanoparticles (Ch-NPs) showed some antioxidant activity in the oxidation of pyrogallol.

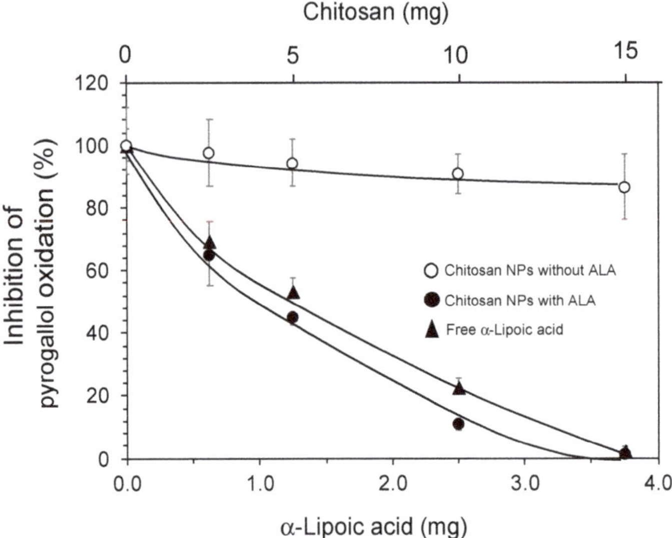

Figure 2. Antioxidant activity of Ch-NPs, CH-ALA-NPs and free ALA. The antioxidant activity was determined by using the inhibition of pyrogallol autooxidation method reported by Marklund and Marklund [54]. The average values from five independent experiments and the standard deviation as error bars are shown.

As mentioned above, the low half-life and bioavailability of ALA, as well as its instability in the stomach, represent disadvantages for its administration as a dietary supplement. Thus, the stability of Ch-ALA-NPs and liberation of ALA in a stomach-like environment (HCl solution at pH 2) were determined. Free ALA was measured via UV-vis spectrophotometry every 5 min for 3 h (Figure 3). A sustained release of ALA was found, but the percentage of released ALA from the NPs was very low according to the total ALA content in the NPs, implying that the Ch-ALA-NPs remained stable and intact in the stomach until further intestinal absorption. Unmodified chitosan is fully soluble at low pH. Nevertheless, the glutaraldehyde crosslinking of chitosan nanoparticles made them stable even in very acidic conditions, thus retaining the ALA inside.

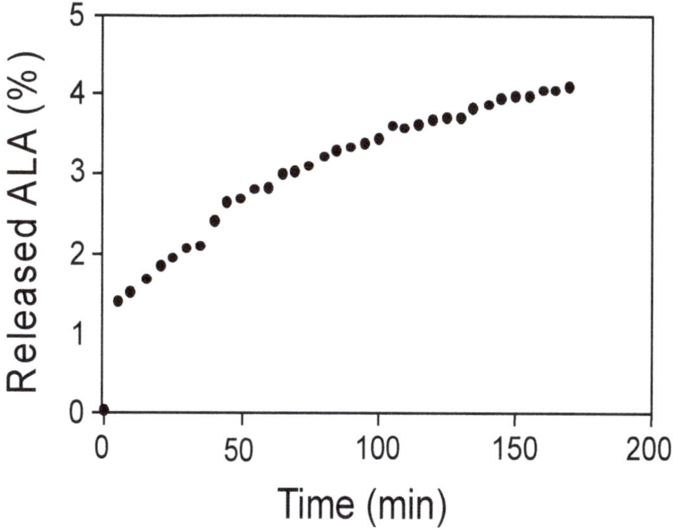

Figure 3. ALA release from Ch-ALA-NPs in a stomach-like environment (HCl, pH 2).

Cell internalization of nanoparticles in 3T3 cells of mouse embryonic fibroblasts was assayed using synthesized fluorescent NPs containing the green fluorescent protein (Ch-GFP-NPs) (Figure 4). These fluorescent chitosan nanoparticles showed hydrodynamic diameter and zeta potential similar to non-labeled nanoparticles (Ch-NPs). A culture of 3T3 cells of mouse embryonic fibroblasts was incubated with different concentrations of Ch-GFP-NPs; after 48 h of incubation, the cells were washed three times with PBS and analyzed by confocal laser scanning microscopy (see Video S1, supplementary information). The cell nuclei were stained with DAPI, and the mitochondria with MitoTracker Red. The confocal analysis showed clear evidence of chitosan nanoparticle cell internalization.

The capacity of chitosan nanoparticles loaded with α-lipoic acid to cross the intestinal barrier was evaluated by using the everted intestine technique. To achieve this, the nanoparticles were labeled with FITC (Ch-ALA-FITC-NPs) to be monitored and quantified. Because the FITC was covalently bonded to the amino groups of glucosamine monomers of chitosan, the zeta potential was reduced to 28 mV (Table 1), but still positive. The everted intestine technique is widely used to evaluate intestinal absorption of nutrients, drugs, and toxicants [55–57]. Everted intestines from rats and chickens were incubated with Ch-ALA-FITC-NPs, and the fluorescence was measured in samples taken from both inside and outside of the everted intestine (Figure 5). Although the absorption rate was low (~0.5%/h), a significant level of fluorescence could be observed compared to the control without NP treatment.

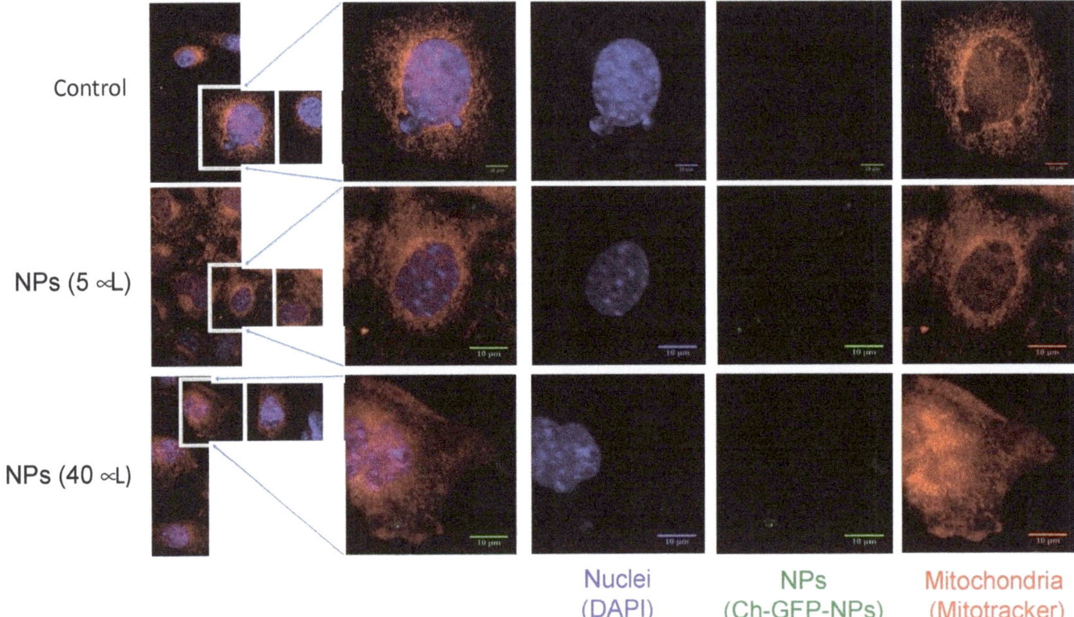

Figure 4. Cell internalization of chitosan nanoparticles containing GFP into 3T3 cells of mouse embryonic fibroblasts. Ch-GFP-NPs were incubated for 48 h and then washed three times. The cell nuclei were stained with DAPI, and the mitochondria with MitoTracker Red. The nanoparticles were detected by emission at 561 nm.

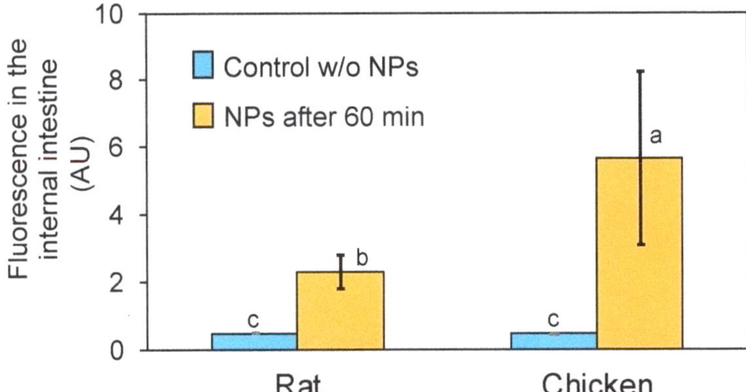

Figure 5. Internalization through the intestinal barrier of Ch-ALA-FITC-NPs assayed in vitro by the everted intestine technique. Fluorescence in arbitrary units (AU) in the control experiments without nanoparticles and after 60 min incubation with Ch-ALA-FITC-NPs. The experiments were carried out at 37 °C, and average values from five independent experiments and the standard deviation as error bars are shown. Significant differences in all determinations were analyzed by one-way ANOVA test, followed by post hoc Tukey rank test. Statistical significance was set at $p < 0.05$. The meaning of a,b,c in figure is the statistical significance expressed as compact letter display.

4. Discussion

α-Lipoic acid is synthesized in human and animal cells at a low rate, and thus it should be consumed as a dietary supplement. However, ALAs' instability in the stomach and low bioavailability represent challenges for efficient administration to the animal organism [40]. To overcome these obstacles, ALA was encapsulated in chitosan nanoparticles. This natural biopolymer is approved by most agencies around the world as safe for use in food and drugs, and it is known to easily achieve a sustained slow release of its cargo as well as increase bioavailability and therapeutic efficiency [47].

Chitosan-based nanoparticles containing ALA were produced (Figure 1 and Table 1). Due to the solubility of chitosan at low pH, the nanoparticles were stabilized by glutaraldehyde crosslinking. However, the glutaraldehyde molecules were covalently conjugated to the glucosamine moieties and not in free form, and thereby reducing their potential toxicity. The antioxidant activity of ALA was slightly higher when encapsulated in chitosan nanoparticles, while the NPs without ALA showed some antioxidant activity (Figure 2). The slight antioxidant capacity of chitosan is well known [58–60].

The chitosan nanoparticles were internalized by mouse embryonic fibroblasts (Figure 4 and Video S1). Chitosan nanoparticle uptake by different cells has been clearly demonstrated [61–63]. The Ch-ALA-NPs showed a positive Zeta potential of 35.7 ± 3.1 at pH of 6. The positive charge promoted the cell internalization rate in different cell lines, increasing the cellular uptake of NPs. In addition, an intracellular trafficking study indicated that positively charged NPs exhibit perinuclear localization and escape from lysosome [60].

Using the everted intestine technique, the capacity of chitosan nanoparticles containing α-lipoic acid to cross the intestinal barrier was demonstrated (Figure 5). Nanoparticles in a size range of 0.1 to 100 nm can quickly enter the digestive tract [64]. Due to their size (84 nm), the chitosan nanoparticles showed the ability to cross the mucosal barrier and interact with the subjacent absorbent epithelial tissue [65,66]. On the other hand, chitosan is a polymer that can help the absorption and increase the permeation capacity of drugs through the paracellular pathway by the reversible opening of the narrow epithelial bonds [67]. In addition, chitosan is muco-adhesive, and thus can adhere to the intestinal mucosa through ionic interactions between its positively charged amino groups and the negative charges found on the surfaces of intestinal cells, promoting intestinal absorption [68]. It is well known that chitosan nanoparticles can increase the drug concentration at the absorption sites [65,66]. In addition, it has been demonstrated that chitosan nanoparticles increase the intestinal absorption in vitro [69] of diverse bioactive compounds including, phenolic compounds [70], catechins and epigallocatechin from green tea [71], resveratrol and cumarin-6 [72].

The results obtained in the present study indicate that the administration of α-lipoic acid in the form of Ch-ALA-NPs as a supplement to a balanced diet might be beneficial for, the prevention and/or treatment of severe diseases such as cardiovascular diseases, diabetes and obesity, amongst others. ALA may play an important role in the treatment and/or prevention of diverse severe diseases including diabetes, Alzheimer's, multiple sclerosis, schizophrenia, and obesity [16,17,21,22,25–28,31,32,35].

5. Conclusions

Chitosan nanoparticles containing α-lipoic acid with antioxidant activity were efficiently synthesized in a size average of 44 nm. The nanoparticles proved to be stable under stomach-like conditions for up to 3 h. Furthermore, the cell internalization of chitosan nanoparticles into fibroblasts cells was demonstrated. Finally, the nanoparticles were able to cross the intestinal barrier as confirmed by the everted intestine technique, suggesting an efficient method to supply ALA. From our results, we concluded that chitosan-based nanoparticles containing ALA were stable in stomach-like conditions and able to cross the intestinal barrier and release their antioxidant cargo. Thus, the use of α-lipoic acid in a nanoparticulate form as a dietary supplement seems to be an attractive method of administration. However, the results shown here are from in vitro experiments; thus, to determine potentially better bioavailability than the native compound, adequately designed pharmacokinetic trials in animal models should be carried out.

Supplementary Materials: The following supporting information can be downloaded at: https://www.mdpi.com/article/10.3390/ani12040417/s1, Video S1: Chitosan-ALA-NPs.

Author Contributions: Conceptualization, R.V.-D., A.S. and O.M.; Data curation, O.M. and R.V.-D.; Formal Analysis, K.Q., O.M. and R.V.-D.; Funding acquisition, A.S. and R.V.-D.; Investigation, K.Q., S.R.-G., S.J.Z.-M. and L.G.-D.; Methodology, A.G.-G., K.Q., C.L.-F. and R.V.-D.; Supervision: R.V.-D., O.M. and A.S.; Writing—original draft, K.Q. and R.V.-D.; Writing—review and editing, R.V.-D. All authors have read and agreed to the published version of the manuscript.

Funding: This work was funded by the Universidad Nacional Autónoma de México (PAPIIT IG200818 and IN209722).

Institutional Review Board Statement: Wistar rats were housed and used at Universidad Nacional Autónoma de México (UNAM) according to regulations of the Mexican government regarding the use of laboratory animals for research purposes (CICUAL-UNAM, NOM-062-ZOO-1999 4.2.2). Animals were sacrificed with CO_2 and by cervical dislocation by trained personnel. This protocol was approved by the Institute's Research Ethics Committee (Comité de Ética en la Investigación, INB-UNAM) with register #065.

Informed Consent Statement: Not applicable.

Data Availability Statement: Not applicable.

Acknowledgments: This work was funded by the Universidad Nacional Autónoma de México (PAPIIT IG200818 and IN209722). We also thank Yareth Silva-Sánchez, undergraduate student, who contributed to studies with everted intestine. We also thank Nydia Hernández Ríos (Unidad de Microscopia Confocal, INB-UNAM) and Alejandro de León Cuevas (LAVIS, UNAM).

Conflicts of Interest: The authors declare no conflict of interest.

References

1. Vergely, C.; Maupoil, V.; Clermont, G.; Bril, A.; Rochette, L. Identification and quantification of free radicals during myocardial ischemia and reperfusion using electron paramagnetic resonance spectroscopy. *Arch. Biochem. Biophys.* **2003**, *420*, 209–216. [CrossRef] [PubMed]
2. Touyz, R.M.; Schiffrin, E.L. Signal transduction mechanisms mediating the physiological and pathophysiological actions of angiotensin II in vascular smooth muscle cells. *Pharmacol. Rev.* **2000**, *52*, 639–672.
3. Sicard, P.; Acar, N.; Grégoire, S.; Lauzier, B.; Bron, A.M.; Creuzot-Garcher, C.; Brétillon, L.; Vergely, C.; Rochette, L. Influence of rosuvastatin on the NAD(P)H oxidase activity in the retina and electroretinographic response of spontaneously hypertensive rats. *Br. J. Pharmacol.* **2007**, *151*, 979–986. [CrossRef] [PubMed]
4. Martin, C.; Dubouchaud, H.; Mosoni, L.; Chardigny, J.M.; Oudot, A.; Fontaine, E.; Vergely, C.; Keriel, C.; Rochette, L.; Leverve, X.; et al. Abnormalities of mitochondrial functioning can partly explain the metabolic disorders encountered in sarcopenic gastrocnemius. *Aging Cell* **2007**, *6*, 165–177. [CrossRef]
5. Figueiredo, P.A.; Mota, M.P.; Appell, H.J.; Duarte, J.A. The role of mitochondria in aging of skeletal muscle. *Biogerontology* **2008**, *9*, 67–84. [CrossRef]
6. Oudot, A.; Martin, C.; Busseuil, D.; Vergely, C.; Demaison, L.; Rochette, L. NADPH oxidases are in part responsible for increased cardiovascular superoxide production during aging. *Free Radic. Biol. Med.* **2006**, *40*, 2214–2222. [CrossRef]
7. Heart Protection Study Collaborative Group. MRC/BHF Heart Protection Study of antioxidant vitamin supplementation in 20,536 high-risk individuals: A randomised placebo controlled trial. *Lancet* **2002**, *360*, 23–33. [CrossRef]
8. Hercberg, S.; Galan, P.; Preziosi, P.; Bertrais, S.; Mennen, L.; Malvy, D.; Roussel, A.M.; Favier, A.; Briançon, S. The SU.VI.MAX Study: A randomized, placebo-controlled trial of the health effects of antioxidant vitamins and minerals. *Arch. Int. Med.* **2004**, *164*, 2335–2342. [CrossRef]
9. Rapola, J.M.; Virtamo, J.; Ripatti, S.; Huttunen, J.K.; Albanes, D.; Taylor, P.R.; Heinonen, O.P. Randomised trial of alphatocopherol and beta-carotene supplements on incidence of major coronary events in men with previous myocardial infarction. *Lancet* **1997**, *349*, 1715–1720. [CrossRef]
10. Kagan, V.E.; Shvedova, A.; Serbinova, E.; Khan, S.; Swanson, C.; Powell, R.; Packer, L. Dihydrolipoic acid—A universal antioxidant both in the membrane and in the aqueous phase: Reduction of peroxyl, ascorbyl and chromanoxyl radicals. *Biochem. Pharmacol.* **1992**, *44*, 1637–1649. [CrossRef]
11. Han, D.; Handelman, G.; Marcocci, L.; Sen, C.K.; Roy, S.; Kobuchi, H.; Tritschler, H.J.; Floh, L.; Packer, L. Lipoic acid increases de novo synthesis of cellular glutathione by improving cystine utilization. *Biofactors* **1997**, *6*, 321–338. [CrossRef] [PubMed]

12. Shay, K.P.; Moreau, R.F.; Smith, E.J.; Smith, A.R.; Hagen, T.M. Alpha-lipoic acid as a dietary supplement: Molecular mechanisms and therapeutic potential. *Biochim. Biophys. Acta (BBA)-Gen. Subj.* **2009**, *1790*, 1149–1160. [CrossRef] [PubMed]

13. Goraca, A.; Huk-Kolega, H.; Piechota, A.; Kleniewska, P.; Ciejka, E.; Skibska, B. Lipoic acid–biological activity and therapeutic potential. *Pharmacol. Rep.* **2011**, *63*, 849–858. [CrossRef]

14. Ou, P.; Tritschler, H.J.; Wolff, S.P. Thioctic (lipoic) acid: A therapeutic metal-chelating antioxidant? *Biochem. Pharmacol.* **1995**, *50*, 123–126. [CrossRef]

15. El Barky, A.R.; Hussein, S.A.; Mohamed, T.M. The potent antioxidant alpha lipoic acid. *J. Plant Chem. Ecophysiol.* **2017**, *2*, id1016.

16. Konrad, D.; Somwar, R.; Sweeney, G.; Yaworsky, K.; Hayashi, M.; Ramlal, T.; Klip, A. The antihyperglycemic drug alpha-lipoic acid stimulates glucose uptake via both GLUT4 translocation and GLUT4 activation: Potential role of p38 mitogen-activated protein kinase in GLUT4 activation. *Diabetes* **2001**, *50*, 1464–1471. [CrossRef]

17. Eason, R.C.; Archer, H.E.; Akhtar, S.; Bailey, C.J. Lipoic acid increases glucose uptake by skeletal muscles of obesediabetic ob/ob mice. *Diabetes Obes. Metabol.* **2002**, *4*, 29–35. [CrossRef] [PubMed]

18. Ghibu, S.; Richard, C.; Vergely, C.; Zeller, M.; Cottin, Y.; Rochette, L. Antioxidant properties of an endogenous thiol: Alpha-lipoic acid, useful in the prevention of cardiovascular diseases. *J. Cardiovasc. Pharmacol.* **2009**, *54*, 391–398. [CrossRef]

19. Hagen, T.M.; Ingersoll, R.T.; Lykkesfeldt, J.; Liu, J.; Wehr, C.M.; Vinarsky, V.; Bartholomew, J.C.; Ames, A.B. (R)-alpha-lipoic acid-supplemented old rats have improved mitochondrial function, decreased oxidative damage, and increased metabolic rate. *FASEB J.* **1999**, *13*, 411–418. [CrossRef]

20. Zhang, W.-J.; Frei, B. Alpha-lipoic acid inhibits TNF-alpha-induced NF-kappaB activation and adhesion molecule expression in human aortic endothelial cells. *FASEB J.* **2001**, *15*, 2423–2432. [CrossRef]

21. Farr, S.A.; Poon, H.F.; Dogrukol-Ak, D.; Drake, J.; Banks, W.A.; Eyerman, E.; Butterfield, D.A.; Morley, J.E. The antioxidants alpha-lipoic acid and N-acetylcysteine reverse memory impairment and brain oxidative stress in aged SAMP8 mice. *J. Neurochem.* **2003**, *84*, 1173–1183. [CrossRef] [PubMed]

22. Lovell, M.A.; Xie, C.; Xiong, S.; Markesbery, W. Protection against amyloid beta peptide and iron/hydrogen peroxide toxicity by alpha lipoic acid. *J. Alzheimers Dis.* **2003**, *5*, 229–239. [CrossRef] [PubMed]

23. Ono, K.; Hirohata, M.; Yamada, M. α-Lipoic acid exhibits anti-amyloidogenicity for β-amyloid fibrils in vitro. *Biochem. Biophys. Res. Comm.* **2006**, *341*, 1046–1052. [CrossRef] [PubMed]

24. Haugaard, N.; Levin, R.M. Regulation of the activity of choline acetyl transferase by lipoic acid. *Mol. Cell. Biochem.* **2000**, *213*, 61–63. [CrossRef] [PubMed]

25. Holmquist, L.; Stauchbury, G.; Berbaum, K.; Muscat, S.; Young, S.; Hager, K.; Engel, J.; Münch, G. Lipoic acid as a novel treatment for Alzheimer's disease and related demenias. *Pharmacol. Ther.* **2007**, *113*, 154–164. [CrossRef]

26. Huerta, A.E.; Navas-Carretero, S.; Prieto-Hontoria, P.L.; Martínez, J.A.; Moreno-Aliaga, M.J. Effects of α-lipoic acid and eicosapentaenoic acid in overweight and obese women during weight loss. *Obesity* **2015**, *23*, 313–321. [CrossRef]

27. Li, N.; Yan, W.; Hu, X.; Huang, Y.; Wang, F.; Zhang, W.; Wang, Q.; Wang, X.; Sun, K. Effects of oral α-lipoic acid administration on body weight in overweight or obese subjects: A crossover randomized, double-blind, placebo-controlled trial. *Clin. Endocrinol.* **2017**, *86*, 680–687. [CrossRef]

28. Escoté, X.; Félix-Soriano, E.; Gayoso, L.; Huerta, A.E.; Alvarado, M.A.; Ansorena, D.; Astiasarán, I.; Martínez, J.A.; Moreno-Aliaga, M.J. Effects of EPA and lipoic acid supplementation on circulating FGF21 and the fatty acid profile in overweight/obese women following a hypocaloric diet. *Food Funct.* **2018**, *9*, 3028–3036. [CrossRef]

29. Romo-Hualde, A.; Huerta, A.E.; González-Navarro, C.J.; Ramos-López, O.; Moreno-Aliaga, M.J.; Martínez, J.A. Untargeted metabolomic on urine samples after α-lipoic acid and/or eicosapentaenoic acid supplementation in healthy overweight/obese women. *Lipids Health Dis.* **2018**, *17*, 103. [CrossRef]

30. Hosseinpour-Arjmand, S.; Amirkhizi, F.; Ebrahimi-Mameghani, M. The effect of alpha-lipoic acid on inflammatory markers and body composition in obese patients with non-alcoholic fatty liver disease: A randomized, double-blind, placebo-controlled trial. *J. Clin. Pharm. Ther.* **2019**, *44*, 258–267. [CrossRef]

31. Khalili, M.; Azimi, A.; Izadi, V.; Eghtesadi, S.; Mirshafiey, A.; Sahraian, M.A.; Motevalian, A.; Norouzi, A.; Sanoobar, M.; Eskandari, G.; et al. Does lipoic acid consumption affect the cytokine profile in multiple sclerosis patients: A double-blind, placebo-controlled, randomized clinical trial. *Neuroimmunomodulation* **2014**, *21*, 291–296. [CrossRef] [PubMed]

32. Fiedler, S.E.; Yadav, V.; Kerns, A.R.; Tsang, C.; Markwardt, S.; Kim, E.; Spain, R.; Bourdette, D.; Salinthone, S. Lipoic acid stimulates cAMP production in healthy control and secondary progressive MS subjects. *Mol. Neurobiol.* **2018**, *55*, 6037–6049. [CrossRef] [PubMed]

33. Loy, B.D.; Fling, B.W.; Horak, F.B.; Bourdette, D.N.; Spain, R.I. Effects of lipoic acid on walking performance, gait, and balance in secondary progressive multiple sclerosis. *Complement. Ther. Med.* **2018**, *41*, 169–174. [CrossRef]

34. Kim, N.W.; Song, Y.M.; Kim, E.; Cho, H.S.; Cheon, K.A.; Kim, S.J.; Park, J.Y. Adjunctive α-lipoic acid reduces weight gain compared with placebo at 12 weeks in schizophrenic patients treated with atypical antipsychotics: A double-blind randomized placebo-controlled study. *Int. Clin. Psychopharmacol.* **2016**, *31*, 265–274. [CrossRef] [PubMed]

35. Sanders, L.L.O.; de Souza Menezes, C.E.; Chaves Filho, A.J.M.; de Almeida Viana, G.; Fechine, F.V.; de Queiroz, M.G.R.; da Cruz Fonseca, S.G.; Vasconcelos, S.M.M.; de Moraes, M.E.A.; Gamam, C.S. α-Lipoic acid as adjunctive treatment for Schizophrenia: An open-label trial. *J. Clin. Psychopharmacol.* **2017**, *37*, 697–701. [CrossRef]

36. Vidović, B.; Milovanović, S.; Stefanović, A.; Kotur-Stevuljević, J.; Takić, M.; Debeljak-Martačić, J.; Pantović, M.; Đorđević, B. Effects of alpha-lipoic acid supplementation on plasma adiponectin levels and some metabolic risk factors in patients with schizophrenia. *J. Med. Food* **2017**, *20*, 79–85. [CrossRef]

37. Díaz-Cruz, A.; Serret, M.; Ramírez, G.; Ávila, E.; Guinzberg, R.; Piña, E. Prophylactic action of lipoic acid on oxidative stress and growth performance in broilers at risk of developing ascites syndrome. *Avian Pathol.* **2003**, *32*, 645–653. [CrossRef]

38. Mora, O.; Álvarez-Alonso, J.; Pérez-Serrano, R.; González-Dávalos, L.; Vargas, C.; Shimada, A.; Piña, E. Lipoic acid enhances broiler metabolic parameters by downregulating gene expression of metabolic enzymes in liver. *Eur. Poult. Sci.* **2017**, *81*, 194.

39. Lodge, J.K.; Youn, H.; Handelman, G. Natural sources of lipoic acid: Determination of lipoyllysine released from protease-digested tissues by high performance liquid chromatography incorporating electrochemical detection. *J. Appl. Nutr.* **1997**, *49*, 3–11.

40. Brufani, M.; Figliola, R. (R)-α-lipoic acid oral liquid formulation: Pharmacokinetic parameters and therapeutic efficacy. *Acta Biomed. Atenei Parm.* **2014**, *85*, 108–115.

41. Bruno, B.J.; Miller, G.D.; Lim, C.S. Basics and recent advances in peptide and protein drug delivery. *Ther. Deliv.* **2013**, *4*, 1443–1467. [CrossRef] [PubMed]

42. Rostro-Alanis, M.J.; Mancera-Andrade, E.I.; Gómez Patiño, M.B.; Arrieta-Baez, D.; Cardenas, B.; Martinez-Chapa, S.O.; Parra, R. Nanobiocatalysis: Nanostructured materials—A minireview. *Biocatalysis* **2016**, *2*, 1–24. [CrossRef]

43. Koyani, R.D.; Vazquez-Duhalt, R. Enzymatic activation of the emerging drug resveratrol. *Appl. Biochem. Biotechnol.* **2017**, *185*, 248–256. [CrossRef]

44. Koyani, R.D.; Andrade, M.; Quester, K.; Gaytan, P.; Huerta-Saquero, A.; Vazquez-Duhalt, R. Surface modification of protein enhances encapsulation in chitosan nanoparticles. *Appl. Nanosci.* **2018**, *8*, 1197–1203. [CrossRef]

45. Alarcón-Payán, D.A.; Koyani, R.D.; Vazquez-Duhalt, R. Chitosan-based biocatalytic nanoparticles for pollutant removal from wastewater. *Enzyme Microb. Technol.* **2017**, *100*, 71–78. [CrossRef] [PubMed]

46. Pavinatto, F.J.; Caseli, L.; Oliveira, O.N., Jr. Chitosan in nanostructured thin films. *Biomacromolecules* **2010**, *11*, 1897–1908. [CrossRef]

47. Safdar, R.; Omar, A.A.; Arunagiri, A.; Regupathi, I.; Thanabalan, M. Potential of Chitosan and its derivatives for controlled drug release applications—A review. *J. Drug Deliv. Sci. Technol.* **2019**, *49*, 642–659. [CrossRef]

48. Gallego, I.; Villate-Beitia, I.; Martinez-Navarrete, G.; Menendez, M.; Lopez-Mendez, T.; Soto-Sanchez, C.; Zarate, J.; Puras, G.; Fernandez, E.; Pedraz, J.L. Non-viral vectors based on cationic niosomes and minicircle DNA technology enhance gene delivery efficiency for biomedical applications in retinal disorders. *Nanomed. Nanotechnol. Biol. Med.* **2019**, *17*, 308–318. [CrossRef]

49. Kamel, M.; El-Sayed, A. Utilization of herpesviridae as recombinant viral vectors in vaccine development against animal pathogens. *Virus Res.* **2019**, *270*, 197648. [CrossRef]

50. Kochhar, S.; Excler, J.L.; Bok, K.; Gurwith, M.; McNeil, M.M.; Seligman, S.J.; Khuri-Bulos, N.; Klug, B.; Laderoute, M.; Robertson, J.S.; et al. Defining the interval for monitoring potential adverse events following immunization (AEFIs) after receipt of live viral vectored vaccines. *Vaccine* **2019**, *37*, 5796–5802. [CrossRef]

51. Mashal, M.; Attia, N.; Martinez-Navarrete, G.; Soto-Sanchez, C.; Fernandez, E.; Grijalvo, S.; Eritja, R.; Puras, G.; Pedraz, J.L. Gene delivery to the rat retina by non-viral vectors based on chloroquine-containing cationic niosomes. *J. Contol. Release* **2019**, *304*, 181–190. [CrossRef] [PubMed]

52. Massaro, M.; Barone, G.; Biddeci, G.; Cavallaro, G.; Di Blasi, F.; Lazzara, G.; Nicotra, G.; Spinella, C.; Spinelli, G.; Riela, S. Halloysite nanotubes-carbon dots hybrids multifunctional nanocarrier with positive cell target ability as a potential non-viral vector for oral gene therapy. *J. Colloid Interface Sci.* **2019**, *552*, 236–246. [CrossRef]

53. Calvo, P.; Remunan-Lopez, R.; Vila-Jato, C.J.L.; Alonso, M.J. Chitosan and chitosan/ethylene oxide–propylene oxide block copolymer nanoparticles as novel carriers for proteins and vaccines. *Pharm. Res.* **1997**, *14*, 1431–1436. [CrossRef] [PubMed]

54. Marklund, S.; Marklund, G. Involvement of the superoxide anion radical in the autoxidation of pyrogallol and a convenient assay for superoxide dismutase. *Eur. J. Biochem.* **1974**, *47*, 469–474. [CrossRef] [PubMed]

55. Daidoji, T.; Ozawa, M.; Sakamoto, H.; Sako, T.; Inoue, H.; Kurihara, R.; Hashimoto, S.; Yokota, H. Slow elimination of nonylphenol from rat intestine. *Drug Metab. Dispos.* **2006**, *34*, 184–190. [CrossRef] [PubMed]

56. Rosas, A.; Vazquez-Duhalt, R.; Tinoco, R.; Shimada, A.; D'Abramo, L.R.; Viana, M.T. Comparative intestinal absorption of amino acids in rainbow trout (*Oncorhynchus mykiss*), totoaba (*Totoaba macdonaldi*), and Pacific bluefin tuna (*Thunnus orientalis*). *Aquac. Nutr.* **2008**, *14*, 481–489. [CrossRef]

57. Ieko, T.; Inoue, S.; Inomata, Y.; Inoue, H.; Fujiki, J.; Iwano, H. Glucuronidation as a metabolic barrier against zearalenone in rat everted intestine. *J. Vet. Med. Sci.* **2020**, *82*, 153–161. [CrossRef]

58. Kim, K.W.; Thomas, R.L. Antioxidative activity of chitosans with varying molecular weights. *Food Chem.* **2007**, *101*, 308–313. [CrossRef]

59. Yen, M.-T.; Yang, J.-H.; Mau, J.-L. Antioxidant properties of chitosan from crab shells. *Carbohydr. Polym.* **2008**, *74*, 840–844. [CrossRef]

60. Zhang, J.; Tan, W.; Wang, G.; Yin, X.; Li, Q.; Dong, F.; Guo, Z. Synthesis, characterization, and the antioxidant activity of N,N,N-trimethyl chitosan salts. *Int. J. Biol. Macromol.* **2018**, *118*, 9–14. [CrossRef]

61. Huang, M.; Ma, Z.; Khor, E.; Lim, L.-Y. Uptake of FITC-chitosan nanoparticles by A549 cells. *Pharm. Res.* **2002**, *19*, 1488–1494. [CrossRef] [PubMed]

62. Loh, J.W.; Yeoh, G.; Saunders, M.; Lim, L.-Y. Uptake and cytotoxicity of chitosan nanoparticles in human liver cells. *Toxicol. Appl. Pharmacol.* **2010**, *249*, 148–157. [CrossRef] [PubMed]
63. Yue, Z.-G.; Wei, W.; Lv, P.-P.; Yue, H.; Wang, L.-Y.; Su, Z.-G.; Ma, G.-H. Surface charge affects cellular uptake and intracellular trafficking of chitosan-based nanoparticles. *Biomacromolecules* **2011**, *12*, 2440–2446. [CrossRef] [PubMed]
64. Bergin, I.; Witzmann, F. Nanoparticle toxicity by the gastrointestinal route: Evidence and knowledge gaps. *Int. J. Biomed. Nanosci. Nanotechnol.* **2013**, *3*, 163–210. [CrossRef]
65. Florence, A.T.; Hussain, N. Transcytosis of nanoparticle and dendrimer delivery systems: Evolving vistas. *Adv. Drug Deliv. Rev.* **2001**, *50*, S69–S89. [CrossRef]
66. Norris, D.A.; Sinko, P.J. Effect of size, surface charge, and hydrophobicity on the translocation of polystyrene microspheres through gastrointestinal mucin. *J. Appl. Polym. Sci.* **1997**, *63*, 1481–1492. [CrossRef]
67. George, M.; Abraham, T.E. Polyionic hydrocolloids for the intestinal delivery of protein drugs: Alginate and chitosan—A review. *J. Control. Release* **2006**, *114*, 1–14. [CrossRef]
68. Smart, J.D. The basics and underlying mechanisms of mucoadhesion. *Adv. Drug Deliv. Rev.* **2005**, *57*, 1556–1568. [CrossRef]
69. Bowman, K.; Leong, K.W. Chitosan nanoparticles for oral drug and gene delivery. *Int. J. Nanomed.* **2006**, *1*, 117–128. [CrossRef]
70. Liang, J.; Yan, H.; Puligundla, P.; Gao, X.; Zhou, Y.; Wan, X. Applications of chitosan nanoparticles to enhance absorption and bioavailability of tea polyphenols: A review. *Food Hydrocoll.* **2017**, *69*, 286–292. [CrossRef]
71. Dube, A.; Ng, K.; Nicolazzo, J.A.; Larson, I. Chitosan nanoparticles enhance the intestinal absorption of the green tea catechins (+)-catechin and (−)-epigallocatechin gallate. *Eur. J. Pharm. Sci.* **2010**, *122*, 662–667. [CrossRef] [PubMed]
72. Bhattacharjee, S.; van Opstal, E.J.; Alink, G.M.; Marcelis, A.T.; Zuilhof, H.; Rietjens, I.M. Surface charge-specific interactions between polymer nanoparticles and ABC transporters in Caco-2 cells. *J. Nanopart. Res.* **2013**, *15*, 1695–1709. [CrossRef]

Article

Antimicrobial Resistance of *Salmonella enteritidis* and *Salmonella typhimurium* Isolated from Laying Hens, Table Eggs, and Humans with Respect to Antimicrobial Activity of Biosynthesized Silver Nanoparticles

Rasha M. M. Abou Elez [1], Ibrahim Elsohaby [2,3,4,*], Nashwa El-Gazzar [5], Hala M. N. Tolba [6], Eman N. Abdelfatah [7], Samah S. Abdellatif [7], Ahmed Atef Mesalam [8,*] and Asmaa B. M. B. Tahoun [7]

[1] Department of Zoonoses, Faculty of Veterinary Medicine, Zagazig University, Zagazig 44511, Egypt; rmmohamed@zu.edu.eg
[2] Department of Animal Medicine, Faculty of Veterinary Medicine, Zagazig University, Zagazig 44511, Egypt
[3] Department of Health Management, Atlantic Veterinary College, University of Prince Edward Island, Charlottetown, PE C1A 4P3, Canada
[4] Department of Infectious Diseases and Public Health, Jockey Club of Veterinary Medicine and Life Sciences, City University of Hong Kong, Kowloon, Hong Kong
[5] Department of Botany and Microbiology, Faculty of Science, Zagazig University, Zagazig 44519, Egypt; ns_elgsazzar@zu.edu.eg
[6] Department of Avian and Rabbit Medicine, Faculty of Veterinary Medicine, Zagazig University, Zagazig 44511, Egypt; halanabil85@gmail.com
[7] Department of Food Control, Faculty of Veterinary Medicine, Zagazig University, Zagazig 44511, Egypt; Enabil@vet.zu.edu.eg (E.N.A.); SSAbdelfattah@vet.zu.edu.eg (S.S.A.); abbadr@vet.zu.edu.eg (A.B.M.B.T.)
[8] Department of Therapeutic Chemistry, National Research Centre (NRC), Dokki, Cairo 12622, Egypt
* Correspondence: ielsohaby@upei.ca (I.E.); Ahmedatefmesalam@hotmail.com or am.mesalam@nrc.sci.eg (A.A.M.)

Citation: Abou Elez, R.M.M.; Elsohaby, I.; El-Gazzar, N.; Tolba, H.M.M.; Abdelfatah, E.N.; Abdellatif, S.S.; Mesalam, A.A.; Tahoun, A.B.M.B. Antimicrobial Resistance of *Salmonella enteritidis* and *Salmonella typhimurium* Isolated from Laying Hens, Table Eggs, and Humans with Respect to Antimicrobial Activity of Biosynthesized Silver Nanoparticles. *Animals* 2021, 11, 3554. https://doi.org/10.3390/ani11123554

Academic Editors: Antonio Gonzalez-Bulnes and Nesreen Hashem

Received: 1 November 2021
Accepted: 10 December 2021
Published: 14 December 2021

Publisher's Note: MDPI stays neutral with regard to jurisdictional claims in published maps and institutional affiliations.

Simple Summary: *Salmonella enterica* are common foodborne pathogens that cause gastrointestinal signs in a wide range of unrelated host species including poultry and humans. The overuse of antibiotics as therapeutic agents and growth promoters in the poultry industry has led to the emergence of multidrug-resistant (MDR) microorganisms. Thus, there is a need to find alternatives to conventional antibiotics. Recently, the biosynthesized silver nanoparticles (AgNPs) have shown an excellent antimicrobial activity. In this study, we investigated the antibacterial, antivirulent, and antiresistant activities of the biosynthesized AgNPs on the MDR and virulent *S. enteritidis* and *S. typhimurium* isolated from laying hens, table eggs, and humans. The obtained results indicated that AgNPs have the potential to be effective antimicrobial agents against MDR *S. enteritidis* and *S. typhimurium* and could be recommended for use in laying hen farms.

Abstract: *Salmonella enterica* is one of the most common causes of foodborne illness worldwide. Contaminated poultry products, especially meat and eggs are the main sources of human salmonellosis. Thus, the aim of the present study was to determine prevalence, antimicrobial resistance profiles, virulence, and resistance genes of *Salmonella* Enteritidis (*S. enteritidis*) and *Salmonella* Typhimurium (*S.* Typhimurium) isolated from laying hens, table eggs, and humans, in Sharkia Governorate, Egypt. The antimicrobial activity of Biosynthesized Silver Nanoparticles (AgNPs) was also evaluated. *Salmonella* spp. were found in 19.3% of tested samples with laying hens having the highest isolation rate (33.1%). *S.* Enteritidis) (5.8%), and *S.* Typhimurium (2.8%) were the dominant serotypes. All isolates were ampicillin resistant (100%); however, none of the isolates were meropenem resistant. Multidrug-resistant (MDR) was detected in 83.8% of the isolates with a multiple antibiotic resistance index of 0.21 to 0.57. Most isolates (81.1%) had at least three virulence genes (*sopB*, *stn*, and *hilA*) and none of the isolates harbored the *pefA* gene; four resistance genes (*blaTEM*, *tetA*, *nfsA*, and *nfsB*) were detected in 56.8% of the examined isolates. The AgNPs biosynthesized by *Aspergillus niveus* exhibit an absorption peak at 420 nm with an average size of 27 nm. AgNPs had a minimum inhibitory concentration of 5 µg/mL against *S. enteritidis* and *S. typhimurium* isolates and a minimum bactericidal

---

concentration of 6 and 8 µg/mL against *S. enteritidis* and *S. typhimurium* isolates, respectively. The bacterial growth and gene expression of *S. enteritidis* and *S. typhimurium* isolates treated with AgNPs were gradually decreased as storage time was increased. In conclusion, this study indicates that *S. enteritidis* and *S. typhimurium* isolated from laying hens, table eggs, and humans exhibits resistance to multiple antimicrobial classes. The biosynthesized AgNPs showed potential antimicrobial activity against MDR *S. enteritidis* and *S. typhimurium* isolates. However, studies to assess the antimicrobial effectiveness of the biosynthesized AgNPs in laying hen farms are warranted.

Keywords: *Salmonella*; antimicrobial agents; virulence genes; resistance genes; silver nanoparticles; expression

1. Introduction

Salmonella enterica are common foodborne pathogens, causing 93.8 million cases of gastroenteritis and 155,000 deaths worldwide and severe economic losses [1]. *Salmonella* Enteritidis (*S. enteritidis*) and *Salmonella* Typhimurium (*S. typhimurium*) are the dominant nontyphoidal serovars that cause mild gastrointestinal signs in a wide range of unrelated host species and severe infections in infants, the elderly, and immunocompromised individuals [2]. Nontyphoidal human salmonellosis is mostly associated with the ingestion of contaminated poultry meats and raw eggs [3]. However, table eggs get contaminated with *Salmonella* via horizontal transmission from the feces of infected laying hens, vertical transmission through the yolk, albumen, or eggshell membranes before oviposition and contamination of the eggshell after oviposition by infected environmental dusts [4].

Antibiotics have been extensively used in developing countries as therapeutic agents and growth promoters in laying hen farms as well as treatment for various human diseases, resulting in the emergence of antimicrobial resistance bacteria [5]. Antibiotic-resistant bacteria can be transmitted to humans either directly through the food chain or indirectly by transferring their antimicrobial resistance genes to human pathogens by mobile genetic elements associated with conjugative plasmids [6]. The presence of various virulence genes in the chromosome and plasmids of *Salmonella* plays a role in the pathogenesis of such bacteria inside the host [7]. *Salmonella* outer proteins (*sops*) and *hilA* virulence genes contribute to the invasion of host epithelial cells [8]. However, the plasmid-encoded fimbriae (*pefA*) gene is mediated through the adherence of *Salmonella* to intestinal epithelial cells [9], whereas the enterotoxin (*stn*) gene is responsible for enterotoxin production and diarrhea in the host [8]. *Salmonella* plasmid virulence (*spvs*) has relevance to *Salmonella* survival and replication inside the host [10]. Therefore, understanding the antibiotic resistance mechanisms will assist in the control and reduction in the spread of resistant bacteria. Furthermore, assessing the distribution of resistance genes in bacterial populations is an additional tool to understand the antimicrobial resistance epidemiology [11]. Thus, the development of an alternatives natural antimicrobial agent is needed.

Nanoparticles are one of the alternatives that could be used as antimicrobial agents in humans and animals [12]. Silver nanoparticles (AgNPs) are among the nanoparticles that have been used for many years in different applications including wound treatment. AgNPs physical, chemical, and biological properties depend on their size and shape [13]. AgNPs have good antimicrobial activity against bacteria, fungi, and viruses, with low cytotoxicity to mammalian cells due to their nanoscale size and various shapes [14]. The mechanism behind AgNPs's actions includes their induction of cell death through generation of reactive oxygen species [15]. Furthermore, there are different factors that make AgNPs suitable for use as antimicrobial: (1) It is simple and safe to synthesize on a large scale with low cost [16]; (2) bacterial resistance to AgNPs is extremely rare; and (3) the surface of AgNPs can be easily modified and synthesized in various shapes [17]. Several studies have reported that AgNPs have antimicrobial activity against a variety of pathogens, including Salmonella enterica, *E. coli* O157:H7, Pseudomonas aeruginosa, Listeria monocytogenes, Klebsiella pneumoniae [18], and mastitis pathogens [19].

This study aimed to (1) determine the prevalence of *Salmonella* spp. in laying hens, table eggs, and humans, in Sharkia Governorate, Egypt; (2) detect the antimicrobial resistance profiles and virulence and resistance genes of *S. enteritidis* and *S. typhimurium* isolates; and (3) prepare and assess the antibacterial, antivirulent, and antiresistant activities of AgNPs biosynthesized by *Aspergillus niveus* (*A. niveus*) on the MDR and virulent *S. enteritidis* and *S. typhimurium* isolates.

2. Materials and Methods

2.1. Sample Collection

A total of 431 samples were collected from laying hen farms, egg retailers, and outpatient clinics in Sharkia Governorate, Egypt, between August 2019 and January 2020. Samples included 166 fecal swabs from laying hens, 165 table eggs, and 100 human stool samples. Aseptic techniques were strictly maintained during sample collection. Table eggs were individually packed in separate sterile plastic bags from egg retailers and brought to the laboratory. All human participants provided verbal or written consent to join the study.

2.2. Salmonella spp. Isolation

Salmonella was isolated from eggs and fecal samples as previously described Im, et al. [20], Yang, et al. [21]. The collected eggs were disinfected with 75% alcohol, the shell was removed and the yolks and whites were mixed. A 25 mL of the sample was added to 225 mL of sterile buffered peptone water (BPW; Oxoid, Basingstoke, Hampshire, UK) and then incubated at 37 °C for 24 h. A 100 μL of incubated BPW suspension was selectively enriched in 10 mL of Rappaport–Vassiliadis medium (Difco, Detroit, MI, USA) and incubated at 42 °C for 24 h. A loopful (10 μL) of enriched Rappaport–Vassiliadis medium was plated onto xylose lysine deoxycholate agar (Difco, Detroit, MI, USA) and incubated at 37 °C for 24 h. The suspected *Salmonella* isolates were kept frozen at −80°C in brain–heart infusion broth (Oxoid, Basingstoke, Hampshire, UK) supplemented with 15% glycerin (Synth®, São Paulo, Brazil). The presumptive colonies were picked up for purification on tryptone soya agar (Oxoid, Basingstoke, Hampshire, UK) and subjected to biotyping [22] and serotyping according to the Kauffmann–White scheme [23] to determine flagellar (H) and somatic (O) antigens using *Salmonella* antiserum (Denka Seiken, Tokyo, Japan).

2.3. Molecular Identification of Salmonella spp.

The extracted DNA was examined by PCR targeting *invA* gene using the QIAamp DNA Mini Kit (Qiagen, Hilden, Germany) according to the manufacturer's guidelines [24]. Isolates identified as *Salmonella* spp. were subjected to identification of *sefA* gene (310 bp) for *S. enteritidis* [25] and *STM4495* gene (915 bp) for *S. typhimurium* [26]. PCR-confirmed *S. enteritidis* and *S.* Typhimurium isolates were screened to detect *sopB* [8], *stn* [9], *pefA* [9], *spvC* [8], and *hilA* [21] virulence genes. *S. typhimurium* ATCC 14028 and *S. enteritidis* ATCC 13076 were used as positive controls. However, *Escherichia coli ATCC 25922 was used as a negative control. The positive and negative controls were donated by the Biotechnology Unit, Reference Laboratory for Veterinary Quality Control on Poultry Production, Animal Health Research Institute, Dokki, Giza, Egypt, and were run alongside the tested isolates.*

2.4. Antimicrobial Susceptibility Test

Antimicrobial resistance of *S. enteritidis* and *S. typhimurium* isolates was determined using the Kirby–Bauer disk diffusion method on Mueller–Hinton Agar (MHA) (Oxoid, Hampshire, UK), in accordance with the guidelines of the Clinical and Laboratory Standards Institute [27]. The antibiotic disks used in this study were ampicillin (AMP, 10 μg), ampicillin/sulbactam (SAM, 20 μg), amoxicillin-clavulanate (AMC, 30 μg), cefotaxime (CTX, 30 μg), imipenem (IPM, 10 μg), gentamicin (GEN, 10 μg), tetracycline (TET, 30 μg), ciprofloxacin (CIP, 5 μg), nalidixic acid (NAL, 30 μg), trimethoprim-sulfamethoxazole (SXT,

25 µg), chloramphenicol (CHL, 30 µg), azithromycin (AZM, 15 µg), nitrofurantoin (NIT, 30 µg), and meropenem (MEM, 10 µg). The zone of inhibition, measured and compared with the world standards [27], was reported as resistant (R), intermediate (I), and sensitive (S). Isolates resistant to ≥3 different antimicrobial classes were considered MDR [28]. For all isolates, the multiple antibiotic resistance (MAR) index was determined using the formula: a/b (where "a" is the number of antimicrobial agents to which an isolate was resistant and "b" is the total number of antimicrobial agents tested) following the protocol designated by Krumperman [29].

S. enteritidis and *S. typhimurium* isolates with phenotypic resistance to specific antimicrobial agents were selected and tested for the presence of relevant resistance genes using PCR. The targeted genes included β-lactams (*blaTEM*) [30], tetracyclines (*tetA* and *tetB*) [31,32], and nitrofurans (*nfsA* and *nfsB*) [33].

2.5. Biosynthesis and Characterization of AgNPs

The AgNPs were biosynthesized from contaminated soil samples containing wastes from ceramics and photographic industries (10th of Ramadan City, Sharkia, Egypt) as described by [34]. Fungal isolate was screened to reduce the AgNo$_3$ solution at 1 mM to AgNPs according to El-Gazzar and Rabie [35]. The available salt of AgNO$_3$ was donated from the NanoTech Egypt (Dreamland, Egypt). The physical properties and concentrations of the biosynthesized AgNPs were investigated.

The extracted DNA of the most potent suspected fungal isolate producing AgNPs was molecularly identified at the Biology Research Unit of the Assiut University using Patho Gene-spin DNA/RNA Extraction Kit (iNtRON Biotechnology, Seongnam, Korea) following the manufacturer's guidelines [36]. Isolates identified as *Aspergillus* spp. were further sequenced to target the same primers (SolGent, Daejeon, South Korea). The obtained sequence was analyzed using the BLAST tool and was placed in the GenBank (accession number MT319815). The nucleotide sequence was aligned with other sequences available at the GenBank to construct a phylogenetic tree using the MegAlign of DNASTAR program package (MegAlign 5.05, DNASTAR Inc., Madison, WI, USA). The recovered fungal isolate was characterized according to DeAlba-Montero, et al. [37] using different tools such as ultraviolet (UV)–visible spectrophotometer (T80 + UV Flash Spectrophotometer, PG Instruments Ltd., Wibtoft, UK); dynamic light scattering (DLS) system; transmission electron microscope (TEM) (JEOL.JEM.1010) at an accelerating voltage of 200 Kv; Fourier transform infrared spectroscopy (FTIR, Thermo Scientific Nicolet 6700 spectrometer) in the range of 400–4000 cm^{-1} and a resolution of 4 cm^{-1}; and zeta potential analyzer (ZEN 1600, Malvern, UK).

2.6. Antimicrobial Activity of Biosynthesized AgNPs

The disk diffusion agar method was used to test the antimicrobial activity of biosynthesized AgNPs against *S. enteritidis* and *S. typhimurium* isolates [38]. A 10 µg of the prepared nanoparticles were paced on standard disks with a diameter of 6 mm (Padtanteb, Qods, Iran). The *S. enteritidis* and *S. typhimurium* isolates were spread on the Mueller–Hinton broth (MHB) (Merck, Darmstadt, Germany). The disks were placed on the agar plate and a sterile blank disk was used as control; then, the inhibition zone was measured after a 24 h incubation at 37 °C.

Minimum inhibitory concentration (MIC) and minimum bactericidal concentration (MBC) values of biosynthesized AgNPs were determined using tube dilution method as previously described Krishnan, et al. [39]. Bacterial inoculums were adjusted to match a 0.5 McFarland turbidity standard. A 100 µL of AgNPs with different concentrations (1, 2, 3, 4, 5, 6, 7, 8, 9, and 10 µg/mL) from each speed (4000, 8000, and 14,000 rpm) and a 100 µL from the tested organism were applied to 5 mL MHB and incubated at 37 °C with shaking for 24 h. Broth media with AgNPs inoculum was used as a positive control. The bacterial growth was monitored by measuring the mean OD at 600 nm.

The MBC was determined by coating the bacterial inoculum with different nanoparticle suspension concentrations into the MHA plate and then incubated at 37 °C for 24 h. The MBC value was defined as the lowest concentration, with no visible growths on the MHA plate.

2.7. Virulence and Resistance Genes Expression

2.7.1. Bacterial Counting

The experiment used 2 MDR *Salmonella* isolates (*S. enteritidis* and *S. typhimurium* isolated during our study and were positive for most virulence and resistance genes). The bacterial inoculums were adjusted to 1.5×10^8 Colony Forming Unite (CFU)/mL. The surface plating method was used for the bacterial count on an agar plate [40]. Bacterial inoculums were inoculated with a final concentration of 5 μg/mL followed by constant stirring to obtain a uniform colloidal nanoparticle suspension. Nanoparticle-free medium was used as a positive control, whereas the bacteria-free medium was used as a negative control. Each inoculated suspension was incubated at 37 °C for 0, 12, 24, 36, and 48 h. Bacterial growth inhibition was determined using the surface plating method by counting the number of CFUs on the plates. Bacteria and nanoparticle mixtures were prepared at different cultivation times. The amplification of the 16S rRNA gene was used to molecularly confirm *Salmonella* colonies. Bacterial cells were counted in triplicate and then the mean values and standard deviations were calculated.

2.7.2. Quantitative Reverse Transcription PCR Analysis of Genes Expression

At each sampling time, 1 volume of the harvested bacterial culture was added to 1 volume of RNAprotect Bacteria Reagent (Qiagen, Hilden, Germany) following the manufacturer's instructions. RNA isolation was implemented using QIAamp RNeasy Mini Kit (Qiagen, Hilden, Germany) according to the manufacturer's guidelines. SYBR Green I-based real-time PCR with the specific primers for *sopB*, *stn*, and *hilA* virulence genes and *blaTEM*, *tetA*, and *nfsA* resistance genes (which are all present in the used 2 *Salmonella* isolates) was performed and the 16S rRNA gene was used as a housekeeping gene [21]. The primers were used in a 25 μL reaction containing 12.5 μL of 2× QuantiTect SYBR Green PCR Master Mix (QIAGEN), 0.25 μL of RevertAid Reverse Transcriptase (200 U/μL) (Thermo Fisher, Waltham, MA, USA), 0.5 μL of each primer (20 pmoL concentration), 8.25 μL of nuclease-free water, and 3 μL of RNA template. The reaction was performed in a Stratagene MX3005P real-time PCR machine where the amplification curves and cycle threshold (Ct) values were determined by Stratagene MX3005P software. The comparative Ct method was used to estimate the variations in RNA gene expression of the different samples by comparing Ct of each sample with that of the positive control. The ΔΔCt method was performed, according to Yuan, et al. [41].

2.8. Statistical Analysis

The R software (R Core Team, 2019; version 3.5.3) was used for the descriptive and statistical analysis. A heatmap was constructed based on the virulence and resistance genes and the antimicrobial susceptibility results using the R package "Complex heatmap" [42]. Nonmetric multidimensional scaling (nMDS) [43] was performed using the "metaMDS" function in "vegan" package to compare the dissimilarity of antimicrobial resistance profiles, using Bray–Curtis distance among isolates across all *Salmonella* spp. and within each species. The "corrplot" function in "corrplot" package was used to assess the correlation between the antimicrobial resistance and the presence of virulence and resistance genes. One-way analysis of variance was used to determine the significant difference between the bacterial counts and the fold change values of the gene's expression at each storage time. Multiple comparisons between the means were assessed by the Tukey's honestly significant difference test. $p < 0.05$ was considered statistically significant.

3. Results

3.1. Salmonella spp. Isolation and Identification

Salmonella spp. were identified in 83 of 431 examined samples (19.3%) and were mostly identified in samples from laying hens (33.1%) followed by humans (22%) and table eggs (3.6%) (Table 1). The most dominant *Salmonella* serotypes were *S. enteritidis* (5.8%) and *S. typhimurium* (2.8%), followed by *S. kentucky* (1.9%), *S. virchow* (1.6%), *S. tamale*, *S. inganda*, *S. wingrove*, *S. bargny* (1.2%, each), *S.* Anatum (0.9%), *S. tsavie*, *S. larochelle* (0.7%, each), and *S. apeyeme* (0.2%). On the other hand, only *S. apeyeme* was isolated from human samples (1 isolate), whereas *S. anatum*, *S. tsavie*, and *S. larochelle* were not identified (Table 1).

Table 1. Serotypes and pathotypes of *Salmonella* spp. isolated from laying hens, table eggs and humans.

Serotypes	Pathotypes	No. (%) of Isolates			Total (n = 431)
		Laying Hens (n = 166)	Table Eggs (n = 165)	Humans (n = 100)	
S.enteritidis	D1; O:1, 9, 12; H:g, m:−	12 (7.2)	5 (3.0)	8 (8.0)	25 (5.8)
S.typhimurium	B; O:1, 4, 5, 12; H:i:1, 2	7 (4.2)	1 (0.6)	4 (4.0)	12 (2.8)
S.kentukey	C3; O:8, 20; H:i:z6	5 (3.01)	0 (0.0)	3 (3.0)	8 (1.9)
S.virchow	C1; O:6, 7, 14; H:r:1, 2	5 (3.01)	0 (0.0)	2 (2.0)	7 (1.6)
S.tamale	C3; O:8, 20; H:Z29:e, n, Z15	4 (2.4)	0 (0.0)	1 (1.0)	5 (1.2)
S.inganda	C1; O:6, 7; H:Z10:1, 5	4 (2.4)	0 (0.0)	1 (1.0)	5 (1.2)
S.wingrove	C2; O:6, 8; H:c:1, 2	4(2.4)	0 (0.0)	1 (1.0)	5 (1.2)
S.bargny	C3; O:8, 20; H:i:1, 5	4 (2.4)	0 (0.0)	1 (1.0)	5 (1.2)
S.anatum	E1; O:3, 10; H:e, h:1, 6	4 (2.4)	0 (0.0)	0 (0.0)	4 (0.9)
S.tsavie	B; O:4,5; H:i:e, n, z15	3 (1.8)	0 (0.0)	0 (0.0)	3 (0.7)
S.larochell	C1; O:6, 7; H:e, h:1, 2	3 (1.8)	0 (0.0)	0 (0.0)	3 (0.7)
S.apeyeme	C3; O:8, 20; H:Z38:−	0 (0.0)	0 (0.0)	1 (1.0)	1 (0.2)
Total		55 (33.1)	6 (3.6)	22 (22)	83 (19.3)

3.2. Antimicrobial Susceptibility Test

The antimicrobial susceptibility profiles of the 25 *S. enteritidis* and 12 *S. typhimurium* isolates against 14 antimicrobial agents are presented in Table 2. All *S. enteritidis* and *S. typhimurium* isolates were resistant to AMP (100%) and sensitive to MEM (100%). In addition, *S. enteritidis* and *S. typhimurium* isolates exhibited high rates of resistance to tetracycline, NIT, and amoxicillin (Figure 1). The MDR was observed in 28 isolates (75.7%) with a MAR index ranging from 0.21 to 0.57.

Table 2. The antimicrobial resistance profile of *S. enteritidis* (n = 25) and *S. typhimurium* (n = 12) isolated from laying hens, table eggs, and humans.

Antimicrobials Class	Antimicrobials	*S. enteritidis* Isolates (%)			*S. typhimurium* Isolates (%)		
		R	I	S	R	I	S
Penicillin	Ampicillin (AMP)	25 (100)	0 (0.0)	0 (0.0)	12 (100)	0 (0.0)	0 (0.0)
	Ampicillin/Sulbactam (SAM)	0 (0.0)	4 (16.0)	21 (84.0)	0 (0.0)	2 (16.7)	10 (83.3)
	Amoxicillin-Clavulanate (AMC)	15 (60.0)	0 (0.0)	10 (40.0)	6 (50.0)	3 (25.0)	3 (25.0)
Cephalosporine	Cefataxime (CTX)	9 (36.0)	0 (0.0)	16 (64.0)	3 (25.0)	0 (0.0)	9 (75.0)
Carbapenems	Imipenem (IPM)	11 (44.0)	0 (0.0)	14 (56.0)	0 (0.0)	2 (16.7)	10 (83.3)
Aminoglycosides	Gentamicin (GEN)	2 (8.0)	0 (0.0)	23 (92.0)	2 (16.7)	0 (0.0)	10 (83.3)
Tetracyclines	Tetracycline (TET)	22 (88.0)	0 (0.0)	3 (12.0)	8 (66.7)	0 (0.0)	4 (33.3)
Quinolones	Ciprofloxacin (CIP)	3 (12.0)	6 (24.0)	16 (64.0)	0 (0.0)	3 (25.0)	9 (75.0)
	Nalidixic acid (NAL)	5 (20.0)	2 (8.0)	18 (72.0)	0 (0.0)	1 (8.3)	11 (91.7)
Sulphonamides	Trimethoprim-Sulfamethoxazole (SXT)	0 (0.0)	5 (20.0)	20 (80.0)	1 (8.3)	2 (16.7)	9 (75.0)
Phenicols	Chloramphenicol (CHL)	4 (16.0)	3 (12.0)	18 (72.0)	0 (0.0)	3 (25.0)	9 (75.0)
Macrolides	Azithromycin (AZM)	5 (20.0)	0 (0.0)	20 (80.0)	4 (33.3)	0 (0.0)	8 (66.7)
Nitrofurans	Nitrofurantoin (NIT)	20 (80.0)	0 (0.0)	5 (20.0)	8 (66.7)	0 (0.0)	4 (33.3)
Carbapenem	Meropenem (MEM)	0 (0.0)	0 (0.0)	25 (100)	0 (0.0)	0 (0.0)	12 (100)

R = resistant, I = intermediate resistance, S = sensitive.

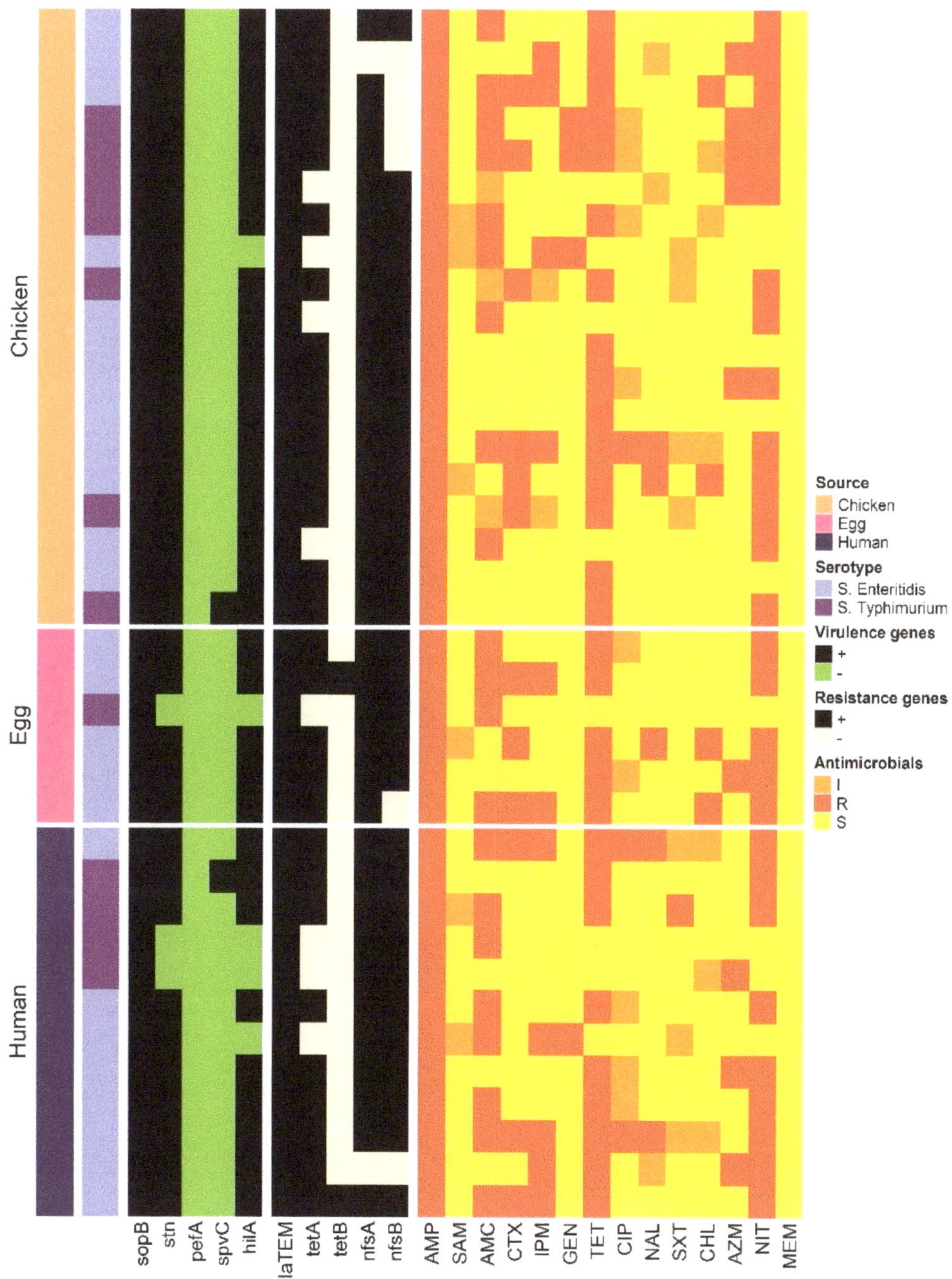

Figure 1. Heat map representation of the virulence and resistance genes and antimicrobial resistance profiles of *S. enteritidis* (*n* = 25) and *S. typhimurium* (*n* = 12) isolated from laying hens, table eggs, and humans.

The nMDS plot (Figure 2) shows that the antimicrobial profiles of *S. enteritidis* and *S. typhimurium* isolated from chickens, eggs, and humans' samples were overlapped. The plot shows no evidence for clustering by isolate source.

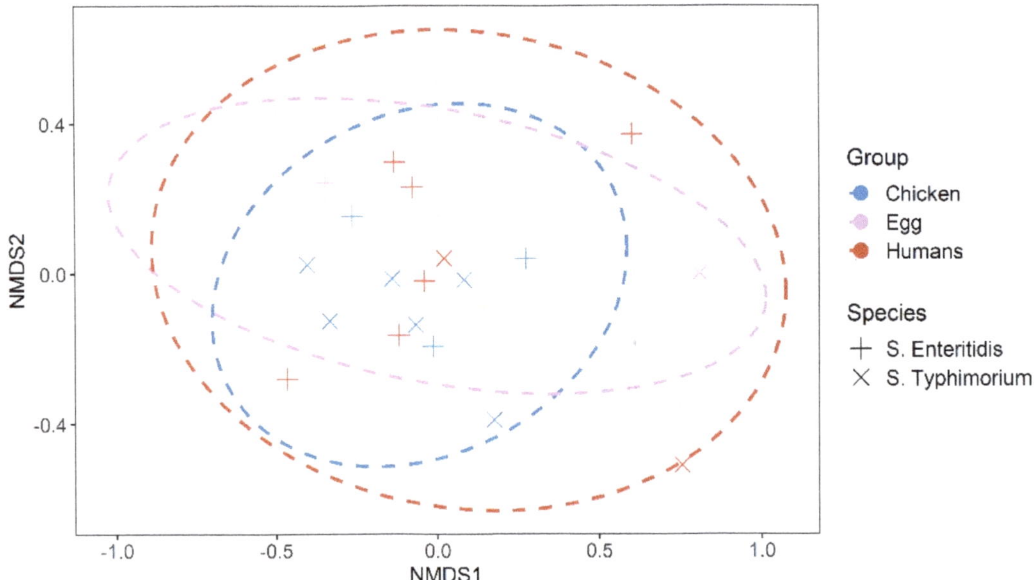

Figure 2. Non-metric multidimensional scaling ordination (NMDS) of antimicrobial-resistant of *S. enteritidis* and *S. typhimurium* isolated from laying hens, table eggs and humans.

3.3. Virulence and Resistance Genes

The presence of virulence and resistance-associated genes in *S. enteritidis* and *S. typhimurium* isolates is presented in Table 3. The *SopB* gene was identified in all *Salmonella* spp. isolates, whereas the *pefA* gene was not detected in any isolate. The *stn* virulence-associated gene was identified in all *S. enteritidis* and 75% of *S. typhimurium* isolates. The *hilA* gene was mostly identified in 94.6% of *Salmonella* isolates; however, the *SpvC* virulence gene was only detected in 5.4% of *S. Typhimurium* isolates. All *Salmonella* isolates (100%) harbored only 1 virulence gene and 34 isolates (91.9%) were positive for >1 virulence gene (Figure 1).

Five resistance genes (*blaTEM*, *tetA*, *tetB*, *nfsA* and *nfsB*) were also identified in *S. enteritidis* and *S. typhimurium* isolates (Table 3). The *blaTEM* gene was identified in all isolates and most isolates harbored *nfsA* (94.6%) and *nfsB* and *tetA* (83.8%, each) resistance genes. However, the *tetB* resistance gene was only recognized in two *S. enteritidis* isolates. Notably, five resistance profiles were exhibited in the study and most isolates (56.8%) were related to a profile carrying four resistance genes (*blaTEM*, *tetA*, *nfsA* and *nfsB*) (Figure 1).

Significant correlations between resistance genes and antimicrobial agents are presented in Figure 3. There were significant positive correlations between resistance genes and their corresponding antimicrobial agents in most cases. The analysis also showed a significant negative correlation between resistance genes and antimicrobial agents other than the corresponding ones.

Table 3. Virulence and resistance genes and antimicrobial resistance patterns of *S. enteritidis* (*n* = 25) and *S. typhimurium* (*n* = 12) isolated from laying hens, table eggs and humans.

Source	No. of Isolates	Virulence Genes					Resistance Genes					No. of Ab	Resistance Patterns	MARIndex
		sopB	*stn*	*pefA*	*spvC*	*hilA*	*blaTEM*	*tetA*	*tetB*	*nfsA*	*nfsB*			
						(I) *S. enteritidis*								
Human	2	+	+	−	−	+	+	+	−	+	+	8	AMP, AMC, CTX, IPM, TET, CIP, NA, NIT	0.57
Chicken	1	+	+	−	−	+	+	+	−	+	+	8	AMP, AMC, CTX, IPM, TET, CIP, NA, NIT	0.57
Chicken	1	+	+	−	−	+	+	+	−	+	−	7	AMP, AMC, CTX, IPM, TET, NA, NIT	0.50
Egg	1	+	+	−	−	+	+	+	−	+	−	7	AMP, AMC, CTX, IPM, TET, NA, NIT	0.50
Egg	1	+	+	−	−	+	+	+	+	+	+	6	AMP, AMC, CTX, IPM, TET, NIT	0.43
Human	1	+	+	−	−	+	+	+	+	+	+	6	AMP, AMC, CTX, IPM, TET, NIT	0.43
Egg	1	+	+	−	−	+	+	+	−	+	+	6	AMP, CTX, IPM, TET, NAL, NIT	0.43
Chicken	1	+	+	−	−	+	+	+	−	+	+	6	AMP, CTX, IPM, TET, NAL, NIT	0.43
Chicken	1	+	+	−	−	+	+	+	−	−	−	5	AMP, IPM, TET, AZM, NIT	0.36
Human	1	+	+	−	−	+	+	+	−	−	−	5	AMP, IPM, TET, AZM, NIT	0.36
Chicken	1	+	+	−	−	+	+	+	−	+	+	4	AMP, AMC, TET, NIT	0.29
Egg	1	+	+	−	−	+	+	+	−	+	+	4	AMP, AMC, TET, NIT	0.29
Human	2	+	+	−	−	+	+	+	−	+	+	4	AMP, AMC, TET, NIT	0.29
Egg	1	+	+	−	−	+	+	+	−	+	+	4	AMP, TET, AZM, NIT	0.29
Chicken	1	+	+	−	−	+	+	+	−	+	+	4	AMP, TET, AZM, NIT	0.29
Human	1	+	+	−	−	+	+	+	−	+	+	4	AMP, TET, AZM, NIT	0.29
Chicken	1	+	+	−	−	−	+	−	−	+	+	4	AMP, AMC, IPM, GEN	0.29
Human	1	+	+	−	−	−	+	−	−	+	+	4	AMP, AMC, IPM, GEN	0.29
Chicken	2	+	+	−	−	+	+	−	−	+	+	3	AMP, AMC, NIT	0.21
Chicken	3	+	+	−	−	+	+	+	−	+	+	2	AMP, AMC	0.14
						(II) *S. typhimurium*								
Chicken	1	+	+	−	−	+	+	+	−	+	−	7	AMP, AMC, CTX, GEN, TET, AZM, NIT	0.50
Chicken	1	+	+	−	−	+	+	+	−	+	−	6	AMP, AMC, GEN, TET, AZM, NIT	0.43
Human	1	+	+	−	−	+	+	+	−	+	+	5	AMP, AMC, TET, SXT, NIT	0.36
Chicken	2	+	+	−	−	+	+	+	−	+	+	4	AMP, CTX, TET, NIT	0.29
Human	1	+	+	−	+	+	+	+	−	+	+	3	AMP, TET, NIT	0.21
Chicken	1	+	+	−	+	+	+	+	−	+	+	3	AMP, TET, NIT	0.21
Chicken	1	+	+	−	−	+	+	−	−	+	+	3	AMP, AZM, NIT	0.21
Chicken	1	+	+	−	−	+	+	+	−	+	+	3	AMP, AMC, TET	0.21
Human	1	+	−	−	−	−	+	−	−	+	+	2	AMP, AMC	0.14
Egg	1	+	−	−	−	−	+	−	−	+	+	2	AMP, AMC	0.14
Human	1	+	−	−	−	−	+	−	−	+	+	2	AMP, AZM	0.14

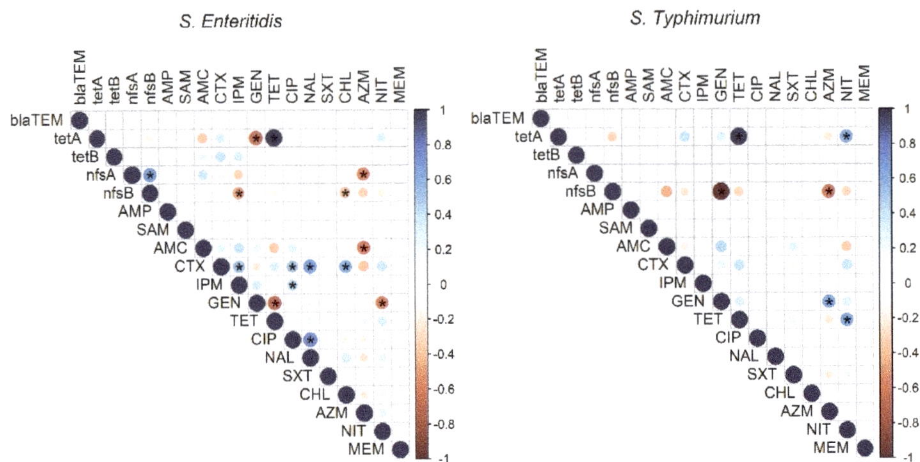

Figure 3. Correlation matrix showing the correlation between resistance phenotypes and genotypes among the examined *S. enteritidis* and *S. typhimurium* isolates recovered from laying hens, table eggs, and humans. The blue colour indicates a positive correlation and red shows a negative correlation. The asterisk (*) indicates significant at $p < 0.001$.

3.4. Characterization of Biosynthesized AgNPs

AgNPs were synthesized from *Aspergillus niveus* (*A. niveus*; MT319815), molecularly confirmed in the study targeting the 28S rRNA region (Figure 4A). AgNPs were oval, cubic, rod shaped, and well distributed without agglomeration at 6.49 nm using TEM microscopy (Figure 4B). The prepared AgNPs were characterized by the absorption peak of AgNPs at 420 nm using the UV–visible spectrum and an average particle size at 27 nm using DLS (Figure 4C). FTIR spectra of AgNPs confirmed the presence of various functional groups at 3451.81, 2081.40, 1634.19, 1384.39, 1117.76, and 614.02 cm^{-1}. The peaks at 3451.81, 2081.40, 1634.19, and 614.02 cm^{-1} correspond to carbonyl residues, alcohol, nitrile, acid chloride, and alkene band C–C stretch in ring of CH3, stretch of alkyl halides and peptide bonds of proteins responsible for the synthesis of the AgNPs (Figure 4D). The bands at 3451.81, 1634.19, and 1384.39 cm^{-1} correspond to the binding vibrations of amide I and amide II of protein and hydroxyl O–H stretch of phenols with amine N–H stretchings. In addition, the band at 1117.76 cm^{-1} refers to one mononuclear aromatic. Dynamic light scattering analysis showed that the average particle size of the prepared AgNPs was at 27 nm (Figure 4E). The stability degree (Zeta Potential) of AgNPs showed a negative charge at -30.4 mv (Figure 4F).

3.5. Antimicrobial Activity of Biosynthesized AgNPs

The disk diffusion method was used to evaluate the antimicrobial activity of *S. enteritidis* and *S. typhimurium* isolates treated with AgNPs. The diameter of inhibition zones of AgNP (10 μg)-treated *S. enteritidis* and *S. typhimurium* isolates were 24 mm and 20 mm, respectively. The MIC for *Salmonella* treated with different concentrations (10, 9, 8, 7, 6, 5, 4, 3, 2, and 1 μg/mL) in this study was 5 μg/mL upon using the dilution method to *S. enteritidis* and *S. typhimurium* isolates, whereas MBC value was 6 μg/mL to *S. enteritidis* isolates and 8 μg/mL to *S. typhimurium* isolates.

The selected MBC value for each strain was evaluated by assessing the bacterial growth (CFU/mL) at time intervals 12, 24, 36, and 48 h (Figure 5). The growth of *S. enteritidis* and *S. typhimurium* significantly ($p < 0.05$) declined at 12 h and reached complete inhibition at 48 h. In parallel, the control plates showed a significant increase in growth.

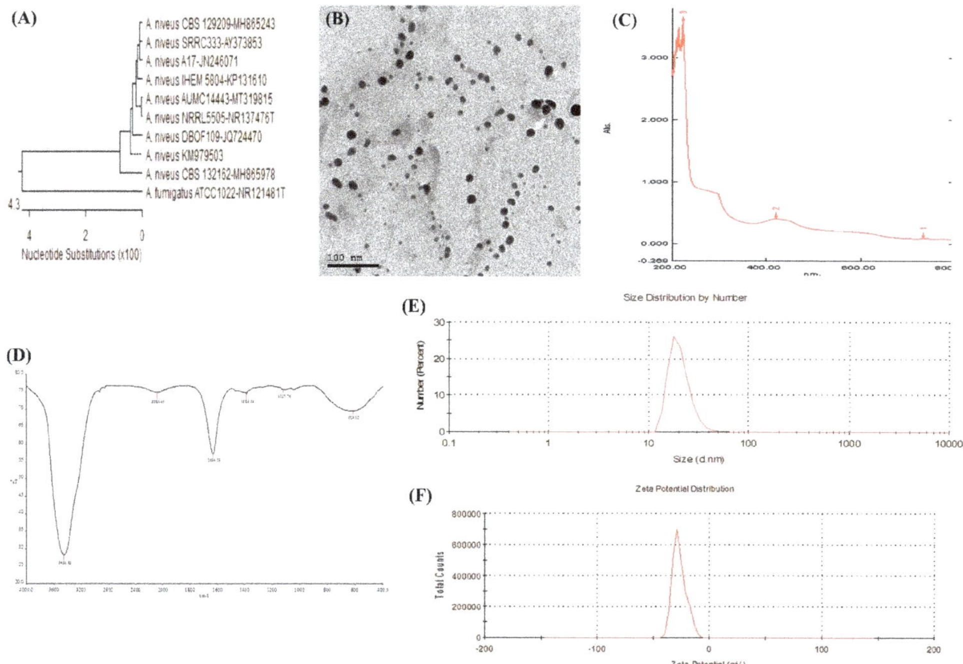

Figure 4. Characterization of AgNPs. (**A**) Phylogenetic analysis of the *Aspergillus niveus* (MT319815) used in the biosynthesis of AgNPs; (**B**) TEM micrographs and size distribution for silver (scale bar: 100 nm); (**C**) UV–Visible spectrum of AgNPs; (**D**) Fourier transform infrared spectrum showing the functional groups on the surface of AgNPs; (**E**) Dynamic light scattering analysis showing the highest peak at 27 nm and (**F**) Zeta-potential of AgNPs.

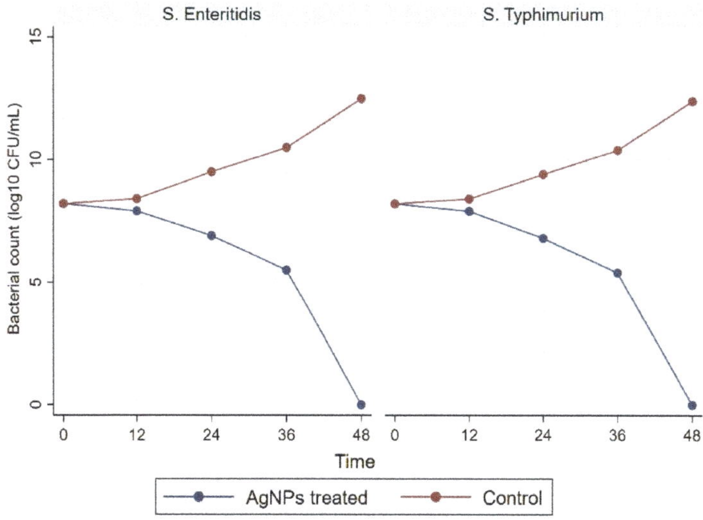

Figure 5. Bacterial growth of multidrug-resistant *S. enteritidis* and *S. typhimurium* treated with AgNPs. Bacterial counts were recorded at 0, 12, 24, 36, and 48 h post-treatment.

3.6. Antivirulent and Antiresistant Activity of Biosynthesized AgNP$_s$

The expression patterns of virulence and resistance genes of *S. enteritidis* and *S. typhimurium* isolates treated with AgNPs at time intervals 6, 12, 24, and 36 h are presented in Table 4. The expression levels of virulence and resistance genes significantly decreased by time. Despite the loss of expression of the virulence genes at 36 h, the resistance genes showed instant expression.

Table 4. Virulence and resistance genes expression of multidrug-resistant *S. enteritidis* and *S. typhimurium* treated with AgNPs.

Storage Time (Hours)	Virulence Genes Expression			Resistance Genes Expression		
	SopB	*stn*	*hilA*	*blaTEM*	*tetA*	*nfsA*
			(I) *S. enteritidis*			
0	1.0 ± 0.0	1.0 ± 0.0	1.0 ± 0.0	1.0 ± 0.0	1.0 ± 0.0	1.0 ± 0.0
6	0.71 ± 0.015	0.73 ± 0.025	0.69 ± 0.015	0.85 ± 0.02	0.78 ± 0.01	0.83 ± 0.006
12	0.45 ± 0.015	0.44 ± 0.015	0.41 ± 0.015	0.69 ± 0.01	0.61 ± 0.01	0.66 ± 0.015
24	0.11 ± 0.001	0.26 ± 0.02	0.15 ± 0.025	0.46 ± 0.01	0.42 ± 0.006	0.47 ± 0.006
36	0.0 ± 0.0	0.0 ± 0.0	0.0 ± 0.0	0.16 ± 0.006	0.11 ± 0.01	0.14 ± 0.006
			(II) *S. typhimurium*			
0	1.0 ± 0.0	1.0 ± 0.0	1.0 ± 0.0	1.0 ± 0.0	1.0 ± 0.0	1.0 ± 0.0
6	0.77 ± 0.02	0.79 ± 0.01	0.76 ± 0.01	0.92 ± 0.015	0.87 ± 0.02	0.94 ± 0.015
12	0.49 ± 0.02	0.47 ± 0.025	0.42 ± 0.01	0.82 ± 0.02	0.56 ± 0.01	0.79 ± 0.01
24	0.23 ± 0.025	0.26 ± 0.025	0.15 ± 0.025	0.59 ± 0.01	0.34 ± 0.015	0.57 ± 0.01
36	0.0 ± 0.0	0.0 ± 0.0	0.0 ± 0.0	0.38 ± 0.015	0.18 ± 0.015	0.36 ± 0.015

4. Discussion

Chicken and eggs are considered one of the main reservoirs for zoonotic pathogens [44]. In this study, 12 different *Salmonella* serovars were identified in the examined samples with *Salmonella* spp. prevalence of 19.3%, and most of them were identified in laying hen samples (33.1%). The observed prevalence in laying hens was comparable to the 32.0% reported recently in Egypt [45] and the 33.3% in Colombia [46], but higher than the 1.02% and 10.4% reported in Spain [47] and Egypt [48], respectively. However, the prevalence of *Salmonella* spp. infection in laying hens was lower than the 59.3% that was reported previously in Korea [20]. Furthermore, the prevalence of *Salmonella* spp. infection in human samples (22%) was higher than the 4% that was reported previously in the same governorate [49]. In this study, the isolation rate of *Salmonella* spp. from egg content was low (3.6%) and lower than the 5.2% reported by Im, Jeong, Kwon, Jeong, Kang, and Lee [20] and the absence of *Salmonella* in egg content reported by Zubair, et al. [50]. The low incidence of *Salmonella* in egg content was caused by the egg's complex system of membrane barriers and the albumen antibacterial effect [51]. Furthermore, *Salmonella* in egg contents may be attributed to contamination of eggshell by *Salmonella* infected poultry feces, which may penetrate the interior of eggs and grow during storage [52].

S. enteritidis and *S. typhimurium* were the most commonly isolated serotypes in this study, which was consistent with previous reports from Egypt [45] and Brazil [53]. The *S. enteritidis* and *S. typhimurium* isolates used in this study showed a high rate of antimicrobial resistance (100%) to at least one antimicrobial agent. Antimicrobial agents (AMP, TET, and NIT) were not effective against *Salmonella* isolates in this study. Our findings were comparable to those reported previously in South India [54] and Ghana [55]. The high rate of resistance reported in this study could be attributed to a number of factors, including the overuse of antimicrobial agents (over-the-counter antibiotics without a prescription) and the inappropriate use of antimicrobial agents as a growth promoter in animals, which could lead to the emergence of resistant bacteria in both animals and humans through direct contact or through the food chain [56]. All *Salmonella* isolates in this study were sensitive to SAM and MEM, which was consistent with previous Egyptian reports [57,58], and was accredited to the limited use of these antimicrobial agents in commercial chicken farms.

Previously, MDR from various sources has been identified in *Salmonella* [57,59], making these antimicrobial agents ineffective in humans and poultry [60]. In the current study, 75.7% of *Salmonella* isolates displayed MDR, which is higher than the prevalence of MDR *Salmonella* isolated from chicken in the USA [61] and China [62,63]. Despite this, the prevalence of MDR *Salmonella* in this study was lower than the 100% previously reported in chicken in Egypt [57] and 92.9% in North India [59]. The resistant isolates were not evenly distributed in laying hens, eggs, and humans. However, these isolates were overlapped, reflecting the high public health impact of MDR isolates and the need for new antimicrobial agents to treat human salmonellosis [64].

The presence of both virulence and antimicrobial resistance genes affects the bacteria' pathogenicity [8]. The emergence of MDR strains of *Salmonella* is primarily due to genetic factors that enhance their survival by retaining drug resistance genes [65]. Of the confirmed *S. enteritidis* and *S. typhimurium* isolates, 81.1% harbored three virulence genes (*sopB*, *stn*, and *hilA*), which was similar to the previous study in Malaysia [66]. None of the isolates harbored *pefA* gene in our study, which was similar to a previous study in Egypt [45]. In contrast to Thung, Radu, Mahyudin, Rukayadi, Zakaria, Mazlan, Tan, Lee, Yeoh, and Chin [66], who reported the absence of *spvC* gene *Salmonella* isolates, the *spvC* was identified in 5.4% of *S. enteritidis* and *S. typhimurium* isolates.

In the present study, *Salmonella* harbors a range of antibiotic resistance genes located on mobile genetic elements that distribute resistance characteristics to other serovars and other different bacteria [59]. All isolates had *blaTEM* resistance genes and most of them had *tetA*, *nfsA*, and *nfsB* resistance genes, which was similar to *Salmonella* isolates from chickens in Egypt [57] and North India [59]. The *TetB* was not identified in previous studies, but *tetA* was the most common, highlighting its resistance role to tetracyclines [59,67].

There is currently a need for more research to find novel materials that inhibit antimicrobial-resistant variants [68]. Therefore, the development of natural antimicrobial agents such as AgNPs may be an alternative way to overcome MDR bacteria. In our study, AgNPs were biosynthesized from molecularly confirmed *A. nivus.* in line with the previous study by Elgazzar and Ismail [69], the peak of AgNPs using UV–visible spectra was witnessed at 420 nm. The small particle size of AgNPs (27 nm) with a single peak indicates the suitable quality of the biosynthesized nanoparticles and ideal inhibitory effects on bacterial growth [70]. The small particle size allows AgNPs to bind to the cell wall and easily penetrate the bacteria cell, enhancing their antimicrobial activity against the bacteria owing to their larger surface area and greater interaction [71]. The TEM microscopy indicated that biosynthesized AgNPs were well dispersed in the solution without agglomeration, explaining that fungal filtrates contain various biomolecules used to cap and stabilize the agglomeration of AgNPs [35]. FTIR measurements of biosynthesized AgNPs have verified different functional groups of biomolecules and capped proteins, excreted by the fungus itself, encapsulated nanoparticles, and increased stability; associated proteins may help to mineralize precursor salts [69]. The higher negative charge of the AgNPs as measured by the zeta potential confirms the synthesized particles' repulsion, resulting in stability and monodispersity of the synthesized AgNPs solution [72].

In this study, a clear zone was found to exist around the AgNP disk, suggesting that the biosynthesized AgNPs possessed a potent antibacterial effect against the growth of MDR bacteria, causing a reduction in bacterial number [73,74]. To evaluate the effect of AgNPs against *Salmonella*, MIC and MBC values were examined in our study, showing the extent of variation with Loo, et al. [74], who reported that the MIC value of *S. enteritidis* and *S. typhimurium* was 3.9 mg/mL (each) and the MBC values for *S. enteritidis* and *S. typhimurium* were 3.9 mg/mL and 7.8 mg/mL, respectively.

The bacterial growth inhibition of AgNPs occurs because AgNPs interfere with sulfur found in biomolecules on the bacterial membrane, attacking the bacterial genome and resulting in bacterial death [75]. In this study, a significant decrease was found in the number of *Salmonella* isolates after different storage time and the growth was completely absent after 48 h of incubation; however, Tanhaeian and Ahmadi [76] reported a decline in the bacterial growth treated with 25 µg/mL and 50 µg/mL of AgNP after 10 h and the growth level dropped to almost zero.

Besides exploring the effect of the biosynthesized AgNPs on the growth inhibition of *Salmonella*, we investigated their impact on the expression of genes that are essential to the virulence and resistance of the pathogen. The results emphasized that biosynthesized AgNPs at a concentration of 5 µg significantly reduce gene expression, as previously reported [76]. The adverse effect of the biosynthesized AgNPs on virulence and resistance genes expression contributes to effective preventative and potent antimicrobial drugs against bacterial infection [76]. AgNPs are broad-spectrum antimicrobial agents that have the same effect on all Gram-negative bacteria strains. However, the concentration of the used AgNPs and bacteria class, nanoparticle size, synthesis method, and physical characteristics of nanoparticles are key factors that can affect the results across studies.

5. Conclusions

In the present study, *S. enteritidis* and *S. typhimurium* were the most common *Salmonella* serovars isolated from laying hens, table eggs, and humans in Egypt. Moreover, most of the recovered *Salmonella* isolates exhibited MDR, which poses a possible risk to consumers in Egypt. The findings of this study support that AgNPs have the potential to be effective antimicrobial agents against MDR *S. enteritidis* and *S. typhimurium* and could be recommended for use in laying hen farms. However, further studies are required to develop and design a safe AgNPs antimicrobial for laying hen farms.

Author Contributions: Conceptualization, R.M.M.A.E.; I.E.; N.E.-G.; H.M.N.T.; E.N.A.; A.A.M.; S.S.A., and A.B.M.B.T.; methodology, R.M.M.A.E.; N.E.-G.; H.M.N.T.; E.N.A.; A.A.M.; S.S.A., and A.B.M.B.T.; software, R.M.M.A.E.; I.E., and N.E.-G.; validation, R.M.M.A.E.; N.E.-G.; H.M.N.T.; E.N.A.; A.A.M.; S.S.A., and A.B.M.B.T.; formal analysis, I.E.; investigation, R.M.M.A.E. and A.B.M.B.T.; resources, N.E.-G.; H.M.N.T.; E.N.A.; A.A.M., and S.S.A.; data curation, R.M.M.A.E.; N.E.-G.; A.A.M., and A.B.M.B.T.; writing—original draft preparation, R.M.M.A.E.; I.E.; N.E.-G., and A.B.M.B.T.; writing—review and editing, R.M.M.A.E.; I.E.; N.E.-G.; H.M.N.T.; E.N.A.; A.A.M.; S.S.A., and A.B.M.B.T.; visualization, R.M.M.A.E.; I.E.; N.E.-G., and A.B.M.B.T.; supervision, R.M.M.A.E.; project administration, R.M.M.A.E. and A.B.M.B.T. All authors have read and agreed to the published version of the manuscript.

Funding: No specific funding was received.

Institutional Review Board Statement: This study was approved by the Institutional Animal Care and Use Committee (IACUC) of Zagazig University (Ref. No.: ZU-IACUC/2/F/7/2020).

Informed Consent Statement: Informed consent was obtained from all subjects involved in the study.

Data Availability Statement: The data presented in this study are available on request from the corresponding author.

Acknowledgments: The authors thank the participants (laying hens farms and patients) who agreed to use their samples in this study. Authors would also thank the Egyptian Knowledge Bank (EKB) for editing and grammar correction of this manuscript.

Conflicts of Interest: The authors declare no conflict of interest.

References

1. Majowicz, S.E.; Musto, J.; Scallan, E.; Angulo, F.J.; Kirk, M.; O'Brien, S.J.; Jones, T.F.; Fazil, A.; Hoekstra, R.M. The global burden of nontyphoidal *Salmonella* gastroenteritis. *Clin. Infect. Dis.* **2010**, *50*, 882–889. [CrossRef]
2. Ceyssens, P.J.; Mattheus, W.; Vanhoof, R.; Bertrand, S. Trends in serotype distribution and antimicrobial susceptibility in *Salmonella enterica* isolates from humans in Belgium, 2009 to 2013. *Antimicrob. Agents Chemother.* **2015**, *59*, 544–552. [CrossRef] [PubMed]
3. Jackson, B.R.; Griffin, P.M.; Cole, D.; Walsh, K.A.; Chai, S.J. Outbreak-associated *Salmonella enterica* serotypes and food commodities, United States, 1998–2008. *Emerg. Infect. Dis.* **2013**, *19*, 1239–1244. [CrossRef]
4. De Reu, K.; Grijspeerdt, K.; Messens, W.; Heyndrickx, M.; Uyttendaele, M.; Debevere, J.; Herman, L. Eggshell factors influencing eggshell penetration and whole egg contamination by different bacteria, including *Salmonella* enteritidis. *Int. J. Food Microbiol.* **2006**, *112*, 253–260. [CrossRef]

5. Li, R.; Lai, J.; Wang, Y.; Liu, S.; Li, Y.; Liu, K.; Shen, J.; Wu, C. Prevalence and characterization of *Salmonella* species isolated from pigs, ducks and chickens in Sichuan Province, China. *Int. J. Food Microbiol.* **2013**, *163*, 14–18. [CrossRef]
6. Heider, L.C.; Funk, J.A.; Hoet, A.E.; Meiring, R.W.; Gebreyes, W.A.; Wittum, T.E. Identification of *Escherichia coli* and *Salmonella enterica* organisms with reduced susceptibility to ceftriaxone from fecal samples of cows in dairy herds. *Am. J. Vet. Res.* **2009**, *70*, 389–393. [CrossRef]
7. Elemfareji, O.I.; Thong, K.L. Comparative Virulotyping of *Salmonella typhi* and *Salmonella enteritidis*. *Indian J. Microbiol.* **2013**, *53*, 410–417. [CrossRef]
8. Huehn, S.; La Ragione, R.M.; Anjum, M.; Saunders, M.; Woodward, M.J.; Bunge, C.; Helmuth, R.; Hauser, E.; Guerra, B.; Beutlich, J. Virulotyping and antimicrobial resistance typing of *Salmonella enterica* serovars relevant to human health in Europe. *Foodborne Pathog. Dis.* **2010**, *7*, 523–535. [CrossRef] [PubMed]
9. Murugkar, H.; Rahman, H.; Dutta, P. Distribution of virulence genes in *Salmonella* serovars isolated from man & animals. *Indian J. Med. Res.* **2003**, *117*, 66–70.
10. Fluit, A.C. Towards more virulent and antibiotic-resistant *Salmonella*? *FEMS Immunol. Med. Microbiol.* **2005**, *43*, 1–11. [CrossRef]
11. Zishiri, O.T.; Mkhize, N.; Mukaratirwa, S. Prevalence of virulence and antimicrobial resistance genes in *Salmonella* spp. isolated from commercial chickens and human clinical isolates from South Africa and Brazil. *Onderstepoort J. Vet. Res.* **2016**, *83*, 1–11. [CrossRef]
12. Shaalan, M.; Saleh, M.; El-Mahdy, M.; El-Matbouli, M. Recent progress in applications of nanoparticles in fish medicine: A review. *Nanomed. Nanotechnol. Biol. Med.* **2016**, *12*, 701–710. [CrossRef]
13. Agnihotri, S.; Mukherji, S.; Mukherji, S. Size-controlled silver nanoparticles synthesized over the range 5–100 nm using the same protocol and their antibacterial efficacy. *RSC Adv.* **2014**, *4*, 3974–3983. [CrossRef]
14. Franci, G.; Falanga, A.; Galdiero, S.; Palomba, L.; Rai, M.; Morelli, G.; Galdiero, M. Silver nanoparticles as potential antibacterial agents. *Molecules* **2015**, *20*, 8856–8874. [CrossRef] [PubMed]
15. Gurunathan, S. Biologically synthesized silver nanoparticles enhance antibiotic activity against Gram-negative bacteria. *J. Ind. Eng. Chem.* **2015**, *29*, 217–226. [CrossRef]
16. Iravani, S.; Korbekandi, H.; Mirmohammadi, S.V.; Zolfaghari, B. Synthesis of silver nanoparticles: Chemical, physical and biological methods. *Res. Pharm. Sci.* **2014**, *9*, 385–406.
17. Khodashenas, B.; Ghorbani, H.R. Synthesis of silver nanoparticles with different shapes. *Arab. J. Chem.* **2019**, *12*, 1823–1838. [CrossRef]
18. El-Gohary, F.A.; Abdel-Hafez, L.J.M.; Zakaria, A.I.; Shata, R.R.; Tahoun, A.; El-Mleeh, A.; Elfadl, E.A.A.; Elmahallawy, E.K. Enhanced Antibacterial Activity of Silver Nanoparticles Combined with Hydrogen Peroxide Against Multidrug-Resistant Pathogens Isolated from Dairy Farms and Beef Slaughterhouses in Egypt. *Infect. Drug Resist.* **2020**, *13*, 3485–3499. [CrossRef]
19. Lange, A.; Grzenia, A.; Wierzbicki, M.; Strojny-Cieslak, B.; Kalińska, A.; Gołębiewski, M.; Radzikowski, D.; Sawosz, E.; Jaworski, S. Silver and copper nanoparticles inhibit biofilm formation by mastitis pathogens. *Animals* **2021**, *11*, 1884. [CrossRef] [PubMed]
20. Im, M.C.; Jeong, S.J.; Kwon, Y.-K.; Jeong, O.-M.; Kang, M.-S.; Lee, Y.J. Prevalence and characteristics of *Salmonella* spp. isolated from commercial layer farms in Korea. *Poult. Sci.* **2015**, *94*, 1691–1698. [CrossRef]
21. Yang, X.; Brisbin, J.; Yu, H.; Wang, Q.; Yin, F.; Zhang, Y.; Sabour, P.; Sharif, S.; Gong, J. Selected lactic acid-producing bacterial isolates with the capacity to reduce *Salmonella* translocation and virulence gene expression in chickens. *PLoS ONE* **2014**, *9*, e93022. [CrossRef] [PubMed]
22. McFaddin, J. *Biochemical Tests for Identification of Medical Bacteria*; Lippicott Williams Wilkins: Philadelphia, PA, USA, 2000.
23. Kauffman, G. Kauffmann white scheme. *J. Acta. Path. Microbiol. Sci.* **1974**, *61*, 385.
24. Oliveira, S.; Rodenbusch, C.; Cé, M.; Rocha, S.; Canal, C. Evaluation of selective and non-selective enrichment PCR procedures for *Salmonella* detection. *Lett. Appl. Microbiol.* **2003**, *36*, 217–221. [CrossRef] [PubMed]
25. Akbarmehr, J.; Salehi, T.Z.; Brujeni, G. Identification of *Salmonella* isolated from poultry by MPCR technique and evaluation of their *hsp groEL* gene diversity based on the PCR-RFLP analysis. *Afr. J. Microbiol. Res.* **2010**, *4*, 1594–1598.
26. Liu, B.; Zhou, X.; Zhang, L.; Liu, W.; Dan, X.; Shi, C.; Shi, X. Development of a novel multiplex PCR assay for the identification of *Salmonella enterica* Typhimurium and Enteritidis. *Food Control* **2012**, *27*, 87–93. [CrossRef]
27. CLSI (Clinical and Laboratory Standards Institute). *Standards for Antimicrobial Disk Susceptibility Test; Approved Standard*; CLSI Document M02-A11; CLSI: Wayne, PA, USA, 2012.
28. Magiorakos, A.P.; Srinivasan, A.; Carey, R.B.; Carmeli, Y.; Falagas, M.E.; Giske, C.G.; Harbarth, S.; Hindler, J.F.; Kahlmeter, G.; Olsson-Liljequist, B.; et al. Multidrug-resistant, extensively drug-resistant and pandrug-resistant bacteria: An international expert proposal for interim standard definitions for acquired resistance. *Clin. Microbiol. Infect.* **2012**, *18*, 268–281. [CrossRef]
29. Krumperman, P.H. Multiple antibiotic resistance indexing of *Escherichia coli* to identify high-risk sources of fecal contamination of foods. *Appl. Environ. Microbiol.* **1983**, *46*, 165–170. [CrossRef]
30. Colom, K.; Pérez, J.; Alonso, R.; Fernández-Aranguiz, A.; Lariño, E.; Cisterna, R. Simple and reliable multiplex PCR assay for detection of *bla TEM*, *bla SHV* and *bla OXA–1* genes in Enterobacteriaceae. *FEMS Microbiol. Lett.* **2003**, *223*, 147–151. [CrossRef]
31. Randall, L.; Cooles, S.; Osborn, M.; Piddock, L.; Woodward, M.J. Antibiotic resistance genes, integrons and multiple antibiotic resistance in thirty-five serotypes of *Salmonella enterica* isolated from humans and animals in the UK. *J. Antimicrob. Chemother.* **2004**, *53*, 208–216. [CrossRef]

32. Sabarinath, A.; Tiwari, K.P.; Deallie, C.; Belot, G.; Vanpee, G.; Matthew, V.; Sharma, R.; Hariharan, H. Antimicrobial resistance and phylogenetic groups of commensal *Escherichia Coli* isolates from healthy pigs in Grenada. *Webmed Cent. Vet. Med.* **2011**, *2*, 1–10.
33. Mottaghizadeh, F.; Haeili, M.; Darban-Sarokhalil, D. Molecular epidemiology and nitrofurantoin resistance determinants from nitrofurantoin non-susceptible *Escherichia coli* isolated from urinary tract infections. *J. Glob. Antimicrob. Resist.* **2019**, *21*, 335–339. [CrossRef]
34. Rabie, G.; El-Abedeen, A.Z.; Bakry, A.A. Biological synthesis of silver nanoparticles using filamentous fungi. *Nano Sci. Nano Technol. J.* **2013**, *7*, 163–171.
35. El-Gazzar, N.S.; Rabie, G.H. Application of silver nanoparticles on *Cephalosporium maydis* in vitro and in vivo. *Egypt. J. Microbiol.* **2018**, *53*, 69–81.
36. White, T.J.; Bruns, T.; Lee, S.; Taylor, J. Amplification and direct sequencing of fungal ribosomal RNA genes for phylogenetics. *PCR Protoc. A Guide Methods Appl.* **1990**, *18*, 315–322.
37. DeAlba-Montero, I.; Guajardo-Pacheco, J.; Morales-Sánchez, E.; Araujo-Martínez, R.; Loredo-Becerra, G.; Martínez-Castañón, G.-A.; Ruiz, F.; Compeán Jasso, M. Antimicrobial properties of copper nanoparticles and amino acid chelated copper nanoparticles produced by using a soya extract. *Bioinorg. Chem. Appl.* **2017**, *2017*, 1064918. [CrossRef] [PubMed]
38. Bauer, A. Antibiotic susceptibility testing by a standardized single disc method. *Am. J. Clin. Pathol.* **1966**, *45*, 149–158. [CrossRef]
39. Krishnan, R.; Arumugam, V.; Vasaviah, S.K. The MIC and MBC of silver nanoparticles against *Enterococcus faecalis*—A facultative anaerobe. *J. Nanomed. Nanotechnol.* **2015**, *6*, 285.
40. Thatcher, F.S.; Clark, D.S. *Micro-Organisms in Foods: Their Significance and Methods of Enumeration*; University of Toronto Press: Toronto, ON, Canada, 1968.
41. Yuan, J.S.; Reed, A.; Chen, F.; Stewart, C.N. Statistical analysis of real-time PCR data. *BMC Bioinform.* **2006**, *7*, 85. [CrossRef]
42. Gu, Z.; Eils, R.; Schlesner, M. Complex heatmaps reveal patterns and correlations in multidimensional genomic data. *Bioinformatics* **2016**, *32*, 2847–2849. [CrossRef]
43. Kruskal, J.B. Multidimensional scaling by optimizing goodness of fit to a nonmetric hypothesis. *Psychometrika* **1964**, *29*, 1–27. [CrossRef]
44. Mughini-Gras, L.; Enserink, R.; Friesema, I.; Heck, M.; van Duynhoven, Y.; van Pelt, W. Risk factors for human salmonellosis originating from pigs, cattle, broiler chickens and egg laying hens: A combined case-control and source attribution analysis. *PLoS ONE* **2014**, *9*, e87933.
45. Elkenany, R.; Elsayed, M.M.; Zakaria, A.I.; El-sayed, S.A.-E.-S.; Rizk, M.A. Antimicrobial resistance profiles and virulence genotyping of *Salmonella enterica* serovars recovered from broiler chickens and chicken carcasses in Egypt. *BMC Vet. Res.* **2019**, *15*, 124. [CrossRef] [PubMed]
46. Rodríguez, R.; Fandiño, C.; Donado, P.; Guzmán, L.; Verjan, N. Characterization of *Salmonella* from commercial egg-laying hen farms in a central region of Colombia. *Avian. Dis.* **2015**, *59*, 57–63. [CrossRef]
47. Lamas, A.; Fernandez-No, I.; Miranda, J.; Vázquez, B.; Cepeda, A.; Franco, C. Prevalence, molecular characterization and antimicrobial resistance of *Salmonella* serovars isolated from northwestern Spanish broiler flocks (2011–2015). *Poult. Sci.* **2016**, *95*, 2097–2105. [CrossRef] [PubMed]
48. Ahmed, H.; Gharieb, R.M.; Mohamed, M.E.; Amin, M.A.; Mohamed, R.E. Bacteriological and molecular characterization of *Salmonella* species isolated from humans and chickens in Sharkia Governorate, Egypt. *Zagazig Vet. J.* **2017**, *45*, 48–61. [CrossRef]
49. Gharieb, R.M.; Tartor, Y.H.; Khedr, M.H. Non-Typhoidal *Salmonella* in poultry meat and diarrhoeic patients: Prevalence, antibiogram, virulotyping, molecular detection and sequencing of class I integrons in multidrug resistant strains. *Gut Pathog.* **2015**, *7*, 34. [CrossRef]
50. Zubair, A.I.; Al-Berfkani, M.I.; Issa, A.R. Prevalence of *Salmonella* species from poultry eggs of local stores in Duhok. *Int. J. Res. Med. Sci.* **2017**, *5*, 2468–2471. [CrossRef]
51. García, C.; Soriano, J.; Benítez, V.; Catalá-Gregori, P. Assessment of *Salmonella* spp. in feces, cloacal swabs, and eggs (eggshell and content separately) from a laying hen farm. *Poult. Sci.* **2011**, *90*, 1581–1585. [CrossRef]
52. Nascimento, V.; Cranstoun, S.; Solomon, S. Relationship between shell structure and movement of *Salmonella* enteritidis across the eggshell wall. *Br. Poult. Sci.* **1992**, *33*, 37–48. [CrossRef]
53. Borges, K.A.; Furian, T.Q.; Borsoi, A.; Moraes, H.L.; Salle, C.T.; Nascimento, V.P. Detection of virulence-associated genes in *Salmonella* Enteritidis isolates from chicken in South of Brazil. *Pesqui. Vet. Bras.* **2013**, *33*, 1416–1422. [CrossRef]
54. Singh, R.; Yadav, A.; Tripathi, V.; Singh, R. Antimicrobial resistance profile of *Salmonella* present in poultry and poultry environment in north India. *Food Control* **2013**, *33*, 545–548. [CrossRef]
55. Andoh, L.A.; Dalsgaard, A.; Obiri-Danso, K.; Newman, M.; Barco, L.; Olsen, J.E. Prevalence and antimicrobial resistance of *Salmonella* serovars isolated from poultry in Ghana. *Epidemiol. Infect.* **2016**, *144*, 3288–3299. [CrossRef] [PubMed]
56. Brown, K.; Uwiera, R.R.; Kalmokoff, M.L.; Brooks, S.P.; Inglis, G.D. Antimicrobial growth promoter use in livestock: A requirement to understand their modes of action to develop effective alternatives. *Int. J. Antimicrob. Agents* **2017**, *49*, 12–24. [CrossRef] [PubMed]
57. Abdeen, E.; Elmonir, W.; Suelam, I.; Mousa, W. Antibiogram and genetic diversity of *Salmonella enterica* with zoonotic potential isolated from morbid native chickens and pigeons in Egypt. *J. Appl. Microbiol.* **2018**, *124*, 1265–1273. [CrossRef]
58. Diab, M.S.; Zaki, R.S.; Ibrahim, N.A.; Abd El Hafez, M.S. Prevalence of Multidrug Resistance Non-Typhoidal *Salmonellae* Isolated from Layer Farms and Humans in Egypt. *World Vet. J.* **2019**, *9*, 280–288. [CrossRef]

59. Sharma, J.; Kumar, D.; Hussain, S.; Pathak, A.; Shukla, M.; Kumar, V.P.; Anisha, P.; Rautela, R.; Upadhyay, A.; Singh, S. Prevalence, antimicrobial resistance and virulence genes characterization of nontyphoidal *Salmonella* isolated from retail chicken meat shops in Northern India. *Food Control* **2019**, *102*, 104–111. [CrossRef]
60. Miranda, J.; Guarddon, M.; Vázquez, B.; Fente, C.; Barros-Velázquez, J.; Cepeda, A.; Franco, C. Antimicrobial resistance in Enterobacteriaceae strains isolated from organic chicken, conventional chicken and conventional turkey meat: A comparative survey. *Food Control* **2008**, *19*, 412–416. [CrossRef]
61. Alali, W.Q.; Thakur, S.; Berghaus, R.D.; Martin, M.P.; Gebreyes, W.A. Prevalence and distribution of *Salmonella* in organic and conventional broiler poultry farms. *Foodborne Pathog. Dis.* **2010**, *7*, 1363–1371. [CrossRef]
62. Bai, L.; Lan, R.; Zhang, X.; Cui, S.; Xu, J.; Guo, Y.; Li, F.; Zhang, D. Prevalence of *Salmonella* isolates from chicken and pig slaughterhouses and emergence of ciprofloxacin and cefotaxime co-resistant S. enterica serovar Indiana in Henan, China. *PLoS ONE* **2015**, *10*, e0144532. [CrossRef]
63. Zhao, X.; Gao, Y.; Ye, C.; Yang, L.; Wang, T.; Chang, W. Prevalence and characteristics of *Salmonella* isolated from free-range chickens in Shandong Province, China. *BioMed Res. Int.* **2016**, *2016*, 8183931. [CrossRef]
64. Lekshmi, M.; Ammini, P.; Kumar, S.; Varela, M.F. The food production environment and the development of antimicrobial resistance in human pathogens of animal origin. *Microorganisms* **2017**, *5*, 11. [CrossRef] [PubMed]
65. Yang, B.; Qu, D.; Zhang, X.; Shen, J.; Cui, S.; Shi, Y.; Xi, M.; Sheng, M.; Zhi, S.; Meng, J. Prevalence and characterization of *Salmonella* serovars in retail meats of marketplace in Shaanxi, China. *Int. J. Food Microbiol.* **2010**, *141*, 63–72. [CrossRef] [PubMed]
66. Thung, T.Y.; Radu, S.; Mahyudin, N.A.; Rukayadi, Y.; Zakaria, Z.; Mazlan, N.; Tan, B.H.; Lee, E.; Yeoh, S.L.; Chin, Y.Z. Prevalence, virulence genes and antimicrobial resistance profiles of *Salmonella* serovars from retail beef in Selangor, Malaysia. *Front. Microbiol.* **2018**, *8*, 2697. [CrossRef]
67. El-Sharkawy, H.; Tahoun, A.; El-Gohary, A.E.-G.A.; El-Abasy, M.; El-Khayat, F.; Gillespie, T.; Kitade, Y.; Hafez, H.M.; Neubauer, H.; El-Adawy, H. Epidemiological, molecular characterization and antibiotic resistance of *Salmonella enterica* serovars isolated from chicken farms in Egypt. *Gut Pathog.* **2017**, *9*, 8. [CrossRef]
68. Abdel-Shafi, S.; Al-Mohammadi, A.-R.; Osman, A.; Enan, G.; Abdel-Hameid, S.; Sitohy, M. Characterization and Antibacterial Activity of 7S and 11S Globulins Isolated from Cowpea Seed Protein. *Molecules* **2019**, *24*, 1082. [CrossRef] [PubMed]
69. El-Gazzard, N.; Ismail, A. The potential use of Titanium, Silver and Selenium nanoparticles in controlling leaf blight of tomato caused by Alternaria alternata. *Biocatal. Agric. Biotechnol.* **2020**, *27*, 101708. [CrossRef]
70. Nakkala, J.R.; Mata, R.; Sadras, S.R. Green synthesized nano silver: Synthesis, physicochemical profiling, antibacterial, anticancer activities and biological in vivo toxicity. *J. Colloid Interface Sci.* **2017**, *499*, 33–45. [CrossRef] [PubMed]
71. Rai, M.; Yadav, A.; Gade, A. Silver nanoparticles as a new generation of antimicrobials. *Biotechnol. Adv.* **2009**, *27*, 76–83. [CrossRef] [PubMed]
72. Shaligram, N.S.; Bule, M.; Bhambure, R.; Singhal, R.S.; Singh, S.K.; Szakacs, G.; Pandey, A. Biosynthesis of silver nanoparticles using aqueous extract from the compactin producing fungal strain. *Process Biochem.* **2009**, *44*, 939–943. [CrossRef]
73. Cavassin, E.D.; de Figueiredo, L.F.P.; Otoch, J.P.; Seckler, M.M.; de Oliveira, R.A.; Franco, F.F.; Marangoni, V.S.; Zucolotto, V.; Levin, A.S.S.; Costa, S.F. Comparison of methods to detect the *in vitro* activity of silver nanoparticles (AgNP) against multidrug resistant bacteria. *J. Nanobiotechnology* **2015**, *13*, 64. [CrossRef]
74. Loo, Y.Y.; Rukayadi, Y.; Nor-Khaizura, M.-A.-R.; Kuan, C.H.; Chieng, B.W.; Nishibuchi, M.; Radu, S. In vitro antimicrobial activity of green synthesized silver nanoparticles against selected gram-negative foodborne pathogens. *Front. Microbiol.* **2018**, *9*, 1555. [CrossRef] [PubMed]
75. Sondi, I.; Siiman, O.; Matijević, E. Synthesis of CdSe nanoparticles in the presence of aminodextran as stabilizing and capping agent. *J. Colloid Interface Sci.* **2004**, *275*, 503–507. [CrossRef] [PubMed]
76. Tanhaeian, A.; Ahmadi, F.S. Inhibitory effect of biologically synthesized silver nanoparticle on growth and virulence of *E. coli*. *Zahedan J. Res. Med. Sci.* **2018**, *20*, e10269.

Article

Anisotropic Silver Nanoparticles Gel Exhibits Antibacterial Action and Reduced Scar Formation on Wounds Contaminated with Methicillin-Resistant *Staphylococcus pseudintermedius* (MRSP) in a Mice Model

Saengrawee Thammawithan [1,2], Oranee Srichaiyapol [1], Pawinee Siritongsuk [1,2], Sakda Daduang [2,3], Sompong Klaynongsruang [1,2], Nuvee Prapasarakul [4] and Rina Patramanon [1,2,*]

1   Department of Biochemistry, Faculty of Science, Khon Kaen University, Khon Kaen 40002, Thailand; th_saengrawee@kkumail.com (S.T.); oranee_sr@kkumail.com (O.S.); pawinee.siri@kkumail.com (P.S.); somkly@kku.ac.th (S.K.)
2   Protein and Proteomics Research Center for Commercial and Industrial Purposes, Khon Kaen University, Khon Kaen 40002, Thailand; sakdad@kku.ac.th
3   Division of Pharmacognosy and Toxicology, Faculty of Pharmaceutical Sciences, Khon Kaen University, Khon Kaen 40002, Thailand
4   Department of Veterinary Microbiology, Faculty of Veterinary Science, Chulalongkorn University, Bangkok 10330, Thailand; nuvee.p@chula.ac.th
*   Correspondence: narin@kku.ac.th; Tel.: +66-84599-9123

Citation: Thammawithan, S.; Srichaiyapol, O.; Siritongsuk, P.; Daduang, S.; Klaynongsruang, S.; Prapasarakul, N.; Patramanon, R. Anisotropic Silver Nanoparticles Gel Exhibits Antibacterial Action and Reduced Scar Formation on Wounds Contaminated with Methicillin-Resistant *Staphylococcus pseudintermedius* (MRSP) in a Mice Model. *Animals* **2021**, *11*, 3412. https://doi.org/10.3390/ani11123412

Academic Editors: Antonio Gonzalez-Bulnes and Nesreen Hashem

Received: 11 October 2021
Accepted: 27 November 2021
Published: 30 November 2021

Publisher's Note: MDPI stays neutral with regard to jurisdictional claims in published maps and institutional affiliations.

Simple Summary: Wound infection in animals with antimicrobial resistant bacteria, especially *Staphylococcus pseudintermedius*, plays an important role in the delay of wound healing. In this work, the antimicrobial and wound healing activities of gels containing anisotropic AgNPs were evaluated on wounds contaminated with Methicillin-resistant *Staphylococcus pseudintermedius* in a mice model. The results show that anisotropic AgNPs gel is effective in eliminating bacteria and preventing pus formation. Furthermore, anisotropic AgNPs gel exhibits improved collagen alignment that supports scar disappearance.

Abstract: *Staphylococcus pseudintermedius* (*S. pseudintermedius*) infected wounds can cause seriously delayed wound healing processes in animals. Antimicrobial agents that have antimicrobial and wound healing efficacy have become an essential tool for overcoming this problem. In our previous study, anisotropic AgNPs have been reported to have antimicrobial efficiency against animal and human pathogens, and could be suitable as antimicrobial agents for infected wounds. Here, antimicrobial and wound healing activities of anisotropic AgNPs gels were assessed *in vivo*. BALB/cAJcl mice wounds were infected by Methicillin-resistant *Staphylococcus pseudintermedius* (MRSP). Then, antibacterial and wound healing activities were evaluated by bacterial cell count, wound contraction, digital capture, and histology. The results show that anisotropic AgNPs gels could eliminate all bacterial cell infected wounds within 7 days, the same as povidone iodine. Wound healing activity was evaluated by wound contraction (%). The results showed 100% wound contraction in groups treated with anisotropic AgNPs gels within 14 days that was not significantly different from povidone iodine and control gel without AgNPs. However, the digital capture of wounds on day 4 showed that anisotropic AgNPs gel prevented pus formation and reduced scar appearance within 21 days. The histology results exhibit improved collagen fiber alignment that supports scar disappearance. In conclusion, these results indicate that anisotropic AgNPs gels are suitable for treating infected wounds. The gel is effective in eliminating bacteria that supports the natural process of wound repair and also causes reduced scar formation.

Keywords: *Staphylococcus pseudintermedius*; wound infection; antimicrobial resistance; alternative antimicrobial agent; nanotechnology; silver nanoparticles; scar reduction

1. Introduction

Bacterial infections can easily occur when animals have a wound that breaks the skin. Contamination at the wound site leads to a delay of wound healing and permits bacteria to spread to other organs. In animals in particular, wounds caused by bites are a common way that animals can get a bacterial infection, because their teeth are covered in bacteria [1]. Therefore, if a wound is caused by bites, they are usually more serious than they look and can easily become infected. Wound infections in animals are usually caused by opportunistic pathogens that are the normal microflora of the skin [2]. Similarly to *Staphylococcus aureus*, which is among normal flora on human skin [3], *S. pseudintermedius* is harmless in healthy individuals, but it is an opportunistic pathogen if an animal gets injured or sick [4]. These bacteria are one of the most common pathogens causing wound infection in animals [5].

S. pseudintermedius is a gram-positive cocci present in the skin and mucosa of animals [6]. Since it is an opportunistic pathogen in most animal species, it is an important pathogen in veterinary medicine. *S. pseudintermedius* has been isolated as a normal inhabitant of the skin, mucosa, nares, mouth, pharynx, forehead, groin and anus of healthy dogs, cats, horses, etc. [7,8]. In fact, 84.7% of all *S. pseudintermedius* isolates originate from canines [5]. *S. pseudintermedius* is a major cause of skin and ear infections, infections of other body tissues, cavities and wound infections in dogs and cats [6,9,10], and has become a pathogenic bacterium of veterinary concern, since a noted property of *staphylococci* is their ability to become resistant to antimicrobials [11,12]. It has been reported that up to 97.8% of Methicillin-resistant *S. pseudintermedius* (MRSP) isolates exhibit multidrug resistance (MDR) to three or more antibiotics routinely used in veterinary medicine [13,14]. Moreover, the bacteria is a zoonotic pathogen that can transmit to and infect humans [15,16]. Therefore, alternative antimicrobial agents to treat MRSP infected wounds are urgently needed in cases of wound infection in animals.

Silver nanoparticles have been used as a topical agent in various formulations for several hundred years; they have been used in applications as diverse as photographics, conductive/antistatic composites, catalysts, biocides and as a wound treatment. The physical, chemical and biological properties of AgNPs depend on their size and shape [17], and in addition, stabilizers are also important to these properties [18,19]. Due to their nanoscale size and the various shapes of AgNPs, they exhibit good antimicrobial action against bacteria, fungi and viruses with low cytotoxicity to mammalian cells [20]. Furthermore, the wound healing activity of AgNPs has been reported in many research studies. AgNPs promote wound healing through anti-inflammatory properties, as indicated by reduced cytokine release, decreased lymphocyte and mast cell infiltration and inhibition of matrix metalloproteinase 2 and 9 [21,22]. In addition to good antimicrobial activity and wound healing properties, AgNPs are also suitable for application as antimicrobial and wound healing agents owing to: (1) Bacterial resistance to AgNPs is extremely rare, due to the presence of multiple bactericidal mechanisms [23]; (2) The fact that the surface of AgNPs can be easily modified [24]; (3) It is easy to synthesize using large scale, inexpensive, safe and reliable methods, including chemical, physical and biological methods [25]; (4) It is easy to synthesize in different shapes (spheres, rods, hexagons, triangles, tubes, plates, cubes) [26].

Owing to AgNPs having various sizes and shapes, they can be divided into two types following their shape; silver nanospheres are spherical in shape, whereas anisotropic AgNPs are non-spherical shapes of AgNPs, such as truncated triangle, hexagon, and rod. Anisotropic AgNPs show various physical, chemical, and biological properties, such as color of colloid solution, conductivity, antimicrobial action and cytotoxicity [27]. Due to their outstanding optical properties and electrical conductivity, anisotropic AgNPs are usually used for catalysts, sensors and imaging [28,29]. However, previous reports show anisotropic AgNPs also have a good antimicrobial effect [30,31]. Moreover, many reports display that the antimicrobial activity of anisotropic AgNPs as being higher than silver nanospheres [32]. Anisotropic AgNPs, such as truncated triangular, display the strongest antibacterial action, compared with spheres [33]. Another report has shown that silver nanoplatelets with triangular or circular shape are more toxic towards *S. aureus* than spherical and rod shapes [34]. In our previous study, anisotropic AgNPs, an appropriate mixture of nanospheres and nanoplates (i.e., circular disk, truncated triangle and hexagon), were synthesized via transformation of silver nanospheres with an optimum concentration of hydrogen peroxide. The anisotropic AgNPs have been presented with antimicrobial efficiency against animal and human pathogens with low cytoxicity to human cell lines [35]. Moreover, we have developed an antimicrobial gel containing AgNPs as an easy way to use an anisotropic AgNP-based antimicrobial formulation for topical use, such as for infected wounds. The gel can inhibit the growth of the bacterial pathogens and shows long-lasting protection against bacteria, compared with povidone iodine. Hence, in this study we evaluated antimicrobial and wound healing activities of gel containing anisotropic AgNPs toward wounds contaminated with MRSP in a mice model to assess the antimicrobial and wound healing efficacy of anisotropic AgNPs gel for alternative applications on infected wounds in animals as well as to expand their application to humans.

2. Materials and Methods

2.1. Animals and Animal Care

The adult female BALB/cAJcl mice of 6 weeks old (17 to 25 g) that were used in this study were supplied by Nomura Siam International Co., Ltd., Bangkok Thailand. Animals were housed in specially designed cages with a size of $26.6 \times 42.5 \times 21$ cm (6 mice per cage), in a room with a controlled temperature (23 ± 2 °C), 30–60% RH, and 12 h light/dark cycle, with specific pathogenic-free conditions, at the Northeast Laboratory Animal Center, Khon Kaen University. The experimental protocol was approved by the animal ethics committee of Khon Kaen University, Khon Kaen, Thailand (Animal Ethics KKU 3/61).

2.2. Bacterial Strain and Culture Media

MRSP MIC 411 (Exudate from dog wound) were kindly provided by the Department of Microbiology, Faculty of Veterinary Science, Chulalongkorn University. The bacteria were identified by the Vitek 2 Compact System (bioMérieux, Marcy l'Etoile, France). The oxacillin-resistant isolates (Minimum inhibitory concentration (MIC) ≥ 4 μg/mL) further underwent mecA gene existence test using the PCR method [36]. The bacteria was streaked on Mueller-Hinton agar (MHA) and incubated at 37 °C for 24 h. A colony was selected and inoculated with 5 mL of Mueller-Hinton broth (MHB, HiMedia Laboratories Pvt. Ltd., Bengaluru, India) at 37 °C overnight. Then, bacteria was sub-cultured in 5 mL of the same medium at 37 °C in a 180-rpm shaker-incubator for 3 h to yield a mid-logarithmic growth phase culture.

2.3. Anisotropic AgNPs Gel Preparation

Tannic acid-stabilized anisotropic AgNPs were provided by our collaborator Prime Nanotechnology Co., Ltd. (Bangkok, Thailand) with a stock concentration of 100 mg/L. Anisotropic AgNPs were synthesized by transformation of silver nanospheres with an optimum concentration of hydrogen peroxide. For anisotropic AgNPs solution (a stock concentration of 100 mg/L) preparation, anisotropic AgNPs were prepared in sterile deionized water. In order to use anisotropic AgNPs in the form of a gel, poly(acrylic acid) (Sigma-Aldrich, St. Louis, MO, USA) was used to prepare the gel. Poly(acrylic acid) was chosen due its ease of preparation. Additionally, they have many advantages, such as good biocompatibility, keeping the wound moist and the ability to absorb many times their weight in water [37–39]. Briefly, 0.2 g poly(acrylic acid) was dissolved in 50 mL of sterile deionized water, followed by the addition 1 mL of glycerol (Sigma-Aldrich, St. Louis, MO, USA). After that, anisotropic AgNPs were added into the solution with final anisotropic AgNPs concentrations of 40 μg/mL (The minimum inhibitory concentration of anisotropic AgNPs against bacteria). Then, 0.2 g of triethanolamine (Sigma-Aldrich, St. Louis, MO, USA) was added to adjust pH and allow poly(acrylic acid) to form a gel. In the final step, sterile deionized water was added into the gel to make 100 g gel, and it was stirred well. The anisotropic AgNPs gel was stored at room temperature (27–30 °C) until use. The antimicrobial activity of anisotropic AgNPs gel was evaluated by diffusion test before application to mice. Briefly, the bacteria at an inoculum of 1×10^7 CFU/mL was swabbed on an MHA plate. A hole with a diameter of 6 mm was punched aseptically with a sterile cork borer. Then, 0.04 mg/g of anisotropic AgNPs gel was filled into the wells at 0.2 g in triplicate and incubated at 37 °C for 24 h. Povidone iodine (Leopard medical brand Co., Ltd., Nakorn Pathom, Thailand) was used as the positive control, and the gel without AgNPs as the negative control. After 24 h of incubation, the inhibition zone was measured with the millimeter (mm) scale.

2.4. Anesthesia and Wounding

All mice were anesthetized with intraperitoneal injection of a ketamine hydrochloride (PubChem, Bethesda, MD, USA) (90 mg/kg) and xylazine (Bimeda, Cambridge, ON, Canada) (10 mg/kg) saline cocktail [40]. The skin areas of the backs of the mice were cleaned with 70% ethanol (Siribuncha Co., Ltd., Nonthaburi, Thailand), povidone iodine (Leopard Medical Brand Co., Ltd., Nakhon Pathom, Thailand) and then 70% ethanol before wound creation. Wounds (8 mm in diameter) were made on the dorsal midline using sterile tissue scissors with curved blades. *S. pseudintermedius* MIC 411 were prepared to 10^7–10^8 CFU/mL in 5 mL of sterile Phosphate-Buffered Saline (PBS). Fifty microliters of the bacterial suspension were added to each wound bed immediately after wound surgery. After 2 days of wound creation, infected mice were divided into 3 groups ($n = 6$) including negative control (gel without AgNPs), positive control (povidone iodine; Leopard Medical Brand Co., Ltd.), and test group (anisotropic AgNPs gel). The mice had one agent applied once a day, every day for 14 days.

2.5. Measurement of Wound Infection and Wound Size

Wound infection was performed using the serial dilution plate count method of bacterial load at the wound site on days 0, 4, 7, 10 and 14 before application of antimicrobial agents. Briefly, the wound sites were swabbed by cotton swab and kept in PBS buffer pH 7.4. Each sample solution was brought to 50 µL for a serial ten-fold dilution plate count (10^{-1}–10^{-8} conc.) with sterile PBS buffer in triplicate. Then, 10 µL of each dilution was dropped on MHA and incubated at 37 °C overnight to count the bacterial colonies formed. In addition, all wounds were digitally photographed on days 1, 4, 7, 10, 14 and 21 to assess wound healing by the naked eye. Moreover, wound size was determined in diameter by using a vernier caliper on days 0, 4, 7, 10 and 14. Determinations were performed for wound area (mm^2) using length and width measurements. Wound contraction was calculated using the following formula:

$$Wound\ contraction\ (\%) = \frac{A_0 - A_t}{A_0} \times 100 \tag{1}$$

where A_0 is the initial wound area and A_t is the area of the wound at the time of image capture [41].

2.6. Histological Analysis

The animals were euthanized via CO_2 inhalation on days 14 or 21, followed by immediately taking tissue from the wound area. The tissues of 0.5×0.5 mm^2 were fixed in 10% neutral buffered formalin (NBF) (RCI Labscan Ltd., Bangkok, Thailand) for 24 h and then embedded in paraffin. Sections were cut at a thickness of 5 µm and stained with hematoxylin and eosin (H&E) (Sigma-Aldrich, St. Louis, MO, USA). Evaluations were made under a light microscope.

2.7. Statistical Analysis

The bacteria load in the wound area and wound contraction data were expressed as the mean ± standard deviation. Analysis of variance with Tukey's test was used for multiple comparisons. A *p*-value of less than 0.05 was considered significant.

3. Results

3.1. Antimicrobial Test of Anisotropic AgNPs Gel

The characteristics of anisotropic AgNPs gel are shown in Figure 1. Control gel (without AgNPs) was colorless, while the gel contained anisotropic AgNPs was dark orange. Anisotropic AgNPs gel remained dark orange and was still stable after storage at room temperature after 1 year (Figure 1c). To evaluate antimicrobial activity of anisotropic AgNPs gel before in vivo testing, an antimicrobial screening test was performed by the diffusion method. As shown in Figure 2, anisotropic AgNPs gel presented an inhibition zone that indicated antimicrobial activity toward the bacteria. The antimicrobial activity of anisotropic AgNPs gel was tested after storage at room temperature for a year. The testing exhibited an inhibition zone at both 24 h and 1 year of storage. The average size of the inhibition zone had no significant differences at 16.37 ± 1.09 mm and 15.67 ± 0.98 mm when the gel was stored for 24 h and 1 year, respectively. This indicated that the anisotropic AgNPs gel still has antimicrobial activity even after storage for 1 year at room temperature.

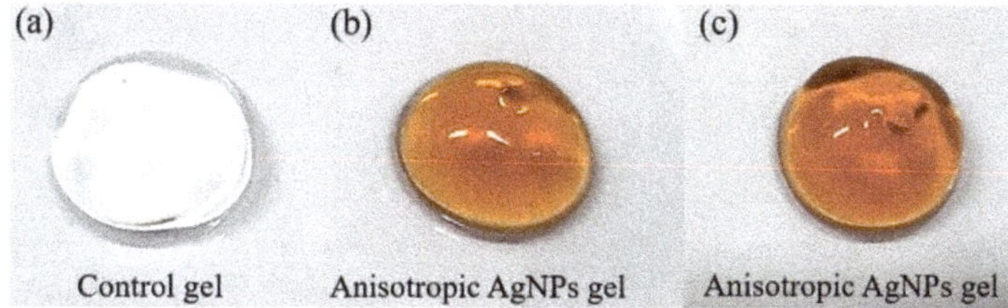

Figure 1. Anisotropic AgNPs containing formulation. Gel without AgNPs (**a**), Gel containing 0.04 mg/g of anisotropic AgNPs stored at room temperature (27–30 °C) for 24 h (**b**) and 1 year (**c**).

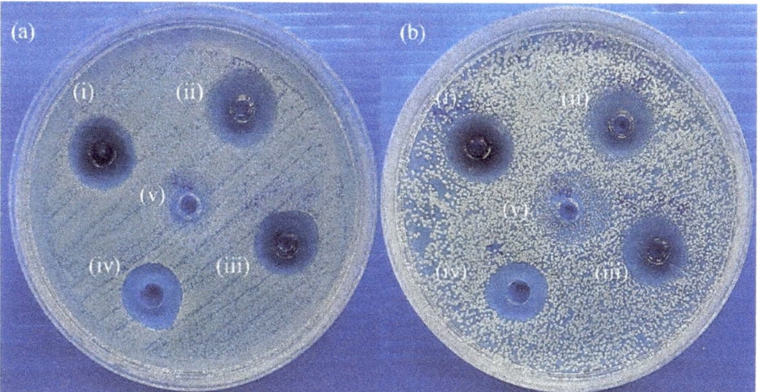

Figure 2. Antimicrobial evaluation of anisotropic AgNPs gel. Antimicrobial test of anisotropic AgNPs gel toward MRSP MIC 411 after preparation for 24 h (**a**), and after storage at room temperature for 1 year (**b**). Inset: anisotropic AgNPs gel (i–iii), povidone iodine (iv), control gel (v).

3.2. Microbial Loads in Wounds

To assess the antimicrobial activity of anisotropic AgNPs gel on wound infection, MRSP were infected with the initial inoculum of approximately 10^7–10^8 CFU/mL (Figure 3). The agents were applied after 48 h of infection. The results show that bacterial cells in the group treated with anisotropic AgNPs gel and povidone iodine were significantly less than the control group on day 4 and day 7 (* $p < 0.05$, ** $p < 0.001$). In the same way, these results show a continuous decrease of bacterial cells on day 0 to day 7 for groups that had anisotropic AgNPs gel applied and povidone iodine. On the other hand, the bacterial cells in the control group (Gel without AgNPs) continuously decreased from day 0 to day 10. These results exhibit that anisotropic AgNPs gel and povidone iodine can eradicate all the bacterial cells within 7 days, while the control group could eliminate all bacterial cells within 10 days.

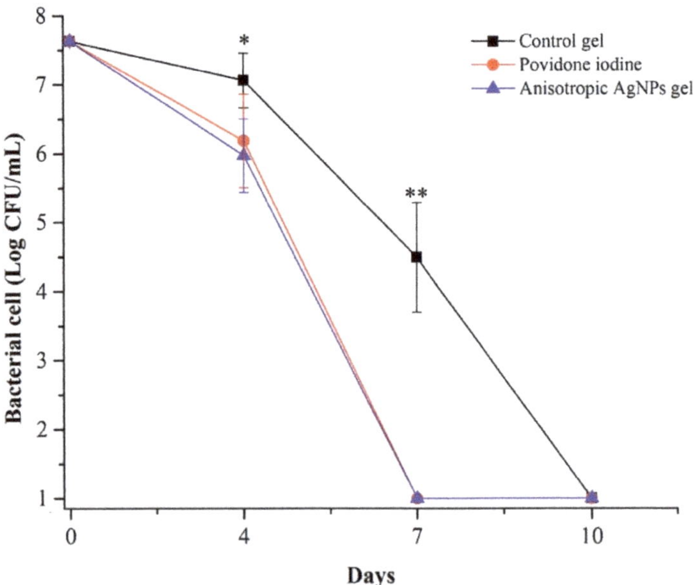

Figure 3. Bacteria load average in wound area. MRSP MIC 411 was loaded into the wound at 10^7–10^8 CFU/mL. The bacteria were evaluated on days 0, 4, 7 and 10. Data represent the mean value ± SD (error bar) of 6 mice ($n = 6$). * $p < 0.05$ and ** $p < 0.001$ are used as significant differences in comparison of the control gel with other groups.

3.3. Wound Closure

We observed the wound healing activity of anisotropic AgNPs gel by quantitative measurement of the wound area on days 0, 4, 7, 10 and 14. The wound area on each day was compared with the initial wound area. Moreover, the wound healing was investigated by capturing digital photographs on days 1, 4, 7, 10, 14 and 21 of the experiment to evaluate wound features and scar formation. In Figure 4, wounds were made up to 8 mm in diameter, followed by adding bacterial cells at 10^7–10^8 CFU/mL. The results show that all groups had an efficiency of wound contraction, which appeared within 14 days. Within all groups, there was not a significant percentage of wound contraction on days 4, 7, 10 and 14. Similarly, digital photographs of wounds showed complete healing on day 14 for all groups. Interestingly, there was only one group that showed no visible scarring on day 21 (Figure 5) and that was the group treated with anisotropic AgNPs gel. From the foregoing, there was no difference in wound contraction observed among all groups within 14 days. However, scars were better in the group treated with anisotropic AgNPs gel. Moreover, it is noteworthy that on day 4 of wound creation, the control group revealed highly purulent wounds, while other groups did not show pus on wounds.

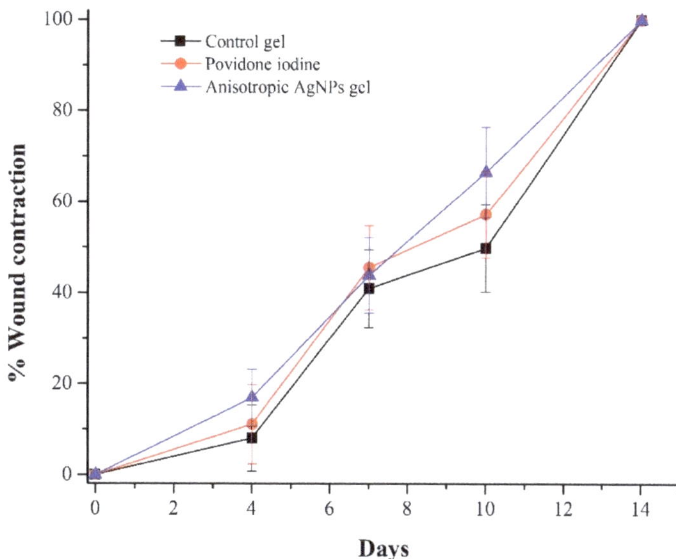

Figure 4. Wound contraction of experimental groups post treatment. Wound contraction (%) = ((wound area on day 0-wound area on day X)/(wound area on day 0)) × 100. Wounds were created 8 mm in diameter. After the wounds were created, they were left to infect with bacteria for 48 h. Control gel (without anisotropic AgNPs), povidone iodine, and anisotropic AgNPs gel were applied to the wounds on day 3 and continuously until the wounds had completely contracted. Wound size was measured on days 4, 7, 10, and 14 to calculate wound contraction. Data represent the mean value ± SD (error bar) for 6 mice ($n = 6$).

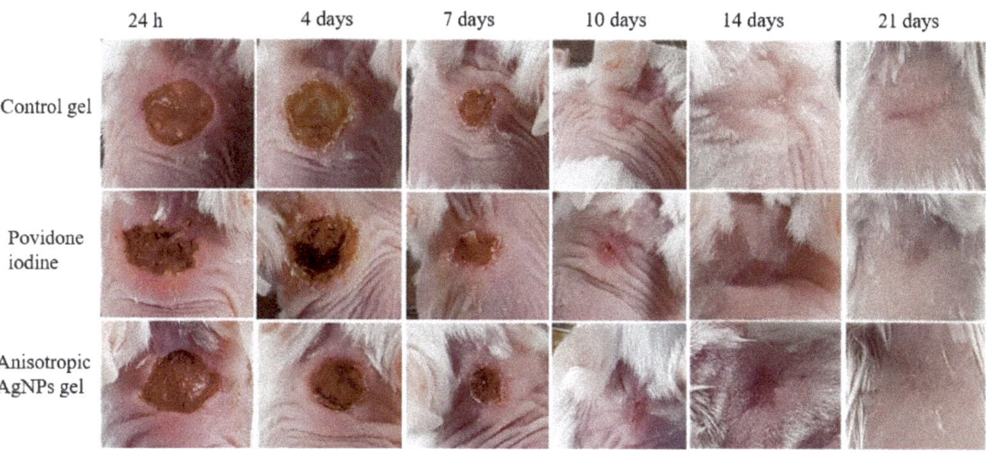

Figure 5. Photographs of wound repair at different times after excision of the wound model in mice. Effects of control gel (without anisotropic AgNPs), povidone iodine and anisotropic AgNPs gel on the healing of MRSP wound infection in mice. Representative images of mice from groups taken on days 1, 4, 7, 10, 14 and 21 after creation of the wound are shown. Mice were given a wound size 8 mm in diameter and bacterial cells added at 10^7–10^8 CFU/mL. After 48 h, the agents were applied to the wound.

3.4. Histology

We evaluated histological sections of wound tissue on days 14 and 21 to support the healing properties of all groups. Hematoxylin and eosin were used to observe cells and tissue such as immune cells, fibroblast and collagen which implied wound status. Normal mouse skin was used as the control that indicated healthy skin. In Figure 6, normal skin showed an integrity of four layers, including epidermis (top layer skin), dermis (next to the top layer), sub-cutaneous fat (next to dermis) and muscle (deepest layer). In addition, there were some fibroblast cells inserted into the collagen fiber in the dermis layer. These are normal characteristics of mice skin. In other groups, including control gel (without AgNPs), povidone iodine and anisotropic AgNPs gel, mice tissue exhibited two layers of epidermis and dermis. They presented a lot of fibroblast cells on day 14. Moreover, the epidermis layer of all groups clearly appeared very swollen, when compared with normal skin.

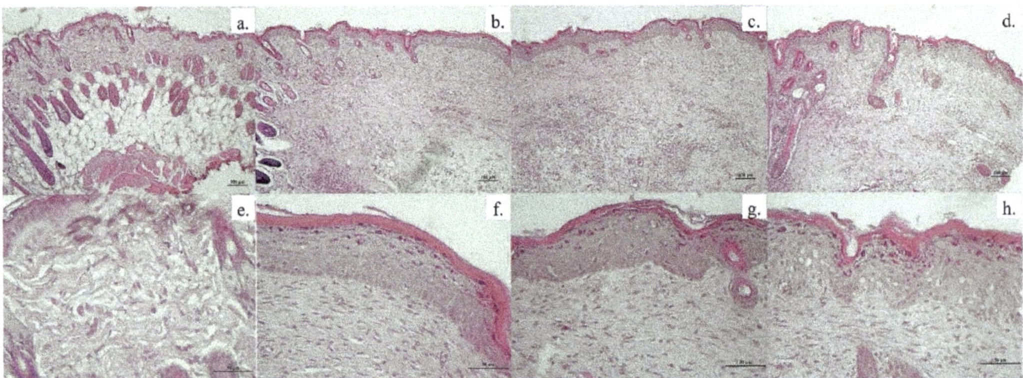

Figure 6. Photomicrographs of histological section of wound tissue of mice. Histology of wounds on day 14, stained with hematoxylin-eosin. Normal tissue (**a,e**), control gel (**b,f**), povidone iodine (**c,g**), anisotropic AgNPs gel (**d,h**) at scale bar of 100× (upper row) and 50× (lower row).

On day 21, as shown in Figure 7, the fibroblast cells in all groups exhibited a decrease when compared to day 14. Moreover, collagen fiber can be observed more clearly. Interestingly, the group treated with anisotropic AgNPs gel displayed collagen fiber that was the most obvious and the most similar to normal skin. In contrast, groups treated with control gel and povidone iodine exhibited too-tight collagen fibers when compared to normal skin. This collagen formation is also related to the amount of fibroblast cells that showed in each group (Figure 7e,f). Such results are consistent with photographs of mice skin without scars on day 21 in the group treated with anisotropic AgNPs gel (Figure 5).

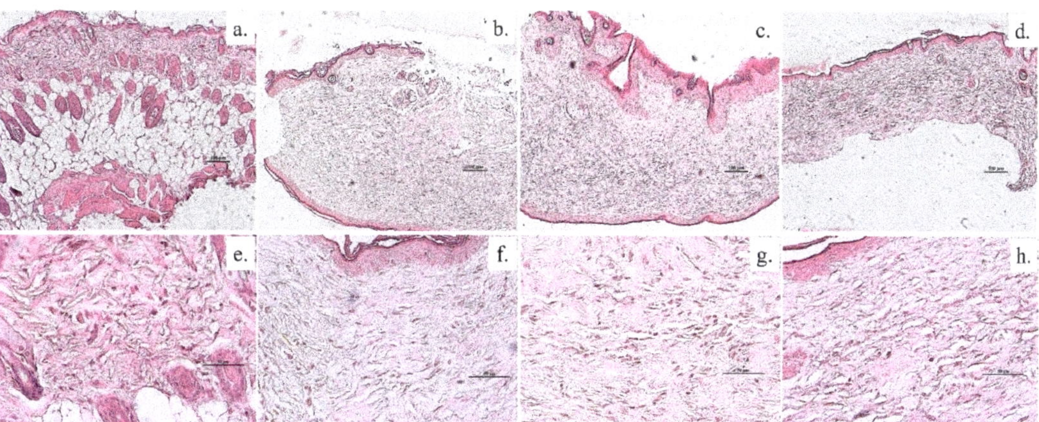

Figure 7. Photomicrographs of histological sections of mice wound tissue to evaluate collagen arrangement. Histology of wound tissue on day 21, stained with hematoxylin-eosin. Normal tissue (**a**,**e**), control gel (**b**,**f**), povidone iodine (**c**,**g**), anisotropic AgNPs gel (**d**,**h**) at scale bar of 100× (upper row) and 50× (lower row).

4. Discussion

S. pseudintermedius is an opportunistic pathogen in most animal species and is an important pathogen in veterinary medicine. Furthermore, the bacteria is a zoonotic pathogen that can transmit to and infect humans, and infection of the bacteria in wounds results in delayed wound healing. In this study, we made a gel that contained anisotropic AgNPs to overcome wound infection. In our previous study, we have shown that anisotropic AgNPs gels have antibacterial activities and prolonged antibacterial activity against animal and human pathogens in vitro [35]. Moreover, anisotropic AgNPs have antimicrobial activity as good as silver nanospheres, but with lower cytotoxicity [35]. Therefore, anisotropic AgNPs were used for this study. In this study, we reported in vivo capabilities of anisotropic AgNPs that appear to promote bacterial killing and improve wound healing, when infected with MRSP in a mice model of a skin wound.

The gel containing anisotropic AgNPs had antimicrobial efficiency and could be stored at room temperature for a year without reduction of antimicrobial efficacy. This result is the same as in a previous study, where AgNPs gel has a good antimicrobial activity and low cytotoxicity [22,42]. To assess the antimicrobial and wound healing efficacy of anisotropic AgNPs gels in vivo, BALB/cAJcl mice were used as a model of wound infection. Mice were given a wound that was infected by methicillin-resistant *S. pseudintermedius*. After 2 days of wound creation, antimicrobial agents were applied and antibacterial, as well as wound healing, activities were determined. The results of antibacterial tests show that the groups treated with anisotropic AgNPs gel can eliminate bacteria within 7 days, the same as povidone iodine, while the control group (gel without anisotropic AgNPs) needed 10 days to remove all bacteria from the wound. Owing to there being no previous reports on the use of AgNPs or AgNPs gel against *S. pseudintermedius* infected wounds in vivo, we can only compare the antibacterial efficacy with the reported results of pathogenic bacteria such as *S. aureus* that share several features with *S. pseudintermedius* [4]. We found that anisotropic AgNPs gel can eliminate *S. pseudintermedius* faster than those observed on *S. aureus* in the past. For example, AgNPs can eliminate all *S. aureus* cells infecting the wound on day 21 after wound induction [40]. In another study, *S. aureus* infected a burn wound, and it was found that AgNPs eliminated bacteria within 17 days after the induction of the burn wound [43]; indeed, the different efficiency at eliminating bacteria might be caused by the bacteria's ability to cause disease. In this case, *S. pseudintermedius* MIC 411 mostly infects dog wounds and exhibits multidrug resistance [36]. These bacteria could contaminate a mice wound, but they might not cause infection similar to an infected dog wound. As shown in Figure 3, the mice's immunity can eliminate bacteria within 10 days without applying antimicrobial agents. However, anisotropic AgNPs gel can support faster bacterial elimination. This point could be an advantage in the case of wound infection in which animals cannot eradicate bacteria by their immune system.

Wound healing efficiency was assessed by the measurement of wound contraction. The results show that all groups were not different on wound healing efficiency, with complete wound contraction within 14 days. In general, mice wounds of size 8 mm in diameter can heal themselves within 14 days [44]. As observed in this study, infection of wound with *S. pseudintermedius* might not delay the wound healing process as shown by the antibacterial test (Figure 3), that mice can remove all bacterial cells within 10 days. The reason is that *S. pseudintermedius* might contaminate the wound, but they cannot replicate and disrupt the wound healing process. Therefore, mice can remove bacteria by their immune system and the results show that wound contraction was not different among all groups; however, even though the wound healing efficiency of anisotropic AgNPs gel was not different from the control group, it was noteworthy that on day 4 of wound creation in the groups treated with anisotropic AgNPs gel, images showed reduction of purulent wounds, similar to the groups treated with povidone iodine, while the control group showed highly purulent wounds. This indicates that anisotropic AgNPs gel prevents pus on wounds, which might be caused by reduction of bacterial cells and inflammation similar to povidone iodine [45]. Our results were similar to a previous report by P. F. Myronov et al., (2019), that AgNPs play an important role in cleaning wounds from purulence by decreasing the microbial contamination and the number of neutrophils and white blood cells [46]. In addition, there are many previous research reports that confirm the wound healing activity of AgNPs by reduction of inflammation through cytokine modulation and inhibition of MMP-2 and MMP-9 [22,47,48]. In another study, AgNPs reduced the levels of the prototypic cytokines, tumor necrosis factor (TNF) and interleukin (IL-6) [49]. Moreover, AgNPs have shown decreased mitochondrial function and induction of apoptosis or apoptosis-like changes of cell morphology [50–52]. However, these points could be studied further in the future to confirm the wound healing action of anisotropic AgNPs.

Besides prevention of purulent wounds, it was clearly noticed in groups treated with anisotropic AgNPs gel that there was negligible scarring on day 21. That anisotropic AgNPs promote scarless wound healing, as found in our study, is similar to the study of Jun Tian et al. (2007); they have demonstrated that AgNPs act directly on dampening the process of inflammation, thus promoting scarless wound healing [47]. In addition, Jaya Jain et al. (2009) reported that, apart from antimicrobial activity, AgNPs inhibited MMPs and promoted scarless wound healing [22]. Moreover, V. Dhapte et al. (2014) proved that AgNPs presented scarless wound recovery without inflammation, because of effective cytokine modulation [53]. The scarless wound healing result corresponds to the histology result; mice given anisotropic AgNPs gel showed collagen fiber similar to normal skin on day 21. Collagen fiber is also important owing to the orientation and arrangement of collagen playing a key role during the remodeling phase, and consequently on the final scar appearance after wound closure. In this case, we could explain our result by the previous study of Karen H.L. Kwan et al. (2011), who pointed to the improvement of collagen properties of AgNPs that led to better fibril alignments in repaired skin, with a close resemblance to normal skin. AgNPs were predominantly responsible for regulating deposition of collagen, and their use resulted in excellent alignment in the wound healing process [54]. The collagen fiber alignment in this study supports scarless wound healing in the group treated with anisotropic AgNPs gel, and could be explained by scientific evidence since it had been previously studied. Nevertheless, the mechanism of collagen fiber alignment could not be elucidated in this study.

Based on the present results, we summarize the wound healing process of anisotropic AgNPs gel against wounds contaminated with MRSP, as illustrated in Figure 8. For the wound after 2 days of applying agents, the number of bacteria in the groups treated with anisotropic AgNPs gel and povidone iodine are reduced, and the wound exhibits no purulence (Figures 3 and 5), whereas the control group shows purulent wounds and exhibits more bacterial cells than other groups. The prevention of purulent wounds in the groups treated with anisotropic AgNPs gel and povidone iodine could be caused by a decrease in bacterial cells. On day 7, all the bacterial cells were eliminated in the groups treated with anisotropic AgNPs gel and povidone iodine, while the control group still had bacterial cell contamination (Figure 3). Next, the wound shows complete contraction within 14 days for all groups (Figure 4). This indicates that anisotropic AgNPs gel might not promote the wound contraction and quicken the healing rate. Interestingly, in the groups treated with anisotropic AgNPs gel, the wound shows scar disappearance as a result of collagen alignment within 21 days (Figures 6 and 7), while the control group and povidone iodine group show scar formation. These imply that anisotropic AgNPs gel could affect the orientation and arrangement of collagen fiber.

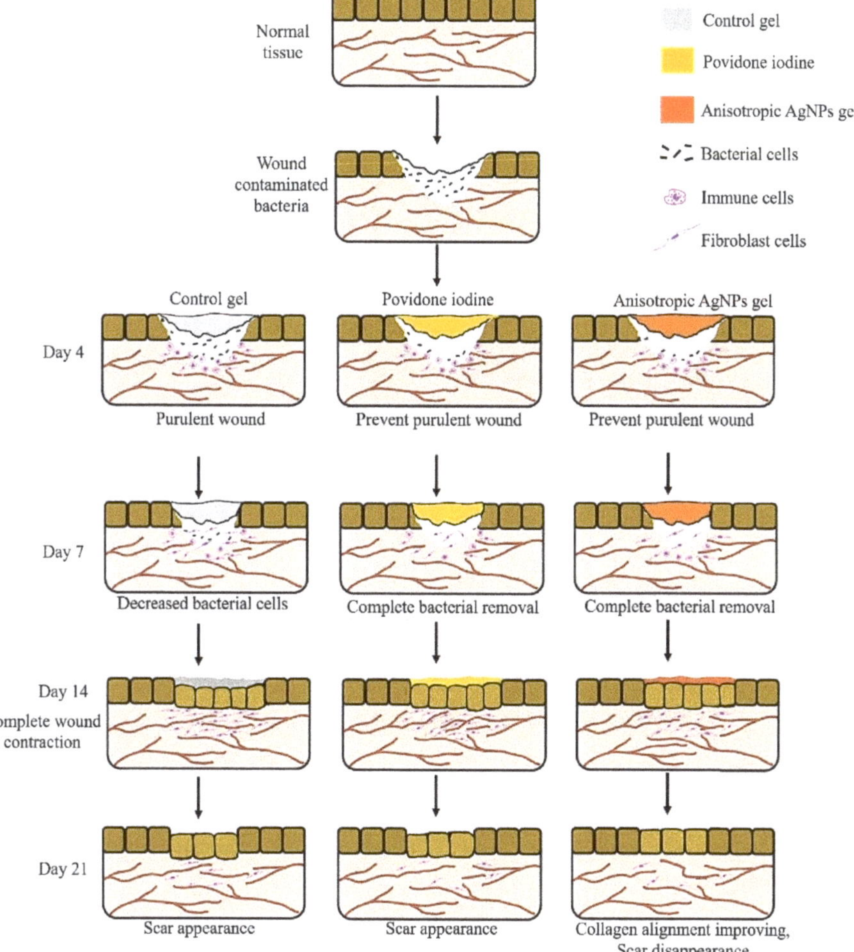

Figure 8. Wound healing process schematic of anisotropic ANPs gel. The anisotropic AgNPs gel reduced wound purulence after applying gel for 2 days, whereas the control gel exhibited high pus formation. After that, anisotropic AgNPs gel can remove all bacterial cells within 7 days and there is complete wound contraction within 14 days. Finally, improving collagen alignment by anisotropic AgNPs gel shows scar disappearance within 21 days.

5. Conclusions

Wound infection in animals such as dogs and cats with antimicrobial resistant bacteria, especially *S. pseudintermedius*, plays an important role in the delay of wound healing. In a previous study, we showed that anisotropic AgNPs have antimicrobial efficacy against animal and human pathogens, and also anisotropic AgNPs have low toxicity to human cells *in vitro*. Therefore, antimicrobial activity and wound healing activity of anisotropic AgNPs *in vivo* were evaluated in this study. Based on the results, even though anisotropic AgNPs gel might not promote the wound contraction and quicken the healing rate, we have proved that anisotropic AgNPs gel can eliminate bacterial cells on infected wounds and reduce pus, as well as showing collagen alignment in repaired skin with a close resemblance to normal skin. This is important evidence to support scarless wound healing in mice, and this research indicates that anisotropic AgNPs gel could be suitable for treating infected wounds. Our findings support the potent antibacterial activities and scar reduction of anisotropic AgNPs gel as antibacterial agents in infected wounds. These gels can prevent the proliferation and colonization of opportunistic bacteria, and finally promote natural wound healing along with reduction of scar occurrence.

Author Contributions: Conceptualization, S.T. and R.P.; methodology, S.T. and O.S.; validation, R.P., S.K. and N.P.; formal analysis, S.T.; investigation, S.T., O.S. and P.S.; resources, R.P. and S.D.; data curation, S.T.; writing—original draft preparation, S.T.; writing—review and editing, R.P. and N.P.; visualization, S.T.; supervision, R.P.; project administration, R.P.; funding acquisition, R.P. and S.D. All authors have read and agreed to the published version of the manuscript.

Funding: This research was funded by Research and Researchers for Industries (RRi) by the Thailand Research Fund (TRF), Bangkok, Thailand, grant number PHD59I0052.

Institutional Review Board Statement: The experimental protocol was approved by the animal ethics committee of Khon Kaen University, Khon Kaen, Thailand (Animal Ethics KKU 3/61).

Informed Consent Statement: Not applicable.

Data Availability Statement: The data presented in this study are available upon request from the corresponding author.

Acknowledgments: We would like to thank Research and Researchers for Industries (RRi) by the Thailand Research Fund (TRF), Bangkok, Thailand for supporting the scholarship. Many thanks to the Northeast Laboratory Animal Center, Khon Kaen University, Khon Kaen, Thailand for providing us with facilities for animal testing. The authors wish to thank our collaborator Prime Nanotechnology Co., Ltd. (Bangkok, Thailand) for kindly giving anisotropic AgNPs. *S. pseudintermedius* MIC 411 were kind gifts from Nuvee Prapasarakul, Department of Veterinary Microbiology Faculty of Veterinary Science, Chulalongkorn University. We would also like to thank Ian Thomas, a retired lecturer from the Department of Physics, Faculty of Science, Khon Kaen University, Khon Kaen, Thailand, for the linguistic correction.

Conflicts of Interest: The authors declare no conflict of interest.

References

1. Oh, C.; Lee, K.; Cheong, Y.; Lee, S.-W.; Park, S.-Y.; Song, C.-S.; Choi, I.-S.; Lee, J.-B. Comparison of the oral microbiomes of canines and their owners using next-generation sequencing. *PLoS ONE* **2015**, *10*, e0131468. [CrossRef]
2. Pompilio, A.; De Nicola, S.; Crocetta, V.; Guarnieri, S.; Savini, V.; Carretto, E.; Di Bonaventura, G. New insights in *Staphylococcus pseudintermedius* pathogenicity: Antibiotic-resistant biofilm formation by a human wound-associated strain. *BMC Microbiol.* **2015**, *15*, 109. [CrossRef]
3. Tong, S.Y.C.; Davis, J.S.; Eichenberger, E.; Holland, T.L.; Fowler, V.G., Jr. Staphylococcus aureus infections: Epidemiology, pathophysiology, clinical manifestations, and management. *Clin. Microbiol. Rev.* **2015**, *28*, 603–661. [CrossRef]
4. Bannoehr, J.; Guardabassi, L. *Staphylococcus pseudintermedius* in the dog: Taxonomy, diagnostics, ecology, epidemiology and pathogenicity. *Vet. Dermatol.* **2012**, *23*, 253–266. [CrossRef]
5. Ruscher, C.; Lübke-Becker, A.; Wleklinski, C.-G.; Soba, A.; Wieler, L.H.; Walther, B. Prevalence of methicillin-resistant *Staphylococcus pseudintermedius* isolated from clinical samples of companion animals and equidaes. *Vet. Microbiol.* **2009**, *136*, 197–201. [CrossRef] [PubMed]

6. Weese, J.S.; van Duijkeren, E. Methicillin-resistant staphylococcus aureus and *Staphylococcus pseudintermedius* in veterinary medicine. *Vet. Microbiol.* **2010**, *140*, 418–429. [CrossRef] [PubMed]

7. Devriese, L.A.; Vancanneyt, M.; Baele, M.; Vaneechoutte, M.; De Graef, E.; Snauwaert, C.; Cleenwerck, I.; Dawyndt, P.; Swings, J.; Decostere, A.; et al. *Staphylococcus pseudintermedius* Sp. Nov., a coagulase-positive species from animals. *Int. J. Syst. Evol. Microbiol.* **2005**, *55*, 1569–1573. [CrossRef] [PubMed]

8. Bardiau, M.; Yamazaki, K.; Ote, I.; Misawa, N.; Mainil, J.G. Characterization of methicillin-resistant *Staphylococcus pseudintermedius* isolated from dogs and cats. *Microbiol. Immunol.* **2013**, *57*, 496–501.

9. Abraham, J.L.; Morris, D.O.; Griffeth, G.C.; Shofer, F.S.; Rankin, S.C. Surveillance of healthy cats and cats with inflammatory skin disease for colonization of the skin by methicillin-resistant coagulase-positive staphylococci and staphylococcus schleiferi Ssp. schleiferi. *Vet. Dermatol.* **2007**, *18*, 252–259. [CrossRef]

10. Griffeth, G.C.; Morris, D.O.; Abraham, J.L.; Shofer, F.S.; Rankin, S.C. Screening for skin carriage of methicillin-resistant coagulase-positive staphylococci and staphylococcus schleiferi in dogs with healthy and inflamed skin. *Vet. Dermatol.* **2008**, *19*, 142–149. [CrossRef] [PubMed]

11. Ferran, A.A.; Liu, J.; Toutain, P.-L.; Bousquet-Mélou, A. Comparison of the in vitro activity of five antimicrobial drugs against *Staphylococcus pseudintermedius* and staphylococcus aureus biofilms. *Front. Microbiol.* **2016**, *7*, 1187. [CrossRef]

12. Jantorn, P.; Heemmamad, H.; Soimala, T.; Indoung, S.; Saising, J.; Chokpaisarn, J.; Wanna, W.; Tipmanee, V.; Saeloh, D. Antibiotic resistance profile and biofilm production of *Staphylococcus pseudintermedius* isolated from dogs in Thailand. *Pharmaceuticals* **2021**, *14*, 592. [CrossRef] [PubMed]

13. Wegener, A.; Broens, E.M.; Zomer, A.; Spaninks, M.; Wagenaar, J.A.; Duim, B. Comparative genomics of phenotypic antimicrobial resistances in methicillin-resistant *Staphylococcus pseudintermedius* of canine origin. *Vet. Microbiol.* **2018**, *225*, 125–131. [CrossRef]

14. Worthing, K.A.; Abraham, S.; Coombs, G.W.; Pang, S.; Saputra, S.; Jordan, D.; Trott, D.J.; Norris, J.M. Clonal diversity and geographic distribution of methicillin-resistant *Staphylococcus pseudintermedius* from australian animals: Discovery of novel sequence types. *Vet. Microbiol.* **2018**, *213*, 58–65. [CrossRef] [PubMed]

15. Van Hoovels, L.; Vankeerberghen, A.; Boel, A.; Van Vaerenbergh, K.; De Beenhouwer, H. First case of *Staphylococcus pseudintermedius* infection in a human. *J. Clin. Microbiol.* **2006**, *44*, 4609–4612. [CrossRef]

16. Robb, A.R.; Wright, E.D.; Foster, A.M.E.; Walker, R.; Malone, C. Skin infection caused by a novel strain of *Staphylococcus pseudintermedius* in a siberian husky dog owner. *JMM Case Rep.* **2017**, *4*, jmmcr005087. [CrossRef] [PubMed]

17. Agnihotri, S.; Mukherji, S.; Mukherji, S. Size-controlled silver nanoparticles synthesized over the range 5–100 Nm using the same protocol and their antibacterial efficacy. *RSC Adv.* **2014**, *4*, 3974–3983. [CrossRef]

18. Zewde, B. A review of stabilized silver nanoparticles—Synthesis, biological properties, characterization, and potential areas of applications. *JSM Nanotechnol. Nanomed.* **2016**, *4*, 1043.

19. Verkhovskii, R.; Kozlova, A.; Atkin, V.; Kamyshinsky, R.; Shulgina, T.; Nechaeva, O. Physical properties and cytotoxicity of silver nanoparticles under different polymeric stabilizers. *Heliyon* **2019**, *5*, e01305. [CrossRef]

20. Franci, G.; Falanga, A.; Galdiero, S.; Palomba, L.; Rai, M.; Morelli, G.; Galdiero, M. Silver nanoparticles as potential antibacterial agents. *Molecules* **2015**, *20*, 8856–8874. [CrossRef]

21. Gunasekaran, T.; Nigusse, T.; Dhanaraju, M.D. Silver nanoparticles as real topical bullets for wound healing. *J. Am. Coll. Clin. Wound Spec.* **2011**, *3*, 82–96. [CrossRef]

22. Jain, J.; Arora, S.; Rajwade, J.M.; Omray, P.; Khandelwal, S.; Paknikar, K.M. Silver nanoparticles in therapeutics: Development of an antimicrobial gel formulation for topical use. *Mol. Pharm.* **2009**, *6*, 1388–1401. [CrossRef] [PubMed]

23. Yin, I.X.; Zhang, J.; Zhao, I.S.; Mei, M.L.; Li, Q.; Chu, C.H. The antibacterial mechanism of silver nanoparticles and its application in dentistry. *Int. J. Nanomed.* **2020**, *15*, 2555–2562. [CrossRef]

24. Qing, Y.; Cheng, L.; Li, R.; Liu, G.; Zhang, Y.; Tang, X.; Wang, J.; Liu, H.; Qin, Y. Potential antibacterial mechanism of silver nanoparticles and the optimization of orthopedic implants by advanced modification technologies. *Int. J. Nanomed.* **2018**, *13*, 3311–3327. [CrossRef]

25. Iravani, S.; Korbekandi, H.; Mirmohammadi, S.V.; Zolfaghari, B. Synthesis of silver nanoparticles: Chemical, physical and biological methods. *Res. Pharm. Sci.* **2014**, *9*, 385–406. [PubMed]

26. Khodashenas, B.; Ghorbani, H.R. Synthesis of silver nanoparticles with different shapes. *Arab. J. Chem.* **2019**, *12*, 1823–1838. [CrossRef]

27. Sajanlal, P.R.; Sreeprasad, T.S.; Samal, A.K.; Pradeep, T. Anisotropic nanomaterials: Structure, growth, assembly, and functions. *Nano Rev.* **2011**, *2*, 5883. [CrossRef] [PubMed]

28. Bhol, P.; Bhavya, M.B.; Swain, S.; Saxena, M.; Samal, A.K. Modern chemical routes for the controlled synthesis of anisotropic bimetallic nanostructures and their application in catalysis. *Front. Chem.* **2020**, *8*, 357. [CrossRef]

29. Thiele, M.; Götz, I.; Trautmann, S.; Müller, R.; Csáki, A.; Henkel, T.; Fritzsche, W. Wet-chemical passivation of anisotropic plasmonic nanoparticles for LSPR-sensing by a silica shell. *Mater. Today Proc.* **2015**, *2*, 33–40. [CrossRef]

30. Osonga, F.J.; Akgul, A.; Yazgan, I.; Akgul, A.; Ontman, R.; Kariuki, V.M.; Eshun, G.B.; Sadik, O.A. Flavonoid-derived anisotropic silver nanoparticles inhibit growth and change the expression of virulence genes in escherichia coli SM10. *RSC Adv.* **2018**, *8*, 4649–4661. [CrossRef]

31. Tang, B.; Wang, J.; Xu, S.; Afrin, T.; Xu, W.; Sun, L.; Wang, X. Application of anisotropic silver nanoparticles: Multifunctionalization of wool fabric. *J. Colloid Interface Sci.* **2011**, *356*, 513–518. [CrossRef] [PubMed]

32. Sangappa, Y.; Latha, S.; Asha, S.; Sindhu, P.; Parushuram, N.; Shilpa, M.; Byrappa, K.; Narayana, B. Synthesis of anisotropic silver nanoparticles using silk fibroin: Characterization and antimicrobial properties. *Mater. Res. Innov.* **2019**, *23*, 79–85. [CrossRef]
33. Pal, S.; Tak, Y.K.; Song, J.M. Does the antibacterial activity of silver nanoparticles depend on the shape of the nanoparticle? A Study of the gram-negative bacterium Escherichia coli. *Appl. Environ. Microbiol.* **2007**, *73*, 1712–1720.
34. Helmlinger, J.; Sengstock, C.; Groß-Heitfeld, C.; Mayer, C.; Schildhauer, T.A.; Köller, M.; Epple, M. Silver nanoparticles with different size and shape: Equal cytotoxicity, but different antibacterial effects. *RSC Adv.* **2016**, *6*, 18490–18501. [CrossRef]
35. Thammawithan, S.; Siritongsuk, P.; Nasompag, S.; Daduang, S.; Klaynongsruang, S.; Prapasarakul, N.; Patramanon, R. A biological study of anisotropic silver nanoparticles and their antimicrobial application for topical use. *Vet. Sci.* **2021**, *8*, 177. [CrossRef]
36. Strommenger, B.; Kettlitz, C.; Werner, G.; Witte, W. Multiplex PCR assay for simultaneous detection of nine clinically relevant antibiotic resistance genes in staphylococcus aureus. *J. Clin. Microbiol.* **2003**, *41*, 4089–4094. [CrossRef]
37. Li, X.; Su, X. Multifunctional smart hydrogels: Potential in tissue engineering and cancer therapy. *J. Mater. Chem. B* **2018**, *6*, 4714–4730. [CrossRef] [PubMed]
38. Wang, H.; Xu, Z.; Zhao, M.; Liu, G.; Wu, J. Advances of hydrogel dressings in diabetic wounds. *Biomater. Sci.* **2021**, *9*, 1530–1546. [CrossRef]
39. Zheng, B.-D.; Ye, J.; Yang, Y.-C.; Huang, Y.-Y.; Xiao, M.-T. Self-healing polysaccharide-based injectable hydrogels with antibacterial activity for wound healing. *Carbohydr. Polym.* **2022**, *275*, 118770. [CrossRef]
40. Adibhesami, M.; Ahmadi, M.; Farshid, A.A.; Sarrafzadeh-Rezaei, F.; Dalir-Naghadeh, B. Effects of silver nanoparticles on staphylococcus aureus contaminated open wounds healing in mice: An experimental study. *Vet. Res. Forum Int. Q. J.* **2017**, *8*, 23–28.
41. Ahmadi, M.; Adibhesami, M. The effect of silver nanoparticles on wounds contaminated with pseudomonas aeruginosa in mice: An experimental study. *Iran. J. Pharm. Res. IJPR* **2017**, *16*, 661–669.
42. Jadhav, K.; Dhamecha, D.; Bhattacharya, D.; Patil, M. Green and ecofriendly synthesis of silver nanoparticles: Characterization, biocompatibility studies and gel formulation for treatment of infections in burns. *J. Photochem. Photobiol. B Biol.* **2016**, *155*, 109–115. [CrossRef]
43. Wasef, L.G.; Shaheen, H.M.; El-Sayed, Y.S.; Shalaby, T.I.A.; Samak, D.H.; Abd El-Hack, M.E.; Al-Owaimer, A.; Saadeldin, I.M.; El-Mleeh, A.; Ba-Awadh, H.; et al. Effects of silver nanoparticles on burn wound healing in a mouse model. *Biol. Trace Elem. Res.* **2020**, *193*, 456–465. [CrossRef] [PubMed]
44. Masson-Meyers, D.S.; Andrade, T.A.M.; Caetano, G.F.; Guimaraes, F.R.; Leite, M.N.; Leite, S.N.; Frade, M.A.C. Experimental models and methods for cutaneous wound healing assessment. *Int. J. Exp. Pathol.* **2020**, *101*, 21–37. [CrossRef]
45. Bigliardi, P.L.; Alsagoff, S.A.L.; El-Kafrawi, H.Y.; Pyon, J.-K.; Wa, C.T.C.; Villa, M.A. Povidone iodine in wound healing: A review of current concepts and practices. *Int. J. Surg.* **2017**, *44*, 260–268. [CrossRef] [PubMed]
46. Myronov, P.; Bugaiov, V.; Tymakova, O.; Maksym, P.; Opanasyuk, A. Cytological examination of experimental purulent wounds in the treatment of silver nanoparticles in ultrasound cavitation. *East. Ukr. Med. J.* **2019**, *7*, 386–395. [CrossRef]
47. Tian, J.; Wong, K.K.Y.; Ho, C.-M.; Lok, C.-N.; Yu, W.-Y.; Che, C.-M.; Chiu, J.-F.; Tam, P.K.H. Topical delivery of silver nanoparticles promotes wound healing. *ChemMedChem* **2007**, *2*, 129–136. [CrossRef]
48. Hebeish, A.; El-Rafie, M.H.; EL-Sheikh, M.A.; Seleem, A.A.; El-Naggar, M.E. Antimicrobial wound dressing and anti-inflammatory efficacy of silver nanoparticles. *Int. J. Biol. Macromol.* **2014**, *65*, 509–515. [CrossRef]
49. Yilma, A.N.; Singh, S.R.; Dixit, S.; Dennis, V.A. Anti-inflammatory effects of silver-polyvinyl pyrrolidone (Ag-PVP) nanoparticles in mouse macrophages infected with live chlamydia trachomatis. *Int. J. Nanomed.* **2013**, *8*, 2421–2432.
50. Braydich-Stolle, L.; Hussain, S.; Schlager, J.J.; Hofmann, M.-C. In vitro cytotoxicity of nanoparticles in mammalian germline stem cells. *Toxicol. Sci.* **2005**, *88*, 412–419. [CrossRef]
51. Hussain, S.M.; Hess, K.L.; Gearhart, J.M.; Geiss, K.T.; Schlager, J.J. In vitro toxicity of nanoparticles in BRL 3A rat liver cells. *Toxicol. Vitr.* **2005**, *19*, 975–983. [CrossRef] [PubMed]
52. Hussain, S.M.; Javorina, A.K.; Schrand, A.M.; Duhart, H.M.; Ali, S.F.; Schlager, J.J. The interaction of manganese nanoparticles with PC-12 cells induces dopamine depletion. *Toxicol. Sci.* **2006**, *92*, 456–463. [CrossRef] [PubMed]
53. Dhapte, V.; Kadam, S.; Moghe, A.; Pokharkar, V. Probing the wound healing potential of biogenic silver nanoparticles. *J. Wound Care* **2014**, *23*, 431–432. [CrossRef] [PubMed]
54. Kwan, K.H.L.; Liu, X.; To, M.K.T.; Yeung, K.W.K.; Ho, C.; Wong, K.K.Y. Modulation of collagen alignment by silver nanoparticles results in better mechanical properties in wound healing. *Nanomedicine* **2011**, *7*, 497–504. [CrossRef] [PubMed]

Article

Impact of Nano-Bromocriptine on Egg Production Performance and Prolactin Expression in Layers

Ahmed Dawod [1], Noha Osman [1], Hanim S. Heikal [1], Korany A. Ali [2], Omaima M. Kandil [3], Awad A. Shehata [4,5], Hafez M. Hafez [6,*] and Hamada Mahboub [1]

[1] Department of Husbandry and Animal Wealth Development, Faculty of Veterinary Medicine, University of Sadat City, Menofia 32897, Egypt; adawod@vet.usc.edu.eg (A.D.); noha.osman@vet.usc.edu.eg (N.O.); hanem.hekal@vet.usc.edu.eg (H.S.H.); hamada.mahboub@vet.usc.edu.eg (H.M.)

[2] Center of Excellence for Advanced Science, Advanced Materials and Nanotechnology Group, Applied Organic Chemistry Department, National Research Centre, Dokki, Giza 12622, Egypt; kornykhlil@gmail.com

[3] Center of Excellence for Embryo and Genetic Resources Conservation Bank, Department of Animal Reproduction and Artificial Insemination, Veterinary Research Division, National Research Center, Dokki , Giza 12622, Egypt; omaima_mk@yahoo.com

[4] Avian and Rabbit Diseases Department, Faculty of Veterinary Medicine, University of Sadat City, Menofia 32897, Egypt; Awad.Shehata@pernaturam.de

[5] Research and Development Section, PerNaturam GmbH, 56290 Gödenroth, Germany

[6] Institute of Poultry Diseases, Free University Berlin, 14195 Berlin, Germany

* Correspondence: hafez.mohamed@fu-berlin.de

Citation: Dawod, A.; Osman, N.; Heikal, H.S.; Ali, K.A.; Kandil, O.M.; Shehata, A.A.; Hafez, H.M.; Mahboub, H. Impact of Nano-Bromocriptine on Egg Production Performance and Prolactin Expression in Layers. *Animals* **2021**, *11*, 2842. https://doi.org/10.3390/ani11102842

Academic Editors: Antonio Gonzalez-Bulnes and Nesreen Hashem

Received: 16 August 2021
Accepted: 24 September 2021
Published: 29 September 2021

Simple Summary: Egg production is one of the most vital axes in the poultry industry. During the late laying period, the egg production continuously decreases, and pauses among the sequence of egg laying increases; however, the feed costs remain constant. Several attempts were carried out to improve the reproductive performance of laying hens by decreasing the prolactin level in the blood; an increase in this hormone initiates the onset of incubation behavior in chickens. In this study, we investigated the potential use of nano-bromocriptine to the improve egg production performance in laying hens. The use of alginate-bromocriptine leads to a significant reduction in the prolactin expression in the pituitary gland, which in turn allows the elongation in sequences and reduction in pauses, as well as the feed per dozen egg in laying hens. Further studies are needed to assess the impacts of nano-bromocriptine on other performance parameters. Thus, the improvement of egg production persistency must also go hand in hand with sustainable egg quality and the maintenance of the birds' health.

Abstract: The current study aimed to investigate the potential use of nano-bromocriptine in improving the laying performance of late laying hens by modulating the prolactin gene expression. A total of 150 NOVOgen brown laying hens aged 70 weeks were randomly allocated into three groups of 50 birds each. The first group was kept as a control, while the second and the third groups were treated with bromocriptine and nano-bromocriptine, respectively, at a dose of 100 µg/kg body weight per week. The pause days, egg production, feed per dozen egg, and Haugh unit were determined on a monthly basis. Also, the relative prolactin gene expression in the pituitary gland was quantified using qPCR and the number of the ovarian follicles was determined after slaughtering at the 84th week of age. It was found that nano-bromocriptine and bromocriptine improved egg laying performance with minimal pause days, reduced feed per dozen egg, and depressed the relative prolactin gene expression; however, nano-bromocriptine treatment was significantly effective compared to bromocriptine. In conclusion, nano-bromocriptine might be beneficial for elongating sequences and reducing pauses.

Keywords: prolactin; sodium alginate; nanotechnology; egg production; qPCR; chickens

1. Introduction

Although the global table eggs production has increased over the past decade to 76.7 million tonnes in 2018 [1], further improvement of egg production performance is urgently needed to fulfill the high demand for animal proteins. This highlights the urgent need to keep the laying persistency with sustainable egg production and quality that goes hand in hand with maintaining health and animal welfare, by considering the bird's physiology, nutritional requirements, management system, reproductive status, and breed selection. Although the extreme efforts which have been done to keep the persistency of egg production, a reduction in egg production accompanied by a deterioration of egg quality are usually common at or around 72 weeks of age [2]. During this period, the egg production continuously decreased and pauses among the sequence of egg laying increased with constant feed costs, causing huge economic losses.

Prolactin belongs to adenohypophysis hormones, is one of the most blamed factors accompanied by a progressive reduction of egg laying performance during the late laying period. This hormone is progressively increased by the time in plasma of late laying hens. It prevents gonadotrophin, which stimulates ovulation as well as estrogen production at the ovarian level [3]. The ovarian tissues in laying hens targeted the prolactin as it expressed the prolactin receptor mRNA [4]. So far, prolactin inhibits estradiol secretion in white ovarian follicles. However, it may improve or depress steroidogenesis in yellow ovarian follicles depended upon the level of prolactin dose, the type of the follicular layer secreting steroids, the order of the ovarian follicle in the ovulation hierarchy, and the phase of the ovulation cycle [5]. Increasing prolactin level during the active stage of laying is accompanied by the appearance of broodiness and cessation of egg production [6]. Several attempts were carried out to improve the reproductive performance of laying hens via decreasing prolactin using recombinant-derived chicken prolactin fusion protein or anti-vasoactive intestinal polypeptide serum to prevent broodiness in laying hens [7,8]. Others tried to depress prolactin chemically using a dopamine agonist, bromocriptine, an ergot derivative [9–11]. The release of prolactin from the adenohypophysis is governed by dopamine as it prevents the stimulatory effect of the vasoactive intestinal peptide on prolactin secretion. Therefore, bromocriptine as a dopamine agonist could be used to overcome the broodiness behaviors in laying hens [8]. Indeed, subcutaneous injection of bromocriptine in laying hens during the 17th to 36th week of life increases the egg production and depresses the prolactin production during the laying cycle up to the 72nd week of age [3].

Nevertheless, due to the low bioavailability of bromocriptine [12], there is a great demand for more convenient, effective, and safe ways for drug delivery. Nano-drug delivery systems could be a promising alternative to extend the half-life time of active principles and to sustain its delivery to the target sites [12–16]. It was found that nanocomposites reduced the dose of the drug and the desired biological activity could be obtained with minimal side effects. Therefore, the present study aimed to synthesize, and characterize, alginate-bromocriptine nanocomposite as well as to assess its efficacy on egg laying performance during the late laying stage. Moreover, its modulatory effect on the prolactin hormone gene expression in the pituitary gland was studied.

2. Materials and Methods

2.1. Preparation of Alginate-Bromocriptine Nano-Composite

The alginate-bromocriptine nano-composite was prepared according to Siddique et al. [12]. Briefly, 100 mg of 2-Bromo-α-ergocryptine (Sigma-Aldrich, St. Louis, MO, USA) was dissolved in 5 mL ethanol and added portion wise to sodium alginate solution, previously prepared by dissolving 0.1 g of sodium alginate (Fisher Scientific, Waltham, MA, USA) in 100 mL distilled water with stirring. The addition process was performed drop by drop over 30 min with stirring and worming at 40 °C. The mixture was then heated at 50 °C under ultra-sonication at 100 W with 35 kHz for 45 min. The obtained nano-composite solution was air-dried then stored at 4 °C, dry, and dark conditions.

The formation and the morphology of alginate-bromocriptine nano-composite were characterized by transmutation electron microscopy (TEM, JEM-2100, JEOL- Tokyo Japan) and the average size of the prepared nano-composite was estimated. The Fourier Transform Infrared Spectrophotometer (FTIR) spectra were analyzed using a spectrometer (JASCO, Tokyo, Japan) in the range of 400–4000 cm^{-1}. Potassium bromide discs (KBr) (5 mg of particles, 100 mg KBr pellets) were used as reference material. The optical properties of the bromocriptine and alginate-bromocriptine nanocomposite were investigated by measuring the UV-visible spectrophotometer (JASCO spectrophotometer, Tokyo, Japan) in the range of 200–800 nm.

2.2. Experimental Assessment of the Efficacy of Bromocriptine and Alginate-Nano-Bromocriptine

2.2.1. Ethical Approval

All procedures of the experiment were conducted under the ethical guidelines of the Institutional Animal Care and Use Committee (IACUC), Faculty of Veterinary Medicine, University of Sadat City, Egypt (Ethical approval number: VUSC-017-1-19).

2.2.2. Experimental Design

One hundred fifty laying hens of NOVOgen brown strain of 70 weeks age were selected from a local commercial layer farm in Egypt based on good feathers, body weight (mean = 1800 ± 150 g), and free from any abnormalities. Birds were randomly allocated into 3 groups of 50 birds per each. Hens were allowed to have 2 weeks acclimatization period. Chickens in the first group were treated with saline and kept as control, while chickens kept in the second and third groups were treated with bromocriptine (2-bromo-alpha-ergocriptine, Sigma-Aldrich, St. Louis, MO, USA) and the prepared nano-bromocriptine, respectively, at a dose of 100 µg/kg BW/week according to Reddy et al. [17] for 12 weeks (from 72nd to 84th week of age). Each group was subdivided into two subgroups; the first subgroup was treated orally, whereas the second subgroup was treated subcutaneously underneath the wing. Hens were kept under the standard laying conditions on the litter floor system of sawdust equipped with automatic drinker and feeding systems. The feed containing 16% crude protein (corn-soybean meal of 16% CP, 2900 kcal ME/kg diet) and 120 g/hen/day was used, fresh-water give at ad libitum. The photoperiod was kept at 16 h of light/day, achieved via an automatic lighting system. The temperature of the house was kept at 24 ± 3 °C.

2.2.3. Egg Production Performance

Eggs were collected and recorded daily from each group at 12 AM. Pause days, egg production%, feed intake, and feed per dozen eggs were daily evaluated for each subgroup. Pause days were daily estimated for each laying hen as the hens were separately reared in deep litter system. Also, twelve eggs were randomly selected from each group/subgroup at 76th, 80th, 84th/week of age to determine the Haugh unit according to Haugh [18] using the formula: Haugh unit $(HU) = 100 \times log \left(H - 1.7W^{0.37} + 7.6 \right)$, where "$H$" is the albumen height and "W" is the egg weight.

At the 84th week of age, all hens were slaughtered by manual cervical dislocation then dressed. The ovary was investigated to estimate the number of normal large yellow follicles (LYF) (>10 mm diameter), small yellow follicles (SYF) (5–10 mm diameter), and large white follicles (LWF) (3–5 mm diameter) according to Renema et al. [19].

2.2.4. Prolactin Gene Expression in Pituitary Gland Using qRT-PCR

Just after slaughtering, the pituitary gland was collected randomly from 5 hens/subgroup. The RNA extraction was carried out using a TRIzol reagent (Invitrogen, Carlsbad, CA, USA) according to the manufacturer and the cDNA was formed from total RNA using Maxima First Strand cDNA Synthesis Kit (Life Technologies, Carlsbad, CA, USA). The PRLE2F: 5′-GTTTGTTTCTGGCGGTTC-3′, and PRLE2R: 5′-AAATTCATTGAATATTTCTGAAG-3′ primers were used to amplify 181 bp fragment of prolactin gene [20], while the QGAPDHF:

5′-CTGCCGTCCTCTCTGGC-3′ and QGAPDHR: 5′-GACAGTGCCCTTGAAGT GT-3′ primers were used as an internal control to amplify 119 bp fragment of glyceraldehyde 3-phosphate dehydrogenase gene [21]. The qPCR was performed in duplicate in a final volume of 25 μL nuclease-free water containing 12.5 μL of Platinum SYBR Green qPCR supermix, 0.05 μL of ROX reference dye (Invitrogen, Carlsbad, CA, USA), 0.2 μM of each primer, 2 μL of cDNA using a 7500 Fast Real-Time PCR System (Applied Biosystems, Foster City, CA, USA). The cycler program was $1\times$ 95 °C/10 min, and $40\times$ (95 °C/30 s, 57 °C/1 min and 72 °C for 30 s). The melting curve of a single peak was determined and used for the gene expression analysis. The normalization was done using the GAPDH as an endogenous control. The relative quantification was carried out using the $2^{-\Delta\Delta CT}$ method [22].

2.3. Statistical Analysis

The data were collected and enrolled into statistical analysis using SAS software (SAS User's Guide: Statistics, Version 8.1 Edition, 2000, SAS Inst. Inc., Cary, NC, USA). Percentage data were subjected to arcsine transformation. Then the data were analyzed using two ways ANOVA of the General Linear Models (GLM) procedures of SAS software according to the following statistical model:

$$Y_{ijk} = \mu + \alpha_i + \beta_j + (\alpha\beta)_{ij} + \epsilon_{ijk}$$

where, Y_{ijk} = overall observation, μ = overall mean, α_i was the treatment effect i = 1, 2, 3 for control, bromocriptine, and nanobromocriptine treatments, β_j was the route of administration j = 1, 2 for oral, and injection rout of administrations, $(\alpha\beta)_{ij}$ was the interaction between treatments and route of administration and ϵ_{ijk} was the random error. Tukey test was used as mean separation test and results were expressed as least square means $\pm$ SE. The level of significance was seated at ($p < 0.05$).

3. Results

3.1. Analysis of Alginate-Bromocriptine

The FTIR spectroscopy of bromocriptine and alginate-bromocriptine composite gave information about chemical bonding to confirm the purity and the formation of the alginate-bromocriptine nanocomposite. The FTIR spectrum displayed bands at 3500 cm^{-1}, corresponding to the OH bending of the hydroxyl group. The stretching vibration appeared as bands at 2300 cm^{-1}, 1500 cm^{-1} and 1200 cm^{-1}, and are attributed to -CO, C=O, and C-O-C bonds, respectively. The FTIR spectroscopy pattern of the alginate-bromocriptine nanocomposite revealed a band at 700 cm^{-1}, corresponding to -CH bending; this confirms a bond formation between alginate and bromocriptine (Figure 1).

The TEM analysis of the prepared alginate-bromocriptine nanocomposite is shown in Figure 2. The alginate-bromocriptine nanocomposite particles are semi-spherical in shape with a size of 20 to 36 nm, and uniformly distributed. The average particle size is up to 33 nm.

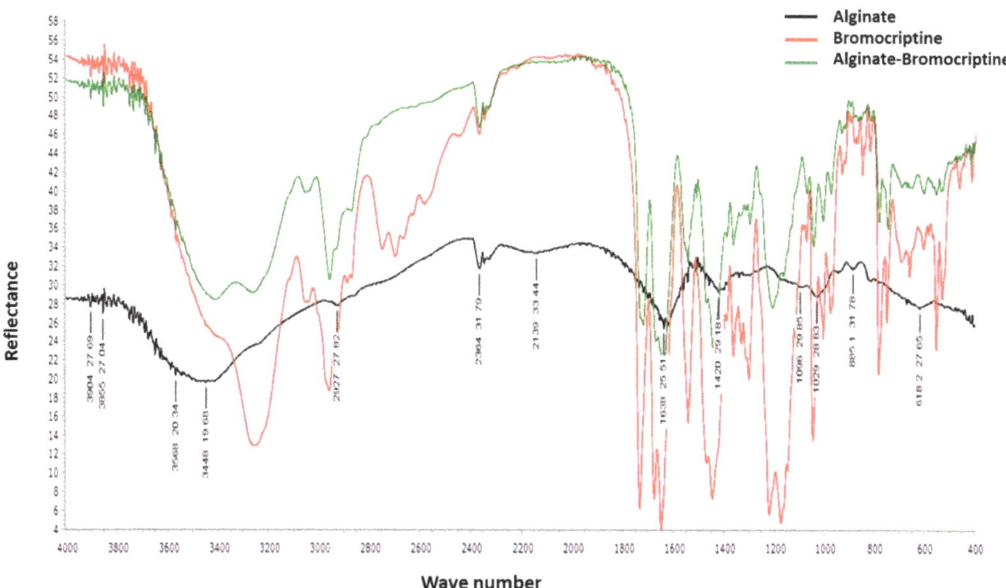

Figure 1. Fourier transform infrared (FTIR) spectra of alginate bromocriptine.

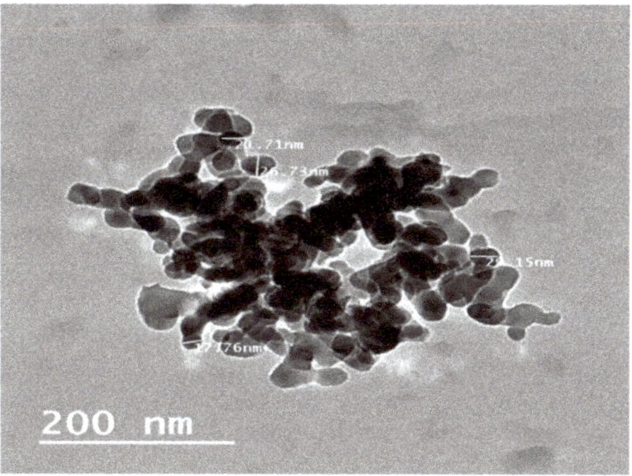

Figure 2. Transmission electron microscope image of alginate-bromocriptine nanocomposite with scale 200 nm at 25 °C.

The optical properties of the bromocriptine and alginate-bromocriptine nanocomposite measured in the wavelength range of 300–900 nm showed a strong absorption band at 366 nm, corresponding to the pure bromocriptine, in addition to a low absorption band at the visible region (Figure 3A,B). Alginate-bromocriptine nanocomposite did not show a significant shift where the strong absorption band appeared at 362 nm. These results revealed a decrease in the bandgap due to the crystallite size of the prepared nanoparticles.

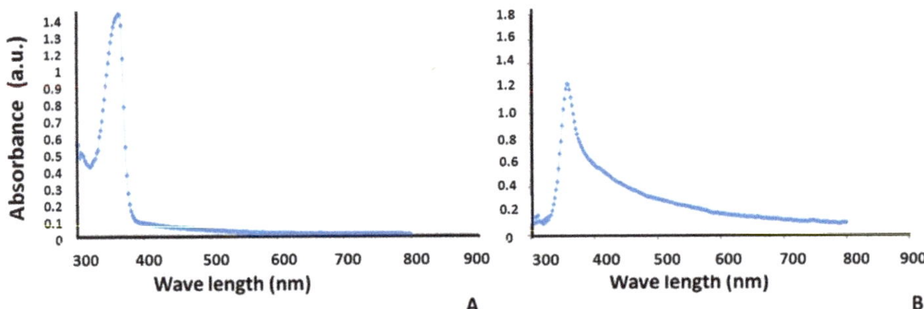

Figure 3. UV-visible absorption spectroscopy for bromocriptine (**A**) and alginate-bromocriptine nanocomposite (**B**).

3.2. Assessment of the Efficacy of Bromocriptine and Nano-Bromocriptine on Egg Production Performance

3.2.1. Pause Days, Egg Production Percentage, and Feed per Dozen Egg

The total number of pause days routinely decreased in nano-bromocriptine treated chickens compared to those treated with bromocriptine and control. This trend appeared clearly over the entire experimental period. The lowest numbers of pause days were 5.38 ± 0.33, 5.93 ± 0.26, and 7.27 ± 0.30 days in chickens treated with alginate-bromocriptine nanocomposite at the 76th, 80th, and 84th week of age, respectively. Furthermore, the oral administration showed the best pause days compared to the injection route. Thus, the lowest number of pause days in the hens treated orally with alginate-bromocriptine nanocomposite were 3.93 ± 0.35, 5.29 ± 0.35, and 6.46 ± 0.44 days at the 76th, 80th, and 84th week of age, respectively (Table 1).

Table 1. Effects of bromocriptine and nano-bromocriptine on pause days and egg production percent of late laying hens.

Parameter		Pause Days (Day)			Egg/Hen/D (%)		
		76th Week	80th Week	84th Week	76th Week	80th Week	84th Week
Treatment	Control ($n = 50$)	11.07 ± 0.26 [a]	10.00 ± 0.25 [a]	11.56 ± 0.34 [a]	61.42 ± 0.95 [c]	61.90 ± 0.78 [c]	51.16 ± 1.42 [b]
	Bromocriptine ($n = 50$)	7.05 ± 0.32 [b]	6.89 ± 0.28 [b]	8.20 ± 0.35 [b]	71.55 ± 1.25 [b]	71.11 ± 1.01 [b]	65.54 ± 1.71 [a]
	Nano-bromocriptine ($n = 50$)	5.38 ± 0.33 [c]	5.93 ± 0.26 [c]	7.27 ± 0.30 [c]	78.33 ± 1.26 [a]	74.54 ± 1.16 [a]	68.13 ± 1.45 [a]
Administration	Orally ($n = 75$)	7.14 ± 0.39 [b]	7.15 ± 0.32 [b]	8.94 ± 0.36	73.37 ± 1.26 [a]	71.04 ± 1.12 [a]	62.41 ± 1.62 [a]
	Injection ($n = 75$)	8.52 ± 0.32 [a]	8.06 ± 0.24 [a]	9.08 ± 0.31	67.49 ± 1.04 [b]	67.33 ± 0.78 [b]	60.82 ± 1.44 [a]
Treatment χ administration	Control (oral, $n = 25$)	11.11 ± 0.37 [a]	10.07 ± 0.37 [a]	11.50 ± 0.48 [a]	61.15 ± 1.34 [d]	61.56 ± 1.12 [c]	51.65 ± 1.99 [c]
	Control (injection, $n = 25$)	11.04 ± 0.38 [a]	9.93 ± 0.35 [a]	11.63 ± 0.48 [a]	61.69 ± 1.37 [d]	62.25 ± 1.09 [c]	50.67 ± 2.05 [c]
	Bromocriptine (oral, $n = 25$)	6.39 ± 0.37 [c]	6.11 ± 0.44 [c]	8.86 ± 0.53 [b]	74.71 ± 1.50 [b]	72.75 ± 1.81 [b]	63.04 ± 2.62 [b]
	Bromocriptine (injection, $n = 25$)	7.71 ± 0.49 [b]	7.68 ± 0.28 [b]	7.54 ± 0.42 [c,d]	68.39 ± 1.87 [c]	69.46 ± 0.84 [b]	68.05 ± 2.12 [a,b]
	Nano-bromocriptine (oral, $n = 25$)	3.93 ± 0.35 [d]	5.29 ± 0.35 [c]	6.46 ± 0.44 [d]	84.26 ± 1.28 [a]	78.80 ± 1.34 [a]	72.52 ± 1.92 [a]
	Nano-bromocriptine (injection, $n = 25$)	6.82 ± 0.41 [c]	6.57 ± 0.36 [c]	8.07 ± 0.34 [b,c]	72.40 ± 1.58 [b,c]	70.29 ± 1.55 [b]	63.75 ± 1.78 [b]
p-value	Treatment	<0.0001	<0.0001	<0.0001	<0.0001	<0.0001	<0.0001
	Administration	<0.0001	<0.0026	<0.7091	<0.0001	<0.0007	<0.3572
	Interaction	<0.0014	<0.0425	<0.0062	<0.0003	<0.0032	<0.0052

Values are presented as least square mean $\pm$ SE. [a–d] Means within the same column for each parameter with different superscripts are statistically different at $p < 0.05$ (Two-way ANOVA, Tukey post-hoc test).

The nano-bromocriptine treated hens recorded the highest egg production percentage (78.33 ± 1.26 and 74.54 ± 1.16) at the 76th and 80th week of life, respectively, compared to the bromocriptine treated (71.55 ± 1.25 and $71.11 \pm 1.11\%$) or control ($61.42 \pm 0.95\%$ and 61.9 ± 0.78) hens (Table 1). However, at the 84th week of age, both nano-bromocriptine (68.13 ± 1.45) and bromocriptine ($65.54 \pm 1.71\%$) treated hens sustained a higher egg production percentage compared to the non-treated hens ($51.16 \pm 1.42\%$). On the other hand, the oral administration of nano-bromocriptine sustained the highest egg production percentage (84.26 ± 1.28, 78.80 ± 1.34 and 72.52 ± 1.92, at 76th, 80th, and 84th week-old, respectively).

3.2.2. Feed Consumption per Dozen Egg, and Haugh Unit

The feed per dozen egg was significantly varied among different bromocriptine forms and control groups (Table 2). Both the bromocriptine and nano-bromocriptine treated hens exhibited a significant reduction in feed per dozen egg compared to the control group over the entire experimental period ($p < 0.0001$). Alginate-bromocriptine nanocomposite treatment significantly reduced the feed per dozen egg (1.87 ± 0.03 and 1.96 ± 0.03 kg, respectively) at the 76th and 80th week of age, followed by bromocriptine treatment (2.06 ± 0.05 and 2.05 ± 0.03 kg, respectively) and control (2.38 ± 0.04 and 2.35 ± 0.03 kg, respectively). The birds that received oral nano-bromocriptine sustained the lowest feed per dozen egg values (1.72 ± 0.03; 1.84 ± 0.03; 2.02 ± 0.06 kg at the 76th, 80th, and 84th week of age, respectively) compared to the other administration methods.

Table 2. Effects of bromocriptine and nano-bromocriptine on feed consumption per dozen egg and Haugh unit of late laying hens.

Parameter		Feed/Dozen Egg (kg)			Haugh Unit		
		76th Week	80th Week	84th Week	76th Week	80th Week	84th Week
Treatment	Control ($n = 50$)	2.38 ± 0.04 [a]	2.35 ± 0.03 [a]	2.94 ± 0.11 [a]	76.70 ± 0.53	77.01 ± 1.19	83.06 ± 1.16 [a]
	Bromocriptine ($n = 50$)	2.06 ± 0.05 [b]	2.05 ± 0.03 [b]	2.29 ± 0.08 [b]	75.63 ± 1.74	77.29 ± 1.39	78.99 ± 1.16 [b]
	Nano-bromocriptine ($n = 50$)	1.87 ± 0.03 [c]	1.96 ± 0.03 [c]	2.16 ± 0.05 [b]	76.24 ± 1.60	75.20 ± 0.84	76.82 ± 1.08 [b]
Administration	Orally ($n = 75$)	2.02 ± 0.04 [b]	2.07 ± 0.03 [b]	2.44 ± 0.08 [a]	75.12 ± 0.58	76.70 ± 0.90	79.21 ± 0.92
	Injection ($n = 75$)	2.18 ± 0.04 [a]	2.16 ± 0.03 [a]	2.48 ± 0.07 [a]	77.26 ± 1.52	76.30 ± 1.03	80.04 ± 0.93
Treatment χ administration	Control (oral, $n = 25$)	2.39 ± 0.05 [a]	2.36 ± 0.04 [a]	2.91 ± 0.15 [a]	76.83 ± 0.72	77.99 ± 2.07	83.29 ± 1.58 [a]
	Control (injection, $n = 25$)	2.37 ± 0.05 [a]	2.33 ± 0.04 [a]	2.96 ± 0.16 [a]	76.58 ± 0.88	76.02 ± 0.80	82.83 ± 1.70 [a,b]
	Bromocriptine (oral, $n = 25$)	1.95 ± 0.04 [b,c]	2.02 ± 0.05 [b]	2.41 ± 0.14 [b]	74.50 ± 0.85	75.75 ± 1.63	75.28 ± 1.77 [b,c]
	Bromocriptine (injection, $n = 25$)	2.17 ± 0.08 [b]	2.08 ± 0.03 [b]	2.17 ± 0.08 [b,c]	76.76 ± 3.49	78.83 ± 2.19	82.71 ± 1.49 [a]
	Nano-bromocriptine (oral, $n = 25$)	1.72 ± 0.03 [c]	1.84 ± 0.03 [c]	2.02 ± 0.06 [c]	74.04 ± 1.08	76.35 ± 0.92	79.06 ± 1.41 [b,c]
	Nano-bromocriptine (injection, $n = 25$)	2.02 ± 0.05 [b]	2.08 ± 0.05 [b]	2.30 ± 0.07 [b,c]	78.44 ± 2.81	74.05 ± 1.28	74.58 ± 1.64 [c]
p-value	Treatment	<0.0001	<0.0001	<0.0001	<0.8726	<0.3834	<0.0007
	Administration	<0.0002	<0.0079	<0.7266	<0.2117	<0.7665	<0.5290
	Interaction	<0.0083	<0.0066	<0.0473	<0.5373	<0.1923	<0.0013

Values are presented as least square mean $\pm$ SE. [a–c] Means within the same column for each parameter with different superscripts are statistically different at $p < 0.05$ (Two-way ANOVA, Tukey post-hoc test).

The Haugh unit possessed no significant differences among different treatments at the 76th and 80th week of age (Table 2). In contrast, the Haugh unit at the 84th week of age was significantly ($p < 0.0007$) increased in the non-treated birds (83.06 ± 1.16) compared to both the bromocriptine (78.99 ± 1.16) and nano-bromocriptine (76.82 ± 1.08) treated groups.

3.2.3. Ovarian Follicles, and Prolactin Gene Expression in the Pituitary Gland Tissue

The number of different types of ovarian follicles varied according to the use of alginate-bromocriptine nanocomposite or bromocriptine during the late laying phase (Table 3). The groups treated with bromocriptine and nano-bromocriptine showed significantly higher numbers of LYF (5.9 ± 0.19 and 6.3 ± 0.15, respectively) compared to the control group (5.33 ± 0.2). However, a non-significant increase in SYF and decrease in LWF were found in chickens treated with nano-bromocriptine.

Table 3. Effects of bromocriptine and nano-bromocriptine on ovarian follicles of late laying hens.

Parameter		Ovarian Follicles		
		LYF	**SYF**	**LWF**
Treatment	Control (n = 50)	5.33 ± 0.20 [b]	5.85 ± 0.63	8.93 ± 1.20
	Bromocriptine (n = 50)	5.90 ± 0.19 [a]	6.40 ± 1.37	6.80 ± 0.81
	Nano-bromocriptine (n = 50)	6.30 ± 0.15 [a]	8.10 ± 0.95	6.80 ± 0.88
Administration	Orally (n = 75)	5.87 ± 0.13	6.73 ± 1.02	7.33 ± 0.87
	Injection (n = 75)	5.82 ± 0.19	6.83 ± 0.68	7.68 ± 0.75
Treatment χ administration	Control (oral, n = 25)	5.40 ± 0.27	6.20 ± 0.97	8.60 ± 1.66
	Control (injection, n = 25)	5.25 ± 0.31	5.50 ± 0.87	9.25 ± 1.97
	Bromocriptine (oral, n = 25)	6.00 ± 0.21	6.60 ± 2.64	6.60 ± 1.60
	Bromocriptine (injection, n = 25)	5.80 ± 0.33	6.20 ± 1.20	7.00 ± 0.63
	Nano-bromocriptine (oral, n = 25)	6.20 ± 0.13	7.40 ± 1.69	6.80 ± 1.39
	Nano-bromocriptine (injection, n = 25)	6.40 ± 0.27	8.80 ± 0.97	6.80 ± 1.24
p-value	Treatment	<0.0021	<0.3429	<0.2705
	Administration	<0.8141	<0.9385	<0.7695
	Interaction	<0.6965	<0.7694	<0.9749

LYF: number of normal large yellow follicles (>10 mm diameter); SYF: number of small yellow follicles (5–10 mm diameter); LWF: number of large white follicles (3–5 mm diameter). Values are presented as least square mean ± SE. [a–b] means within the same column for each parameter with different superscripts are statistically different at $p < 0.05$ (Two-way ANOVA, Tukey post-hoc test).

Additionally, the expression of the prolactin gene was determined in the pituitary gland tissues of late laying hens (Figure 4). The findings of the prolactin gene expression showed obvious depression in response to the alginate-bromocriptine nanocomposite treated group. Both bromocriptine and nano-bromocriptine exhibited a significant depression of prolactin gene expression in treated birds. Furthermore, the findings revealed that the depression of prolactin gene expression was enhanced via the injection route of administration in bromocriptine-treated birds ($p < 0.05$). However, there were no significant differences between the two routes in the groups treated with nano-bromocriptine.

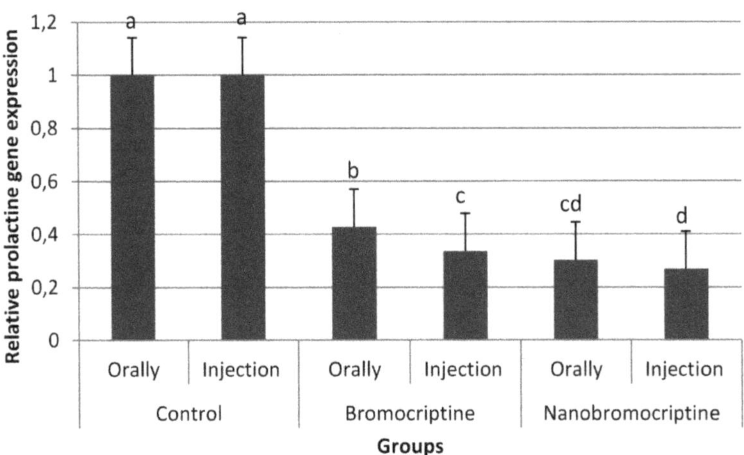

Figure 4. Quantitative gene expression of prolactin gene in pituitary after either bromocriptine or nano-bromocriptine administration. The qRT-PCR data were normalized relative to the expression of glyceraldehyde-3-phosphate dehydrogenase (GAPDH) as endogenous control and calculated using the $2^{-\Delta\Delta CT}$ method (n = 5). Different superscripts indicated significant differences at $p < 0.05$. Error bars indicate the standard error of the mean.

4. Discussion

Eggs are important food for the human population because of their inexpensive price and high nutrient value. There is a huge demand for animal protein supply that could be covered with eggs [23]. In the present investigation, we tried to modulate the level of the prolactin gene expression in hens during their late laying phase using a minimal dose of bromocriptine and nano-bromocriptine.

The alginate nanocomposites have recently received much attention as drug delivery nanoparticles [14]. It has efficient biodegradability, biocompatibility, and mucoadhesiveness nature making them superior natural polymers that are ready to use as a drug delivery nanocomposite for target tissue [24]. The formulation used is highly advantageous for maximizing the effectiveness of the drug on target tissues. In our study, the alginate-bromocriptine nano-composite was synthesized and characterized. The prepared particle size ranged from 20 to 36 nm. The size of nanoparticles is a critical criterion for the crossing of the mucosal tissue barriers and the enhancement of the cellular uptake [25]. The size of most nanoparticles ranged between 50–250 nm [26,27]. However, the average size of alginate-bromocriptine nanocomposite could reach 20 nm [12]. The biological activity of bromocriptine and the prepared nano-bromocriptine was carried out in the 70 week old NOVOgen brown hens. Our findings revealed that alginate-bromocriptine nano-composite improved the egg laying performance in late laying hens, thus treated hens sustained the highest egg production, the lowest pause days, and the lowest feed per dozen egg compared to the negative control and bromocriptine treated birds. This could be due to rapid gut absorption and bioavailability of the alginate-bromocriptine nano-composite compared to bromocriptine. It was reported that nano-drugs could be accumulated in the target tissue with longer blood circulation time and binding properties [28]. Moreover, the drug nano-particles crossed efficiently from blood vessels and lymphatics. Furthermore, nano-drugs had positive charges that could be attracted to the negative charges of mucin, this property may improve the transportation and absorption of the drug through the epithelial membranes [29].

Furthermore, bromocriptine increased significantly egg laying performance compared with control, in terms of egg production percentage, reduction of pause days, and feed per dozen egg. The same findings were proved in earlier studies [3,10,11,17,30,31]. This could be due to the lower circulating prolactin in the bromocriptine and alginate-bromocriptine nanocomposite treated hens, thus the increase of prolactin above its physiological range resulting in a subsequent decrease in circulating gonadotropins, regression of the ovarian functions, ending of the reproductive phase of the laying hens, and altered it to the brooding phase [32]. However, there is one significant limitation in this study: it was not possible to determine both plasma level of luteinizing hormone (LH) and sex steroids.

Interestingly, significant increases in LYF numbers were reported in bromocriptine and alginate-bromocriptine nanocomposite treated birds. This may explain the higher egg production percentages that were observed in these groups. These findings could be attributed to the low level of the circulating prolactin hormone in treated birds. Prolactin depressed the development of ovarian follicles to reach the final stage through interference with the follicular steroidogenesis in avian species [33]. However, the outcomes suggested that the Haugh unit was significantly decreased in response to bromocriptine and alginate-bromocriptine nanocomposite treatment. In contrast, Banu et al. reported that bromocriptine treatment did not affect the Haugh unit of the eggs of laying hens [31]. This outcome may belong to high egg production sustained in such groups compared with control.

Oral administration may enhance the alginate-bromocriptine nanocomposite effects. This could be due to the high absorption of alginate-bromocriptine nanocomposite particles in the chicken gut compared with bromocriptine. Caster et al. found that oral and intravenous routes were the most common routes of administration of the nano-drugs rather than the transdermal route [34]. Furthermore, the oral administration of bromocriptine is rapidly and incompletely absorbed in animals. However, this rapidly absorbed portion is highly metabolized in the liver, so its bioavailability is rapidly declined [35].

Nano-drug had many advantages properties, such as improved solubility, efficacy, tissue selectivity, and reduced toxicity compared to the conventional drugs [34,36–38]. This could explain why the alginate-bromocriptine nanocomposite depressed the relative prolactin gene expression in the pituitary gland compared with bromocriptine. Moreover, the injectable route of administration of bromocriptine sustained lower prolactin gene expression compared with the oral route. Linearly, it was reported that injection of bromocriptine frequently induced a prolonged normoprolactinemia compared to oral administration [39]. In contrast, there were no significant differences between the two different routes of alginate-bromocriptine nanocomposite administration, highlighting the high absorption and bioavailability of nanoform.

5. Conclusions

Nano-bromocriptine could be used to improve the egg production performance in late laying hens. The same effects were obtained with bromocriptine but with lower efficacy than its nanoform. Further studies about its impact on other performance and health parameters such as carcass quality, blood parameters, LH plasma concentration and sex steroids as well as liver functions are in progress.

Author Contributions: Conceptualization, A.D., N.O., A.A.S., H.M.H. and H.M.; methodology, A.D., N.O., K.A.A. and O.M.K.; software, A.D. and N.O.; validation, A.D., N.O. and H.M.; formal analysis, A.D., N.O. and A.A.S.; investigation, A.D., N.O., A.A.S. and H.M.H.; data curation, A.D. and N.O.; writing—original draft preparation, A.D., N.O., A.A.S. and H.M.H.; writing—review and editing, A.D., K.A.A., O.M.K., A.A.S. and H.M.H.; visualization, A.D. and A.A.S.; supervision, A.D., H.S.H. and H.M.H. All authors have read and agreed to the published version of the manuscript.

Funding: This research received no external funding.

Institutional Review Board Statement: The study was conducted according to the guidelines of the Declaration of Helsinki, and approved by the Institutional Animal Care and Use Committee (IACUC), Faculty of Veterinary Medicine, University of Sadat City, Egypt (Ethical approval number: VUSC-017-1-19).

Informed Consent Statement: Not applicable.

Data Availability Statement: The data sets used during the current study are available from the corresponding author on reasonable request.

Acknowledgments: The authors acknowledge the great support of Alaa Eldean Mostafa, Dean of the Faculty of Veterinary Medicine, Sadat City University, Egypt during conduction of the this study methodology.

Conflicts of Interest: The authors declare no conflict of interest.

References

1. FAO. FAOSTAT. Available online: http://www.fao.org/faostat/en/ (accessed on 15 July 2021).
2. Bain, M.M.; Nys, Y.; Dunn, I.C. Increasing persistency in lay and stabilising egg quality in longer laying cycles. What are the challenges? *Br. Poult. Sci.* **2016**, *57*, 330–338. [CrossRef]
3. Reddy, I.; David, C.; Sarma, P.; Singh, K. Modulation of prolactin hormone and intersequence pause days in domestic chickens. *Vet. Rec.* **2001**, *149*, 590. [CrossRef]
4. Ohkubo, T.; Tanaka, M.; Nakashima, K.; Talbot, R.; Sharp, P. Prolactin receptor gene expression in the brain and peripheral tissues in broody and nonbroody breeds of domestic hen. *Gen. Comp. Endocrinol.* **1998**, *109*, 60–68. [CrossRef]
5. Hrabia, A.; Paczoska-Eliasiewicz, H.; Rząsa, J. Effect of prolactin on estradiol and progesterone secretion by isolated chicken ovarian follicles. *Folia Biol.* **2004**, *52*, 197–203, (In Kraków). [CrossRef] [PubMed]
6. Sharp, P.J.; Dawson, A.; Lea, R.W. Control of luteinizing hormone and prolactin secretion in birds. *Comp. Biochem. Physiol. Part C Pharmacol. Toxicol. Endocrinol.* **1998**, *119*, 275–282. [CrossRef]
7. Li, W.; Liu, Y.; Yu, Y.; Huang, Y.; Liang, S.; Shi, Z. Prolactin plays a stimulatory role in ovarian follicular development and egg laying in chicken hens. *Domest. Anim. Endocrinol.* **2011**, *41*, 57–66. [CrossRef]

8. Sharp, P.; Sterling, R.; Talbot, R.; Huskisson, N. The role of hypothalamic vasoactive intestinal polypeptide in the maintenance of prolactin secretion in incubating bantam hens: Observations using passive immunization, radioimmunoassay and immunohistochemistry. *J. Endocrinol.* **1989**, *122*, 5–13. [CrossRef] [PubMed]
9. Badakhshan, Y.; Mazhari, M. Effect of epinephrine and bromocriptine on ovary folliculogenesis and productive ratits of laying hens. *Iran. J. Appl. Anim. Sci.* **2014**, *4*, 421–424.
10. Reddy, I.; David, C.; Raju, S. Chemical control of prolactin secretion and it's effects on pause days, egg production and steroid hormone concentration in Girirani birds. *Int. J. Poult. Sci.* **2006**, *5*, 685–692.
11. Parvez, M.M.I.R.; Rashid, B.; Hasan, M.; Mobarak, H.; Roy, K.K.; Haque, M.A. Effect of serum from laying hen and antiprolactin drug on egg production of indigenous chicken in Bangladesh. *Asian Australas. J. Biosci. Biotechnol.* **2017**, *2*, 51–54.
12. Siddique, Y.H.; Khan, W.; Fatima, A.; Jyoti, S.; Khanam, S.; Naz, F.; Ali, F.; Singh, B.R.; Naqvi, A.H. Effect of bromocriptine alginate nanocomposite (BANC) on a transgenic Drosophila model of Parkinson's disease. *Dis. Models Mech.* **2016**, *9*, 63–68.
13. El-Sissi, A.F.; Mohamed, F.H.; Danial, N.M.; Gaballah, A.Q.; Ali, K.A. Chitosan and chitosan nanoparticles as adjuvant in local Rift Valley Fever inactivated vaccine. *3 Biotech* **2020**, *10*, 1–11. [CrossRef]
14. Hassanein, E.M.; Hashem, N.M.; El-Azrak, K.E.-D.M.; Gonzalez-Bulnes, A.; Hassan, G.A.; Salem, M.H. Efficiency of GnRH–loaded chitosan nanoparticles for inducing LH secretion and fertile ovulations in protocols for artificial insemination in rabbit does. *Animals* **2021**, *11*, 440. [CrossRef]
15. Wang, X.; Chi, N.; Tang, X. Preparation of estradiol chitosan nanoparticles for improving nasal absorption and brain targeting. *Eur. J. Pharm. Biopharm.* **2008**, *70*, 735–740. [CrossRef]
16. Zhang, Q.; Jiang, X.; Jiang, W.; Lu, W.; Su, L.; Shi, Z. Preparation of nimodipine-loaded microemulsion for intranasal delivery and evaluation on the targeting efficiency to the brain. *Int. J. Pharm.* **2004**, *275*, 85–96. [CrossRef]
17. Reddy, I.; David, C.; Raju, S. Effect of suppression of plasma prolactin on luteinizing hormone concentration, intersequence pause days and egg production in domestic hen. *Domest. Anim. Endocrinol.* **2007**, *33*, 167–175. [CrossRef] [PubMed]
18. Haugh, R. The Haugh unit for measuring egg quality. *U. S. Egg Poult. Mag.* **1937**, *43*, 522–555.
19. Renema, R.; Robinson, F.; Melnychuk, V.; Hardin, R.; Bagley, L.; Emmerson, D.; Blackman, J. The use of feed restriction for improving reproductive traits in male-line large white Turkey hens.: 2. Ovary morphology and laying traits. *Poult. Sci.* **1995**, *74*, 102–120. [CrossRef]
20. Jiang, R.-S.; Xu, G.-Y.; Zhang, X.-Q.; Yang, N. Association of polymorphisms for prolactin and prolactin receptor genes with broody traits in chickens. *Poult. Sci.* **2005**, *84*, 839–845. [CrossRef] [PubMed]
21. Bhattacharya, T.; Chatterjee, R.; Sharma, R.; Niranjan, M.; Rajkumar, U. Associations between novel polymorphisms at the 5′-UTR region of the prolactin gene and egg production and quality in chickens. *Theriogenology* **2011**, *75*, 655–661. [CrossRef] [PubMed]
22. Livak, K.J.; Schmittgen, T.D. Analysis of relative gene expression data using real-time quantitative PCR and the $2^{-\Delta\Delta CT}$ method. *Methods* **2001**, *25*, 402–408. [CrossRef] [PubMed]
23. Molnár, S.; Szőllősi, L. Sustainability and quality aspects of different table egg production systems: A literature review. *Sustainability* **2020**, *12*, 7884. [CrossRef]
24. Choukaife, H.; Doolaanea, A.A.; Alfatama, M. Alginate nanoformulation: Influence of process and selected variables. *Pharmaceuticals* **2020**, *13*, 335. [CrossRef] [PubMed]
25. Panyam, J.; Labhasetwar, V. Biodegradable nanoparticles for drug and gene delivery to cells and tissue. *Adv. Drug Deliv. Rev.* **2003**, *55*, 329–347. [CrossRef]
26. Danaei, M.; Dehghankhold, M.; Ataei, S.; Hasanzadeh Davarani, F.; Javanmard, R.; Dokhani, A.; Khorasani, S.; Mozafari, M. Impact of particle size and polydispersity index on the clinical applications of lipidic nanocarrier systems. *Pharmaceutics* **2018**, *10*, 57. [CrossRef]
27. Gan, Q.; Wang, T.; Cochrane, C.; McCarron, P. Modulation of surface charge, particle size and morphological properties of chitosan–TPP nanoparticles intended for gene delivery. *Colloids Surf. B Biointerfaces* **2005**, *44*, 65–73. [CrossRef]
28. Win, K.Y.; Feng, S.-S. Effects of particle size and surface coating on cellular uptake of polymeric nanoparticles for oral delivery of anticancer drugs. *Biomaterials* **2005**, *26*, 2713–2722. [CrossRef]
29. Roger, E.; Lagarce, F.; Garcion, E.; Benoit, J.-P. Biopharmaceutical parameters to consider in order to alter the fate of nanocarriers after oral delivery. *Nanomedicine* **2010**, *5*, 287–306. [CrossRef]
30. Reddy, I.; David, C.; Sarma, P.; Singh, K. The possible role of prolactin in laying performance and steroid hormone secretion in domestic hen (Gallus domesticus). *Gen. Comp. Endocrinol.* **2002**, *127*, 249–255. [CrossRef]
31. Banu, M.N.; Rashid, M.B.; Hasan, M.M.; Aziz, F.B.; Islam, M.R.; Haque, M.A. Effect of anti-prolactin drug and peppermint on broodiness, laying performance and egg quality in indigenous hens. *Asian J. Med Biol. Res.* **2016**, *2*, 547–554. [CrossRef]
32. Crisostomo, S.; Guémené, D.; Garreau-Mills, M.; Morvan, C.; Zadworny, D. Prevention of incubation behavior expression in turkey hens by active immunization against prolactin. *Theriogenology* **1998**, *50*, 675–690. [CrossRef]
33. Dajee, M.; Fey, G.H.; Richards, J.S. Stat 5b and the orphan nuclear receptors regulate expression of the α2-macroglobulin (α2M) gene in rat ovarian granulosa cells. *Mol. Endocrinol.* **1998**, *12*, 1393–1409. [CrossRef] [PubMed]
34. Caster, J.M.; Patel, A.N.; Zhang, T.; Wang, A. Investigational nanomedicines in 2016: A review of nanotherapeutics currently undergoing clinical trials. *Wiley Interdiscip. Rev. Nanomed. Nanobiotechnol.* **2017**, *9*, e1416. [CrossRef] [PubMed]

35. Vautier, S.; Lacomblez, L.; Chacun, H.; Picard, V.; Gimenez, F.; Farinotti, R.; Fernandez, C. Interactions between the dopamine agonist, bromocriptine and the efflux protein, P-glycoprotein at the blood–brain barrier in the mouse. *Eur. J. Pharm. Sci.* **2006**, *27*, 167–174. [CrossRef] [PubMed]

36. Sainz, V.; Conniot, J.; Matos, A.I.; Peres, C.; Zupančič, E.; Moura, L.; Silva, L.C.; Florindo, H.F.; Gaspar, R.S. Regulatory aspects on nanomedicines. *Biochem. Biophys. Res. Commun.* **2015**, *468*, 504–510. [CrossRef]

37. Bobo, D.; Robinson, K.J.; Islam, J.; Thurecht, K.J.; Corrie, S.R. Nanoparticle-based medicines: A review of FDA-approved materials and clinical trials to date. *Pharm. Res.* **2016**, *33*, 2373–2387. [CrossRef]

38. Havel, H.A. Where are the nanodrugs? An industry perspective on development of drug products containing nanomaterials. *AAPS J.* **2016**, *18*, 1351–1353. [CrossRef]

39. Ciccarelli, E.; Grottoli, S.; Miola, C.; Avataneo, T.; Lancranjan, I.; Camanni, F. Double blind randomized study using oral or injectable bromocriptine in patients with hyperprolactinaemia. *Clin. Endocrinol.* **1994**, *40*, 193–198. [CrossRef] [PubMed]

Article

Modified *Spirulina maxima* Pectin Nanoparticles Improve the Developmental Competence of In Vitro Matured Porcine Oocytes

Pantu-Kumar Roy [1], Ahmad-Yar Qamar [1,2], Bereket-Molla Tanga [1,3], Seonggyu Bang [1], Gyeonghwan Seong [1], Xun Fang [1], Ghangyong Kim [1], Shan-Lakmal Edirisinghe [1], Mahanama De Zoysa [1], Do-Hyung Kang [4,5], Islam M. Saadeldin [1] and Jongki Cho [1,*]

[1] College of Veterinary Medicine, Chungnam National University, Daejeon 34134, Korea; vetpantu88@gmail.com (P.-K.R.); ahmad.qamar@uvas.edu.pk (A.-Y.Q.); tanga@o.cnu.ac.kr (B.-M.T.); bangsk97@o.cnu.ac.kr (S.B.); 202050377@o.cnu.ac.kr (G.S.); fx2442@o.cnu.ac.kr (X.F.); gykim1007@gmail.com (G.K.); shan.lakmal09011@gmail.com (S.-L.E.); mahanama@cnu.ac.kr (M.D.Z.); islamms@zu.edu.eg (I.M.S.)
[2] Department of Clinical Sciences, College of Veterinary and Animal Sciences, Jhang 35200, Pakistan
[3] Faculty of Veterinary Medicine, Hawassa University, Hawassa 05, Ethiopia
[4] Jeju Marine Research Center, Korea Institute of Ocean Science and Technology (KIOST), Jeju 63349, Korea; dohkang@kiost.ac.kr
[5] Department of Ocean Science, University of Science and Technology (UST), Jeju 63349, Korea
* Correspondence: cjki@cnu.ac.kr; Tel.: +82-42-821-6788

Citation: Roy, P.-K.; Qamar, A.-Y.; Tanga, B.-M.; Bang, S.; Seong, G.; Fang, X.; Kim, G.; Edirisinghe, S.-L.; De Zoysa, M.; Kang, D.-H.; et al. Modified *Spirulina maxima* Pectin Nanoparticles Improve the Developmental Competence of In Vitro Matured Porcine Oocytes. *Animals* **2021**, *11*, 2483. https://doi.org/10.3390/ani11092483

Academic Editors: Antonio Gonzalez-Bulnes and Nesreen Hashem

Received: 13 July 2021
Accepted: 20 August 2021
Published: 24 August 2021

Publisher's Note: MDPI stays neutral with regard to jurisdictional claims in published maps and institutional affiliations.

Simple Summary: Poor in vitro embryo development is a major obstacle in porcine assisted reproduction. In the current study, we utilized modified *Spirulina maxima* pectin nanoparticles as a supplement to improve porcine in vitro maturation medium. Results showed that modified *Spirulina maxima* pectin nanoparticles at 2.5 μg/mL improved oocyte maturation in form of first polar body extrusion, reduced oxidative stress, and increased the developmental competence of the oocytes after parthenogenetic activation and somatic cell nuclear transfer. Moreover, the relative transcripts quantification showed significant increase in the pluripotency-associated transcripts in the resultant cloned embryos after modified *Spirulina maxima* pectin nanoparticles supplementation. Therefore, we provide an optimum in vitro maturation condition to improve the in vitro embryo production in porcine.

Abstract: Molecular approaches have been used to determine metabolic substrates involved in the early embryonic processes to provide adequate culture conditions. To investigate the effect of modified *Spirulina maxima* pectin nanoparticles (MSmPNPs) on oocyte developmental competence, cumulus–oocyte complexes (COCs) retrieved from pig slaughterhouse ovaries were subjected to various concentrations of MSmPNPs (0, 2.5, 5.0, and 10 μg/mL) during in vitro maturation (IVM). In comparison to the control, MSmPNPs-5.0, and MSmPNPs-10 groups, oocytes treated with 2.5 μg/mL MSmPNPs had significantly increased glutathione (GSH) levels and lower levels of reactive oxygen species (ROS). Following parthenogenetic activation, the MSmPNPs-2.5 group had a considerably higher maturation and cleavage rates, blastocyst development, total cell number, and ratio of inner cell mass/trophectoderm (ICM:TE) cells, when compared with those in the control and all other treated groups. Furthermore, similar findings were reported for the developmental competence of somatic cell nuclear transfer (SCNT)-derived embryos. Additionally, the relative quantification of POU5F1, DPPA2, and NDP52 mRNA transcript levels were significantly higher in the MSmPNPs-2.5 group than in the control and other treated groups. Taken together, the current findings suggest that MSmPNP treatment alleviates oxidative stress and enhances the developmental competence of porcine in vitro matured oocytes after parthenogenetic activation and SCNT.

Keywords: *Spirulina maxima* pectin; nanoparticles; porcine; embryos; development

1. Introduction

Nanotechnology is a promising technique owing to its increasing applicability in economically areas, such as agriculture, industry, medicine, and public health [1,2]. Nanotechnology has been applied to improve animal production and health using different approaches [3,4]. Nanoparticles (NPs) have been used in disease diagnosis, drug administration, animal nutrition, reproduction, and food safety [5,6]. Because of their unique and distinctive physicochemical properties that differ significantly from bulk materials of the same composition, nanomaterials are being created for use in a wide range of commercial goods worldwide. There are some physicochemical properties give synthetic NPs features and a higher surface reactivity than their counterparts of the regular bulk materials; such as the minute size (increased surface area and size distribution), purity, surface reactivity, solubility, shape, and aggregation [7,8].

NPs, such as lipid core NPs, supplemented during in vitro maturation (IVM), have previously been shown to improve oocyte quality, embryo cleavage, and blastocyst rates [9]. Furthermore, chitosan NPs efficiently prevent oxidative damage to oocytes [10]. In a previous study, we examined the effects of chitosan NP supplementation during IVM on porcine oocyte developmental competence and pre-implantation development in parthenogenetic and cloned embryos [11]. NPs containing various antioxidant materials can scavenge reactive oxygen species (ROS) and thus protect cellular molecules, such as lipids, proteins, and DNA, from oxidative stress, [11–13]. Therefore, many NPs are currently being used to improve the entire process of in vitro embryo production (IVP), such as gametes cryopreservation, oocyte in vitro maturation, and embryo culture [12,14–17], but some of these NPs exert toxic effects [18–25]. To overcome these toxic effects, plant-derived nanoparticles are biologically safe and applicable for improving the quality of oocytes and subsequent embryo development [1,26].

Despite the chemical activation of in vitro fertilization (IVF), the therapeutic properties of plant extracts or their secondary derivatives on the regulation of folliculogenesis have been extensively studied under both in vitro and in vivo conditions [27]. In particular, plant polysaccharides have shown developmental competence of in vitro matured mouse oocytes by protective effects, such as reducing endoplasmic reticulum (ER) stress, preventing cell death, and activating both phosphatidylinositol 3-kinase (PI3K)/ Akt (protein kinase B, AKT) and mitogen-activated protein kinase (MAPK3/1) signaling [28].

Pectin is a cell wall ingredient in terrestrial plants, and feeds incorporated polysaccharides from livestock animals. The three major pectic polysaccharides are homogalacturonan (HG), rhamnogalacturonan-I (RG-I), and rhamnogalacturonan-II (RG-II) [29]. In clinical studies, pectin has been shown to have a broad range of immunomodulatory activities due to different molecular characteristics, such as source, extraction technique, degree of esterification (DE), degree of acetylation, and chemical modifications [30,31]. Among the other farm animals, pectin has shown a beneficial impact on pig nutrition and health contexts, which was performed in vivo, ex vivo, and in vitro studies [32]. The development of reproductive performance using crude extracts of microalgae has been widely discussed in livestock animals [33–36]. Microalgae *Spirulina* is a photosynthetic cyanobacterium that has been used to improve reproductive efficiency in in vivo bovine studies [33]. Recently, *S. maxima*-based pectin (SmP), and their modified two products namely, modified SmP (MSmP) and its nanoparticles (MSmPNPs) were investigated for different bioactivities in in vitro and in vivo studies [37–39]. Modification of SmP remarkably enhanced their functionality via physicochemical properties.

The production of cloned and transgenic pigs is a crucial step in xenotransplantation [40]. For successful production of these pigs, oxidative stress of recipient oocytes must be decreased during in vitro maturation (IVM), which can reprogram the gene-modified donor cells appropriately [41–43]. The goal of this study was to determine how MSmPNPs supplementation during IVM affected porcine oocyte developmental competence and pre-implantation development in parthenogenetic and cloned embryos. Consequently, this study was conducted to investigate the effects of MSmPNP supplementation on porcine oocyte maturation to reveal novel functional properties that enhance porcine IVP through parthenogenesis and SCNT. Intracellular glutathione (GSH) and ROS levels in the oocytes, pre-implantation embryo development, and the expression of a reprogramming-related gene were investigated.

2. Materials and Methods

2.1. Chemicals and Reagents

Unless otherwise stated, all chemicals and reagents were acquired from Sigma-Aldrich (St. Louis, MO, USA).

2.2. Preparation of MSmPNPs

S. maxima based MSmP was provided by the Jeju Marine Research Institute, Korea Institute of Ocean Science and Technology (KIOST), Jeju Special Self-Governing Province, Republic of Korea. The particle size of MSmP was further reduced by mechanical sonication. Briefly, MSmP was dissolved in deionized water and sonicated under an amplitude of 30%, 10:10 s pulse at 4 °C for 30 min (Sonics & Materials. Inc. Newtown, CT, USA). Sonicated MSmP was centrifuged at 3500 rpm for 15 min to collect MSmPNPs in the supernatant [39]. The average particle sizes of MSmp and MSmPNPs are 542.4 nm and 78.6 nm, respectively. Zeta potential of MSmp and MSmPNPs are −22.8 mV and −19.8 mV, respectively.

2.3. Oocyte In Vitro Maturation and MSmPNPs Treatment

The procedure for collecting porcine oocytes and IVM was performed according to our previous investigations [44]. Briefly, apparently healthy porcine ovaries were obtained within 4 h of slaughter at a local abattoir. Follicular fluid from 3–8 mm in diameter was aspirated into a 15 mL conical tube (Corning, Acton, MA, USA) using a 10 mL syringe and an 18-gauge needle. The fluid was rinsed with HEPES-buffered Tyrode's (TLH) medium (119 mM NaCl, 5 mM KCl, 25 mM HEPES buffer, 2 mM CaCl2, 2 mM MgCl2, 6 g/liter glucose, adjust pH to 7.4 with NaOH) containing 0.05% (*w/v*) polyvinyl alcohol (PVA) after the oocytes settled for 5 min. A stereomicroscope was used to select cumulus–oocyte complexes (COCs) with at least three layers of compact cumulus cells and homogenous ooplasm. COCs were washed three times with TLH–PVA and then with Dulbecco's phosphate-buffered saline (DPBS; Gibco, Life Technologies, Grand Island, NY, USA) modified with 0.4% bovine serum albumin (BSA) (mDPBS). The IVM medium consisted of TCM-199 (Gibco) supplemented with 10% (*v/v*) of porcine follicular fluid (pFF), 0.6mM cysteine, 0.91mM sodium pyruvate, 75 μg/mL kanamycin, 10 ng/mL epidermal growth factor (EGF), 1 μg/mL insulin, 10 IU/mL human chorionic gonadotrophin (hCG; Intervet International BV, Boxmeer, Holland), and 10 IU/mL pregnant mare serum gonadotrophin (PMSG). COCs were then incubated in IVM medium (50 COCs/500 μL) at 39 °C in a humidified environment containing 5% CO_2. COCs were moved into hormone-free IVM medium after 22 h of incubation and cultured in four-well multi dishes (SPL, Pocheon, South Korea) for another 22 h under the same conditions. The oocytes in the control group were not supplemented to MSmPNPs during IVM, while the remaining oocytes were separated into three groups and treated with varied doses of MSmPNPs (2.5, 5.0, and 10 μg/mL) for the first 22 h. Gentle pipetting of COCs in 0.5 mg/mL hyaluronidase solution (Catalog no. H7630, in PBS) in PBS was used to denude the in vitro matured oocytes. The oocyte morphology and appearance of the polar body in the perivitelline space were used to evaluate oocyte maturation, where oocytes in the metaphase II (MII) stage showed the first polar bodies with a dark distinct cytoplasm.

2.4. Measurement of Intracellular GSH and ROS Levels

The levels of intracellular GSH and ROS in in vitro-matured oocytes were measured as previously described [11,45]. The intensity of green fluorescence of 2,7-dichlorodihydrofluorescein diacetate was used to estimate intracellular ROS levels (H_2DCFDA; Invitrogen Corporation, Carlsbad, CA, USA). After 44 h of IVM, intracellular GSH levels were measured using

CellTracker^{TM} Blue, which contains the blue fluorescent dye 4-chloromethyl-6,8-difluoro-7-hydroxycoumarin (CMF$_2$HC, Invitrogen Corporation). Twenty oocytes from each group were incubated in TLH-PVA supplemented with 10 µM H$_2$DCFDA and 10 µM Cell Tracker Blue in the dark for 30 min. Finally, oocytes were analyzed using an epifluorescence microscope with UV filters (×200 magnification; Leica Application Suite X; Leica Microsystems, Wetzlar, Germany) (460 nm for ROS and 370 nm for GSH) (Figure 1a). The intensity of fluorescence was measured, and the photos were stored as TIFF graphic files for subsequent examination. The fluorescence intensity of the oocytes was standardized to that of the control oocytes using ImageJ software (version 1.41; National Institutes of Health, Bethesda, MD, USA).

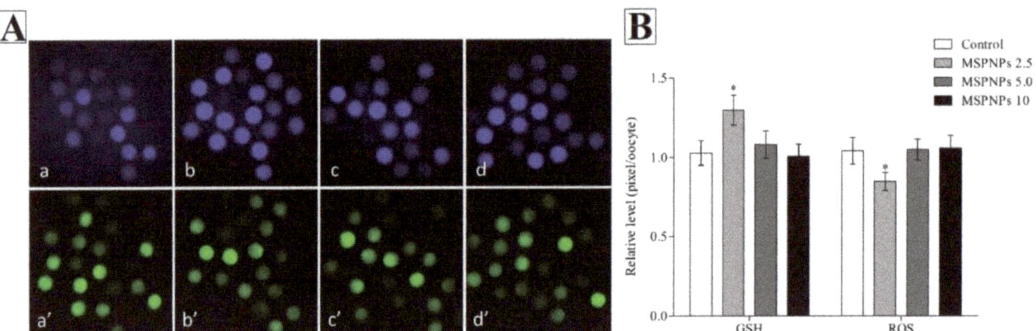

Figure 1. Evaluation of glutathione and reactive oxygen species of in vitro matured porcine oocytes at different concentrations of MSPNPs. (**A**) Oocytes were stained with (**a–d**) Cell Tracker Blue and (**a'–d'**) 2′,7′-dichlorodihydrofluorescein diacetate (H2DCFDA) to detect intracellular levels of glutathione (GSH) and reactive oxygen species (ROS), respectively, whereas (**a,a'**) control and (**b,b'**) MSPNPs-2.5, (**c,c'**) MSPNPs-5.0, and (**d,d'**) MSPNPs-10 matured oocytes. (**B**) Effects on intracellular levels of GSH and ROS in vitro matured porcine oocytes. * Indicates that there is a significant difference in the GSH and ROS levels between each group ($p < 0.05$). GSH samples, n = 60; ROS samples, n = 60. The experiment was independently replicated three times.

2.5. In Vitro Embryo Production

Oocytes were parthenogenetically activated (PA) or used as karyoplasts for SCNT, with some modifications to our previous studies [45]. For PA, COCs were cultured in IVM medium 22 h and hormone-free IVM medium for 22 h and were then exposed to 0.1% (*w/v*) hyaluronidase. Cumulus cells were then and repeatedly pipetted gently. The mature, good-quality oocytes were parthenogenetically triggered with two 60 µsec direct current (DC) pulses of 120 V/cm in 280 mM mannitol solution with 0.01 mM CaCl$_2$ and 0.05 mM MgCl$_2$ using a BTX 2001 Electro-cell Manipulator (BTX Harvard Apparatus, San Diego, CA, USA). For SCNT, a primary culture of donor cells was made from the kidney cells of an aborted cloned male pig at 50 days of gestation, which was chopped into small pieces and centrifuged three times. The culture medium comprised Dulbecco's Modified Eagle Medium (Gibco) with 10% (*v/v*) fetal bovine serum in a 60 mm tissue culture plate until a monolayer of cells was established with 70–80% confluency. For 48–72 h, donor cells in the G0/G1 stage of the cell cycle were synchronized by serum starvation. Donor cells were prepared by resuspending trypsinized cultured cells prior to nuclear transfer with 0.4% (*w/v*) BSA (TLH). Mature COCs were denuded as mentioned above. The denuded oocytes were then incubated in 5 µg/mL Hoechst 33342 medium for 15 min. A 17 µm beveled glass pipette was used to enucleate and extract polar bodies from metaphase II oocytes. A single cell was injected into the perivitelline space. In a 280 mM mannitol solution with a low Ca^{+2} concentration (0.001 mM), reconstructed SCNT oocytes were electrofused with two pulses of DC at 160 V/cm for 40 µs, followed by an AC of 2 V/cm for 2 s, using a BTX 2001 Electro-cell Manipulator (Harvard Apparatus, San Diego, CA, USA). Presumptive zygotes were activated in a 280 mM mannitol solution containing 0.01 mM $CaCl_2$ and 0.05 mM $MgCl_2$, after 30 min of fusing with two pulses of DC at 120 V/cm for 60 µs. Both PA and SCNT embryos were post-activated for 4 h with 10 µg/mL cytochalasin B and 6-dimethylaminopurine after electrical activation [46]. Activated oocytes were then washed three times in an in vitro culture medium (porcine zygote medium-5; IFP, Higashine, Yamagata, Japan) and 20 oocytes cultured in 25 µL droplets, covered with pre-warmed mineral oil, and the embryos were then cultured at 39 °C in a humidified atmosphere (5% O_2, 5% CO_2, and 90% N_2). Cleavage and blastocyst formation were measured on days two and six for embryo development and blastocyst formation, respectively. Cell numbers were counted on day six to determine the total cell number, the inner cell mass (ICM), and trophectoderm (TE) expression in accordance with the differential staining methodology described in our previous studies [45].

2.6. Analysis of mRNA Transcript Expression by Relative Quantitative Real-Time Polymerase Chain Reaction (qRT-PCR)

A relative quantitative polymerase chain reaction (qRT-PCR) was used to analyze the mRNA expression of genes involved in nuclear reprogramming and pluripotency (*POU5F1*, *NDP52*, and *DPPA2*). The primer sequences are listed in Table 1. TRIzol reagent (Invitrogen Corporation) was used to extract total RNA from six-day-old blastocysts from the untreated (control) and treated groups [47]. Reverse transcription 2X RT Pre-Mix (BioFACT Co., Ltd., Daejeon, Korea) and oligo dT primers were used to generate complementary DNA (cDNA) from 300 ng of total RNA (Neoprobe). The absorbance of a diluted RNA samples was measured at 260 and 280 nm by NanoDrop Spectrophotometer (Thermo Fisher Scientific, Pittsburgh, PA, USA). Each RNA sample consisted of 0.5µg/µL. The following reaction settings were used for RT-qPCR: denaturation at 95 °C for 15 min and 20 s, followed by 40 cycles of annealing and extension at 60 °C for 40 s (BIOFACT Co., Ltd., Daejeon, Korea). The expression of each target gene was measured in comparison to that of the internal control gene (β-actin). The threshold cycle (Ct) at a constant fluorescence intensity was used for relative quantification of gene expression using the $2^{-(\Delta Ct\,sample\,-\,\Delta Ct\,control)}$ method [48]. Each value was normalized to that of β-actin to determine the normalized arbitrary values for each gene.

Table 1. S Specific primers used for gene expression analysis by qPCR.

Genes	Gene Full Name	Sequences (5′-3′)	Product Size (bp)	NCBI Accession No.
β-actin	Beta actin	F: CCC TGG AGA GCT ACG AG R: TCC TTC CTG ATG TCC ACG TC	172	XM_003124280.5
POU5F1	POU class 5 homeobox 1	F: AGT GAG AGG CAA CCT GGA GA R: TCG TTG CGA ATA GTC ACT GC	166	XM_021097869.1
NDP52	Nuclear domain 10 protein	F: TGC TGA GTT ACA TGG GTC TGG R: ACC AAG GTC TGA TTT GCA GGT	182	XM_003131552.4
DPPA2	Developmental pluripotency associated 2	F: TGA GAG AGG GGA AAA GAC CAA R: TGG CAG AAA GGT CTC AAC AGA	151	XM_003358822.4

2.7. Experimental Design

The effects of supplementing porcine oocytes with or without 0, 2.5, 5.0, and 10 µg/mL MSmPNPs during IVM were examined in Experiment 1. Intracellular GSH and ROS levels were measured after 44 h of IVM. The effects of MSmPNP treatment during IVM of oocytes were investigated on the developmental competence of parthenogenetic and cloned embryos in Experiment 2. The effect of MSmPNP supplementation on the number of parthenogenetic and cloned blastocysts was investigated in Experiment 3. The effects of MSmPNP supplementation during IVM reprogramming-related genes (POU5F1, DPPA2, and NDP52) and the control gene (β-actin) in the cloned blastocysts acquired in Experiment 2 were investigated in Experiment 4.

2.8. Statistical Analysis

Origin software (version 8.1; OriginLab Corporation, Northampton, MA, USA) was used to analyze the data. All data are reported as mean $\pm$ standard error of the mean (SEM), with a probability (p) value of <0.05, regarded as statistically significant. The generalized linear model technique and one-way analysis of variance (ANOVA) were used to assess data on oocyte maturation, blastocyst development rates in PA and cloned embryos, cell number, GSH, ROS, and gene expression. Duncan's multiple range test was used to establish significance.

3. Results

3.1. GSH and ROS Intracellular Levels Treated with/without MSmPNPs

Following oocyte maturation, the levels of intracellular GSH and ROS were measured. Mature oocytes in the MSmPNPs-2.5 group had significantly higher intracellular GSH levels ($p < 0.05$)) compared to the control, MSmPNPs-5.0 and MSmPNPs-10 groups, (Figure 1b). Furthermore, intracellular ROS levels in the MSmPNPs-2.5 group were considerably lower ($p < 0.05$)) than those in the control and other MSmPNP-treated groups.

3.2. The Effect of MSmPNPs on the Developmental Competence of PA Embryos

To improve the maturation, cleavage, and blastocyst development rate, varying amounts of MSmPNPs were added to in vitro maturation media. The results showed that the MSmPNPs-2.5 group significantly increased ($p < 0.05$) maturation (91.0 $\pm$ 1.0% vs. 86.5 $\pm$ 1.3% vs. 80.0 $\pm$ 0.8% vs. 83.5 $\pm$ 1.8%, respectively) compared to MSmPNPs-5.0, MSmNPS-10, and control groups (Table 2). The MSmPNPS-2.5 group displayed a significantly increased cleavage (90.5 $\pm$ 0.8% vs. 86.4 $\pm$ 1.3% vs. 80.3 $\pm$ 1.1% vs. 85.1 $\pm$ 1.3%, respectively) and blastocysts rate (34.5 $\pm$ 1.4% vs. 29.6 $\pm$ 1.7% vs. 24.9 $\pm$ 1.1% vs. 29.2 $\pm$ 1.0%, respectively) than other groups.

Table 2. Effect of in vitro maturation of porcine oocytes with different concentrations of MSmPNPs on the in vitro development rate of PA embryos.

Conc. of MSmPNPs (µg/mL)	No. of COCs	No. of Matured Oocytes (%)	No. of Embryos (Mean $\pm$ SEM)		
			Cultured	Cleaved (%)	Blastocyst (%)
0 (Control)	200	167 (83.5 $\pm$ 1.8) [b]	161	137 (85.1 $\pm$ 1.3) [b]	47 (29.2 $\pm$ 1.0) [b]
2.5	200	182 (91.0 $\pm$ 1.0) [a]	168	152 (90.5 $\pm$ 0.8) [a]	58 (34.5 $\pm$ 1.4) [a]
5.0	200	173 (86.5 $\pm$ 1.3) [b]	162	140 (86.4 $\pm$ 1.3) [a,b]	48 (29.6 $\pm$ 1.7) [a,b]
10	200	160 (80.0 $\pm$ 0.8) [b]	157	126 (80.3 $\pm$ 1.1) [c]	39 (24.9 $\pm$ 1.1) [b]

[a,b] Values in the same column with different superscript letters are significantly different ($p < 0.05$). Number of replicates ($n = 4$).

3.3. MSmPNPs Effects on Cell Number of PA Embryos

Table 3 shows a comparison of the total number of cells, ICM, TE, and ICM:TE ratio in PA embryos between the MSmPNP-treated groups and the control group. Compared to the control group (42.8 $\pm$ 1.2%), the MSmPNPs-2.5 group (48.7 $\pm$ 1.4%) showed a significant ($p < 0.05$) increase in the total cell number than the MSmPNP$_S$-5.0 (44.0 $\pm$ 1.1%) and MSmPNPS-10 (40.2 $\pm$ 1.0%) groups. These results revealed that higher concentrations of MSmPNPs reduced the total cell number, the TE, ICM, and ICM:TE, while MSmPNPs-2.5 groups showed a significantly increased in these parameters.

Table 3. The effects of in vitro maturation of porcine oocytes with different concentrations of MSmPNPs on different cell number of in vitro PA blastocysts.

Conc. of MSmPNPs (µg/mL)	No. of Cells (Mean ± SEM)			Ratio (%) of ICM: TE
	Total Cells	**TE (%)**	**ICM (%)**	
0 (Control)	42.8 ± 1.2 [b,c]	33.8 (78.7 ± 1.0) [a]	9.0 (21.3 ± 1.0) [b]	27.9 ± 1.7 [b]
2.5	48.7 ± 1.4 [a]	36.6 (74.8 ± 0.8) [b]	12.0 (25.2 ± 0.8) [a]	34.5 ± 1.4 [a]
5.0	44.0 ± 1.1 [b]	33.4 (75.6 ± 0.7) [b,c]	10.7 (24.4 ± 0.7) [a,b]	32.8 ± 1.2 [a,b]
10	40.2 ± 1.0 [c]	32.0 (79.6 ± 0.9) [a]	8.2 (20.4 ± 0.9) [b]	26.1 ± 1.5 [b]

[a–c] Values in the same column with different superscript letters are significantly different ($p < 0.05$). Number of replicates ($n = 4$).

3.4. MSmPNPs Effects on the Developmental Competence of Cloned Embryos

Cloned embryos produced from MSmPNPs-2.5-treated oocytes had higher development rates than the other groups ($p < 0.05$) (Table 4). Oocytes in the MSmPNPs-2.5 group had significantly higher in maturation (89.5 ± 1.8%), cleavage (88.5 ± 1.6%), and blastocysts rates (31.1 ± 1.1%) than those in the control (83.0 ± 1.5, 84.7 ± 1.0, and 24.8 ± 0.9%, respectively), MSmPNPs-5.0 (86.5 ± 1.3, 87.4 ± 2.1, and 25.9 ± 0.9%, respectively), and MSmPNPs-10 (80.5 ± 1.4, 82.6 ± 1.3, and 22.7 ± 1.4%, respectively) treated groups.

Table 4. Effect of in vitro maturation of porcine oocytes with different concentrations of MSmPNPs on in vitro development rate of SCNT embryos.

Conc. of MSmPNPs (µg/mL)	No. of COCs	No. of Matured Oocytes (%)	No. of Embryos (Mean ± SEM)		
			Cultured	**Cleaved (%)**	**Blastocyst (%)**
0 (Control)	200	166 (83.0 ± 1.5) [b]	137	116 (84.7 ± 1.0)	34 (24.8 ± 0.9) [b]
2.5	200	179 (89.5 ± 1.8) [a]	148	131 (88.5 ± 1.6)	46 (31.1 ± 1.1) [a]
5.0	200	173 (86.5 ± 1.3) [b]	143	125 (87.4 ± 2.1)	37 (25.9 ± 0.9) [b]
10	200	161 (80.5 ± 1.4) [b]	132	109 (82.6 ± 1.3)	30 (22.7 ± 1.4) [b]

[a,b] Values in the same column with different superscript letters are significantly different ($p < 0.05$). Number of replicates ($n = 4$).

3.5. MSmPNPs Effects on Cloned Cell Number

Table 5 compares the total cell number, ICM, TE expression, and ICM:TE ratio of cloned embryos. In the MSmPNPs-2.5-treated oocytes group, the overall cell number (48.9 ± 1.5%), TE (74.0 ± 0.9%), ICM (26.0 ± 0.9%), and ICM:TE (35.7 ± 1.8%) were significantly higher ($p < 0.05$) than those in the control and other treated groups.

Table 5. Effect of in vitro maturation of porcine oocytes with different concentrations of MSmPNPs on different cell number of SCNT blastocysts.

Conc. of MSmPNPs (µg/mL)	No. of Cells (Mean ± SEM)			Ratio (%) of ICM: TE
	Total Cells	**TE (%)**	**ICM (%)**	
0 (Control)	41.9 ± 1.9 [b,c]	33.2 (78.7 ± 1.2) [a]	8.8 (21.3 ± 1.2) [b]	27.9 ± 2.2 [b]
2.5	48.9 ± 1.5 [a]	36.4 (74.0 ± 0.9) [b]	12.5 (26.0 ± 0.9) [a]	35.7 ± 1.8 [a]
5.0	44.4 ± 1.0 [b]	34.6 (77.7 ± 1.0) [a]	9.7 (22.3 ± 1.0) [b]	29.0 ± 1.7 [b]
10	39.2 ± 2.5 [c]	31.0 (78.5 ± 1.1) [a]	8.2 (21.5 ± 1.1) [b]	27.6 ± 1.7 [b]

[a–c] Values in the same column with different superscript letters are significantly different ($p < 0.05$). Number of replicates ($n = 4$).

3.6. MSmPNPs Effects on Reprogramming-Related Gene Expression

qRT-PCR analysis of cloned blastocyst mRNA transcripts supported the benefits of MSmPNPs-2.5 and MSmPNPs-5.0 supplementation during porcine oocyte IVM on the development of resultant embryos. The expression levels of the reprogramming-related genes, *POU5F1*, *DPPA2*, and *NDP52*, were significantly higher in cloned blastocysts produced from MSmPNPs-2.5-and MSmPNPs-5.0-treated oocytes than in MSmPNPs-10 and untreated control blastocysts. Moreover, the relative expression levels of these genes in the MSmPNPs-2.5 group were significantly higher than those in the other groups (Figure 2).

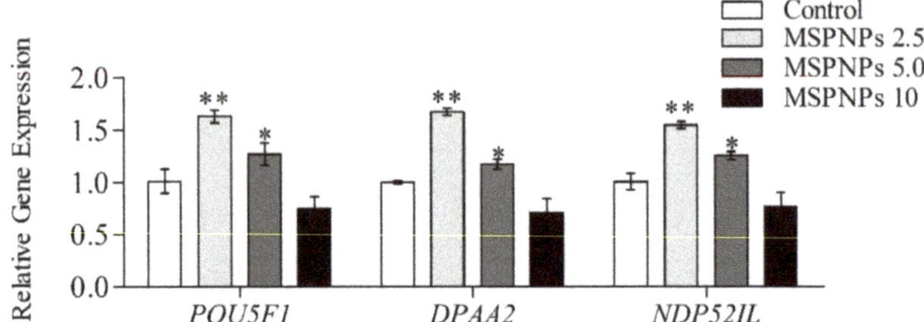

Figure 2. Relative expression of *POU5F1*, *DPPA2*, and *NDP52* mRNA transcripts in cloned blastocysts obtained from oocytes supplemented with varying concentrations of MSmPNPs (0, 2.5, 5.0, and 10 µg/mL) in maturation media. Each sample consisted of five blastocysts and three times replicate. * Indicates that there is a significant difference in mRNA values between each group ($p < 0.05$). ** Indicates that there is a significant difference in mRNA values at $p < 0.01$.

4. Discussion

Given the importance of genetically modified pigs in biomedical research, an ideal environment for porcine preimplantation embryo development in vitro is required. Interdisciplinary investigations between biology and nanotechnology may solve some problems associated with biological applications, particularly IVP in pigs [2]. To provide proper culture conditions during IVP, omics technologies are being applied to discover the metabolic substrates that are activated during early embryonic processes. Therefore, combining omics and nanotechnology to develop a suitable culture medium for improving porcine IVP would increase the number of pigs that are valid for transgenic technologies. However, the use of NPs showed a considerable increase in the risk of toxic effects to the animals, because they can cross the placental barrier, leading to embryo damage [18,49].

In this experiment, we used MSmPNPs as a component to improve porcine IVM and subsequent preimplantation embryo development. Recent study showed that the biliprotein phycocyanin, derived from *Spirulina platensis*, enhanced the developmental competence of the oocytes from obese female mice [50]. Additionally, IVM supplementation with phycocyanin improved the embryonic development of parthenogenetic and cloned embryos in porcine [51]. The natural polysaccharide pectin has gained increasing attention owing to its physicochemical and biomedical activities [52]. It exerts antioxidant activity in experimental, in vitro free radical scavenging [52–57], and in vivo models [58]. Furthermore, Zhang et al. reported that lower doses (5–25 µM) of naturally occurring phenolic compounds (rosmarinic acid) attenuated intracellular ROS levels in porcine oocytes and cumulus cells during IVM [59]. Moreover, low doses (100 µM) of the polyphenolic compound procyanidin, which is derived from plant sources, also decreased ROS production and apoptosis, while promoting the quality of oocytes and PA embryo development [60]. Therefore, it is ideal to investigate the anti-oxidative effects of MSmPNPs on the critical stage of IVP in porcine IVM and subsequent embryo development.

The results showed that MSmPNPs mediate the reduction in ROS levels and increase in GSH, which are favorable conditions during IVM [61,62]. Although ROS are generated during different cellular metabolic processes, in vitro-matured oocytes are more sensitive to oxidative stress [63]. Therefore, alleviating such negative effects of oxidative stress will enhance mitochondrial function and improve oocyte maturation and cleavage capabilities [64]. Therefore, we hypothesized that pectin-like biomaterials extracted from *Spirulina maxima* may have positive anti-oxidative and free radical scavenging effects during oocyte IVM.

Previous studies have shown that SmP and SmPNPs enhance wound healing [17], disease resistance, and stress tolerance [65] by reducing oxidative stress. There is no previous report on the use of *Spirulina* pectins to improve the quality of developing oocytes. Thus, to our knowledge, this is the first study to explore the effects of marine *S. maxima* pectin on oocyte maturation and developmental competence in oocytes *in vitro*. According to our findings, the optimum concentration of MSmPNPs to effectively neutralize IVM-derived ROS in porcine oocytes was 2.5 µg/mL. Compared to the control and other treatment groups, matured oocytes in the MSmPNPs 2.5-treated group had significantly higher intracellular GSH levels and lower ROS levels. Increased intracellular GSH levels have been linked to improved cytoplasmic maturation, embryonic development, and offspring production [3]. Furthermore, GSH levels are important for maintaining the cellular redox; a lack of GSH can lead to apoptotic stimuli in mature ovarian follicles [66,67]. In the present study, 2.5 µg/mL of MSmPNPs effectively reduced intracellular ROS levels and enhanced the non-enzymatic antioxidant (GSH) level in oocytes and cumulus cells at the end of IVM.

Interestingly, IVM medium containing 2.5 µg/mL of MSmPNPs markedly improved the quality of porcine developmental competence of PA and SCNT embryos, as indicated by enhanced hatching and total cell counts of blastocysts. Moreover, qRT-PCR analysis further confirmed the upregulated expression pattern of selected reprogramming and pluripotency-related genes (*POU5F1*, *DPAA2*, and *NDP52*) at lower doses of MSmPNPs (2.5 and 5 µg/mL). From the given results, it appears that increasing the concentrations of the MSmPNPs tends to be toxic to the IVM environment and the IVM conditions favors the reduced and optimal concentration, which is 2.5 µg/mL. *PPOU5F1* (or *OCT4)* is known as a master gene for pluripotency, as its expression controls early embryonic stem cells and development [68]. *DPPA2* (or *PESCRG1)* acts as a transcription factor, is involved in the maintenance of the active epigenetic status of these genes, and maintains the pluripotency of stem cells [69–71]. Moreover, *NDP52* (or *CALCOCO2)* is an *OCT4*-related gene, and its expression is limited to the pluripotent cells of the early embryo and the germline (blastocysts, epiblasts, and purified primordial germ cells) and regulates embryonic stem cell pluripotency and early blastocyst development [72,73]. Collectively, the activation of *POU5F1*, *DPPA2*, and *NDP52* improved the proper reprogramming of donor nuclei after reactivation of genes [74]. Overall, the results confirmed that excess ROS neutralization by antioxidant properties of MSmPNPs could effectively improve oocyte maturation and subsequent embryonic developmental competence of cloned and parthenogenetic embryos. Further investigations are required to study the effects on in vitro fertilization-derived embryos. This finding provides novel IVM conditions through the integration of innovative nanomaterials from marine *Spirulina*.

5. Conclusions

Oocytes treated with MSmPNPs during IVM resulted in a higher rate of pre-implantation porcine parthenogenetic and cloned embryos development. The highest impacts were seen at a supplementation of 2.5 µg/mL MSmPNPs and had a constructive impact on oocyte quality and embryonic development and embryo eminence by increasing the levels of intracellular GSH while reducing ROS levels, as well as increasing the expression of the pluripotency-associated genes *POU5F1*, *DPPA2*, and *NDP52* in the resultant blasto-cysts. This finding provides novel IVM conditions through the integration of innovative nanomaterials from marine *Spirulina*.

Author Contributions: Conceptualization, P.-K.R., M.D.Z., D.-H.K., and J.C.; methodology, P.-K.R., X.F., and G.K.; investigation, P.-K.R., G.K., and M.D.Z.; data curation, P.-K.R., A.-Y.Q., B.-M.T., G.S., I.M.S., and J.C.; writing—original draft preparation, P.-K.R., B.-M.T., S.B., S.-L.E., I.M.S., and M.D.Z.; writing—review and editing, P.-K.R., A.-Y.Q., M.D.Z., D.-H.K., I.M.S., and J.C.; project administration, M.D.Z. and J.C.; funding acquisition, D.-H.K. and J.C. All authors have read and agreed to the published version of the manuscript.

Funding: This research was supported by a grant from the Korea Health Technology R&D Project through the Korea Health Industry Development Institute, funded by the Ministry of Health and Welfare, Republic of Korea (grant number: HI13C0954), and from a research funds (PE99922) of Jeju Marine Research Institute, Korea Institute of Ocean Science and Technology (KIOST), Republic of Korea.

Institutional Review Board Statement: All experiments were approved and performed under the guidelines of the Institutional Animal Care and Use Committee of Chungnam National University (IACUC approval no.: CNU-00624).

Acknowledgments: We thank all the members of our laboratory for their technical support and helpful discussions.

Conflicts of Interest: The authors declare no conflict of interest.

References

1. Saadeldin, I.M.; Khalil, W.A.; Alharbi, M.G.; Lee, S.H. The Current Trends in Using Nanoparticles, Liposomes, and Exosomes for Semen Cryopreservation. *Animals* **2020**, *10*, 2281. [CrossRef]
2. Lucas, C.G.; Chen, P.R.; Seixas, F.K.; Prather, R.S.; Collares, T. Applications of omics and nanotechnology to improve pig embryo production in vitro. *Mol. Reprod. Dev.* **2019**, *86*, 1531–1547. [CrossRef]
3. Neculai-Valeanu, A.S.; Ariton, A.M.; Mădescu, B.M.; Rîmbu, C.M.; Creangă, Ş. Nanomaterials and Essential Oils as Candidates for Developing Novel Treatment Options for Bovine Mastitis. *Animals* **2021**, *11*, 1625. [CrossRef] [PubMed]
4. Abo-Al-Ela, H.G.; El-Kassas, S.; El-Naggar, K.; Abdo, S.E.; Jahejo, A.R.; Al Wakeel, R.A. Stress and immunity in poultry: Light management and nanotechnology as effective immune enhancers to fight stress. *Cell Stress Chaperones* **2021**, *26*, 457–472. [CrossRef]
5. Abdelnour, S.A.; Alagawany, M.; Hashem, N.M.; Farag, M.R.; Alghamdi, E.S.; Hassan, F.U.; Bilal, R.M.; Elnesr, S.S.; Dawood, M.A.O.; Nagadi, S.A.; et al. Nanominerals: Fabrication Methods, Benefits and Hazards, and Their Applications in Ruminants with Special Reference to Selenium and Zinc Nanoparticles. *Animals* **2021**, *11*, 1916. [CrossRef]
6. Hashem, N.M.; Gonzalez-Bulnes, A. Nanotechnology and Reproductive Management of Farm Animals: Challenges and Advances. *Animals* **2021**, *11*, 1932. [CrossRef] [PubMed]
7. Guisbiers, G.; Mejía-Rosales, S.; Leonard Deepak, F. Nanomaterial Properties: Size and Shape Dependencies. *J. Nanomater.* **2012**, *2012*, 1–2. [CrossRef]
8. Jeevanandam, J.; Barhoum, A.; Chan, Y.S.; Dufresne, A.; Danquah, M.K. Review on nanoparticles and nanostructured materials: History, sources, toxicity and regulations. *Beilstein J. Nanotechnol.* **2018**, *9*, 1050–1074. [CrossRef]
9. Lucas, C.G.; Remiao, M.H.; Bruinsmann, F.A.; Lopes, I.A.R.; Borges, M.A.; Feijo, A.L.S.; Basso, A.C.; Pohlmann, A.R.; Guterres, S.S.; Campos, V.F.; et al. High doses of lipid-core nanocapsules do not affect bovine embryonic development in vitro. *Toxicol. Vitr.* **2017**, *45*, 194–201. [CrossRef] [PubMed]
10. Abdel-Halim, B.R. Protective effect of Chitosan nanoparticles against the inhibitory effect of linoleic acid supplementation on maturation and developmental competence of bovine oocytes. *Theriogenology* **2018**, *114*, 143–148. [CrossRef] [PubMed]
11. Roy, P.K.; Qamar, A.Y.; Fang, X.; Kim, G.; Bang, S.; De Zoysa, M.; Shin, S.T.; Cho, J. Chitosan nanoparticles enhance developmental competence of in vitro-matured porcine oocytes. *Reprod. Domest. Anim.* **2021**, *56*, 342–350. [CrossRef]
12. Bakhtari, A.; Nazari, S.; Alaee, S.; Kargar-Abarghouei, E.; Mesbah, F.; Mirzaei, E.; Molaei, M.J. Effects of Dextran-Coated Superparamagnetic Iron Oxide Nanoparticles on Mouse Embryo Development, Antioxidant Enzymes and Apoptosis Genes Expression, and Ultrastructure of Sperm, Oocytes and Granulosa Cells. *Int. J. Fertil. Steril.* **2020**, *14*, 161–170. [CrossRef]
13. Shehata, A.M.; Salem, F.M.S.; El-Saied, E.M.; Abd El-Rahman, S.S.; Mahmoud, M.Y.; Noshy, P.A. Zinc Nanoparticles Ameliorate the Reproductive Toxicity Induced by Silver Nanoparticles in Male Rats. *Int. J. Nanomed.* **2021**, *16*, 2555–2568. [CrossRef]
14. Abbasi, Y.; Hajiaghalou, S.; Baniasadi, F.; Mahabadi, V.P.; Ghalamboran, M.R.; Fathi, R. Fe$_3$O$_4$ magnetic nanoparticles improve the vitrification of mouse immature oocytes and modulate the pluripotent genes expression in derived pronuclear-stage embryos. *Cryobiology* **2021**, *100*, 81–89. [CrossRef] [PubMed]
15. El-Naby, A.-s.A.-H.H.; Ibrahim, S.; Hozyen, H.F.; Sosa, A.S.A.; Mahmoud, K.G.M.; Farghali, A.A. Impact of nano-selenium on nuclear maturation and genes expression profile of buffalo oocytes matured in vitro. *Mol. Biol. Rep.* **2020**, *47*, 8593–8603. [CrossRef] [PubMed]
16. Ariu, F.; Bogliolo, L.; Pinna, A.; Malfatti, L.; Innocenzi, P.; Falchi, L.; Bebbere, D.; Ledda, S. Cerium oxide nanoparticles (CeO$_2$ NPs) improve the developmental competence of in vitro-matured prepubertal ovine oocytes. *Reprod. Fertil. Dev.* **2017**, *29*, 1046. [CrossRef] [PubMed]
17. Li, W.; Zhou, X.; Dai, J.; Zhang, D.; Liu, B.; Wang, H.; Xu, L. Effect of hydroxyapatite nanoparticles on MII-stage porcine oocytes vitrification and the study of its mechanism. *J. Biomed. Eng. = Shengwu Yixue Gongchengxue Zazhi* **2013**, *30*, 789–793.
18. Huang, C.H.; Yeh, J.M.; Chan, W.H. Hazardous impacts of silver nanoparticles on mouse oocyte maturation and fertilization and fetal development through induction of apoptotic processes. *Environ. Toxicol.* **2018**, *33*, 1039–1049. [CrossRef]
19. Santacruz-Márquez, R.; González-De los Santos, M.; Hernández-Ochoa, I. Ovarian toxicity of nanoparticles. *Reprod. Toxicol.* **2021**, *103*, 79–95. [CrossRef]
20. Lin, Y.H.; Zhuang, S.X.; Wang, Y.L.; Lin, S.; Hong, Z.W.; Liu, Y.; Xu, L.; Li, F.P.; Xu, B.H.; Chen, M.H.; et al. The effects of graphene quantum dots on the maturation of mouse oocytes and development of offspring. *J. Cell. Physiol.* **2019**, *234*, 13820–13831. [CrossRef]
21. Taylor, U.; Tiedemann, D.; Rehbock, C.; Kues, W.A.; Barcikowski, S.; Rath, D. Influence of gold, silver and gold–Silver alloy nanoparticles on germ cell function and embryo development. *Beilstein J. Nanotechnol.* **2015**, *6*, 651–664. [CrossRef]

22. Karimipour, M.; Zirak Javanmard, M.; Ahmadi, A.; Jafari, A. Oral administration of titanium dioxide nanoparticle through ovarian tissue alterations impairs mice embryonic development. *Int. J. Reprod. Biomed.* **2018**, *16*, 397–404. [CrossRef]

23. Lei, R.; Bai, X.; Chang, Y.; Li, J.; Qin, Y.; Chen, K.; Gu, W.; Xia, S.; Zhang, J.; Wang, Z.; et al. Effects of Fullerenol Nanoparticles on Rat Oocyte Meiosis Resumption. *Int. J. Mol. Sci.* **2018**, *19*, 699. [CrossRef]

24. Wang, R.; Song, B.; Wu, J.; Zhang, Y.; Chen, A.; Shao, L. Potential adverse effects of nanoparticles on the reproductive system. *Int. J. Nanomed.* **2018**, *13*, 8487–8506. [CrossRef] [PubMed]

25. Liu, J.; Zhao, Y.; Ge, W.; Zhang, P.; Liu, X.; Zhang, W.; Hao, Y.; Yu, S.; Li, L.; Chu, M.; et al. Oocyte exposure to ZnO nanoparticles inhibits early embryonic development through the γ-H2AX and NF-κB signaling pathways. *Oncotarget* **2017**, *8*, 42673–42692. [CrossRef] [PubMed]

26. Ismail, A.A.; Abdel-Khalek, A.E.; Khalil, W.A.; Yousif, A.I.; Saadeldin, I.M.; Abomughaid, M.M.; El-Harairy, M.A. Effects of mint, thyme, and curcumin extract nanoformulations on the sperm quality, apoptosis, chromatin decondensation, enzyme activity, and oxidative status of cryopreserved goat semen. *Cryobiology* **2020**, *97*, 144–152. [CrossRef] [PubMed]

27. Mbemya, G.T.; Vieira, L.A.; Canafistula, F.G.; Pessoa, O.D.L.; Rodrigues, A.P.R. Reports on in vivo and in vitro contribution of medicinal plants to improve the female reproductive function. *Reprod. Clim.* **2017**, *32*, 109–119. [CrossRef]

28. Yang, L.; Lei, L.; Zhao, Q.; Gao, Z.; Xu, X. Lycium barbarum polysaccharide improves the development of mouse oocytes vitrified at the germinal vesicle stage. *Cryobiology* **2018**, *85*, 7–11. [CrossRef]

29. Willats, W.G.; McCartney, L.; Mackie, W.; Knox, J.P. Pectin: Cell biology and prospects for functional analysis. *Plant Mol. Biol.* **2001**, *47*, 9–27. [CrossRef]

30. de Moura, F.A.; Macagnan, F.T.; Dos Santos, L.R.; Bizzani, M.; de Oliveira Petkowicz, C.L.; da Silva, L.P. Characterization and physicochemical properties of pectins extracted from agroindustrial by-products. *J. Food Sci. Technol.* **2017**, *54*, 3111–3117. [CrossRef]

31. Fracasso, A.F.; Perussello, C.A.; Carpiné, D.; Petkowicz, C.L.O.; Haminiuk, C.W.I. Chemical modification of citrus pectin: Structural, physical and rheologial implications. *Int. J. Biol. Macromol.* **2018**, *109*, 784–792. [CrossRef] [PubMed]

32. Wiese, M. The potential of pectin to impact pig nutrition and health: Feeding the animal and its microbiome. *FEMS Microbiol. Lett.* **2019**, *366*, fnz029. [CrossRef] [PubMed]

33. Mizera, A.; Kuczaj, M.; Szul, A. Impact of the Spirulina maxima extract addition to semen extender on bovine sperm quality. *Ital. J. Anim. Sci.* **2019**, *18*, 601–607. [CrossRef]

34. Barkallah, M.; Slima, A.B.; Elleuch, F.; Fendri, I.; Pichon, C.; Abdelkafi, S.; Baril, P. Protective Role of Spirulina platensis Against Bifenthrin-Induced Reprotoxicity in Adult Male Mice by Reversing Expression of Altered Histological, Biochemical, and Molecular Markers Including MicroRNAs. *Biomolecules* **2020**, *10*, 753. [CrossRef]

35. Lee, A.V.; You, L.; Oh, S.Y.; Li, Z.; Fisher-Heffernan, R.E.; Regnault, T.R.H.; de Lange, C.F.M.; Huber, L.; Karrow, N.A. Microalgae supplementation to late gestation sows and its effects on the health status of weaned piglets fed diets containing high- or low-quality protein sources. *Vet. Immunol. Immunopathol.* **2019**, *218*, 109937. [CrossRef]

36. Senosy, W.; Kassab, A.Y.; Mohammed, A.A. Effects of feeding green microalgae on ovarian activity, reproductive hormones and metabolic parameters of Boer goats in arid subtropics. *Theriogenology* **2017**, *96*, 16–22. [CrossRef] [PubMed]

37. Edirisinghe, S.L.; Dananjaya, S.H.S.; Nikapitiya, C.; Liyanage, T.D.; Lee, K.A.; Oh, C.; Kang, D.H.; De Zoysa, M. Novel pectin isolated from Spirulina maxima enhances the disease resistance and immune responses in zebrafish against Edwardsiella piscicida and Aeromonas hydrophila. *Fish Shellfish Immunol.* **2019**, *94*, 558–565. [CrossRef] [PubMed]

38. Edirisinghe, S.L.; Rajapaksha, D.C.; Nikapitiya, C.; Oh, C.; Lee, K.-A.; Kang, D.-H.; De Zoysa, M. Spirulina maxima derived marine pectin promotes the in vitro and in vivo regeneration and wound healing in zebrafish. *Fish Shellfish Immunol.* **2020**, *107*, 414–425. [CrossRef]

39. Chandrarathna, H.; Liyanage, T.D.; Edirisinghe, S.L.; Dananjaya, S.H.S.; Thulshan, E.H.T.; Nikapitiya, C.; Oh, C.; Kang, D.H.; De Zoysa, M. Marine Microalgae, Spirulina maxima-Derived Modified Pectin and Modified Pectin Nanoparticles Modulate the Gut Microbiota and Trigger Immune Responses in Mice. *Mar. Drugs* **2020**, *18*, 175. [CrossRef]

40. Park, D.S.; Kim, S.; Koo, D.-B.; Kang, M.-J. Current Status of Production of Transgenic Livestock by Genome Editing Technology. *J. Anim. Reprod. Biotechnol.* **2019**, *34*, 148–156. [CrossRef]

41. Roy, P.K.; Qamar, A.Y.; Fang, X.; Hassan, B.M.S.; Cho, J. Effects of cobalamin on meiotic resumption and developmental competence of growing porcine oocytes. *Theriogenology* **2020**, *154*, 24–30. [CrossRef]

42. Lin, T.; Lee, J.E.; Kang, J.W.; Oqani, R.K.; Cho, E.S.; Kim, S.B.; Il Jin, D. Melatonin supplementation during prolonged in vitro maturation improves the quality and development of poor-quality porcine oocytes via anti-oxidative and anti-apoptotic effects. *Mol. Reprod. Dev.* **2018**, *85*, 665–681. [CrossRef] [PubMed]

43. Kim, J.-W.; Park, H.-J.; Yang, S.-G.; Koo, D.-B. Anti-oxidative effects of exogenous ganglioside GD1a and GT1b on embryonic developmental competence in pigs. *J. Anim. Reprod. Biotechnol.* **2020**, *35*, 347–356. [CrossRef]

44. Cho, J.; Kim, G.; Qamar, A.Y.; Fang, X.; Roy, P.K.; Tanga, B.M.; Bang, S.; Kim, J.K.; Galli, C.; Perota, A.; et al. Improved efficiencies in the generation of multigene-modified pigs by recloning and using sows as the recipient. *Zygote* **2021**, 1–8. [CrossRef]

45. Roy, P.K.; Qamar, A.Y.; Tanga, B.M.; Fang, X.; Kim, G.; Bang, S.; Cho, J. Enhancing Oocyte Competence With Milrinone as a Phosphodiesterase 3A Inhibitor to Improve the Development of Porcine Cloned Embryos. *Front. Cell Dev. Biol.* **2021**, *9*, 647616. [CrossRef] [PubMed]

46. Roy, P.K.; Kim, G.; Fang, X.; Hassan, B.; Soysa, M.D.; Shin, S.T.; Cho, J.K. Optimization of post-activation systems to improve the embryonic development in porcine parthenogenesis and somatic cell nuclear transfer. *J. Embryo Transf.* **2017**, *32*, 95–104. [CrossRef]

47. Kim, G.; Roy, P.K.; Fang, X.; Hassan, B.M.; Cho, J. Improved preimplantation development of porcine somatic cell nuclear transfer embryos by caffeine treatment. *J. Vet. Sci.* **2019**, *20*, e31. [CrossRef] [PubMed]

48. Livak, K.J.; Schmittgen, T.D. Analysis of Relative Gene Expression Data Using Real-Time Quantitative PCR and the $2^{-\Delta\Delta CT}$ Method. *Methods* **2001**, *25*, 402–408. [CrossRef] [PubMed]

49. Wang, Z.; Wang, Z. Nanoparticles induced embryo–fetal toxicity. *Toxicol. Ind. Health* **2020**, *36*, 181–213. [CrossRef]

50. Wen, X.; Han, Z.; Liu, S.-J.; Hao, X.; Zhang, X.-J.; Wang, X.-Y.; Zhou, C.-J.; Ma, Y.-Z.; Liang, C.-G. Phycocyanin Improves Reproductive Ability in Obese Female Mice by Restoring Ovary and Oocyte Quality. *Front. Cell Dev. Biol.* **2020**, *8*, 1208. [CrossRef]

51. Liang, S.; Guo, J.; Jin, Y.X.; Yuan, B.; Zhang, J.B.; Kim, N.H. C-Phycocyanin supplementation during in vitro maturation enhances pre-implantation developmental competence of parthenogenetic and cloned embryos in pigs. *Theriogenology* **2018**, *106*, 69–78. [CrossRef]

52. Wathoni, N.; Yuan Shan, C.; Yi Shan, W.; Rostinawati, T.; Indradi, R.B.; Pratiwi, R.; Muchtaridi, M. Characterization and antioxidant activity of pectin from Indonesian mangosteen (*Garcinia mangostana* L.) rind. *Heliyon* **2019**, *5*, e02299. [CrossRef] [PubMed]

53. Smirnov, V.V.; Golovchenko, V.V.; Vityazev, F.V.; Patova, O.A.; Selivanov, N.Y.; Selivanova, O.G.; Popov, S.V. The Antioxidant Properties of Pectin Fractions Isolated from Vegetables Using a Simulated Gastric Fluid. *J. Chem.* **2017**, *2017*, 1–10. [CrossRef]

54. Hosseini Abari, A.; Amini Rourani, H.; Ghasemi, S.M.; Kim, H.; Kim, Y.-G. Investigation of antioxidant and anticancer activities of unsaturated oligo-galacturonic acids produced by pectinase of Streptomyces hydrogenans YAM1. *Sci. Rep.* **2021**, *11*, 8491. [CrossRef]

55. Ogutu, F.O.; Mu, T.-H. Ultrasonic degradation of sweet potato pectin and its antioxidant activity. *Ultrason. Sonochem.* **2017**, *38*, 726–734. [CrossRef] [PubMed]

56. Chen, R.; Jin, C.; Tong, Z.; Lu, J.; Tan, L.; Tian, L.; Chang, Q. Optimization extraction, characterization and antioxidant activities of pectic polysaccharide from tangerine peels. *Carbohydr. Polym.* **2016**, *136*, 187–197. [CrossRef]

57. Kang, H.J.; Jo, C.; Kwon, J.H.; Son, J.H.; An, B.J.; Byun, M.W. Antioxidant and Cancer Cell Proliferation Inhibition Effect of Citrus Pectin-Oligosaccharide Prepared by Irradiation. *J. Med. Food* **2006**, *9*, 313–320. [CrossRef]

58. Li, T.; Li, S.; Dong, Y.; Zhu, R.; Liu, Y. Antioxidant activity of penta-oligogalacturonide, isolated from haw pectin, suppresses triglyceride synthesis in mice fed with a high-fat diet. *Food Chem.* **2014**, *145*, 335–341. [CrossRef]

59. Zhang, Y.; Guo, J.; Nie, X.W.; Li, Z.Y.; Wang, Y.M.; Liang, S.; Li, S. Rosmarinic acid treatment during porcine oocyte maturation attenuates oxidative stress and improves subsequent embryo development in vitro. *PeerJ* **2019**, *7*, e6930. [CrossRef]

60. Gao, W.; Jin, Y.; Hao, J.; Huang, S.; Wang, D.; Quan, F.; Ren, W.; Zhang, J.; Zhang, M.; Yu, X. Procyanidin B1 promotes in vitro maturation of pig oocytes by reducing oxidative stress. *Mol. Reprod. Dev.* **2021**, *88*, 55–66. [CrossRef] [PubMed]

61. Choi, J.-Y.; Kang, J.-T.; Park, S.-J.; Kim, S.-J.; Moon, J.-H.; Saadeldin, I.M.; Jang, G.; Lee, B.-C. Effect of 7,8-Dihydroxyflavone as an Antioxidant on In Vitro Maturation of Oocytes and Development of Parthenogenetic Embryos in Pigs. *J. Reprod. Dev.* **2013**, *59*, 450–456. [CrossRef] [PubMed]

62. Kang, J.-T.; Moon, J.H.; Choi, J.-Y.; Park, S.J.; Kim, S.J.; Saadeldin, I.M.; Lee, B.C. Effect of Antioxidant Flavonoids (Quercetin and Taxifolin) on In Vitro Maturation of Porcine Oocytes. *Asian-Australas. J. Anim. Sci.* **2016**, *29*, 352–358. [CrossRef] [PubMed]

63. Guerin, P. Oxidative stress and protection against reactive oxygen species in the pre-implantation embryo and its surroundings. *Hum. Reprod. Update* **2001**, *7*, 175–189. [CrossRef] [PubMed]

64. Somfai, T.; Kaneda, M.; Akagi, S.; Watanabe, S.; Haraguchi, S.; Mizutani, E.; Dang-Nguyen, T.Q.; Geshi, M.; Kikuchi, K.; Nagai, T. Enhancement of lipid metabolism with L-carnitine during in vitro maturation improves nuclear maturation and cleavage ability of follicular porcine oocytes. *Reprod. Fertil. Dev.* **2011**, *23*, 912–920. [CrossRef]

65. Rajapaksha, D.C.; Edirisinghe, S.L.; Nikapitiya, C.; Dananjaya, S.; Kwun, H.J.; Kim, C.H.; Oh, C.; Kang, D.H.; De Zoysa, M. Spirulina maxima Derived Pectin Nanoparticles Enhance the Immunomodulation, Stress Tolerance, and Wound Healing in Zebrafish. *Mar. Drugs* **2020**, *18*, 556. [CrossRef] [PubMed]

66. Devine, P.J.; Perreault, S.D.; Luderer, U. Roles of Reactive Oxygen Species and Antioxidants in Ovarian Toxicity1. *Biol. Reprod.* **2012**, *86*, 1–10. [CrossRef]

67. Circu, M.L.; Yee Aw, T. Glutathione and apoptosis. *Free Radic. Res.* **2009**, *42*, 689–706. [CrossRef]

68. Kim, S.J.; Koo, O.J.; Park, H.J.; Moon, J.H.; da Torre, B.R.; Javaregowda, P.K.; Kang, J.T.; Park, S.J.; Saadeldin, I.M.; Choi, J.Y.; et al. Oct4 overexpression facilitates proliferation of porcine fibroblasts and development of cloned embryos. *Zygote* **2014**, *23*, 704–711. [CrossRef]

69. Sato, Y.; Kobayashi, H.; Higuchi, T.; Shimada, Y.; Era, T.; Kimura, S.; Eto, Y.; Ida, H.; Ohashi, T. Disease modeling and lentiviral gene transfer in patient-specific induced pluripotent stem cells from late-onset Pompe disease patient. *Mol. Ther. Methods Clin. Dev.* **2015**, *2*, 15023. [CrossRef]

70. Eckersley-Maslin, M.A. Keeping your options open: Insights from Dppa2/4 into how epigenetic priming factors promote cell plasticity. *Biochem. Soc. Trans.* **2020**, *48*, 2891–2902. [CrossRef]

71. Lim, P.S.L.; Meshorer, E. Dppa2 and Dppa4 safeguard bivalent chromatin in order to establish a pluripotent epigenome. *Nat. Struct. Mol. Biol.* **2020**, *27*, 685–686. [CrossRef] [PubMed]

72. Hellweg, C.E.; Shinde, V.; Srinivasan, S.P.; Henry, M.; Rotshteyn, T.; Baumstark-Khan, C.; Schmitz, C.; Feles, S.; Spitta, L.F.; Hemmersbach, R.; et al. Radiation Response of Murine Embryonic Stem Cells. *Cells* **2020**, *9*, 650. [CrossRef] [PubMed]
73. Bortvin, A.; Eggan, K.; Skaletsky, H.; Akutsu, H.; Berry, D.L.; Yanagimachi, R.; Page, D.C.; Jaenisch, R. Incomplete reactivation of Oct4-related genes in mouse embryos cloned from somatic nuclei. *Development* **2003**, *130*, 1673–1680. [CrossRef]
74. You, J.; Lee, J.; Kim, J.; Park, J.; Lee, E. Post-fusion treatment with MG132 increases transcription factor expression in somatic cell nuclear transfer embryos in pigs. *Mol. Reprod. Dev.* **2010**, *77*, 149–157. [CrossRef] [PubMed]

Article

In Vitro and *In Vivo* Assessment of Dietary Supplementation of Both Natural or Nano-Zeolite in Goat Diets: Effects on Ruminal Fermentation and Nutrients Digestibility

Amr El-Nile [1], Mahmoud Elazab [1], Hani El-Zaiat [2,3], Kheir El-Din El-Azrak [2], Alaa Elkomy [1,4], Sobhy Sallam [2,*] and Yosra Soltan [2]

[1] Department of Livestock Research, Arid Land Cultivation Research Institute, City of Scientific Research and Technological Applications, New Borg El-Arab, Alexandria 21934, Egypt; amr_elneel88@yahoo.com (A.E.-N.); melazab@srtacity.sci.eg (M.E.); alaa_elkomy@yahoo.com (A.E.)

[2] Department of Animal and Fish Production, Faculty of Agriculture, Alexandria University, Alexandria 21545, Egypt; hm_elzaiat@yahoo.com (H.E.-Z.); kheir_elazrak@yahoo.com (K.E.-D.E.-A.); uosra_eng@yahoo.com (Y.S.)

[3] Department of Animal and Veterinary Sciences, College of Agricultural and Marine Sciences, Sultan Qaboos University, P.O. Box 34, Al-Khod 123, Oman

[4] Faculty of Desert and Environmental Agriculture, Matrouh University, Matrouh 51512, Egypt

* Correspondence: s_sallam@yahoo.com or soubhy.salam@alexu.edu.eg; Tel.: +20-1001729576

Citation: El-Nile, A.; Elazab, M.; El-Zaiat, H.; El-Azrak, K.E.-D.; Elkomy, A.; Sallam, S.; Soltan, Y. *In Vitro* and *In Vivo* Assessment of Dietary Supplementation of Both Natural or Nano-Zeolite in Goat Diets: Effects on Ruminal Fermentation and Nutrients Digestibility. *Animals* **2021**, *11*, 2215. https://doi.org/10.3390/ani11082215

Academic Editors: Antonio Gonzalez-Bulnes, Nesreen Hashem and Byeng Ryel Min

Received: 13 May 2021
Accepted: 5 July 2021
Published: 27 July 2021

Publisher's Note: MDPI stays neutral with regard to jurisdictional claims in published maps and institutional affiliations.

Simple Summary: Increasing fibrous feed digestibility while reducing methane (CH_4) emission through manipulating rumen fermentation patterns to improve animal performance is the most critical challenge in the animal nutrition field. Nanotechnology has revolutionized the commercial application of nano-sized minerals in medicine, engineering, information, environmental technology, pigments, food, electronics appliances, biological and pharmaceutical applications, and many more. Therefore, animal nutrition scientists also resorted to using minerals and clays such as zeolite with different forms in feeding animals and evaluate this additive in animal performance. The natural zeolite clay is known for its high cation exchange capacity and adsorption characteristics that can modify ruminal fluid viscosity and binding capacity with ammonia (NH_3-N). After evaluating the addition of zeolite *in vivo* and *in vitro*, results indicated that zeolite (natural and nano forms) maintained rumen pH, increased protozoa numbers, and improved propionate production. Medium supplementation level of the natural form of zeolite at 20 g/kg dry matter (DM) was the most efficient dose in reducing CH_4 production, while the zeolite nano-form supplemented at 0.4 g/kg DM was the most effective dose in improving the organic matter (OM) degradability and reducing the NH_3-N concentration compared to the control.

Abstract: This study aimed to evaluate *in vitro* and *in vivo* dietary supplementation with different levels of natural or nano-zeolite forms on rumen fermentation patterns and nutrient digestibility. In the *in vitro* experiment, a basal diet (50% concentrate: 50% forage) was incubated without additives (control) and with natural zeolite (10, 20, 30 g/kg DM) or nano-zeolite (0.2, 0.3, 0.4, 0.5, 1.0 g/kg DM) for 24 h to assess their effect on ruminal fermentation, feed degradability, and gas and methane production using a semi-automatic system of *in vitro* gas production (GP). The most effective doses obtained from the *in vitro* experiment were evaluated *in vivo* using 30 Barki goats (26 ± 0.9 SE kg body weight). Goats were allocated into three dietary treatments (n = 10/treatment) as follows: control (basal diet without any supplementations), natural zeolite (20 g/kg DM diet), and nano-zeolite (0.40 g/kg DM diet). The *in vitro* results revealed that only the nano-zeolite supplementation form quadratically (p = 0.004) increased GP, and the level of 0.5 g/kg DM had the highest GP value compared to the control. Both zeolite forms affected the CH_4 production, linear, and quadratic reductions (p < 0.05) in CH_4 (mL/g DM), consistent with linear increases in truly degraded organic matter (TDOM) (p = 0.09), and propionate molar proportions (p = 0.007) were observed by nano zeolite treatment, while the natural form of zeolite resulted in a linear CH_4 reduction consistent with a linear decrease (p = 0.004) in NH_3-N, linear increases in TDOM (p = 0.09), and propionate molar proportions (p = 0.004). Results of the *in vivo* experiment demonstrated that the nutrient digestibility was similar among all treatments. Nano zeolite enhanced ($p < 0.05$) the total short-chain fatty acids and butyrate concentrations, while both zeolite forms decreased ($p < 0.001$) NH_3-N compared to the control. These results suggested that both zeolite supplementation forms favorably modified the rumen fermentation in different patterns.

Keywords: zeolite; nano-zeolite; *in vitro* gas production; digestibility; goat; methane emission; clay minerals

1. Introduction

The application of feed additives in ruminant rations is one solution to improve the animal's performance via manipulating ruminal fermentation patterns and improving nutrients utilization. Microbial fermentation of the dietary organic matter results in loss of gross energy and nitrogen. Enteric CH_4 emission in ruminants represents a loss of up to 15% of gross energy of feeds; also, 75–85% of the nitrogen consumed by ruminants is excreted in the feces and urine [1]. Therefore, enhancing fibrous feed digestibility, reducing CH_4 emission, and nitrogen excretion by ruminants have to improve their performance [2].

Natural zeolite clay is composited of crystalline aluminosilicates and characterized by a high cation exchange capacity, high sorbent property that can modify ruminal fluid viscosity and binding capacity with NH_3-N; therefore, it has been extensively used as a potential feed additive [3]. It also can capture ammonium ions, reducing the rate of their release and absorption from the rumen wall, and act as adsorbents for mycotoxins [4]. Besides, clinoptilolite of zeolite can enhance microbial ruminal fermentation by regulating ruminal pH to act as a pH-buffering agent [5].

The literature reported that zeolite supplementation levels had been examined ranging from 1% to 9% of DM of ruminant diets [6–8]. Dietary supplementation with zeolite clay exhibited positive effects on nutrients digestion and growth performance of sheep [9]. Furthermore, zeolite positively affected animal health status and performance due to its characteristic sorbent properties that modify the ruminal environment [10,11]. Nanoclays and other nano-particles have been shown to specifically absorb mycotoxins through the gastrointestinal tract of ruminants [12]. Nanotechnology is one of the most promising applications of the twenty-first century. It can create new materials with unique properties, which change the physical and chemical characteristics of the molecules/element to have the potential to revolutionize agriculture sectors and has given birth to the new area of agro-nanotechnology, particularly in livestock production. Size reduction of materials to the nano range can increase their adsorption, absorption, and cation exchange capacity [13]. Comparative research studies of nano and natural zeolite supplementations on rumen fermentation patterns and nutrient digestibility are limited. Therefore, we hypothesized that the effects of nano zeolite on ruminal microbial activity might differ from its natural form. Therefore, the objective of this study was to investigate the *in vitro* dose–response effects of natural and nano-zeolite supplementations on ruminal antimethanogenic activity, fermentation end-products, and nutrient degradation. The most effective doses of both zeolite forms were evaluated *in vivo* for ruminal fermentation characteristics and nutrient digestibility.

2. Materials and Methods

The study was carried out at the Advanced Laboratory of Animal Nutrition and experimental farm Faculty of Agriculture, Alexandria University and Laboratory of Livestock Research Department of Arid Land Cultivation Research Institute, the City of Scientific Research and Technological Applications, Alexandria. All procedures following protocols were approved and authorized by the Institutional Animal Care and Use Committee of the Alexandria University (ALEXU-IACUC/08-19-05-14-2-22).

2.1. Experimental Feed Additives

Natural zeolite was commercially purchased from A & O trading company, Giza, Egypt. Zeolite is composed of a microporous arrangement of silica and alumina tetrahedra (Clinoptilolite) with general formula (Ca, K_2, Na_2, Mg)$_4$ Al$_8$ Si$_{40}$ O$_{96}$. 24H$_2$O. The chemical composition and physical properties of zeolite in its natural form are according to the zeolite datasheet by the A & O trading company (Table 1).

Table 1. Chemical composition and physical properties of the natural form of zeolite.

Item	Zeolite Characteristics
Chemical composition	
SiO_2	650.0–713.0 g/kg
Al_2O_3	115.0–131.0 g/kg
CaO	27.0–52.0 g/kg
K_2O	22.0–34.0 g/kg
Fe_2O_3	7.00–19.0 g/kg
MgO	6.00–12.0 g/kg
Na_2O	2.00–13.0 g/kg
TiO_2	1.00–3.00 g/kg
Si/Al ratio	4.80–5.40
Physical properties	
Softing point	1260 °C
Melting point	1340 °C
Flow temperature	1420 °C
Specific gravity	2200–2440 kg/m^3
Volume density	1600–1800 kg/m^3
Porosity	24–32%
Compactness	70%
Whitens	70%
Appearance	Gray-green

The nano-zeolite powder was prepared mechanically by a high-energy planetary ball mill (Retsch PM, Germany) [14]. The mechanical route was performed in a period of 6 h with a reverse rotation speed of 300 rpm and vial rotation speed of 600 rpm with a ball to powder ratio of 9:1 mass/mass. The particle size of the obtained nano zeolite was measured by N_5 submicron particle size analyzer (BECKMAN COULTER, Brea, CA, USA), with a range of 3 nm–5 µm of particle size.

To detect the distribution size and shape of zeolite nano-particles, the scanning electron microscope (SEM; Jeol JSM-6360 LA, 3-1-2 Musashino, Akishima, Tokyo, Japan) and transmission electron microscope (TEM; JEOL JEM-2100, 3-1-2 Musashino, Akishima, Tokyo, Japan) were used to provide three-dimensional images, which are very useful for understanding the morphological characters of the tested nanoparticles [15]. The sample was coated with gold to improve the imaging of the sample. The SEM was operated at a vacuum of the order of 10, and the accelerating voltage of the microscope was kept in the range of 10–20 kV (Figure 1). The TEM nano-particles' shape and size were prepared by dropping approximately 10–15 µL of a dilute sample of ZnO-NPs on the top of the carbon-coated copper grid and left in the hood to dry (Figure 2). The particle size mean was 60.2 nm of the nano zeolite.

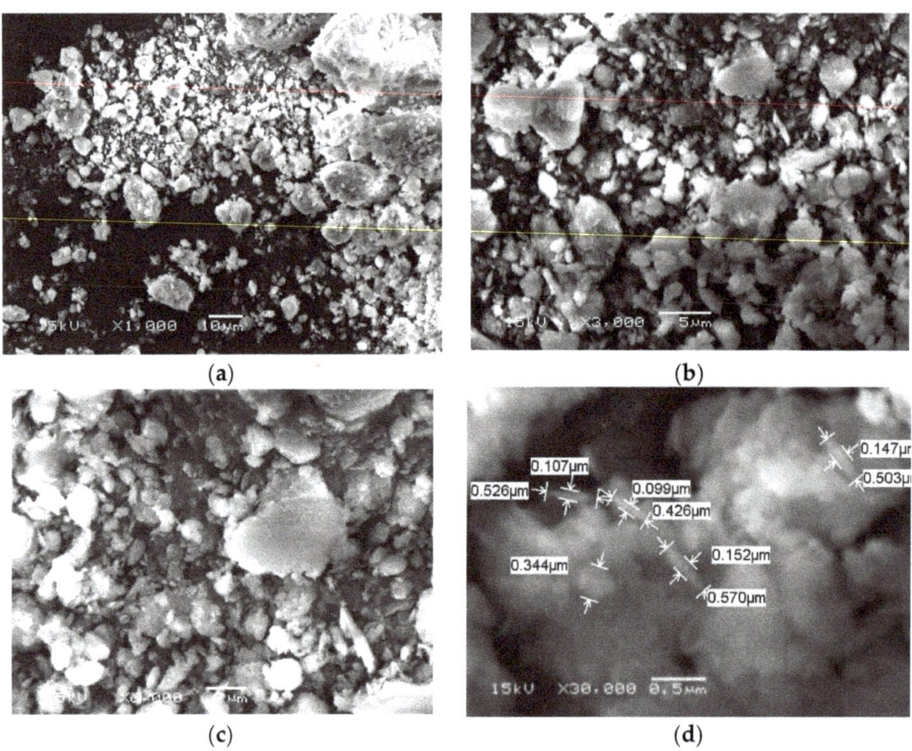

Figure 1. The surface morphology of the nano-zeolite by scanning electron microscope (SEM): (**a**) SEM with X1000; (**b**) SEM with X3000; (**c**) SEM with X6000; (**d**) SEM with X30000.

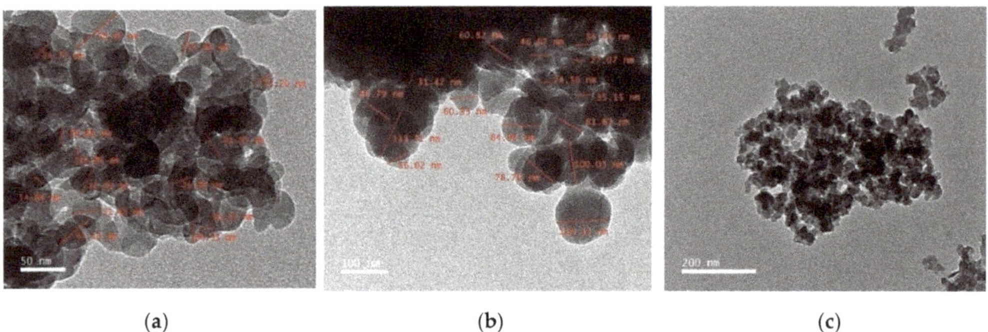

Figure 2. The nano-particles size and shape of the nano-zeolite by transmission electron microscope (TEM): (**a**) TEM with 50 nm; (**b**) TEM with 100 nm; (**c**) TEM with 200 nm.

To identify the functional groups of the prepared nano-zeolite form, the Fourier Transform Infra-Red Spectroscopy (FTIR) analysis was performed using an infrared spectrometer (Shimadzu FTIR-8400S, Nakagyo-ku, Kyoto, Japan) by employing the KBr pellet technique [16], as shown in Figure 3.

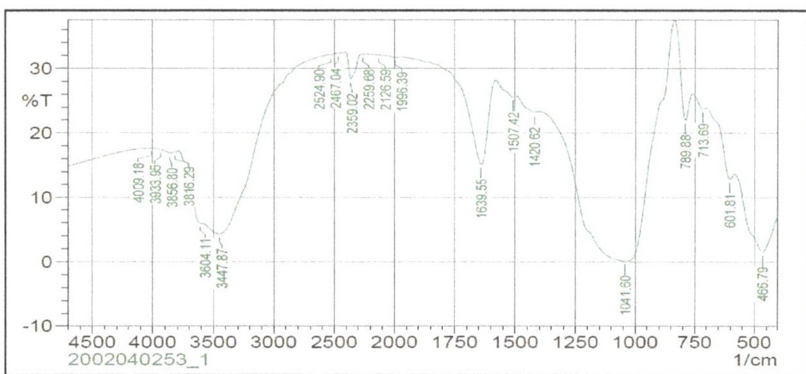

Figure 3. Fourier transform infrared spectroscopy (FTIR) spectra for the experimental nano-zeolite.

The surface charge of the nano-zeolite was measured by zeta potential analysis using a Malvern ZETASIZER Nano series (Malvern, Worcestershire, England, United Kingdom) [17], under the following circumstances: temperature (°C) 25.0, count Rate (kcps) 347.4, measurement position (mm) 2.00, and attenuator 7.00. The zeta potential of the prepared nano-zeolite was −5.85 (mv), zeta deviation and conductivity were 63.8 (mV) and 0.00165 (mS/cm), respectively, as presented in Figure 4.

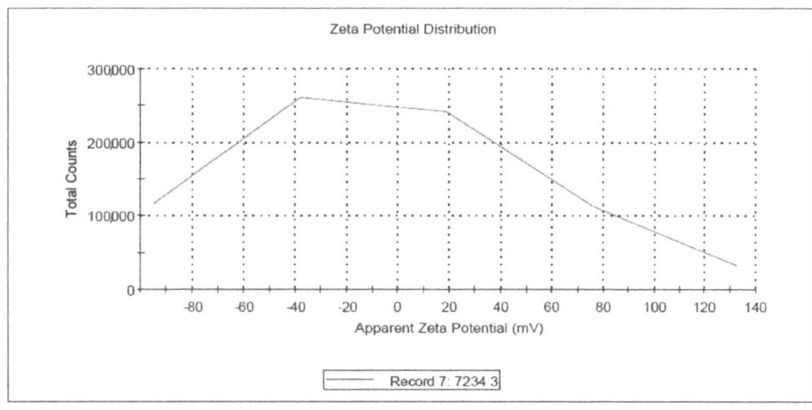

Figure 4. Zeta potential distribution of the experimental nano-zeolite.

2.2. Basal Diet

The experimental basal diet (used in the *in vitro* and *in vivo* experiments) consisted of (g/kg DM) 500 g concentrate and 500 g berseem hay (*Trifolium alexandrinum*) formulated as a total mixed ration (TMR) diet to meet the nutrient requirements of lactating goats [18]. The AOAC [19] analytical procedures were used for dry matter (DM), organic matter (OM), crude protein (CP as 6.25 × N; by Kjeldahl technique), and ether extract (EE). Cell wall ingredients (neutral detergent fiber (NDF), acid detergent fiber (ADF), and lignin contents (ADL)) were determined sequentially by an Ankom 200 fiber analyzer unit (ANKOM Technology Corporation, Macedon, NY, USA) and expressed exclusive of residual ash as described by Van Soest et al. [20]. Concentrations of hemicellulose were calculated as NDF—ADF, and cellulose as ADF—ADL.

The major ingredients and chemical composition of the experimental diet are presented in Table 2.

Table 2. Ingredients and chemical composition of experimental basal diet used in the *in vitro* and *in vivo* experiments.

Item	Basal Diet (g/kg Dry Matter)
Ingredients	
Berseem clover hay	500
Ground yellow corn	345
Soybean meal	150
Mineral and vitamin mixture [1]	5.00
Chemical composition (g/kg DM)	
Organic matter	924
Crude protein	131
Ether extract	20.0
Neutral detergent fiber	718
Acid detergent fiber	343
Acid detergent lignin	60
Hemicellulose	375
Cellulose	283

[1] Each kg contained: 45.8 g dicalcium phosphate, 15 g magnesium sulfate, 6.15 g ferrous sulfate, 0.393 g potassium iodide, 0.753 g copper sulfate, 0.248 g cobalt sulfate, 0.373 g zinc sulfate, 0.02 g slinat sodium.

2.3. The In Vitro Experiment

The experimental treatments consisted of control (basal diet without supplementation), five supplemental doses of the nano-zeolite (0.2, 0.3, 0.4, 0.5, and 1 g/kg DM basal diet), and three doses of the natural zeolite (10, 20, and 30 g/kg DM), and were evaluated *in vitro*.

2.3.1. Gas Production Procedure

Method of the semi-automatic system of GP equipped with pressure transducer and a data logger (Pressure Press Data GN200, Sao Paulo, Brazil) as described by Bueno et al. [21] and adapted by Soltan et al. [22] was used to evaluate the dose–response effects of the experimental supplementations.

Rumen contents were collected freshly from adult fasted slaughtered of three Egyptian buffalo steers at the slaughterhouse of Faculty of Agriculture Alexandria University. The slaughtered animals were fed *ad libitum* a diet consisting of 50:50 commercial concentrate mixture: clover hay (*Trifolium alexandrinum* L.) and had free access to fresh water. Rumen contents were collected and kept separately in pre-warmed containers (39 °C) under anaerobic conditions. To prepare the rumen inocula ($n = 3$), the rumen content of each animal was blended for 10 s, squeezed through three layers of cheesecloth, and kept in a water bath (39 °C) under CO_2 until inoculation took place. The different ruminal inocula were used to prevent the unusual effects of rumen environmental conditions [23,24].

Four analytical repetitions (4 bottles/inoculum/treatment) were used; two for the fermentation parameters and protozoal count, and the other two were for the determination of truly degraded organic matter (TDOM). Similarly, blank bottles (rumen fluid and buffer solution), and internal standard bottles (rumen inoculum, buffer solution, and clover hay) were prepared to correct for the sensitivity induced by the inocula [24,25].

Samples (0.5 g) of the experimental supplemented diets were weighed into numbered bottles and were incubated with 45 mL of diluted rumen fluid (15 mL mixed rumen fluid + 30 mL of Menkes buffered medium) in 120 mL incubation bottles [24,25]. Bottles were then sealed immediately with 20 mm butyl septum stoppers (Bellco Glass Inc., Vineland, NJ, USA), mixed, and incubated in a forced-air oven (FLAC STF-N 52 Lt, Treviglio, Italy) at 39 °C for 24 h. The gas head-space pressure of all bottles was recorded at 3, 6, 9, 12, 24 h incubation using a pressure transducer and a data logger (Pressure Press Data GN200, Piracicaba, Sao Paulo, Brazil). The pressure of GP in all bottles at each measuring time was converted into volumes to calculate the total accumulative gas produced through 24 h [22].

For CH_4 determination through 24 h, one mL of gas of the bottle head-space was sampled by a syringe (med Dawliaico, Assiut, Egypt) at each gas pressure measuring time and accumulated in a 5 mL vacutainer tubes (BD Vacutainer® Tubes, Franklin Lakes, NJ, USA). Methane concentration was determined using a gas chromatograph (Model 7890, Agilent Technologies, Inc., CO 80537, Santa Clara, CA, USA); the separation conditions were in detail described by Soltan et al. [22]. The amounts of CH_4 produced were calculated according to Longo et al. [26]. Net values of both GP and CH_4 were corrected for the corresponding blank values.

2.3.2. Rumen Degradability

At the end of the incubation, all bottles were put in cold water (4 °C) to stop the microbial fermentation process. Determination of TDOM was carried out according to Blümmel et al. [27] by immediate addition of neutral detergent solution (70 mL) without heat-stable α-amylase and incubated in a forced-air oven at 105 °C for 3 h. The remains were filtered in clean pre-weighed crucibles, washed with hot water, and dried at 105 °C for 16 h, and allowed to be burned at 550 °C for 4 h. The TDOM values were calculated from the difference between the amounts of the incubated OM and those remaining non-degraded. The portioning factor (PF) was calculated as the ratio of TDOM (mg) and gas volume (mL) [27].

2.3.3. Rumen Fermentation Characteristics

Rumen pH was determined using a pH meter (GLP 21 model; CRISON, Barcelona, Spain) in all fermentation bottles. Protozoal count was microscopically determined and differentiated by Digital Zoom Video microscope (LCD 3D, GiPPON; Wanchai, Hong Kong) following the procedure described by Dehority et al. [28].

Individual short-chain fatty acids (SCFAs) concentrations were determined according to Palmquist and Conrad [29] and adapted to Soltan et al. [22] using gas chromatography (Thermo fisher scientific, Inc., TRACE1300, Rodano, Milan, Italy) fitted with an AS3800 autosampler and equipped with a capillary column HP-FFAP (19091F-112; 0.320 mm o.d., 0.50 μm i.d., and 25 m length; J & W Agilent Technologies Inc., Palo Alto, CA, USA). A mixture of known concentrations of individual SCFAs was used as an external standard (Sigma Chemie GmbH, Steinheim, Germany) to calibrate the integrator. Concentrations of ruminal NH_3-N were measured colorimetrically using a commercial lab kit (Biodiagnostic kits, Giza, Egypt) [30].

2.4. In Vivo Experiment

2.4.1. Animals and Experimental Design

Based on the *in vitro* assay results, the most effective level of both natural and nano-zeolite was selected to evaluate their responses on apparent nutrients digestibility. Thirty female non-lactating Barki goats were randomly divided into three dietary treatments ($n = 10$/treatment) according to initial body weight (26 ± 0.9 kg SE bodyweight) as follows: control (the same control basal diet that was used in the *in vitro* experiment), natural zeolite (20 g/kg DM), and nano-zeolite (0.40 g/kg DM). Animals were fed their experimental diets *ad libitum*. Zeolite supplementation was orally administrated to ensure the complete dose was received.

Goats were fed twice daily at 08:00 and 16:00 and allowed free access to fresh water throughout the experimental period. Animals were adapted to the experimental diets for 15 days, followed by 7 days as a collection period.

2.4.2. Rumen Fermentation Parameters

Samples of rumen fluid (~30 mL) were collected using an esophageal probe 3 h after the morning feeding. The first 15 mL of the ruminal sample was discarded to avoid saliva contamination; all samples were then strained through three layers of cheesecloth and immediately subjected to ruminal pH using the same portable digital pH meter that was used in the *in vitro* assay. Ruminal individual SCFAs, total protozoa numbers, and NH_3-N concentration were analyzed as described previously in the *in vitro* experiment.

2.4.3. Apparent Nutrients Digestibility

Fresh fecal samples (~40 g each) were obtained daily from each goat at 09:00 and 17:00, about 1 h post-feeding. Apparent nutrient digestibility was determined in which acid-insoluble fiber was used as an internal marker based on the relative concentrations of these nutrients in the feed and feces [20]. These samples were pooled per goat and stored at −20 °C for later analysis. At the end of this period, all the fecal samples were dried in a forced-air oven at 60 °C for 72 h, ground to pass through a 1 mm screen, and chemically analyzed for DM, OM, EE, NDF, and ADF as described previously.

2.5. Statistical Analyses

All results were analyzed using the general linear model procedure (PROC GLM) procedure of SAS [31]. The *in vitro* gas production experiment was performed in one run for all treatments. The analytical replicates were averaged before statistical analysis, with each inoculum being the statistical replicate; thus, the statistical number of replications of treatments (n = 3) are the true replications. Orthogonal contrast statements were designed to test the linear and quadratic responses of each dependent variable to the increasing concentrations of nano or natural zeolite. The results of *in vivo* experiment were subjected to analysis of variance using the following statistical model as Yi = μ + Ti + ei, where Yi = observations mean, μ = overall mean, Tj = treatment effect, and ei = residual error. Differences between the treatments were considered significant at (p < 0.05), and trends were accepted if (p < 0.10). Tukey's procedure for multiple comparisons was used to detect differences among means of the *in vivo* experiment.

3. Results

3.1. In Vitro Experiment

The effects of different levels of natural and nano-zeolite forms on ruminal GP, CH_4, TDOM, and partitioning factors are presented in Table 3. The GP increased quadratically (p = 0.004) with increasing doses of nano-zeolite supplementations, while the natural zeolite did not affect the GP values. Linear reductions (p < 0.05) in CH_4 production (related to the incubated DM and TDOM) consistent with tended increases (p = 0.09) in TDOM were observed by both zeolite form supplementations. The most significant CH_4 reductions (49 and 15%) were achieved by supplementations of 20 g/kg DM natural zeolite, and 0.4 g/DM kg nano zeolite, respectively, compared to the control. Neither nano nor the natural form of zeolite supplementation affected the partitioning factor.

Table 3. Supplementation effects of natural and nano-zeolite forms on ruminal gas production (GP), methane, truly degraded organic matter (TDOM), and partitioning factor through 24 h incubation period (*in vitro* experiment).

Treatment	GP (mL/g DM Incubated)	Methane (mL/g DM Incubated)	Methane (mL/g TDOM)	TDOM (g/kg)	Partitioning Factor (mg TDOM/mL GP)
Control	133	7.7	10.8	709	1.10
Nano zeolite (g/kg DM)					
0.20	143	10.51	14.33	730	1.11
0.30	129	10.12	13.80	733	1.13
0.40	142	6.74	9.13	756	1.18
0.50	153	6.94	9.90	734	1.09
1.00	141	7.32	10.2	742	0.97
Contrast 1					
SEM	1.40	0.22	0.38	8.19	0.01

Table 3. *Cont.*

Treatment	GP (mL/g DM Incubated)	Methane		TDOM (g/kg)	Partitioning Factor (mg TDOM/mL GP)
		(mL/g DM Incubated)	(mL/g TDOM)		
Linear	0.36	0.01	0.05	0.09	0.56
Quadratic	0.004	0.04	0.10	0.46	0.86
Natural zeolite (g/kg DM)					
10	138	4.99	7.40	710	1.15
20	137	3.98	5.50	741	1.13
30	140	5.15	7.25	711	1.16
Contrast 2					
SEM	0.42	0.06	0.11	2.45	0.002
Linear	0.25	0.002	0.001	0.09	0.62
Quadratic	0.36	0.16	0.24	0.36	0.50

Contrast 1 = effects of control (0 supplementation g/kg DM) compared with nano zeolite supplementations, and Contrast 2 = effects of control (0 supplementation g/kg DM) compared with natural zeolite supplementations. SEM: standard error of the mean.

The *in vitro* effects of natural and nano-zeolite forms on rumen protozoal count are presented in Table 4. Increases in the total protozoal count consistent with increases in *Diplodinium* sp. and *Epidinium* sp. were observed by nano zeolite (quadratic effect; $p < 0.05$) and natural zeolite (linear effect, $p < 0.01$) supplementations. Only natural zeolite supplementation increased linearly ($p = 0.001$) and quadratically ($p = 0.02$) the *Eudiplodinium* sp., while no effects were observed by nano zeolite treatments. Similarly, *Isotricha* sp. tended to be increased (linearly, $p = 0.09$, and quadratically $p = 0.05$) with the increasing levels of the natural zeolite, while neither *Entodinium* nor *Ophryscolex* sp. was affected by both zeolite supplementations.

Table 4. Supplementation effects of natural and nano-zeolite forms on ruminal protozoal count through 24 h incubation period (*in vitro* experiment).

Treatment	Protozoal Count ($\times 10^5$/mL)						
	Diplodinium	*Entodinium*	*Epidinium*	*Eudiplodinium*	*Isotricha*	*Ophryscolex*	Total
Control	8.49	1.42	0.412	0.150	0.26	0.150	10.9
Nano zeolite (g/kg DM)							
0.20	10.4	1.20	0.07	0.11	0.15	0.22	12.1
0.30	8.62	1.12	0.15	0.37	0.15	0.11	10.5
0.40	12.9	1.53	0.07	0.30	0.41	0.15	15.4
0.50	10.0	1.46	0.01	0.67	0.30	0.07	12.5
1.00	11.5	1.39	0.03	0.90	0.41	0.11	14.3
Contrast 1							
SEM	0.39	0.143	0.045	0.082	0.098	0.052	0.44
Linear	0.86	0.28	0.01	0.18	0.49	0.71	0.67
Quadratic	0.009	0.75	0.03	0.31	0.68	0.28	0.05
Natural zeolite (g/kg DM)							
10	11.7	1.20	0.07	0.11	0.94	0.11	14.2
20	13.5	1.01	0.11	0.83	0.71	0.30	16.5
30	11.6	1.20	0.04	0.22	0.37	0.26	13.7
Contrast 2							
SEM	0.124	0.043	0.013	0.024	0.029	0.015	0.1334
Linear	<0.001	0.20	0.04	0.001	0.09	0.21	<0.001
Quadratic	0.36	0.95	0.13	0.03	0.05	0.27	0.61

Contrast 1 = effects of control (0 supplementation g/kg DM) compared with nano zeolite supplementations, and Contrast 2 = effects of control (0 supplementation g/kg DM) compared with natural zeolite supplementations. SEM: standard error of the mean.

Quadratic increases ($p < 0.05$) in total SCFAs concentrations and acetate molar proportions by the natural zeolite, while no effects were observed by the nano zeolite form. Both zeolite forms linearly enhanced ($p < 0.05$) and tended to increase quadratically ($p < 0.001$) propionate to molar proportions. Ratio of C2:C3 declined linearly ($p = 0.01$) by nano-zeolite, and quadratic ($p = 0.02$) by natural zeolite supplementation. The ruminal pH was not affected by dietary levels of nano or natural zeolite, while only natural zeolite linearly decreased ($p = 0.004$) the NH_3-N concentration (Table 5).

Table 5. Supplementation effects of natural or nano-zeolite on ruminal total short-chain fatty acids (SCFAs) concentration, molar proportions of individual SCFAs, pH, and ammonia nitrogen (NH_3-N) concentrations through 24 h incubation period (*in vitro* experiment).

Item	SCFAs (% of Total SCFAs)							Total SCFAs (mM)	pH	NH_3-N (mg/100 mL)
	Acetate	Propionate	Butyrate	Iso-butyrate	Valerate	Iso-Valerate	C2:C3			
Control	59.1	22.2	11.6	0.297	2.02	3.31	2.66	90.0	6.34	20.7
Nano zeolite (g/kg DM)										
0.20	58.4	24.9	11.1	0.276	1.96	3.30	2.34	103	6.31	21.1
0.30	58.7	24.8	10.9	0.282	1.95	3.37	2.37	99.0	6.32	19.3
0.40	58.6	24.4	11.3	0.309	1.97	3.26	2.36	94.0	6.26	16.3
0.50	60.3	24.7	10.2	0.209	1.71	2.85	2.44	95.0	6.25	19.8
1.00	59.7	23.1	11.4	0.310	1.99	3.39	2.59	89.0	6.32	20.6
Contrast 1										
SEM	0.631	0.316	0.622	0.033	0.074	0.146	0.041	3.150	0.012	0.475
Linear	0.71	0.007	0.56	0.82	0.68	0.87	0.01	0.29	0.54	0.27
Quadratic	0.65	0.06	0.86	0.82	0.88	0.91	0.07	0.24	0.49	0.33
Natural zeolite (g/kg DM)										
10	63.7	21.6	9.88	0.18	1.61	3.03	2.82	83.0	6.36	20.3
20	59.4	24.8	10.7	0.21	1.75	2.94	2.39	100	6.33	17.9
30	58.2	23.4	12.6	0.32	1.95	3.40	2.49	103	6.34	19.2
Contrast 2										
SEM	0.19	0.09	0.186	0.010	0.022	0.044	0.012	0.94	0.004	0.14
Linear	0.86	0.001	0.64	0.43	0.16	0.47	0.10	0.14	0.83	0.004
Quadratic	0.03	0.08	0.421	0.44	0.11	0.82	0.02	0.05	0.15	0.13

Contrast 1 = effects of control (0 supplementation g/kg DM) compared with nano zeolite supplementations, and Contrast 2 = effects of control (0 supplementation g/kg DM) compared with natural zeolite supplementations. C2:C3 = acetate to propionate ratio. SEM: standard error of the mean.

3.2. In Vivo Experiment

The effects of zeolite type supplementation on ruminal fermentation characteristics and protozoal count are shown in Table 6. Nano-zeolite increased total SCFAs ($p = 0.021$) and butyrate ($p = 0.001$) concentrations compared to other treatments, while it decreased ($p = 0.03$) valeric molar proportion compared with the natural form of zeolite. Goats fed natural zeolite had an increase ($p = 0.05$) in ruminal pH compared with goats fed the control diet, while no differences were observed between both zeolite forms on ruminal pH.

Both natural and nano-zeolite forms declined ($p < 0.001$) NH_3-N concentration compared with the control. Moreover, both nano and natural zeolite increased ($p < 0.001$) ruminal *Isotrica* sp. populations compared with the control, while no differences were detected among the experimental treatments on the other protozoal populations.

Table 6. Supplementation effects of natural or nano-zeolite on goat rumen fermentation parameters and protozoal count 3 h post-feeding (*in vivo* experiment).

Items	Treatments			SEM	*p*-Value
	Control	Natural Zeolite	Nano Zeolite		
Total SCFAs, mM	72.3 [b]	74.3 [b]	86.8 [a]	2.48	0.02
SCFAs (% of total SCFAs)					
Acetic	64.3	65.8	61.6	0.93	0.20
Propionic	17.1	15.7	16.0	0.56	0.50
Isobutyric	2.13	2.30	1.93	0.10	0.41
Butyric	9.92 [b]	10.2 [b]	14.3 [a]	0.66	0.001
Isovaleric	4.37	3.68	4.45	0.42	0.76
Valeric	2.24 [ab]	2.46 [a]	1.74 [b]	0.12	0.03
C2:C3	3.83	4.25	3.95	0.19	0.69
pH	5.50 [b]	5.96 [a]	5.72 [ab]	0.07	0.005
NH_3-N, (mg/100 mL)	6.28 [a]	5.16 [b]	3.90 [c]	0.30	0.001>
Protozoa, ($\times 10^5$/mL)					
Diplodinium	10.33	12.3	11.4	0.39	0.113
Entodinium	1.37	1.50	1.11	0.09	0.212
Epidinium	1.07	1.13	1.17	0.11	0.932
Eudiplodinium	0.57	0.70	0.63	0.08	0.793
Isotrica	0.23 [c]	0.77 [a]	0.50 [b]	0.06	<0.001
Ophryscolex	0.17	0.13	0.40	0.06	0.123
Total	13.7	16.8	15.3	0.65	0.147

[a,b,c] Means within a row without a common superscript letter differ significantly at $p < 0.05$. SCFAs = short chain fatty acids concentration. % of total SCFAs = molar proportions of individual SCFAs. NH_3-N = ammonia. C2/C3 = acetate to propionate ratio. SEM = standard error of the mean.

The digestibility coefficients of DM, OM, CP, and EE are shown in Table 7. Natural and nano zeolite supplemented diets did not affect the DMI and nutrients digestibility.

Table 7. Supplementation effects of natural or nano-zeolite on dry matter intake (DMI) and apparent nutrients digestibility of goats (*in vivo* experiment).

Items	Treatments			SEM	*p*-Value
	Control	Natural Zeolite	Nano Zeolite		
DMI (g/day)	1167	1184	1179	13.21	0.904
Digestibility (g/kg)					
Dry matter	443	446	445	0.13	0.56
Organic matter	435	459	445	0.61	0.31
Ether extract	556	565	608	1.42	0.30
Crude protein	387	386	425	0.90	0.12
Neutral detergent fiber	429	436	425	0.43	0.62
Acid detergent fiber	355	362	341	0.66	0.45
Hemicellulose	505	513	513	0.44	0.764
Cellulose	472	467	447	0.75	0.389

4. Discussion

Both TEM and SEM images of the experimental nano zeolite indicated that the mechanical grinding of the natural zeolite reduced their particle size and successfully presented in the nano-scale. The zeta potential of nano zeolite was negative charges that favorably enhance the affinity of palygorskite with cationic matters, e.g., cationic dyes, and then enhance the adsorption capacity. Results of the FTIR of the nano form of zeolite indicated the high efficiency of the performed nano-particles; 20 well-defined peaks of zeolite functional groups were observed, and 12 of them were found in higher frequencies at 2126–4009, while only four peaks appeared in the lower frequency range (from 466 to 789). These physico-chemical properties of the zeolite nano-form may result in different effects in rumen fermentation compared to its natural form. Increases in GP values caused by nano-zeolite addition may indicate the higher efficiency of nano-zeolite to improve the ruminal microbial fermentation than the natural zeolite; this can be due to the large surface areas, large capacity for cation exchange, and high activities caused by the size–quantization effect [32]. Rumen CH_4 production is strongly related to microbial fermentation extent; therefore, enhancements in GP and nutrient degradability can increase rumen CH_4 emission [33]. Thus, such increases in the total GP caused by the nano zeolite may partly explain the low efficiency of the nano-zeolite to reduce CH_4 production compared to the normal form. Reductions in CH_4 production caused by nano or natural forms confirmed the anti-methanogenic activity of zeolite in this study. Zeolite may act as an alkalinizer and has a high capacity for H+ exchange at different pH ranges [34,35]. Therefore, zeolite can reduce CH_4 emission by affecting rumen H+ exchange capacity and can also affect all the end fermentation characteristics.

Most common CH_4 inhibitors may adversely affect the ruminal nutrient degradability and/or microbial fermentation at doses that achieve desirable CH_4 reduction [22]. In the *in vitro* study, CH_4 reduction consistent with increases in TDOM and GP caused by nano zeolite supplementations may indicate that both zeolite types might benefit the alteration of ruminal fermentation pattern towards less CH_4 production without adverse effect on feed degradability. This can be due to the catalytic activity of zeolite nano-particles which can increase some digestive fiber enzymes (such as amylase, α-amylase) to improve OM degradability [36]. Moreover, the literature reported that enhancing the rumen nutrient degradability is a typical action of zeolite through the buffering effect and maintaining the ruminal pH from the rapid decrease [36]. In the current study, both zeolite forms enhanced the TDOM *in vitro* and the pH *in vivo*, while no differences were detected in the nutrient digestibility *in vivo*. The reasons for this phenomenon are not clear, but it seems that the activity of zeolite is more efficient in the rumen than in the post-ruminal digestive tract parts. The current results are in line with Galindo et al. [37], who reported that zeolite could provide favorable conditions for the increase of cellulolytic rumen bacteria and subsequently increase the ruminal degradable organic matter. The lacking effects of zeolite supplementation on apparent nutrient digestibility are consistent with what reported by Câmara et al. [38], as total tract DM digestion was unaffected when zeolite supplemented at levels of 30–50 g zeolite/kg of dietary DM, while McCollum et al. [39] observed enhancements in ruminal digestion of OM and starch with supplementation of 25 g zeolite/kg of finishing diet.

Reduction in CH_4 can be achieved indirectly by decreasing protozoal abundance [22], but results of the current study indicated that CH_4 reduction was consistent with increases in the total protozoal count, which is mainly related to the significant increases in *Diplodinium* sp., the highest number of protozoal species naturally found in the typical rumen conditions of ruminants. This may partly explain the high TDOM caused by both zeolite types, as *Diplodinium* sp. is known for the high efficiency for cellulose degradation, and consequently, H+ abundance [33–40]. Therefore, the current results indicated that the CH_4 reduction was a result of the high rumen H+ exchange capacity of zeolite.

According to some previous studies [40,41], ruminal pH stability provides more favorable environmental conditions for more microbial proliferation. In the present study, the observed increase in total protozoal abundance may be due to the practical stability of rumen pH within the normal range associated with the available energy (as SCFAs production) and nitrogen (as adequate NH_3-N concentration) for more microbial protein synthesis. This explanation agrees with Dschaak et al. [42], who reported that the great affinity of zeolites for holding water and osmotically active cations could enhance ruminal microbial fermentation and osmotic activity that can regulate pH in the rumen by buffering against hydrogen ions of organic acids. Our results also confirmed that rumen pH plays an important role in the survival of rumen -ciliated protozoa [40,41].

The SCFAs patterns of both zeolite forms declared the ability of nano zeolite to modify the microbial fermentation activity differently from its natural form. The *in vitro* experiment revealed that natural zeolite quadratically enhanced acetate concentration; consequently, the total SCFAs (as acetate is the main contributor of total SCFAs), while these were not caused by the nano form of zeolite. Additionally, the nano form of zeolite enhanced butyric concentration in the *in vivo* experiment compared with the natural zeolite form. These differences may confirm our suggested hypothesis that performing the nano form of the zeolite may affect their efficiency as feed additive differently from its natural form.

Results of the *in vitro* assay showed that both zeolite forms enhanced propionate molar proportions concentration. These results, alongside decreases in acetate to a propionate ratio, might be due to shifting SCFAs production pattern from acetate toward more propionate production, which may explain that the fermentation process occurred in a more efficient manner where more hydrogen ion (H^+) may be used by ruminal microbes to synthesize SCFAs (propionate) rather than CH_4.

Additionally, differences in fermentation patterns were observed by the *in vivo* and *in vitro* experiments using the same experimental dose. It seems that the time of collection of the ruminal samples (3 h post-feeding) of the *in vivo* assay, rather than the nutritive buffering solution used in the *in vitro* assays, may affect the obtained results.

Both zeolite forms decreased NH_3-N concentration in the *in vivo* assay, while it occurred only by the natural form in the *in vitro* experiment. Zeolite, as a cation exchanger, is capable of exchanging and holding the ammonium ion before its release by the sodium ion (Na^+) present in the saliva that was entering the rumen [43–45]. In this regard, zeolite additive could exhibit a higher potential to sink hydrogen through its cation exchange capacity, which might be another possible explanation for zeolite-buffering properties. Lower ruminal NH_3-N concentration with the addition of nano-zeolite indicated that zeolite was able to capture NH_3 through the character of cation exchange capacity [46].

The modulation of rumen fermentation patterns that occurred by both zeolite forms may be nutritionally advantageous for lactating and growing ruminants through enhancing ruminal OM degradability and propionate production [47].

5. Conclusions

The nano transformation of the natural zeolite positively affected the physico-chemical properties of the natural zeolite. Zeolite, whether in its natural or nano-form, was able to maintain rumen pH while reducing NH3-N concentration and affecting CH4 production without adverse effects on the apparent nutrient digestibility. Zeolites as clay minerals play a role in improving the rumen environment and fermentation end-products because of their buffering role. In both experiments, nano zeolite modified the SCFAs pattern differently from the natural zeolite. These results may suggest that the consideration of zeolite as a modifier of rumen fermentation was not only dose-dependent but also particle-size-dependent.

Author Contributions: Conceptualization, S.S.; Methodology, A.E.-N., M.E. and K.E.-D.E.-A.; Investigation, A.E.-N., M.E., H.E.-Z. and K.E.-D.E.-A.; Data curation, A.E.-N., M.E., Y.S. and S.S.; Resources, S.S.; Visualization, A.E.-N. and M.E.; Writing—review and editing, A.E.-N., M.E., H.E.-Z., S.S. and Y.S.; Supervision, M.E., K.E.-D.E.-A., A.E., S.S. and Y.S.; Project administration, S.S.; Formal analysis, S.S. All authors have read and agreed to the published version of the manuscript.

Funding: This research received no external funding.

Institutional Review Board Statement: The study was carried out at the Advanced Laboratory of Animal Nutrition and experimental farm at the Faculty of Agriculture, Alexandria University, and the Laboratory of Livestock Research Department of Arid Land Cultivation Research Institute, the City of Scientific Research and Technological Applications, Alexandria. All procedures followed protocols approved and authorized by the Institutional Animal Care and Use Committee of Alexandria University, Egypt (ALEXU-IACUC/08-19-05-14-2-22).

Data Availability Statement: Not applicable.

Conflicts of Interest: The authors declare no conflict of interest.

References

1. Grabherr, H.; Spolders, M.; Furll, M.; Flachwosky, G. Effect of several doses of zeolite A on feed intake, energy metabolism and on mineral metabolism in dairy cows around calving. *J. Anim. Physiol. Anim. Nutr.* **2009**, *93*, 221–236. [CrossRef] [PubMed]
2. Connor, E. Invited review: Improving feed efficiency in dairy production: Challenges and possibilities. *Animal* **2015**, *9*, 395–408. [CrossRef]
3. Baglieri, A.; Reyneri, A. Organically modified clays as binders of fumonisins in feedstocks. *J. Environ. Sci. Health* **2013**, *48*, 776–783. [CrossRef]
4. Jaynes, W.F.; Zartman, R.E.; Hudnall, W.H. Aflatoxin B1 adsorption by clays from water and corn meal. *Appl. Clay Sci.* **2007**, *36*, 197–205. [CrossRef]
5. Valpotić, H.; Gračner, D.; Turk, R.; Đuričić, D.; Vince, S.; Folnožić, I.; Lojkić, M.; Žura Žaja, I.; Bedrica, L.; Maćešić, N.; et al. Zeolite clinoptilolite nanoporous feed additive for animals of veterinary importance: Potentials and limitations. *Period. Biol.* **2017**, *119*, 159–172. [CrossRef]
6. Trckova, M.; Matlova, L.; Dvorska, L.; Pavlik, I. Kaolin, bentonite and zeolites as feed supplements for animals: Health advantages and risks. *Vet. Med. Czech.* **2004**, *49*, 389–399. [CrossRef]
7. Cole, N.A.; Todd, R.W.; Parker, D.B. Use of fat and zeolite to reduce ammonia emissions from beef cattle feed yards. In Proceedings of the International Symposium on Air Quality and Waste Management for Agriculture, (M31), Broomfield, CO, USA, 16–19 September 2007; ASABE: Broomfield, CO, USA, 2007; p. 41.
8. Ghaemnia, L.; Bojarpour, M.; Mirzadeh, K.; Chaji, M.; Eslami, M. Effects of different levels of zeolite on digestibility and some blood parameters in Arabic lambs. *J. Anim. Vet. Adv.* **2010**, *9*, 779–781. [CrossRef]
9. Goodarzi, M.; Nanekarani, S. The effects of calcic and potassic clinoptilolite on ruminal parameters in Lori breed sheep. *APCBEE Procedia.* **2012**, *4*, 140–145. [CrossRef]
10. Karatzia, M.A.; Pourliotis, K.; Katsoulos, P.D.; Karatzias, H. Effects of in-feed inclusion of clinoptilolite on blood serum concentrations of aluminium and inorganic phosphorus and on ruminal pH and volatile fatty acid concentrations in dairy cows. *Biol. Trace Elem. Res.* **2011**, *142*, 159–166. [CrossRef]
11. Spotti, M.; Fracchiolla, M.L.; Arioli, F.; Caloni, F.; Pompa, G. Aflatoxin B 1 binding to sorbents in bovine ruminal fluid. *Vet. Res. Commun.* **2005**, *29*, 507–515. [CrossRef]
12. Kuzma, J.; Romanchek, J.; Kokotovich, A. Upstream oversight assessment for agrifood nanotechnology: A case studies approach. *Risk Anal.* **2008**, *28*, 1081–1098. [CrossRef]
13. Mahler, G.J.; Esch, M.B.; Tako, E.; Southard, T.L.; Archer, S.D.; Glahn, R.P.; Shuler, M.L. Oral exposure to polystyrene nano-particles affects iron absorption. *Nat. Nanotechnol.* **2012**, *7*, 264–271. [CrossRef]
14. Zhang, D. Processing of advanced materials using high-energy mechanical milling. *Prog. Mater. Sci.* **2004**, *49*, 537–560. [CrossRef]
15. Zhang, X.F.; Liu, Z.G.; Shen, W.; Gurunathan, S. Silver nanoparticles: Synthesis, characterization, properties, applications, and therapeutic approaches. *Int. J. Mol. Sci.* **2016**, *17*, 1534. [CrossRef]
16. Thomas, P.; Turnbull, J.; Roberts, J.; Patel, A.; Brain, A. Response to: 'Failed supraglottic airway': An algorithm for suboptimally placed supraglottic airway devices based on videolaryngoscopy. *Br. J. Anaesth.* **2017**, *119*, 1243. [CrossRef]
17. Varenne, F.; Botton, J.; Merlet, J.C.; Vachon, J.; Geiger, S.; Infante, I.C.; Chehimi, M.M.; Vauthier, C. Standardization and validation of a protocol of zeta potential measurements by electrophoretic light scattering for nanomaterial characterization. *Colloids Surf.* **2015**, *486*, 218–231. [CrossRef]
18. NRC. *National Research Council, Nutrient Requirements of Small Ruminants*, 7th ed.; National Academy Press: Washington, DC, USA, 2001.
19. AOAC. *Official Methods of Analysis*, 20th ed.; Association of Official Agricultural Chemists: Arlington, VA, USA, 2006.
20. Van Soest, P.V.; Robertson, J.B.; Lewis, B. Methods for dietary fiber, neutral detergent fiber, and nonstarch polysaccharides in relation to animal nutrition. *J. Dairy Sci.* **1991**, *74*, 3583–3597. [CrossRef]
21. Bueno, I.C.S.; Cabral Filhoa, S.L.S.; Gobboa, S.P.; Louvandininb, H.; Vittia, D.M.S.S.; Abdallaa, A.L. Influence of inoculum source in a gas production method. *Anim. Feed Sci. Technol.* **2005**, *123*, 95–105. [CrossRef]
22. Soltan, Y.A.; Hashem, N.M.; Morsy, A.S.; El-Azrak, K.M.; Nour El-Din, A.; Sallam, S.M. Comparative effects of *Moringa oleifera* root bark and monensin supplementations on ruminal fermentation, nutrient digestibility and growth performance of growing lambs. *Anim. Feed Sci. Technol.* **2018**, *235*, 189–201. [CrossRef]
23. Salama, H.S.A.; El-Zaiat, H.M.; Sallam, S.M.A.; Soltan, Y.A. Agronomic and qualitative characterization of multi-cut berseem clover (*Trifolium alexandrinum* L.) cultivars. *J. Sci. Food Agric.* **2020**, *100*, 3857–3865. [CrossRef] [PubMed]

24. Soltan, Y.A.; Morsy, A.S.; Sallam, S.M.A.; Louvandini, H.; Abdalla, A.L. Comparative *in vitro* evaluation of forage legumes (prosopis, acacia, atriplex, and leucaena) on ruminal fermentation and methanogenesis. *J. Anim. Feed Sci.* **2012**, *21*, 759–772. [CrossRef]

25. El-Zaiat, H.M.; Abdalla, A.L. Potentials of patchouli (*Pogostemon cablin*) essential oil on ruminal methanogenesis, feed degradability, and enzyme activities *in vitro*. *Environ. Sci. Pollut. Res.* **2019**, *26*, 30220–30228. [CrossRef] [PubMed]

26. Longo, C.; Bueno, I.C.S.; Nozella, E.F.; Goddoy, P.B.; Cabral-Filho, S.L.S.; Abdalla, A.L. The influence of head-space and inoculum dilution on *in vitro* ruminal methane measurements. *Int. Congr. Ser.* **2006**, *1293*, 62–65. [CrossRef]

27. Blümmel, M.; Makkar, H.P.S.; Becker, K. *In vitro* gas production: A technique revisited. *J. Anim. Physiol. Anim. Nutr.* **1997**, *77*, 24–34. [CrossRef]

28. Dehority, B.A.; Damron, W.S.; McLaren, J.B. Occurrence of the rumen ciliate *Oligoisotricha bubali* in domestic cattle (*Bos taurus*). *Appl. Environ. Microbiol.* **1983**, *45*, 1394–1397. [CrossRef] [PubMed]

29. Palmquist, D.L.; Conrad, H.R. Origin of plasma fatty acids in lactating cows fed high grain or high fat diets. *J. Dairy Sci.* **1971**, *54*, 1025–1033. [CrossRef]

30. Konitzer, K.; Voigt, S. Direktbestimmung von Ammonium in Blut-und Gewebsextrakten mit der Phenol-Hypochlorit-reaktion. *Clin. Chim. Acta* **1963**, *8*, 5–11. [CrossRef]

31. SAS. *SAS User's guide: Statistics*; Version 9; SAS Institute Inc.: Cary, NC, USA, 2002.

32. Hua, M.; Zhang, S.; Pan, B.; Zhang, W.; Lv, L.; Zhang, Q. Heavy metal removal from water/wastewater by nano-sized metal oxides: A review. *J. Hazard. Mater.* **2012**, *212*, 317–331. [CrossRef]

33. Boadi, D.; Benchaar, C.; Chiquette, J.; Massé, D. Mitigation strategies to reduce enteric methane emissions from dairy cows: Update review. *Can. J. Anim. Sci.* **2004**, *84*, 319–335. [CrossRef]

34. Sarker, N.C.; Keomanivong, F.; Borhan, M.; Rahman, S.; Swanson, K. *In vitro* evaluation of nano zinc oxide (nZnO) on mitigation of gaseous emissions. *J. Anim. Sci. Technol.* **2018**, *60*, 27. [CrossRef]

35. Adegbeye, M.J.; Elghandour, M.M.; Barbabosa-Pliego, A.; Monroy, J.C.; Mellado, M. Nano-particles in equine nutrition: Mechanism of action and application as feed additives. *J. Equine Vet. Sci.* **2019**, *78*, 29–37. [CrossRef]

36. Galindo, J.A.; Elias, A.; Cardero, J. The addition zeolite to silage diets. 1. Effect of zeolite level on rumen cellulolisis of cows fed silage. *Cuban J. Agric. Sci.* **1982**, *16*, 277.

37. Câmara, L.R.A.; Valadares, S.C.; Leão, M.I.; Valadares, R.F.D.; Dias, M.; Gomide, A.P.C.; Barros, A.C.W.; Nascimento, V.A.; Ferreira, D.J.; Faé, J.T.; et al. Zeolite in the diet of beef cattle. *Arq. Bras. Med. Vet. Zootec.* **2012**, *64*, 631–639. [CrossRef]

38. McCollum, F.T.; Galyean, M.L. Effects of clinoptilolite on rumen fermentation, digestion and feedlot performance in beef steers fed high concentrate diets. *J. Anim. Sci.* **1983**, *56*, 517. [CrossRef] [PubMed]

39. Dehority, B.A. *Rumen Microbiology*; Nottingham University Press: Nottingham, UK, 2003.

40. Sweeney, T.F.; Cervantes, A.; Bull, L.S.; Hemken, R.W. *Zeo-Agriculture. Use of Natural Zeolites in Agriculture and Aquaculture*; Pond, W.G., Mumpton, F.A., Eds.; Westview Press Inc.: Boulder, CO, USA, 1984; p. 183.

41. Marden, J.P.; Julien, C.; Monteils, V.; Auclair, E.; Moncoulon, R.; Bayourthe, C. How does live yeast differ from sodium bicarbonate to stabilize ruminal pH in highyielding dairy cows? *J. Dairy Sci.* **2008**, *91*, 3528. [CrossRef] [PubMed]

42. Dschaak, C.M.; Eun, J.-S.; Young, A.J.; Stott, R.D.; Peterson, S. Effects of supplementation of natural zeolite on intake, digestion, ruminal fermentation, and lactational performance of dairy cows. *Prof. Anim. Sci.* **2010**, *26*, 647–654. [CrossRef]

43. Mumpton, F.A. La roca magica: Uses of Natural Previous Zeolites in Agriculture and Industry. *Proc. Natl. Acad. Sci. USA* **1999**, *96*, 3463–3470. [CrossRef]

44. Erwanto, I.R.; Zakaria, W.A.; Prayuwidayati, M. The use of ammoniated zeolite to improve rumen metabolism in ruminant. *Anim. Prod. Sci.* **2012**, *13*, 138–142.

45. Kardaya, D.; Sudrajat, D.; Dihansih, E. Efficacy of dietary urea-impregnated zeolite in improving rumen fermentation characteristics of local lamb. *Media Paternak.* **2012**, *35*, 207–213. [CrossRef]

46. Khachlouf, K.; Hamed, H.; Gdoura, R.; Gargouri, A. Effects of zeolite supplementation on dairy cow production and ruminal parameters–a review. *Ann. Anim. Sci.* **2018**, *18*, 857–877. [CrossRef]

47. Soltan, Y.A.; Morsy, A.S.; Hashem, N.M.; Sallam, S.M. *Boswellia sacra* resin as a phytogenic feed supplement to enhance ruminal fermentation, milk yield, and metabolic energy status of early lactating goats. *Anim. Feed Sci. Technol.* **2021**, *277*, 114963. [CrossRef]

Review

The Mechanistic Action of Biosynthesised Silver Nanoparticles and Its Application in Aquaculture and Livestock Industries

Catrenar De Silva [1], Norazah Mohammad Nawawi [2,3], Murni Marlina Abd Karim [4,5], Shafinaz Abd Gani [1], Mas Jaffri Masarudin [6], Baskaran Gunasekaran [7] and Siti Aqlima Ahmad [1,8,*]

[1] Department of Biochemistry, Faculty of Biotechnology and Biomolecular Sciences, Universiti Putra Malaysia UPM, Serdang 43400, Selangor, Malaysia; catrenar94@gmail.com (C.D.S.); shafinaz_abgani@upm.edu.my (S.A.G.)

[2] Institute of Bio-IT Selangor, Universiti Selangor, Jalan Zirkon A7/A, Seksyen 7, Shah Alam 40000, Selangor, Malaysia; norazah@unisel.edu.my

[3] Centre for Foundation and General Studies, Universiti Selangor, Jalan Timur Tambahan, Bestari Jaya 45600, Selangor, Malaysia

[4] Department of Aquaculture, Faculty of Agriculture, Universiti Putra Malaysia UPM, Serdang 43400, Selangor, Malaysia; murnimarlina@upm.edu.my

[5] Laboratory of Sustainable Aquaculture and Aquatic Sciences, Port Dickson 71050, Negeri Sembilan, Malaysia

[6] Department of Cell and Molecular Biology, Faculty of Biotechnology and Biomolecular Sciences, Universiti Putra Malaysia UPM, Serdang 43400, Selangor, Malaysia; masjaffri@upm.edu.my

[7] Department of Biotechnology, Faculty of Applied Sciences, UCSI University, Taman Connaught, Kuala Lumpur 56000, Malaysia; baskaran@ucsiuniversity.edu.my

[8] Laboratory of Bioresource Management, Institute of Tropical Forestry and Forest Products (INTROP), Universiti Putra Malaysia UPM, Serdang 43400, Selangor, Malaysia

* Correspondence: aqlima@upm.edu.my

Citation: De Silva, C.; Nawawi, N.M.; Abd Karim, M.M.; Abd Gani, S.; Masarudin, M.J.; Gunasekaran, B.; Ahmad, S.A. The Mechanistic Action of Biosynthesised Silver Nanoparticles and Its Application in Aquaculture and Livestock Industries. *Animals* **2021**, *11*, 2097. https://doi.org/10.3390/ani11072097

Academic Editors: Antonio Gonzalez-Bulnes and Nesreen Hashem

Received: 11 May 2021
Accepted: 6 July 2021
Published: 14 July 2021

Publisher's Note: MDPI stays neutral with regard to jurisdictional claims in published maps and institutional affiliations.

Simple Summary: Silver nanoparticles (AgNPs) have been employed in various fields of studies due to their impeccable ability as an antibacterial, antifungal and antiviral agent. AgNPs are generally synthesised via three methods, which are the chemical, physical and biological methods, with the biological method being the preferred method recently due to the capability of synthesizing nanoparticles of various shapes and sizes to enhance antimicrobial activity and reduced environmental effects. In recent years, AgNPs have been employed in the aquaculture, livestock and poultry industries to combat pathogens. This review is centred on the cytotoxic mechanistic action of AgNPs, which contributes to its application against pathogens in the aquaculture, livestock and poultry industries.

Abstract: Nanotechnology is a rapidly developing field due to the emergence of various resistant pathogens and the failure of commercial methods of treatment. AgNPs have emerged as one of the best nanotechnology metal nanoparticles due to their large surface-to-volume ratio and success and efficiency in combating various pathogens over the years, with the biological method of synthesis being the most effective and environmentally friendly method. The primary mode of action of AgNPs against pathogens are via their cytotoxicity, which is influenced by the size and shape of the nanoparticles. The cytotoxicity of the AgNPs gives rise to various theorized mechanisms of action of AgNPs against pathogens such as activation of reactive oxygen species, attachment to cellular membranes, intracellular damage and inducing the viable but non-culturable state (VBNC) of pathogens. This review will be centred on the various theorized mechanisms of actions and its application in the aquaculture, livestock and poultry industries. The application of AgNPs in aquaculture is focused around water treatment, disease control and aquatic nutrition, and in the livestock application it is focused on livestock and poultry.

Keywords: biotechnology; nanotechnology; toxicity; aquaculture; livestock; poultry

---

1. Introduction

Nanotechnology has gained attention since the 1980s and has developed rapidly in the years to follow, which has been proven impeccable in combating various pathogens and human diseases in various fields such as medicine, pharmaceuticals and the food industry [1,2]. The most common application of nanotechnology is the generation of nanoparticles such as organic, non-organic, polymer-based, non-polymeric and metallic nanoparticles. In recent years, metallic nanoparticles have gained attention in the field of nanotechnology due to their ability to offer a range of surface modifications, which contributes to the nanoparticles being more stable, biocompatible and confers specific functionalities needed for applications in various fields. Researchers have successfully synthesised metallic nanoparticles via physical, chemical and biological synthesis methods, of which biological synthesis is the most preferable method due to its eco-friendly and low-cost approach compared to physical and chemical means that generate nanoparticles at a high-cost rate from the use of various types of machinery and chemicals, thus resulting in environmental issues. The biological synthesis method generates nanoparticles of various shapes and sizes [2,3], and the primary mode of action of nanoparticles that gives them their multidisciplinary functions is their small size (1–100 nm) and their shape (spherical, triangular or rod).

Silver nanoparticles (AgNPs) are one of the many nanoparticles that have been employed in various fields. AgNPs are the nano-sized version of bulk silver that possesses the same antibacterial activity as their large counterpart, but with better efficiency and effectiveness due to their small size and shape, which contribute to their large surface-to-volume ratio, enhancing their antibacterial potential. Over the years, AgNPs have gained attention as one of the most successful nanoparticles that have been applied in many fields, demonstrating the highest success as antibacterial agents. However, the application of AgNPs in aquaculture and livestock is relatively new; there are in vitro results, yet in vivo results are lacking [1]. The application of AgNPs in aquaculture and livestock in crucial to combat pathogens that pose a threat to the industry, resulting is loss of product and leading to economic losses.

Due to this major issue, study into silver as an antibacterial agent has been accelerating over the years, specifically research on AgNPs. AgNPs are promising antibacterial agents, are one of a kind and can significantly change their physical, chemical and biological properties due to their surface-to-volume ratio [4]. Among various methods of synthesis, the biological means of synthesis is by far the simplest; it is non-toxic, dependable, rapid and produces nanoparticles of well-defined size and shape under optimum conditions [5]. Silver nanoparticles trigger cell death via a primary action of cytotoxicity. The action of silver nanoparticles is complex and centred around their size and shape. This review will be centred on the mechanistic action of AgNPs and its application in the aquaculture and livestock industries.

2. Properties of AgNPs Based on Size and Shape

The toxicity of AgNPs depends primarily on their size and shape; other factors such as surface charge, functionalisation and core structure are also crucial for the biological action of the nanoparticles [6,7]. These factors play a crucial part in the mechanism of action towards bacterial cells in terms of cellular uptake, cellular activation and intracellular distribution [6,7]. Researchers have argued over how size and shape play a key role in AgNP toxicity towards bacterial cells, and it was concluded that small nanoparticles with sharp edges or facets are able to better penetrate and kill bacterial cells [8,9].

2.1. Size of AgNPs (Size Dependent)

The main characteristics studied were particle size and antibacterial activity. It was found that the antimicrobial activity of nanosilver was closely related to size. In their study, Morones et al. [8] concluded that AgNPs, mainly of smaller size ranging between 1 and 10 nm, attached to the cell membrane of target bacteria and interfered with cell functions, primarily penetration and bacterial respiration. The nanoparticles were able to penetrate the bacterial cell and cause severe damage to intracellular mechanisms, leading to cell death by interacting with bacterial DNA. Morones et al. [8] also summarized that AgNPs of smaller size released silver ions, which has an added advantage to antibacterial effects of nanosilver. The study by Feng et al. [10] supported that the interaction of silver ions with thiol proteins interrupted the enzymatic activity of bacterial cells. Meanwhile, Sotiriou et al. [11] argued that nanosilver of smaller size releases silver ions faster as compared to larger nanoparticles, which leads to higher toxicity due to more effective silver ions. El-Nour et al. [12] also reported that the smaller the size of silver nuclei, the higher the antibacterial potential displayed. Table 1 shows the strains of pathogenic bacteria on which antibacterial testing was done and the specific nanoparticle size responsible in killing the bacterial strains.

Table 1. Summary of antibacterial testing done on various strains of pathogenic bacteria and the size of the AgNPs found to be most effective in killing the bacterial strains based on literature.

Bacterial Strain	AgNP Size	Author
Escherichia coli	7 nm	[13]
Oral pathogenic bacteria: *Aggregatibacter actinomycetemcomitans, Fusobacterium nucleatum, Streptococcus mitis, Streptococcus mutans, Streptococcus sanguinis.* Aerobic: *Escherichia coli*	5 nm	[14]
Escherichia coli	55 nm	[15]
Foodborne pathogenic bacteria: *Bacillus cereus* ATCC 13061, *Listeria monocytogenes* ATCC 19115, *Staphylococcus aureus* ATCC 49444, *Escherichia coli* ATCC 43890, and *Salmonella Typhimurium* ATCC 43174 Candida species: *C. albicans* KACC 30003 and KACC 30062, *C. glabrata* KBNO6P00368, *C. geochares* KACC 30061, and *C. saitoana* KACC 41238	31.18 nm, 35.74 nm and 69.14 nm	[16]
Gram-positive bacteria: *Streptococcus* sp., *Bacillus* sp., *Staphylococcus* sp. Gram-negative bacteria: *Shigella* sp., *Escherichia coli, Pseudomonas aeruginosa* and *Klebsiella* sp. Fungus: *Candida* sp.	8.8 nm to 21.4 nm	[17]
Pseudomonas fluorescens MTCC 1749, *Proteus mirabilis* MTCC 425, *Escherichia coli* MTCC 1610, *Bacillus cereus* and *Staphylococcus aureus* MTCC 2940	18 nm to 100 nm and 49 nm to 153 nm	[18]
Gram-negative :*Escherichia coli* O157:H7 Gram-positive: *Listeria monocytogenes*	8 nm to 15 nm	[19]

2.2. Shape of AgNPs (Shape Dependent)

In recent years, studies have focused on nanoparticle size and how size is the primary means for nanoparticles to penetrate the bacterial cell and kill target bacteria. However, more researchers have ventured into studies related to the shape of the nanoparticles and how shape and size are correlated in the mechanism of antibacterial action of nanoparticles. The antibacterial activity of nanosilver with different shapes has been discussed by Pal et al. [20] and Dong et al. [9]. The most common shape of nanoparticles is circular. Since venturing into nanoparticles-related research, researchers have discussed how circular-shaped nanoparticles can easily penetrate bacterial cells by passing through protein channels on the plasma membrane. Huang et al. [21] were able to optimize and synthesize AgNPs of circular shape, which were effective against the plant pathogen *Bipolaris maydis*. However, recent research suggests triangular AgNPs exhibit a higher inhibition activity against bacteria compared to circular AgNPs due to the presence of a basal plane, which gives the triangular AgNPs a stronger antibacterial activity against bacteria at a high atom density [20]. Dong et al. [9] added to this by synthesising triangular nanoparticles with sharper vertexes and edges and tested the antibacterial potential against bacteria with success. Thus, they hypothesised that a geometrically triangular Nano prism, having a very sharp apex and sharp edges, would better penetrate and damage bacterial cells [9]. Figure 1 shows various shapes of nanoparticles with sharp vertexes. AgNPs with different shapes would also have diverse effects on the bacterial cell. Besides spherical and triangular shapes, nanoparticles can be synthesised as cubes, platelets, ovals, hexagons and rods.

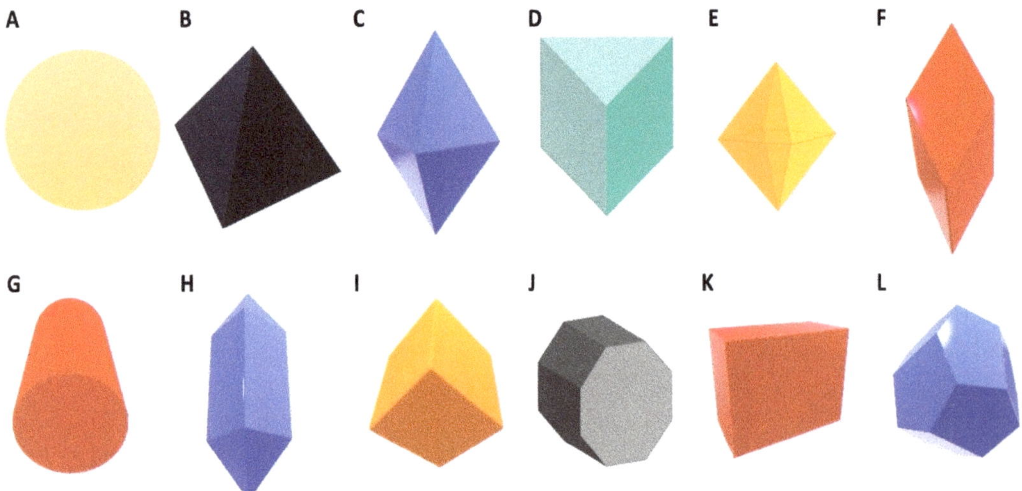

Figure 1. The various shapes of synthesized nanoparticles (adapted from Jo et al. [22] and Walters and Parkin [23]).

Numerous methods have been developed to synthesize nanosilver in various shapes. AgNPs can be structurally synthesized in the form of cubes, rods, platelets, pyramids and bipyramids [24,25]. The difference in shape results is the difference in efficacy of the AgNPs. As reported by Dong et al. [9], shapes with edges, specifically sharp edges, are able to penetrate the bacterial cell easier compared to spheres, which have no edges. Hong et al. [15] reported that AgNPs in the form of cubes were most effective against *Escherichia coli* as compared to spheres and wires. Pal et al. [20] reported comparable results for rod, circular and triangular AgNPs against *Escherichia coli*. However, there are reports of no antibacterial activity from AgNPs in the form of cubes, spheres and triangles against *Staphylococcus aureus* [26]. The published results obtained from research into shape-dependent nanoparticle antibacterial activity have been inconsistent; therefore, further research is required to fill the existing gaps [25].

Nanoparticle size and shape work in correlation; the smaller the nanoparticle and the sharper the edges, or the more edges present on the nanoparticle, the easier the penetration into the bacterial cell. The combination of these two factors makes AgNPs an effective antibacterial agent [9].

3. Mechanism of Action of AgNPs

The mechanism of action of AgNPs on bacterial cells has been yet to be fully understood. Researchers have theorised on possible mechanisms that may be related to changes of the morphology and structure of the bacterial cell caused by the size of nanoparticles in relation to the large surface-to-volume ratio and their shape [27]. These physiochemical factors play a crucial role in the antibacterial action of AgNPs [28,29]. The small size of the nanoparticles provides better interaction with the bacterial cell and ease in penetrating the bacterial cell [27]. However, for the specific size and shape production of AgNPs, factors including temperature and pH are important [30]. Application of heat over a time period can increase the size of nanoparticles as well as change their shape, as observed by Mokhena et al. [31] after heating AgNPs at 90 °C for 3 h, which resulted in the increase in the size of the nanoparticles from 28 nm to 30 nm while the shape of the nanoparticles changed from spherical to an irregular shape. Meanwhile, further heating for 48 h resulted in the formation of a mixture of rod-like nanoparticles with sizes between 76 nm and 121 nm as well as spherical nanoparticles of sizes between 28 nm and 56 nm. Huang et al. [32] were able to synthesise AgNPs of sizes ranging from 8 nm to 24 nm via heating up to 80 °C for 15 min.

Another theory argued by researchers is that the antimicrobial action of AgNPs is similar to silver ions due to the former being oxidised into the latter [8,33–35]. Xiu et al. [35] reported that the antimicrobial action of AgNPs is oxygen-dependent and ineffective in the absence of oxygen, primarily being that the conversion of AgNPs to silver ions occurs in the presence of oxygen, and the defined molecular toxicants are silver ions. They then tested the antimicrobial action of nanosilver by synthesizing it in anaerobic conditions, and they observed the lack of toxicity of the AgNPs due to the reduction in the formation of silver ions.

3.1. Production of Reactive Oxygen Species (ROS)

Cell death is induced by ROS. ROS causes cell death in two separate ways: apoptosis or necrosis. The apoptotic course is activated by the ROS through caspases, which are the killers of apoptosis [36]. The activation of caspases is determined as the point of no return in apoptosis [37]. The function of caspases in the apoptosis pathway is to cleave DNA of the bacteria, which is an onset of apoptosis.

Silver ions produced from the oxidation of AgNPs catalyse the production of ROS [38,39]. Metal nanoparticles have been reported to induce a significant rise in ROS in cells inducing toxicity related to oxidative stress [40–42]. Oxidative stress can be triggered by disruption on the respiratory chain of the targeted bacteria or by the silver nanoparticles [43]. ROS possibly result from the interaction of silver ions with the thiol enzyme group during the inhibition of the respiratory chain via respiratory enzymes [44]. As a major factor affecting oxidative stress, ROS could cause damage to bacterial cell macromolecules such as protein synthesis and alteration, inhibition of enzymes, lipid synthesis as well as oxidation and damage of DNA and RNA of the bacterial cell [45]. At high levels, ROS can cause cell death, whereas severe damage or mutation to the DNA of the bacterial cells can occur at low levels.

Mats et al. [46] reported that ROS may react directly with the DNA or protein of bacterial cells or lipids of the bacterial cell producing malondialdehyde, which is a marker for oxidative stress that can, in turn, react with bacterial DNA, protein or lipids and cause cell damage or death. An early study by Messner and Imlay [47] reported that oxidative stress triggered by high temperatures of *Escherichia coli* treated with nanosilver resulted in the formation of ROS via the autoxidation of NADH dehydrogenase II in the respiratory chain, leading to cellular damage. AgNPs can directly interact with essential enzymes, induce nitrogen reactive species production and induce programmed cell death [29,48]. The full mechanism requires extensive research to be conducted. Until then, researchers can only theorise on the possible mechanisms of action. Figure 2 illustrates a simplified diagram of the possible mechanism of action towards Gram-negative and Gram-positive bacterial cells.

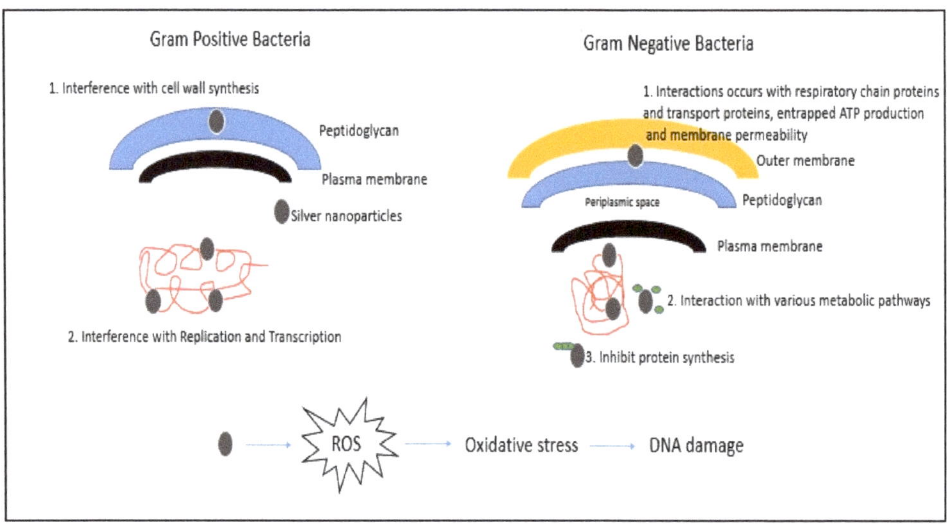

Figure 2. The activation of ROS by AgNPs and the possible mechanism of action towards Gram-negative and Gram-positive bacterial cells (adapted from Pandey et al. [49]).

3.2. Attachment of AgNPs to the Cell Membrane of Bacterial Cells

AgNPs have the ability to adhere to the cell wall and penetrate it, which in turn causes changes to the structure of the cell membrane, including permeability and leading to cell death [33]. The AgNPs attach to the cell membrane by means of electrostatic charges, where the positive surface charge of the nanoparticle electrostatically adheres to the negative charge of the cell membrane, thus facilitating nanoparticle membrane attachment [50]. Upon interaction, an evident morphological change is observed and can be characterised by cytoplasm shrinkage and cell membrane rupture [50]. Raffi et al. [51] reported that, via transmission electron microscope (TEM) analysis, complete cell membrane disruption of *Escherichia coli* was observed upon minutes of exposure to AgNPs. Pits were observed around the areas of AgNPs damage induced by the AgNPs during attachment with the bacterial cell membrane [33].

Another factor is that AgNPs can interact with sulphur-containing proteins present on the cell wall, causing cell wall damage [52]. This in turn affects the lipid bilayer and permeability of the cell membrane, which affect the bacterial cells' ability to regulate transport through the membrane [53]. A study has been done on *E. coli* to examine this factor. Gram-negative bacteria have an outer membrane outside the peptidoglycan layer, which is absent in Gram-positive bacteria. The outer membrane serves as a selectively permeable layer that only allows the entry of selective components for cell growth, whereas the layer serves as a protective barrier for the bacterial cell to prevent the entry of harmful substances that may temper bacterial cell activity and lead to cell death. The outer membrane is made up of liposaccharide (LPS) molecules with a small portion of membrane proteins. The LPS layer serves as a selectively permeable membrane for Gram-negative bacterial cells [45]. The presence of AgNPs may increase the permeability of the membrane, hence causing a reduction in intake of sugar and protein by the bacterial cell that in turn results in bacterial growth inhibition [54]. Schreurs and Rosenberg [53], in an early study, reported that silver can reduce the uptake and release of phosphate ions in *E. coli*. The transport and release of potassium ions can be altered from the bacterial cell. An increase in membrane permeability can also cause the leakage of cellular contents such as proteins, ions, reducing sugars and, in some cases, ATP molecules [55–57]. Figure 3 displays the pathway of cell death caused by increased permeability of the cell membrane based on the literature of Rai et al. [27].

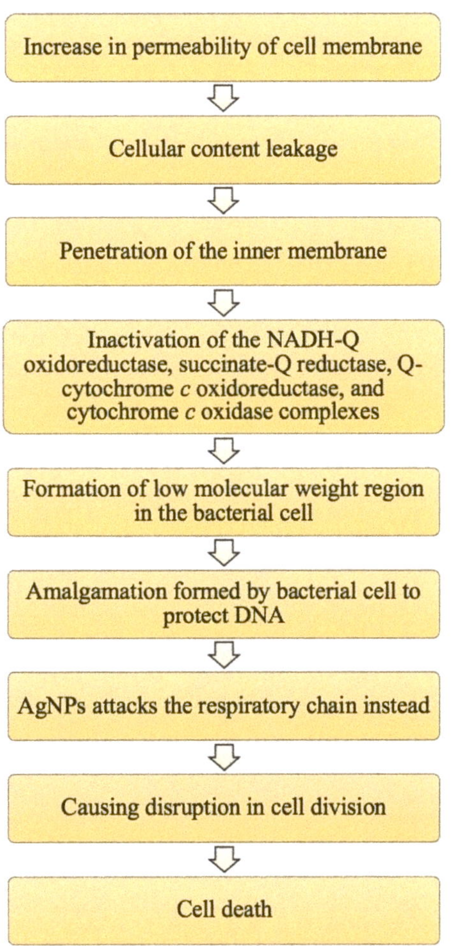

Figure 3. The summary of the pathway of cell death caused by an increase in membrane permeability due to the presence of AgNPs (adapted from Rai et al. [27]).

3.3. Damage to Intracellular Components of Bacterial Cells

Once the AgNPs adhere to the cell membrane of bacterial cells, they can penetrate into the bacterial membrane and enter the cell, which affects important cell functioning [58,59]. Smaller-sized nanoparticles affect the intracellular structures of bacterial cells at a faster rate than bigger nanoparticles due to their large surface-area-to-volume ratio [60]. Once penetrated by the bacteria, nanosilver can interact with cell components such as proteins, lipids, enzymes and DNA. The interaction between the silver nanoparticles and the cellular components can lead to severe damage to bacterial cells such as dysfunction and eventually bacterial cell death. Interactions between AgNPs and bacterial cell ribosomes cause the denaturation of ribosomes that can inhibit protein synthesis [8,61,62]. Figure 4 illustrates intracellular damage of the cell caused by AgNPs.

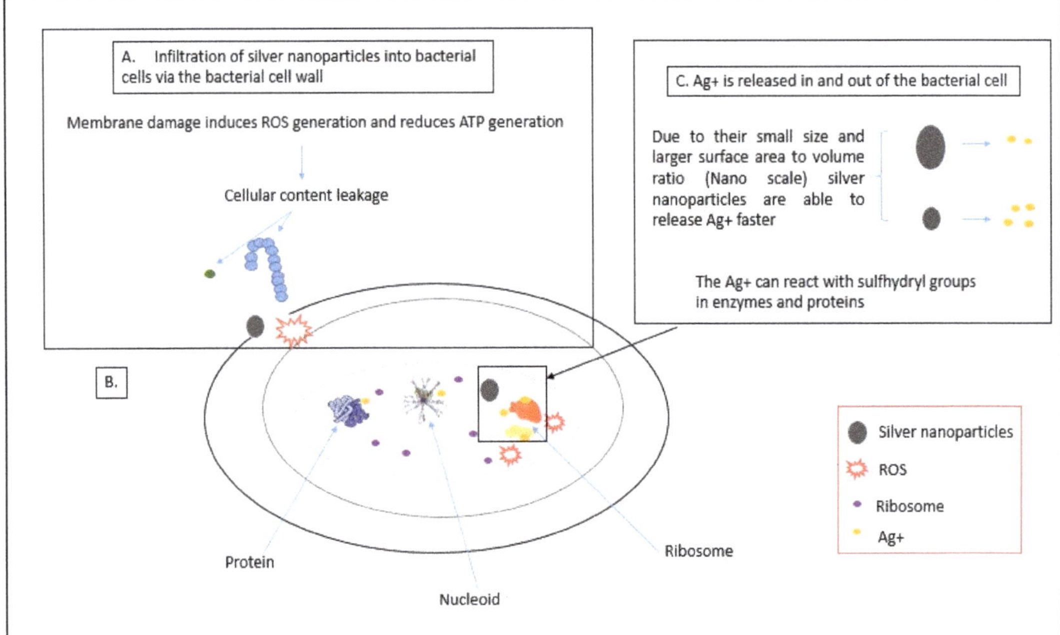

Figure 4. Intracellular damage of the cell caused by AgNPs. (**A**) Process of AgNPs penetration into bacterial cell and disruptions caused on the membrane; (**B**) process of ribosome denaturation by AgNPs; (**C**) interaction of AgNPs with proteins and enzymes, adapted from Qing et al. [63].

AgNPs can also directly attach to functional groups of proteins, resulting in deactivation. For example, silver ions have been shown to bind to thiol groups of proteins present in the membrane of bacterial cells, thus forming stable bonds that result in the deactivation of protein molecules, leading to inhibition of ion transport across the membrane and transmembrane ATP generation [62,64]. Lok et al. [34] observed that nanosilver in its pure form or in the form of silver ions can alter the 3D structure of protein molecules present in bacterial cells; it interferes with the disulfide bonds in protein molecules and blocks the active binding side on protein molecules, leading to defects in the function of the bacterial cells. AgNPs have also been reported by Bhattacharya and Mukherjee [65] to interact and block sugar metabolism, which affects the glycolytic pathway of the bacterial cell and cause bacterial cell death. This was done by the interaction of AgNPs with the enzyme phosphomannose isomerase, which mediates the isomerisation of mannose-6-phosphate into fructose-6-phosphate that plays a key role as an intermediate in the pathway.

Direct interaction of AgNPs with DNA of the bacterial cells can lead to devastating effects in terms of cell division [66,67]. Klueh et al. [64] revealed that silver ions can interact with the nucleoside of the nucleic acid, imbed between the purine and pyrimidine base pairs and cause disruption of the double helix structure of DNA by disrupting the hydrogen bonds between the base pair of the anti-parallel DNA strand, which may block the transcription of genes by the bacterial cells [8]. Silver nanoparticles may also cause the DNA helix to become more condensed, which results in the bacterial cells losing their replication capacity [10].

3.4. Inducing the Viable But Non-Culturable (VBNC) State

Silver ions can have a viable but non-culturable (VBNC) effect on cells. The phenomenon by which bacterial cells are alive but unable to grow in standard conditions is known as the viable but non-culturable (VNCB) state [68,69]. In this stage, bacterial cells are unable to carry out cell division due to the inhibition of key factors such as uptake and utilisation of important substrates when the AgNPs are attached and cause structural and morphological changes to the bacterial cell membrane. These bacterial cells are metabolically active, but they show a reduced rate of nutrient transport, respiration and synthesis of macromolecules [45]. In some cases, bacterial cells in this state can stay alive for a certain period; however, they eventually die due to lack of nutrients and cell dysfunction. Jung et al. [61] reported that silver ions were able to inhibit the growth of *Escherichia coli* and *Staphylococcus aureus* and eventually led to bacterial cell death after 2 hours of exposure to silver ions. Although the exact mechanism of action for AgNPs has not been identified, results by various researchers have shown possible mechanisms required thorough investigation, observation and results.

4. Application of AgNPs in Aquaculture, Livestock and Poultry Industries

The application of AgNPs in aquaculture and livestock and poultry industries plays a crucial role in contributing to an increase in economic value through the reduction in issues related to aquatic and livestock diseases. Application of AgNPs in aquaculture can increase the survival rates and yield of aquatic life in ponds through water treatment; in livestock breeding, AgNPs can improve animal immunity through the reduction in the use of antibiotics and increased production of poultry [1]. Figure 5 illustrates the summary of the application of AgNPs in aquaculture, livestock and poultry industries.

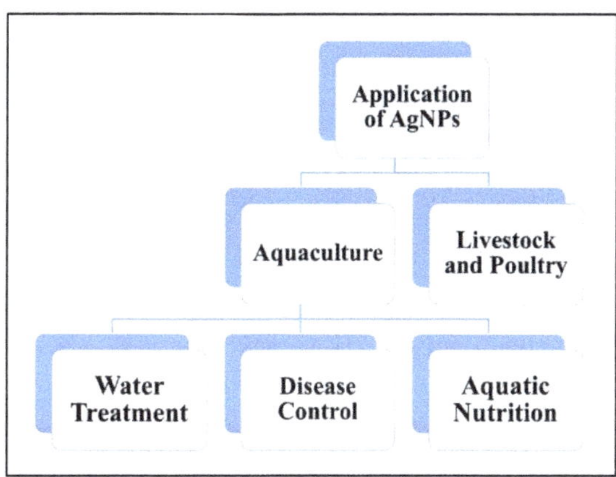

Figure 5. The applications of AgNPs in aquaculture (water treatment, disease control and aquatic nutrients), livestock and poultry industries.

4.1. Application of AgNPs in Aquaculture

The aquaculture industry is one of the rising industries in recent years that has to keep up with the demand for seafood required for the global population. Although the industry is relatively new as compared to farming and fishing industries, it has gained much attention. The industry faces a relatively serious issue of fish and shellfish-related diseases caused by pathogenic bacteria and viruses, which have caused a lot of economic losses to counter this issue. Besides diseases, water quality is also an issue that is correlated with the occurrence of diseases in aquaculture products. Although methods are available for treating fish and shellfish-related diseases, the methods do not stop the issue at hand but rather offer a temporary solution. Moreover, resistance to commercially available antibiotics has also occurred. Thus, a reliable and effective solution is required to overcome this problem and increase the production of aquaculture products. Nanotechnology, specifically AgNPs, offers an excellent solution and has been proven effective in in vitro testing and small-scale in vivo applications to overcome three major concerns, which are water treatment, disease control and aquatic nutrients [70,71].

4.1.1. AgNPs in Water Treatment

Due to the large-scale production of the aquaculture industry, a large land area and high water quantities are required to keep up with supply and demand. As a result, aquaculture imposes an environmental risk to areas surrounding the aquaculture farm regardless of freshwater, seawater, brackish ponds or seawater cages. Water source is a limited and valuable source, and its use in aquaculture puts a strain on water resources. The contamination of ponds with unconsumed food, excretory by-products, chemicals and antibiotics generates pollution of pond water resulting in bacterial, fungal and viral infections. The common solution is continuous water changes, which causes a reduction in the water source as hundreds of cubic meters of water are needed per day depending on pond size and pollution of the surrounding environment with the polluted wastewater [72,73]. Thus, nanotechnology can offer a more cost-effective and efficient solution.

AgNPs offer a means to solve water pollution issues via remediation and water treatment using a filter system. Pradeep [74] reviewed the use of nanoparticles in water treatment, which showed success in eliminating bacterial and viral pathogens from water sources. However, another study concluded that the use of AgNPs directly into ponds reduced fungal infection on rainbow trout (*Oncorhynchus mykiss*), yet also caused the reduction in chloride and potassium in blood plasma in juvenile rainbow trout, suggesting that AgNPs need to be applied in other means to only combat bacterial and viral infections without causing harm on aquaculture products [75]. An effective solution to this problem is the use of AgNPs in filter systems. In one study, AgNPs were combined with zeolite and incorporated into filter systems with varying zeolite and AgNPs concentrations (the control was zeolite alone in a semi-recirculating system. Water was treated with the filter systems, and fertilised ovules of rainbow trout were added, with results showing a 5% increase in survival rate of fingerlings compared to the control test [75]. This proves that AgNPs are excellent water treatment agents and are best used as a filter system. Figure 6 shows a simplified diagram based on the study done by Sarkheil et al. [76] of the application of AgNPs in a water filtration system with a silver absorbent layer against *Vibrio* sp. strain Persian1 infecting Pacific White Shrimp. Two filter systems were designed, one with a silver absorbent and one without a silver absorbent. The results indicate that the filter system with the silver absorbent showed a significantly higher bacterial removal as compared to the filter system without the silver absorbent after 2, 6 and 12 h, respectively.

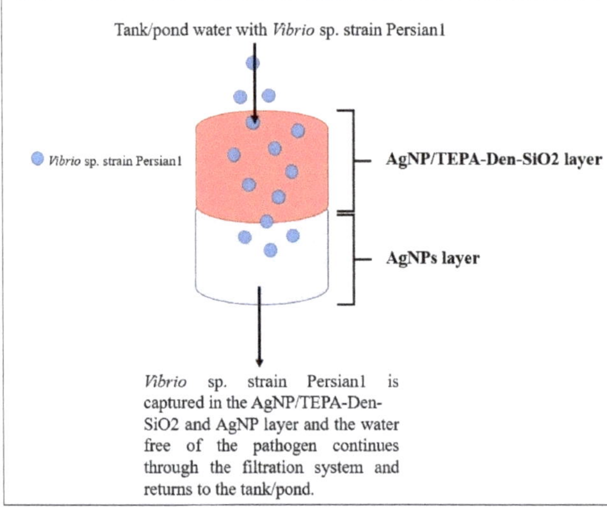

Figure 6. Water filtration system employing AgNPs with a silver absorbent layer (adapted from Sarkheil et al. [76]).

4.1.2. AgNPs in Disease Control

Bacterial, fungal and viral infections are major problems in the aquaculture industry. The commercially available method is the use of antibiotics to treat the infections; however, excessive application over time results in the emergence of resistant strains, making antibiotic treatments unsuccessful [77]. The most commonly used antibiotic based on data from 25 countries is tetracycline. Bacterial pathogens are also resistant to a wide range of antibiotics. The most common resistant bacterial strains are *Aeromonas salmonicida*, *Photobacterium damselae*, *Yersinia ruckeri*, *Listeria* sp., *Vibrio* sp., *Pseudomonas* sp. and *Edwardsiella* sp., which affect both freshwater and saltwater fish as well as shellfish species. AgNPs have been selected as an alternative to combat pathogens and produce efficient and effective results.

The use of biologically synthesised AgNPs has shown positive results in in vitro studies. AgNPs synthesised from tea plants showed 70% inhibition of *Vibrio harveyi* in infected Indian white prawns (*Feneropenaeus indicus*) at a concentration of 10 μg/mL [78]. Sivaramasamy et al. [79] synthesised AgNPs using *Bacillus subtilis* as a nanofactory and tested its antibacterial potential on *Vibrio parahaemolyticus* and *Vibrio harveyi* on white leg shrimp, showing a 90% inhibition. In another study, AgNPs encapsulated with starch at 10 mg AgNPs concentration was used to treat *Ichthyophthirius multifiliis* and *Aphanomyces invadans* fungal infections in fish species, and a recovery rate of three days was demonstrated, indicating inhibition or death of the fungal pathogens [80,81]. AgNPs were also able to be formulated into a vaccine to treat the white spot syndrome, a viral infection in shrimps, with success [82]. In another study, researchers were able to increase the mortality rate of shrimp (*Litopenaeus vannamei*) against the white spot syndrome virus by 50% with the application of the patented Agrovit-4® (Novosibirsk, Russia; patent 2427380), a PVP-coated spheroid AgNP, at a dose of 1000 ng. A minimal dose of 10 μg/g was shown to be effective in protecting the shrimp, and an increase of 16% in mortality rate was observed as compared to the positive control, which had no survival [83–85]. The application of Argovit® on the other hand showed an increased mortality rate of 70–80% (various doses) against the white spot syndrome virus in shrimp [83].

4.1.3. AgNPs in Aquatic Nutrition

The application of AgNPs in aquatic nutrition is fairly new yet employs the same technique of using AgNPs as nanocarriers to increase absorption of nutrients in fish and shellfish via encapsulation, targeted delivery and controlled release, or as cargo to a nanocarrier depending on their necessity. As a nanocarrier, AgNPs offer better adsorption and delivery of nutraceuticals required by aquatic species for growth. Biologically synthesised AgNPs are the best choice of nanoparticles due to their biodegradable ability at the end of the delivery pathway [86]. Alishahi et al. [87] reported that the application of nanotechnology in fish food pellets incorporating vitamin C resulted in better absorption of the nutrient in rainbow trout compared to the control with regular fish food pellets. This paves a platform for the further application of AgNPs in aquatic nutrition as it offers a reduced cost in fish food due to the smaller feeding portions and higher nutrient impacts, thus also reducing food wastage and reducing toxic effects of AgNPs.

4.2. Application of AgNPs in Livestock and Poultry

The livestock and poultry industry are the largest industries in the food sector due to the global demand for meat. Like the agricultural and aquaculture industries, the livestock and poultry industry also face a threat, which is diseases that affect the breeding and growth of livestock and poultry [1]. Nanotechnology offers a possible solution to the problem, as commercially available methods such as antibiotics have failed to counter the problem due to the emergence of resistance in bacterial pathogens. Colloidal silver has been used since the early 1950s as an additive in animal feed and showed improved growth and reduction in infections, yet the use of colloidal silver was stopped and replaced with a cheaper alternative known as antibiotics [88]. In recent years, in vitro research has been done on the use of AgNPs to kill pathogenic bacterial cells and as a feed additive for antibacterial effects to promote animal growth.

4.2.1. Application of Silver Nanoparticles in the Livestock Industry

The livestock industry plays a major role in contributing to economic growth; however, over the past decade the quantity and quality of livestock products have reduced due to the emergence of resistant strains of pathogenic microorganisms failing commercial methods in combating said pathogens. Foodborne pathogens *Escherichia coli, Staphylococcus aureus, Streptococcus agalactiae, Pseudomonas aeruginosa, Salmonella enterica* and *Klebsiella pneumoniae* are some species that cause diseases that result in increased mortality rates in livestock such as cows, goats, sheep and pigs [89]. AgNPs offer an effective and efficient solution to existing issues of resistance primarily due to the mechanism of action of AgNPs, which plays a crucial role in the treatment of pathogenic bacteria. The primary mechanism of action of AgNPs is their toxicity, via the release of silver ions, and activation of oxygen reactive species that trigger oxidative stress; both put stress on the bacterial components, such as protein synthesis and DNA formation, and result in deactivation or cellular death.

Sondi and Salopek-Sondi [33] and Jung et al. [61] observed cellular death via inhibitory effects of AgNPs on *Escherichia coli*. Meanwhile, Li et al. [90] and Kim et al. [91] observed similar effects of AgNPs against *Staphylococcus aureus*. AgNPs have also been applied as an additive in animal feed, where Fondevila et al. [92] observed the reduction in coliforms in pig microbiota and an increase in pig growth rate when the AgNPs colloid was used as a supplement in the feed. The results indicated the death of harmful microorganisms (coliforms) and no effects on healthy gut bacteria of the pigs. They also reported that an increase in the dose of AgNPs from 0 to 40 mg/kg resulted in an increase in the growth rate of weaned pigs in a period of 28 to 56 days [92]. In another study done by Kalinska et al. [93], in vitro results of the application of AgNPs as a potential treatment/prevention against mastitis in dairy cows and goats, caused by pathogenic species such as *Escherichia coli, Staphylococcus aureus, Enterobacter cloacae, Enterococcus faecalis, Streptococcus agalactiae* and *Candida albicans*, showed strong antibacterial and antifungal activity in pure AgNPs and with a combination of silver and copper nanoparticle complex (AgCuNP). AgNPs showed a stronger antibacterial and antifungal activity as compared to the AgCuNP complex with pathogen viability of $p < 0.01$.

4.2.2. Application of Silver Nanoparticles in the Poultry Industry

The poultry industry faces the tremendous loss of products primarily due to pathogenic infections brought on by pollution of water sources used in poultry farms as well as improper handling, which in turn causes a decrease in growth rate and increase in mortality. AgNPs have proven successful in combating pathogenic infections and reducing the mortality rate in poultry. AgNPs can be introduced to target organisms via various methods such as in feed and water source, by injecting target organisms and by tropical applications such as spraying or dipping in the AgNPs colloid for eggs. Figure 7 illustrates the basic methods of application of AgNPs on poultry.

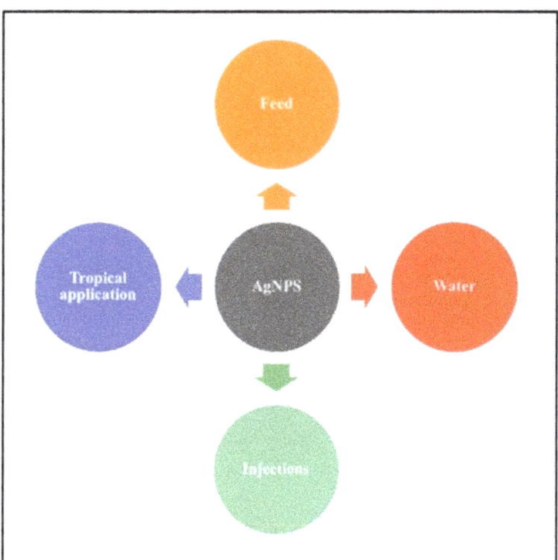

Figure 7. Methods of application of AgNPs on target organisms.

Studies have shown that AgNPs can increase the weight of poultry by acting as a growth promoter as well as an effective antibacterial agent by targeting the cell membrane and causing DNA damage via direct and indirect routes against pathogens that infect poultry. Authors in [27,94,95] reported that groups of quail that consumed AgNPs in feed and water had an increase in body weight from 40 g to 98.9–102.2 g after 12 days of the study. AgNPs with a size range of 1 to 100 nm at lower concentrations (<70 ppm) have shown to be effective in combating pathogens and induce growth of poultry with minimal to no toxic effects as compared to higher concentrations of the same size of AgNPs [96].

Sawosz et al. [94] reported that the application of 50 ppm of colloidal AgNPs to chicken embryos caused no effects on the growth, development and mortality of chicken embryos after 48 h and after 20 days. Biochemical tests of blood serum of the chicken embryos showed no effects to the liver. However, increased mineral content was noticed, indicating that AgNPs influence bone mineralization. In another study, chicken eggs were treated with 50 ppm colloidal AgNPs. The results obtained indicated no pathogenic infection or spoilt eggs, but a pathogenic study revealed the effects of AgNPs on hatchlings due to oxidative stress [97]. A recent study done by Kumar et al. [98] showed that AgNPs were successful in reducing the rate of mortality and increase in growth of chicken in two farms in India when applied in poultry drinking water to combat *Escherichia coli* infections. The nanoparticles were of average size, 15 nm, with a minimum inhibitory concentration of 50 mg/L. The research also proved that the poultry used in this research was safe for human consumption, with a hazard quotient of 0.34, which falls in the non-toxic range.

In terms of the use of AgNPs in livestock and poultry production, more in vivo and large-scale research is required, as the existing studies are in vitro, but results are promising in the use of AgNPs as an alternative to antibiotic since the ban of many antibiotics in livestock and poultry industries.

5. Toxicity of AgNPs

Toxicity is the root mode of action of AgNPs against pathogens. Toxicity of AgNPs, which is size- and shape-dependent, can inhibit the vital cellular activity of the pathogens at lower concentrations of AgNPs, thus lowering their effects on target organisms. At high concentrations, AgNPs can cause cellular death by interrupting vital cellular activity, causing changes that lead to cellular damage and eventually cellular death [99]. This ability of AgNPs makes them an effective tool against various pathogens and broadens their application in aquaculture and livestock production.

The advantage of the AgNPs toxicity mechanism is their ability to penetrate target organisms with ease due to their large surface-to-volume ratio. This allows them to be excellent antibacterial, antifungal and antiviral agents as they can travel through cellular membranes with ease and increase their bioavailability, thus being able to employ the 'Trojan Horse' mechanism [100]. The AgNPs toxicity mechanism also allows for applications as a drug delivery tool where AgNPs can serve as a carrier to ensure that the drugs reach target cells and are released on-site [101].

Although AgNPs offer great benefits, their effects on animal and human health are still in question. The primary disadvantage is the bioaccumulation of AgNPs in aquatic and livestock products. This, in the long run, causes organ damage in animals, and it results in possible human organ damage when consumed, owing to their small size that easily passes through the blood–organ barrier [1]. Another issue is that AgNPs are difficult to remove using common methods of rinsing. They can accumulate in the environment, which may cause changes to the size, shape, surface area and crystalline structure due to environmental factors such as temperature, pH and light intensity. This results in AgNPs of different sizes and shapes with different toxic effects [1,102].

Research into AgNPs effects needs to be conducted further, and in vitro and in vivo applications need to done to understand the extent of negative effects posed by AgNPs to better improve the application process. At the moment, it can be said that the toxicity of AgNPs is unpredictable despite their excellent contribution in the field of nanotechnology.

6. Conclusions

The mechanistic action of AgNPs offers an effective and efficient action against pathogens as compared to commercial methods. The application of nanotechnology, primarily AgNPs, in aquaculture and livestock sectors has been proven to solve problems faced by the respective sectors. AgNPs offer a long-term and effective solution in disease control, which in our opinion is the root of the majority of issues faced by the respective industries. An effective solution leads to an increase in aquaculture and livestock production with minimal to no side effects, ensuring the quality of products and, thus, increasing the economic growth of the country as well as the global economy.

Author Contributions: Conceptualization, S.A.A., C.D.S., N.M.N.; Writing—original draft preparation, C.D.S.; Writing—review and editing, S.A.A., N.M.N., M.M.A.K., S.A.G., M.J.M., B.G.; Supervision, S.A.A., N.M.N., M.M.A.K., S.A.G., M.J.M., B.G. All authors have read and agreed to the published version of the manuscript.

Funding: This project was financially supported by Universiti Putra Malaysia and Malaysia Ministry of Higher Education (MOHE) under the research grant attached to N.M. Nawawi, S.A. Ahmad, and M.M. Abd Karim (FRGS/1/2020/STG01/UNISEL/02/4). The authors also thank the Universiti Putra Malaysia for granting a personal scholarship (GRF) to C. De Silva.

Institutional Review Board Statement: Not applicable.

Data Availability Statement: Not applicable.

Acknowledgments: The authors would like to thank Universiti Putra Malaysia and Universiti Selangor.

Conflicts of Interest: The authors declare no conflict of interest.

References

1. Huang, S.; Wang, L.; Liu, L.; Hou, Y.; Li, L. Nanotechnology in agriculture, livestock, and aquaculture in China. A review. *Agron. Sustain. Dev.* **2015**, *35*, 369–400. [CrossRef]
2. Ghidan, A.Y.; Al-Antary, T.M.; Awwad, A.M. Aphidicidal potential of green synthesized magnesium hydroxide nanoparticles using *Olea europaea* leaves extract. *J. Agric. Biol. Sci.* **2017**, *12*, 293–301.
3. Ghidan, A.Y.; Al-Antary, T.M.; Salem, N.M.; Awwad, A.M. Facile green synthetic route to the zinc oxide (ZnONPs) nanoparticles: Effect on green peach aphid and antibacterial activity. *J. Agric. Biol. Sci.* **2017**, *9*, 131–138. [CrossRef]
4. Franci, G.; Falanga, A.; Galdiero, S.; Palomba, L.; Rai, M.; Morelli, G.; Galdiero, M. Silver nanoparticles as potential antibacterial agents. *Molecules* **2015**, *20*, 8856–8874. [CrossRef]

5. Zhang, X.-F.; Liu, Z.-G.; Shen, W.; Gurunathan, S. Silver Nanoparticles: Synthesis, characterization, properties, applications, and therapeutic approaches. *Int. J. Mol. Sci.* **2016**, *17*, 1534. [CrossRef]
6. Shang, L.; Nienhaus, K.; Jiang, X.; Yang, L.; Landfester, K.; Mailänder, V.; Simmet, T.; Nienhaus, G.U. Nanoparticle interactions with live cells: Quantitative fluorescence microscopy of nanoparticle size effects. *Beilstein J. Nanotechnol.* **2014**, *5*, 2388–2397. [CrossRef] [PubMed]
7. Kuhn, D.A.; Vanhecke, D.; Michen, B.; Blank, F.; Gehr, P.; Petri-Fink, A.; Rothen-Rutishauser, B. Different endocytotic uptake mechanisms for nanoparticles in epithelial cells and macrophages. *Beilstein J. Nanotechnol.* **2014**, *5*, 1625–1636. [CrossRef]
8. Morones, J.R.; Elechiguerra, J.L.; Camacho, A.; Holt, K.; Kouri, J.B.; Ramírez, J.T.; Yacaman, M.J. The bactericidal effect of silver nanoparticles. *Nanotechnology* **2005**, *16*, 2346–2353. [CrossRef]
9. Van Dong, P.; Ha, C.H.; Binh, L.T.; Kasbohm, J. Chemical synthesis and antibacterial activity of novel-shaped silver nanoparticles. *Int. Nano Lett.* **2012**, *2*, 9. [CrossRef]
10. Feng, Q.L.; Wu, J.; Chen, G.Q.; Cui, F.Z.; Kim, T.N.; Kim, J.O. A mechanistic study of the antibacterial effect of silver ions on Escherichia coli and *Staphylococcus aureus*. *J. Biomed. Mater. Res.* **2000**, *52*, 662–668. [CrossRef]
11. Sotiriou, G.A.; Pratsinis, S.E. Engineering nanosilver as an antibacterial, biosensor and bioimaging material. *Curr. Opin. Chem. Eng.* **2011**, *1*, 3–10. [CrossRef]
12. El-Nour, K.M.A.; Eftaiha, A.; Al-Warthan, A.; Ammar, R.A. Synthesis and applications of silver nanoparticles. *Arab. J. Chem.* **2010**, *3*, 135–140. [CrossRef]
13. Martínez-Castañón, G.A.; Niño, N.; Martinez-Gutierrez, F.; Martínez-Mendoza, J.R.; Ruiz, F. Synthesis and antibacterial activity of silver nanoparticles with different sizes. *J. Nanoparticle Res.* **2008**, *10*, 1343–1348. [CrossRef]
14. Lu, Z.; Rong, K.; Li, J.; Yang, H.; Chen, R. Size-dependent antibacterial activities of silver nanoparticles against oral anaerobic pathogenic bacteria. *J. Mater. Sci. Mater. Med.* **2013**, *24*, 1465–1471. [CrossRef] [PubMed]
15. Hong, X.; Wen, J.; Xiong, X.; Hu, Y. Shape effect on the antibacterial activity of silver nanoparticles synthesized via a microwave-assisted method. *Environ. Sci. Pollut. Res.* **2016**, *23*, 4489–4497. [CrossRef] [PubMed]
16. Patra, J.K.; Baek, K.-H. Antibacterial activity and synergistic antibacterial potential of biosynthesized silver nanoparticles against foodborne pathogenic bacteria along with its anticandidal and antioxidant effects. *Front. Microbiol.* **2017**, *8*, 167. [CrossRef]
17. Soliman, H.; Elsayed, A.; Dyaa, A. Antimicrobial activity of silver nanoparticles biosynthesised by *Rhodotorula* sp. strain ATL72. *Egypt. J. Basic Appl. Sci.* **2018**, *5*, 228–233. [CrossRef]
18. Mathivanan, K.; Selva, R.; Chandirika, J.U.; Govindarajan, R.; Srinivasan, R.; Annadurai, G.; Duc, P.A. Biologically synthesized silver nanoparticles against pathogenic bacteria: Synthesis, calcination and characterization. *Biocatal. Agric. Biotechnol.* **2019**, *22*, 101373. [CrossRef]
19. Yu, Z.; Wang, W.; Kong, F.; Lin, M.; Mustapha, A. Cellulose nanofibril/silver nanoparticle composite as an active food packaging system and its toxicity to human colon cells. *Int. J. Biol. Macromol.* **2019**, *129*, 887–894. [CrossRef] [PubMed]
20. Pal, S.; Tak, Y.K.; Song, J.M. Does the antibacterial activity of silver nanoparticles depend on the shape of the nanoparticle? A study of the gram-negative bacterium *Escherichia coli*. *Appl. Environ. Microbiol.* **2007**, *73*, 1712–1720. [CrossRef]
21. Huang, W.; Li, H.; Duan, H.; Bi, Y.; Wu, D.; Yu, H. Optimized biosynthesis and antifungal activity of *Osmanthus fragrans* leaf extract-mediated silver nanoparticles. *Int. J. Agric. Biol.* **2017**, *19*, 668–672. [CrossRef]
22. Jo, D.H.; Kim, J.H.; Lee, T.G.; Kim, J.H. Size, surface charge, and shape determine therapeutic effects of nanoparticles on brain and retinal diseases. *Nanomedicine* **2015**, *11*, 1603–1611. [CrossRef]
23. Walters, G.; Parkin, I.P. The incorporation of noble metal nanoparticles into host matrix thin films: Synthesis, characterisation and applications. *J. Mater. Chem.* **2009**, *19*, 574–590. [CrossRef]
24. Xia, Y.; Xia, X.; Wang, Y.; Xie, S. Shape-controlled synthesis of metal nanocrystals. *MRS Bull.* **2013**, *38*, 335–344. [CrossRef]
25. Helmlinger, J.; Heise, M.; Heggen, M.; Ruck, M.; Epple, M. A rapid, high-yield and large-scale synthesis of uniform spherical silver nanoparticles by a microwave-assisted polyol process. *RSC Adv.* **2015**, *5*, 92144–92150. [CrossRef]
26. Actis, L.; Srinivasan, A.; Lopez-Ribot, J.; Ramasubramanian, A.K.; Ong, J. Effect of silver nanoparticle geometry on methicillin susceptible and resistant *Staphylococcus aureus*, and osteoblast viability. *J. Mater. Sci. Mater. Med.* **2015**, *26*, 1–7. [CrossRef]
27. Rai, M.; Yadav, A.; Gade, A. Silver nanoparticles as a new generation of antimicrobials. *Biotechnol. Adv.* **2009**, *27*, 76–83. [CrossRef]
28. Gatoo, M.A.; Naseem, S.; Arfat, M.Y.; Dar, A.M.; Qasim, K.; Zubair, S. Physicochemical properties of nanomaterials: Implication in associated toxic manifestations. *BioMed Res. Int.* **2014**, *2014*, 498420. [CrossRef]
29. Beyth, N.; Houri-Haddad, Y.; Domb, A.; Khan, W.; Hazan, R. Alternative antimicrobial approach: Nano-antimicrobial materials. *Evid. Based Complement. Altern. Med.* **2015**, *2015*, 246012. [CrossRef]
30. Raman, G.; Park, S.J.; Sakthivel, N.; Suresh, A.K. Physico-cultural parameters during AgNPs biotransformation with bactericidal activity against human pathogens. *Enzym. Microb. Technol.* **2017**, *100*, 45–51. [CrossRef]
31. Mokhena, T.; Luyt, A. Electrospun alginate nanofibres impregnated with silver nanoparticles: Preparation, morphology and antibacterial properties. *Carbohydr. Polym.* **2017**, *165*, 304–312. [CrossRef]
32. Huang, W.; Yan, M.; Duan, H.; Bi, Y.; Cheng, X.; Yu, H. synergistic antifungal activity of green synthesized silver nanoparticles and epoxiconazole against *Setosphaeria turcica*. *J. Nanomater.* **2020**, *2020*, 1–7. [CrossRef]
33. Sondi, I.; Salopek-Sondi, B. Silver nanoparticles as antimicrobial agent: A case study on *E. coli* as a model for Gram-negative bacteria. *J. Colloid Interface Sci.* **2004**, *275*, 177–182. [CrossRef]

34. Lok, C.-N.; Ho, C.-M.; Chen, R.; He, Q.-Y.; Yu, W.-Y.; Sun, H.; Tam, P.K.-H.; Chiu, J.-F.; Che, C.-M. Silver nanoparticles: Partial oxidation and antibacterial activities. *JBIC J. Biol. Inorg. Chem.* **2007**, *12*, 527–534. [CrossRef] [PubMed]

35. Xiu, Z.-M.; Ma, J.; Alvarez, P.J.J. Differential effect of common ligands and molecular oxygen on antimicrobial activity of silver nanoparticles versus silver ions. *Environ. Sci. Technol.* **2011**, *45*, 9003–9008. [CrossRef]

36. Fadeel, B.; Ahlin, A.; Henter, J.I.; Orrenius, S.; Hampton, M.B. Involvement of caspases in neutrophil apoptosis: Regulation by reactive oxygen species. *Blood* **1998**, *92*, 4808–4818. [CrossRef] [PubMed]

37. Cohen, G.M. Caspases: The executioners of apoptosis. *Biochem. J.* **1997**, *326*, 1–16. [CrossRef] [PubMed]

38. Ebrahiminezhad, A.; Davaran, S.; Rasoul-Amini, S.; Barar, J.; Moghadam, M.; Ghasemi, Y. Synthesis, characterization and anti-listeria monocytogenes effect of amino acid coated magnetite nanoparticles. *Curr. Nanosci.* **2012**, *8*, 868–874. [CrossRef]

39. Ebrahiminezhad, A.; Rasoul-Amini, S.; Kouhpayeh, A.; Davaran, S.; Barar, J.; Ghasemi, Y. Impacts of amine functionalized iron oxide nanoparticles on HepG2 cell line. *Curr. Nanosci.* **2014**, *11*, 113–119. [CrossRef]

40. Gurr, J.-R.; Wang, A.S.S.; Chen, C.-H.; Jan, K.-Y. Ultrafine titanium dioxide particles in the absence of photoactivation can induce oxidative damage to human bronchial epithelial cells. *Toxicology* **2005**, *213*, 66–73. [CrossRef]

41. Olmedo, D.G.; Tasat, D.R.; Guglielmotti, M.; Cabrini, R. Effect of titanium dioxide on the oxidative metabolism of alveolar macrophages: An experimental study in rats. *J. Biomed. Mater. Res. Part A* **2005**, *73*, 142–149. [CrossRef] [PubMed]

42. Limbach, L.K.; Wick, P.; Manser, P.; Grass, R.N.; Bruinink, A.; Stark, W.J. Exposure of engineered nanoparticles to human lung epithelial cells: Influence of chemical composition and catalytic activity on oxidative stress. *Environ. Sci. Technol.* **2007**, *41*, 4158–4163. [CrossRef] [PubMed]

43. Nathan, C.; Cunningham-Bussel, A. Beyond oxidative stress: An immunologist's guide to reactive oxygen species. *Nat. Rev. Immunol.* **2013**, *13*, 349–361. [CrossRef]

44. Matsumura, Y.; Yoshikata, K.; Kunisaki, S.-I.; Tsuchido, T. Mode of bactericidal action of silver zeolite and its comparison with that of silver nitrate. *Appl. Environ. Microbiol.* **2003**, *69*, 4278–4281. [CrossRef] [PubMed]

45. Ebrahiminezhad, A.; Raee, M.J.; Manafi, Z.; Jahromi, A.S.; Ghasemi, Y. Ancient and novel forms of silver in medicine and biomedicine. *J. Adv. Med Sci. Appl. Technol.* **2016**, *2*, 122–128. [CrossRef]

46. Matés, J.M.; Jiménez, F.M.S. Role of reactive oxygen species in apoptosis: Implications for cancer therapy. *Int. J. Biochem. Cell Biol.* **2000**, *32*, 157–170. [CrossRef]

47. Messner, K.; Imlay, J.A. The identification of primary sites of superoxide and hydrogen peroxide formation in the aerobic respiratory chain and sulfite reductase complex of *Escherichia coli*. *J. Biol. Chem.* **1999**, *274*, 10119–10128. [CrossRef] [PubMed]

48. Blecher, K.; Nasir, A.; Friedman, A. The growing role of nanotechnology in combating infectious disease. *Virulence* **2011**, *2*, 395–401. [CrossRef]

49. Pandey, J.K.; Swarnkar, R.K.; Soumya, K.K.; Dwivedi, P.; Singh, M.K.; Sundaram, S.; Gopal, R. Silver nanoparticles synthesized by pulsed laser ablation: As a potent antibacterial agent for human enteropathogenic Gram-positive and Gram-negative bacterial strains. *Appl. Biochem. Biotechnol.* **2014**, *174*, 1021–1031. [CrossRef]

50. Nalwade, A.R.; Jadhav, A.A. Biosynthesis of silver nanoparticles using leaf extract of *Datura alba* Nees. and evaluation of their antibacterial activity. *Arch. Appl. Sci. Res.* **2013**, *5*, 45–49.

51. Raffi, M.; Hussain, F.; Bhatti, T.M.; Akhter, J.I.; Hameed, A.; Hasan, M.M. Antibacterial characterization of silver nanoparticles against *E. Coli* ATCC-15224. *J. Mater. Sci. Technol.* **2008**, *24*, 192–196.

52. Ghosh, S.; Patil, S.; Ahire, M.; Kitture, R.; Kale, S.; Pardesi, K.; Cameotra, S.S.; Bellare, J.; Dhavale, D.D.; Jabgunde, A.; et al. Synthesis of silver nanoparticles using *Dioscorea bulbifera* tuber extract and evaluation of its synergistic potential in combination with antimicrobial agents. *Int. J. Nanomed.* **2012**, *7*, 483–496. [CrossRef]

53. Schreurs, W.J.; Rosenberg, H. Effect of silver ions on transport and retention of phosphate by *Escherichia coli*. *J. Bacteriol.* **1982**, *152*, 7–13. [CrossRef]

54. Li, J.J.; Muralikrishnan, S.; Ng, C.-T.; Yung, L.-Y.L.; Bay, B.-H. Nanoparticle-induced pulmonary toxicity. *Exp. Biol. Med.* **2010**, *235*, 1025–1033. [CrossRef]

55. Lok, C.-N.; Ho, C.-M.; Chen, R.; He, Q.-Y.; Yu, W.-Y.; Sun, H.; Tam, P.K.-H.; Chiu, J.-F.; Che, C.M. Proteomic analysis of the mode of antibacterial action of silver nanoparticles. *J. Proteome Res.* **2006**, *5*, 916–924. [CrossRef] [PubMed]

56. Kim, S.-H.; Lee, H.-S.; Ryu, D.-S.; Choi, S.-J.; Lee, D.-S. Antibacterial activity of silver-nanoparticles against *Staphylococcus aureus* and *Escherichia coli*. *Korean J. Microbiol. Biotechnol.* **2011**, *39*, 77–85.

57. Li, J.; Rong, K.; Zhao, H.; Li, F.; Lu, Z.; Chen, R. Highly selective antibacterial activities of silver nanoparticles against *Bacillus subtilis*. *J. Nanosci. Nanotechnol.* **2013**, *13*, 6806–6813. [CrossRef] [PubMed]

58. Habash, M.B.; Park, A.J.; Vis, E.C.; Harris, R.J.; Khursigara, C.M. Synergy of silver nanoparticles and aztreonam against *Pseudomonas aeruginosa* PAO1 biofilms. *Antimicrob. Agents Chemother.* **2014**, *58*, 5818–5830. [CrossRef] [PubMed]

59. Singh, B.R.; Singh, B.N.; Singh, A.; Khan, W.A.; Naqvi, A.H.; Singh, H.B. Mycofabricated biosilver nanoparticles interrupt *Pseudomonas aeruginosa* quorum sensing systems. *Sci. Rep.* **2015**, *5*, 1–14. [CrossRef]

60. Khalandi, B.; Asadi, N.; Milani, M.; Davaran, S.; Abadi, A.J.N.; Abasi, E.; Akbarzadeh, A. A review on potential role of silver nanoparticles and possible mechanisms of their actions on bacteria. *Drug Res. (Stuttg)* **2016**, *67*, 70–76. [CrossRef]

61. Jung, W.K.; Koo, H.C.; Kim, K.W.; Shin, S.; Kim, S.H.; Park, Y.H. Antibacterial activity and mechanism of action of the silver ion in *Staphylococcus aureus* and *Escherichia coli*. *Appl. Environ. Microbiol.* **2008**, *74*, 2171–2178. [CrossRef] [PubMed]

62. Rai, M.; Deshmukh, S.; Ingle, A.; Gade, A. Silver nanoparticles: The powerful nanoweapon against multidrug-resistant bacteria. *J. Appl. Microbiol.* **2012**, *112*, 841–852. [CrossRef]

63. Qing, Y.; Cheng, L.; Li, R.; Liu, G.; Zhang, Y.; Tang, X.; Wang, J.; Liu, H.; Qin, Y. Potential antibacterial mechanism of silver nanoparticles and the optimization of orthopedic implants by advanced modification technologies. *Int. J. Nanomed.* **2018**, *13*, 3311–3327. [CrossRef]

64. Klueh, U.; Wagner, V.; Kelly, S.; Johnson, A.; Bryers, J.D. Efficacy of silver-coated fabric to prevent bacterial colonization and subsequent device-based biofilm formation. *J. Biomed. Mater. Res.* **2000**, *53*, 621–631. [CrossRef]

65. Bhattacharya, R.; Mukherjee, P. Biological properties of "naked" metal nanoparticles. *Adv. Drug Deliv. Rev.* **2008**, *60*, 1289–1306. [CrossRef]

66. Hsueh, Y.-H.; Lin, K.-S.; Ke, W.-J.; Hsieh, C.-T.; Chiang, C.-L.; Tzou, D.-Y.; Liu, S.-T. The antimicrobial properties of silver nanoparticles in *Bacillus subtilis* are mediated by released Ag+ ions. *PLoS ONE* **2015**, *10*, e0144306. [CrossRef] [PubMed]

67. Kumar, N.; Das, S.; Jyoti, A.; Kaushik, S. Synergistic effect of silver nanoparticles with doxycycline against *Klebsiella pneumoniae*. *Int. J. Pharm. Sci.* **2016**, *8*, 183–186.

68. Oliver, J.D. The viable but non-culturable state in bacteria. *J. Microbiol.* **2005**, *43*, 93–100.

69. Li, L.; Mendis, N.; Trigui, H.; Oliver, J.D.; Faucher, S.P. The importance of the viable but non-culturable state in human bacterial pathogens. *Front. Microbiol.* **2014**, *5*, 258. [CrossRef]

70. Zhou, X.; Wang, Y.; Gu, Q.; Li, W. Effects of different dietary selenium sources (selenium nanoparticle and selenomethionine) on growth performance, muscle composition and glutathione peroxidase enzyme activity of crucian carp (*Carassius auratus gibelio*). *Aquaculture* **2009**, *291*, 78–81. [CrossRef]

71. Meneses, J.; Hamdan, A.; Del Carmen Monroy Dosta, M.; Mejía, J.; Bustos Martinez, J. Silver nanoparticles applications (AgNPS) in aquaculture. *Int. J. Fish. Aquat.* **2018**, *6*, 5–11.

72. Martinez-Porchas, M.; Martinez-Cordova, L.R. World Aquaculture: Environmental Impacts and Troubleshooting Alternatives. *Sci. World J.* **2012**, *2012*, 389623. [CrossRef] [PubMed]

73. Pillay, T.V. R *Aquaculture and the Environment*, 2nd ed.; Blackwell Publishing Ltd.: Oxford, UK, 2008; p. 208. [CrossRef]

74. Pradeep, T.; Anshup. Noble metal nanoparticles for water purification: A critical review. *Thin Solid Films* **2009**, *517*, 6441–6478. [CrossRef]

75. Johari, S.A.; Kalbassi, M.R.; Soltani, M.; Yu, I.J. Toxicity comparison of colloidal silver nanoparticles in various life stages of rainbow trout (*Oncorhynchus mykiss*). *Iran. J. Fish. Sci.* **2013**, *12*, 76–95.

76. Sarkheil, M.; Sourinejad, I.; Mirbakhsh, M.; Kordestani, D.; Johari, S.A. Application of silver nanoparticles immobilized on TEPA-Den-SiO2 as water filter media for bacterial disinfection in culture of Penaeid shrimp larvae. *Aquac. Eng.* **2016**, *74*, 17–29. [CrossRef]

77. Pelgrift, R.Y.; Friedman, A. Nanotechnology as a therapeutic tool to combat microbial resistance. *Adv. Drug Deliv. Rev.* **2013**, *65*, 1803–1815. [CrossRef]

78. Vaseeharan, B.; Ramasamy, P.; Chen, J.C. Antibacterial activity of silver nanoparticles (AgNps) synthesized by tea leaf extracts against pathogenic *Vibrio harveyi* and its protective efficacy on juvenile *Feneropenaeus indicus*. *Lett. Appl. Microbiol.* **2010**, *50*, 352–356. [CrossRef] [PubMed]

79. Sivaramasamy, E.; Zhiwei, W.; Li, F.; Xiang, J. Enhancement of vibriosis resistance in *Litopenaeus vannamei* by supplementation of biomastered silver nanoparticles by *Bacillus subtilis*. *J. Nanomed. Nanotechnol.* **2016**, *7*, 2. [CrossRef]

80. Barakat, K.M.; El-Sayed, H.S.; Gohar, Y.M. Protective effect of squilla chitosan–silver nanoparticles for *Dicentrarchus labrax* larvae infected with *Vibrio anguillarum*. *Int. Aquat. Res.* **2016**, *8*, 179–189. [CrossRef]

81. Daniel, S.C.G.K.; Sironmani, T.A.; Dinakaran, S. Nano formulations as curative and protective agent for fish diseases: Studies on red spot and white spot diseases of ornamental gold fish *Carassius auratus*. *Int. J. Fish. Aquat.* **2016**, *4*, 255–261.

82. Rajeshkumar, S.; Venkatesan, C.; Sarathi, M.; Sarathbabu, V.; Thomas, J.; Basha, K.A.; Hameed, A.S. Oral delivery of DNA construct using chitosan nanoparticles to protect the shrimp from white spot syndrome virus (WSSV). *Fish Shellfish. Immunol.* **2009**, *26*, 429–437. [CrossRef]

83. Juarez-Moreno, K.; Mejia-Ruiz, C.H.; Díaz, F.; Reyna-Verdugo, H.; Re, A.D.; Vazquez-Felix, E.F.; Sánchez-Castrejón, E.; Mota-Morales, J.D.; Pestryakov, A.; Bogdanchikova, N. Effect of silver nanoparticles on the metabolic rate, hematological response, and survival of juvenile white shrimp *Litopenaeus vannamei*. *Chemosphere* **2017**, *169*, 716–724. [CrossRef] [PubMed]

84. Ochoa-Meza, A.R.; Álvarez-Sánchez, A.R.; Romo-Quiñonez, C.R.; Barraza, A.; Magallón-Barajas, F.J.; Chávez-Sánchez, A.; García-Ramos, J.C.; Toledano-Magaña, Y.; Bogdanchikova, N.; Pestryakov, A.; et al. Silver nanoparticles enhance survival of white spot syndrome virus infected *Penaeus vannamei* shrimps by activation of its immunological system. *Fish Shellfish. Immunol.* **2019**, *84*, 1083–1089. [CrossRef]

85. Romo-Quiñonez, C.R.; Sanchez, A.R.A.; Álvarez-Ruiz, P.; Chávez-Sánchez, M.C.; Bogdanchikova, N.; Pestryakov, A.; Mejia-Ruiz, C.H. Evaluation of a new Argovit as an antiviral agent included in feed to protect the shrimp *Litopenaeus vannamei* against White Spot Syndrome Virus infection. *PeerJ* **2020**, *8*, e8446. [CrossRef] [PubMed]

86. Rather, M.A.; Sharma, R.; Gupta, S.; Ferosekhan, S.; Ramya, V.L.; Jadhao, S.B. Chitosan-nanoconjugated hormone nanoparticles for sustained surge of gonadotropins and enhanced reproductive output in female fish. *PLoS ONE* **2013**, *8*, e57094. [CrossRef] [PubMed]

87. Alishahi, A.; Mirvaghefi, A.; Tehrani, M.; Farahmand, H.; Koshio, S.; Dorkoosh, F.; Elsabee, M.Z. Chitosan nanoparticle to carry vitamin C through the gastrointestinal tract and induce the non-specific immunity system of rainbow trout (*Oncorhynchus mykiss*). *Carbohydr. Polym.* **2011**, *86*, 142–146. [CrossRef]

88. Fondevila, M. Potential Use of Silver Nanoparticles as an Additive in Animal Feeding. In *Silver Nanoparticles*; Perez, D.P., Ed.; InTech: London, UK, 2010; Available online: https://www.intechopen.com/books/silver-nanoparticles/potential-use-of-silver-nanoparticles-as-an-additive-in-animal-feeding (accessed on 13 July 2021).

89. El-Gohary, F.A.; Abdel-Hafez, L.J.M.; Zakaria, A.I.; Shata, R.R.; Tahoun, A.; El-Mleeh, A.; Elfadl, E.A.A.; Elmahallawy, E.K. Enhanced antibacterial activity of silver nanoparticles combined with hydrogen peroxide against multidrug-resistant pathogens isolated from dairy farms and beef slaughterhouses in Egypt. *Infect. Drug Resist.* **2020**, *13*, 3485–3499. [CrossRef]

90. Li, Y.; Leung, P.; Yao, L.; Song, Q.; Newton, E. Antimicrobial effect of surgical masks coated with nanoparticles. *J. Hosp. Infect.* **2006**, *62*, 58–63. [CrossRef] [PubMed]

91. Kim, J.S.; Kuk, E.; Yu, K.N.; Kim, J.-H.; Park, S.J.; Lee, H.J.; Kim, S.H.; Park, Y.K.; Park, Y.H.; Hwang, C.-Y.; et al. Antimicrobial effects of silver nanoparticles. *Nanomed. Nanotechnol. Biol. Med.* **2007**, *3*, 95–101. [CrossRef] [PubMed]

92. Fondevila, M.; Herrer, R.; Casallas, M.; Abecia, L.; Ducha, J. Silver nanoparticles as a potential antimicrobial additive for weaned pigs. *Anim. Feed. Sci. Technol.* **2009**, *150*, 259–269. [CrossRef]

93. Kalińska, A.; Jaworski, S.; Wierzbicki, M.; Gołębiewski, M. Silver and copper nanoparticles—an alternative in future mastitis treatment and prevention? *Int. J. Mol. Sci.* **2019**, *20*, 1672. [CrossRef]

94. Sawosz, F.; Marta, G.; Marlena, Z.; Niemiec, T.; Boena, O.; Chwalibog, A. Nanoparticles of silver do not effect growth, development and DNA oxidative damage in chicken embryo. *Arch. Geflugelkd.* **2009**, *73*, 208–213.

95. Sawosz, F.; Pineda, L.; Hotowy, A.; Hyttel, P.; Sawosz, E.; Szmidt, M.; Niemiec, T.; Chwalibog, A. Nano-nutrition of chicken embryos. The effect of silver nanoparticles and glutamine on molecular responses, and the morphology of pectoral muscle. *Baltic J. Comp. Clin. Syst. Biol.* **2012**, *2*, 29–45. [CrossRef]

96. Raieszadeh, H.; Noaman, V.; Yadegari, M. Echocardiographic assessment of cardiac structural and functional indices in broiler chickens treated with silver nanoparticles. *Sci. World J.* **2013**, *2013*, 931432. [CrossRef] [PubMed]

97. Chmielowiec-Korzeniowska, A.; Tymczyna, L.; Dobrowolska, M.; Banach, M.; Nowakowicz-Debek, B.; Bryl, M.; Drabik, A.; Tymczyna-Sobotka, M.; Kolejko, M. Silver (Ag) in tissues and eggshells, biochemical parameters and oxidative stress in chickens. *Open Chem.* **2015**, *13*, 1269–1274. [CrossRef]

98. Kumar, I.; Bhattacharya, J.; Das, B.K.; Lahiri, P. Growth, serum biochemical, and histopathological responses of broilers administered with silver nanoparticles as a drinking water disinfectant. *3 Biotech* **2020**, *10*, 1–12. [CrossRef]

99. Stensberg, M.C.; Wei, Q.; McLamore, E.S.; Porterfield, D.M.; Wei, A.; Sepúlveda, M.S. Toxicological studies on silver nanoparticles: Challenges and opportunities in assessment, monitoring and imaging. *Nanomedicine* **2011**, *6*, 879–898. [CrossRef] [PubMed]

100. Choi, O.; Hu, Z. Size dependent and reactive oxygen species related nanosilver toxicity to nitrifying bacteria. *Environ. Sci. Technol.* **2008**, *42*, 4583–4588. [CrossRef]

101. Jamil, B.; Javed, R.; Qazi, A.S.; Syed, M.A. Nanomaterials: Toxicity, risk managment and public perception. In *Nanomaterials: Ecotoxicity, Safety, and Public Perception*; Rai, M., Biswas, J.K., Eds.; Springer: Cham, Switzerland, 2018; pp. 283–304. [CrossRef]

102. Ferdous, Z.; Nemmar, A. Health impact of silver nanoparticles: A review of the biodistribution and toxicity following various routes of exposure. *Int. J. Mol. Sci.* **2020**, *21*, 2375. [CrossRef]

Article

Antibacterial Potential of Biosynthesized Zinc Oxide Nanoparticles against Poultry-Associated Foodborne Pathogens: An In Vitro Study

Hidayat Mohd Yusof [1], Nor'Aini Abdul Rahman [1,2,*], Rosfarizan Mohamad [1,2], Uswatun Hasanah Zaidan [3] and Anjas Asmara Samsudin [4]

[1] Department of Bioprocess Technology, Faculty of Biotechnology and Biomolecular Sciences, Universiti Putra Malaysia, Serdang 43400, Selangor, Malaysia; hidayatmy@gmail.com (H.M.Y.); farizan@upm.edu.my (R.M.)
[2] Bioprocessing and Biomanufacturing Research Centre, Faculty of Biotechnology and Biomolecular Sciences, Universiti Putra Malaysia, Serdang 43400, Selangor, Malaysia
[3] Department of Biochemistry, Faculty of Biotechnology and Biomolecular Sciences, Universiti Putra Malaysia, Serdang 43400, Selangor, Malaysia; uswatun@upm.edu.my
[4] Department of Animal Science, Faculty of Agriculture, Universiti Putra Malaysia, Serdang 43400, Selangor, Malaysia; anjas@upm.edu.my
*   Correspondence: nor_aini@upm.edu.my; Tel.: +60-3-9769-1945; Fax: +60-3-9769-7590

Citation: Mohd Yusof, H.; Abdul Rahman, N.; Mohamad, R.; Hasanah Zaidan, U.; Samsudin, A.A. Antibacterial Potential of Biosynthesized Zinc Oxide Nanoparticles against Poultry-Associated Foodborne Pathogens: An In Vitro Study. *Animals* **2021**, *11*, 2093. https://doi.org/10.3390/ani11072093

Academic Editors: Antonio Gonzalez-Bulnes and Nesreen Hashem

Received: 25 April 2021
Accepted: 7 June 2021
Published: 14 July 2021

Publisher's Note: MDPI stays neutral with regard to jurisdictional claims in published maps and institutional affiliations.

Simple Summary: The overuse of antibiotics in the poultry industry has led to the emergence of multidrug-resistant microorganisms. Thus, there is a need to find an alternative to conventional antibiotics. Recently, zinc oxide nanoparticles (ZnO NPs) have gained much attention due to their excellent antibacterial activity. In addition, ZnO NPs is an essential trace mineral in poultry diets. In this sense, incorporating ZnO NPs into poultry can promote growth and performance while serving as an alternative antibacterial agent to control diseases. Therefore, this study aimed to assess the in vitro antibacterial activity and antibacterial mechanisms of ZnO NPs against poultry-associated foodborne pathogens (*Salmonella* spp., *Escherichia coli*, and *Staphylococcus aureus*). The obtained findings demonstrated effective antibacterial actions against the tested microorganisms. The nanotechnology approach could represent a new tool for combating pathogens in the poultry industry.

Abstract: Since the emergence of multidrug-resistant bacteria in the poultry industry is currently a serious threat, there is an urgent need to develop a more efficient and alternative antibacterial substance. Zinc oxide nanoparticles (ZnO NPs) have exhibited antibacterial efficacy against a wide range of microorganisms. Although the in vitro antibacterial activity of ZnO NPs has been studied, little is known about the antibacterial mechanisms of ZnO NPs against poultry-associated foodborne pathogens. In the present study, ZnO NPs were successfully synthesized using *Lactobacillus plantarum* TA4, characterized, and their antibacterial potential against common avian pathogens (*Salmonella* spp., *Escherichia coli*, and *Staphylococcus aureus*) was investigated. Confirmation of ZnO NPs by UV-Visual spectroscopy showed an absorption band center at 360 nm. Morphologically, the synthesized ZnO NPs were oval with an average particle size of 29.7 nm. Based on the dissolution study of Zn^{2+}, ZnO NPs released more ions than their bulk counterparts. Results from the agar well diffusion assay indicated that ZnO NPs effectively inhibited the growth of the three poultry-associated foodborne pathogens. The minimum inhibitory concentration (MIC) and minimum bactericidal concentration (MBC) were assessed using various concentrations of ZnO NPs, which resulted in excellent antibacterial activity as compared to their bulkier counterparts. *S. aureus* was more susceptible to ZnO NPs compared to the other tested bacteria. Furthermore, the ZnO NPs demonstrated substantial biofilm inhibition and eradication. The formation of reactive oxygen species (ROS) and cellular material leakage was quantified to determine the underlying antibacterial mechanisms, whereas a scanning electron microscope (SEM) was used to examine the morphological changes of tested bacteria treated with ZnO NPs. The findings suggested that ROS-induced oxidative stress caused membrane damage and bacterial cell death. Overall, the results demonstrated that

ZnO NPs could be developed as an alternative antibiotic in poultry production and revealed new possibilities in combating pathogenic microorganisms.

Keywords: antibacterial; antibiotic; mechanisms; poultry; reactive oxygen species; zinc oxide nanoparticles

1. Introduction

The unrestricted overuse of antibiotics promotes multidrug resistance in microorganisms, compromising human and animal health. Hence, there is a pressing need to develop alternatives to traditional antimicrobials that are more effective with new action mechanisms. One of the current alternatives to combat multidrug-resistant pathogens is nanobiotics, also known as nanoparticles (NPs) with antimicrobial properties. Nanoparticles currently gained much attention due to their unique characteristics, which offer many explicit properties for biomedical applications [1]. Inorganic NPs, such as gold (AuNPs), silver (AgNPs), zinc oxide (ZnO NPs), and selenium NPs (SeNPs), have profound applications in medical and biological fields such as medical diagnosis, biosensor, and personal care products and have been widely explored as antibacterial agents [2–4] due to their distinctive physicochemical and biological properties over their bulk phase.

Among the NPs, ZnO NPs are specifically vital as efficient metal oxide NPs and exhibited various antibacterial properties against a wide range of microorganisms, including Gram-positive and Gram-negative bacteria, as well as major foodborne pathogens. The properties of ZnO NPs, such as extensive surface area, biocompatibility, biodegradability, semiconductor behavior, and UV light barrier, render their vast application. In addition, ZnO NPs have been utilized as antimicrobial agents in food packaging [5], topical creams, and antibiotic agents in animal feed due to their strong bactericidal effect associated with their small particles and higher surface energies [1].

Zinc (Zn) is an important trace element for poultry required for the growth, health, and metabolic function of the body [6]. However, the Zn content in the raw material diet in poultry feed is too low to meet the poultry requirements. Thus, Zn supplementation in the form of inorganic Zn has been widely used. However, the major drawbacks of using inorganic Zn is its poor bioavailability and utilization rate [7], causing the feed manufacturers to apply a greater amount of dietary Zn (100 to 120 mg/kg feed) to achieve the maximum performance of poultry, which is above the recommended standard (40 mg/kg) by the National Research Council (NRC, 1994) [8]. Nonetheless, the extensive use of high-dose dietary zinc oxide (ZnO) may affect the stability of other trace elements [9,10] and lead to excess Zn in excreta, causing environmental contamination. With the emergence of nanotechnology, which is associated with the smallest particle size ranging from 1 to 100 nm, the bioavailability and absorption efficiency of Zn were improved. Therefore, the effects of ZnO NPs have been prominently studied as a feed supplement in the poultry industry to improve the bioavailability of Zn in the body [9–11] and reduce the excretion of Zn in feces. Aside from promoting poultry growth and performance, the supplementation of ZnO NPs also received great attention for its antibacterial effects. Furthermore, many studies have shown the effectiveness of ZnO NPs in controlling the gut microbiota in poultry and livestock [12–15], demonstrating its potential use as an antibiotic. In this sense, introducing ZnO NPs into poultry feed could potentially serve as an antibiotic, as well as a Zn supply.

Several challenges emerged with the global exponential growth of broiler meat consumption, including infectious diseases caused by poor biosecurity and husbandry practices [16]. Microbial diseases caused by pathogens mainly contribute to the high mortality in the poultry industry. In poultry production, different kinds of antibiotics are used to control diseases and promote body growth performance. However, frequent use of antibiotics in the poultry industry contributed to major health threats in the society associated with antibiotic residue in meat, eggs, and other animal products [16]. *Salmonella, Escherichia coli,* and *Staphylococcus aureus* are major poultry product contaminants associated with human foodborne pathogen illness [17,18] and a major cause of opportunistic human and animal infections. These pathogenic bacteria commonly colonize poultry and could also lead to serious consequences if not treated. *Salmonella*, for example, is one of the most prevalent foodborne pathogens that can spread from contaminated poultry products such as meat and eggs to humans, resulting in salmonellosis [19]. Therefore, lowering the burden of poultry-associated foodborne pathogens in the gut of poultry could help reduce the contamination of poultry products as well as diseases caused by these pathogens. As more pathogenic bacteria develop resistance to conventional antibiotics, nanotechnology offers new paths for antibacterial drug design.

Scientific evidence demonstrated the potential of ZnO NPs as an alternative to conventional antibiotics in livestock farming. Yausheva et al. [12] conducted an intestinal microbiome assessment on the supplementation of ZnO NPs in broiler chicken. The results show that the ZnO NPs supplementation resulted in a decreased number of pathogenic microorganisms in broiler chicken cecum. Likewise, Wang et al. [20] reported that ZnO NPs supplementation in piglets diet could reduce diarrhea caused by some common bacterial such as *E. coli* and *Salmonella* than high dietary ZnO. The study suggested that ZnO NPs could be used as a feed antibiotic and to replace high dietary ZnO levels. However, the exact antimicrobial mechanisms of ZnO NPs are not completely understood, and the most common antibacterial action of ZnO NPs is attributed through multifaceted mechanisms, including the generation of ions from the surface of NPs [1]. The ions will bind to electron donor groups on the bacterial cell surface and subsequently damage the bacterial cell membrane [21]. Besides, the intrinsic physicochemical properties of ZnO NPs allow for the generation of reactive oxygen species (ROS), which induces oxidative stress and cell death [22].

The biological synthesis of metal nanoparticles (NPs) has many advantages compared to physical and chemical synthesis due to their eco-friendliness, biocompatible properties, and low cost. Although chemical approaches outperform biological alternatives in terms of production rate and NPs size control, chemically synthesized NPs are less biocompatible due to the usage of toxic chemicals for capping and stabilizing agents. Triethylamine, thioglycerol, and ethylenediaminetetraacetic acid (EDTA) [23] are commonly employed as agents in the capping and stabilizing process to control the size of NPs and prevent their aggregation. Nonetheless, these chemicals may reside or bind with the final product of NPs, and their presence is considered to have an adverse impact when used in biological applications. This issue does not arise when NPs are synthesized using biological synthesis routes such as by plant extract or microorganisms. Biological synthesis does not require external chemical sources for reducing and stabilizing, as the reactions are mediated by the biological molecules present within the system [6,24]. Moreover, the production of ZnO NPs using biological molecules is of interest for sustainable production and biomedical applications, in line with the green chemistry prospect. A previous study by Darvishi et al. [25] compared the cytotoxicity effect between biosynthesized (walnut extract) and chemically synthesized ZnO NPs on human skin fibroblasts and found that the biosynthesized ZnO NPs were less toxic to the cells. They discovered that the hazardous capping agent in chemically synthesized ZnO NPs causes cytotoxicity, while biomolecules from walnut extract serve as a capping agent, reducing toxicity.

The present work involves the biological synthesis of ZnO NPs using *Lactobacillus plantarum* TA4 from our previous study [26]. The biosynthesized ZnO NPs were analyzed by UV-Visible spectroscopy, and the size and shape morphology was determined using a high-resolution transmission electron microscope (HR-TEM). To date, studies related to the efficacy of ZnO NPs against poultry-associated foodborne pathogens are scarce. Hence, the present study attempted to assess the in vitro antibacterial potential of ZnO NPs against poultry-associated foodborne pathogen isolates of *Salmonella*, *E. coli*, and *S. aureus* for their potential application in combating pathogens in the poultry industry. The in vitro antibacterial potential of ZnO NPs was investigated through the determination of the inhibition zone using the agar well diffusion method. The minimum inhibitory concentration (MIC) and minimum bactericidal concentration (MBC) and time-killing assay of ZnO NPs were also determined. Trypan blue exclusion assay, quantification of protein, reducing sugar, ROS formation, and cell morphological changes using a scanning electron microscope (SEM) were performed to investigate the antibacterial mechanisms of ZnO NPs.

2. Materials and Methods

2.1. Bacterial Strain and Media

Three poultry-associated foodborne pathogenic isolates, i.e., *Salmonella* spp., *E. coli*, and *S. aureus* isolated from the small intestine of broiler chicken, were obtained from the Department of Animal Science, Universiti Putra Malaysia. The isolates were grown overnight in nutrient broth at 37 °C. A previously isolated zinc-tolerant *L. plantarum* TA4 was used for the biosynthesis of ZnO NPs [27]. The strain was grown in the de Man, Rogosa and Sharpe (MRS) (Oxoid™, Basingstoke, UK) broth and incubated for 24 h at 37 °C on an orbital shaker at 150 rpm. After the incubation, the supernatant was collected by centrifugation at 10,000× g for 5 min.

2.2. Zinc Oxide Nanoparticles Preparation and Characterization

The established procedure from our previous study [26] was followed for the biosynthesis of ZnO NPs with some modifications. Briefly, 20 mL of supernatant from zinc-tolerant *L. plantarum* TA4 was mixed with 80 mL of 500 mM aqueous zinc nitrate ($Zn(NO_3)_2 \cdot 6H_2O$) solution. The mixture was continuously stirred with a magnetic stirrer for 4 h at room temperature. The formation of ZnO NPs was confirmed by the appearance of white coalescence in the reaction mixture. Subsequently, the biosynthesized ZnO NPs were separated using centrifugation (10,000× g for 20 min) and washed with dH_2O and ethanol multiple times, and drying overnight at 100 °C to obtain the white powder of ZnO NPs.

ZnO NPs were characterized by physicochemical methods. The surface plasmon resonance (SPR) was characterized by UV-Visual spectroscopy (Uviline 9400, Secomam, Alès, France) in the range of 300 to 600 nm and 1 nm resolution. The morphology of ZnO NPs was observed by HR-TEM (JEM-2100F, JEOL, Tokyo, Japan). The average particle size of ZnO NPs was determined by measuring at least 300 particles using ImageJ software (National Institute of Health, Bethesda, MD, USA).

2.3. Determination of Zn^{2+} Dissolution

The concentrations of dissolved Zn^{2+} from ZnO NPs were measured following the method described by Haque et al. [28]. A 100 µg/mL of ZnO NPs or bulk ZnO was suspended in 0.85% NaCl. The suspensions were incubated at room temperature for 24 h on an orbital shaker at 150 rpm. About 1 mL of sample was withdrawn at intervals of 0, 2, 4, 6, 8, 10, 12, and 24 h and centrifuged (20,000× g for 10 min). The zinc concentration in the media was determined by inductively coupled plasma atomic emission spectroscopy (ICP-OES) (Optima 3700, Perkin Elmer, Waltham, MA, USA).

2.4. In Vitro Antibacterial Activity of Biosynthesized Zinc Oxide Nanoparticles

2.4.1. Agar Well Diffusion Method

Agar well diffusion method was used to investigate the antibacterial activity of biosynthesized ZnO NPs against poultry-associated foodborne pathogens following the procedure outlined in our previous study [26]. Briefly, the bacterial strains were grown until they reached the 0.5 McFarland turbidity standards and a lawn of bacterial strain was made by spreading them uniformly onto a nutrient agar (NA) (Merck, Darmstadt, Germany) plate using a sterile cotton swab. A sterile cork borer of 6 mm in diameter was used to make the wells, and about 100 µL of ZnO NPs (at concentrations of 1000, 2000, 3000, 4000, and 5000 µg/mL) were filled into respective wells. The agar plates were incubated for 24 h at 37 °C, and the diameter (mm) of inhibitory activity shown by a clear zone around each well was measured with a ruler. Bulk ZnO served as a negative control. The experiments were carried out in triplicate.

2.4.2. Minimum Inhibitory Concentration (MIC) and Minimum Bactericidal Concentration (MBC) of Biosynthesized Zinc Oxide Nanoparticles

The MIC is the lowest concentration of a particular antibacterial agent that could inhibit bacterial growth. The MIC of biosynthesized ZnO NPs against the bacteria was examined using 2,3,5-triphenyl tetrazolium chloride (TTC) (Merck, Darmstadt, Germany) in a 96-well microtiter plate, according to the Clinical and Laboratory Standards Institute (CLSI) guidelines [29] with some modifications as a previously described method by Ashengroph et al. [30]. Briefly, the bacterial culture was grown until they reach a 0.5 McFarland standard. Then, 10 μL of bacterial suspension was pipetted to the wells containing 140 μL of nutrient broth containing various concentrations of ZnO NPs (10 to 5000 μg/mL). Nutrient broth without ZnO NPs was served as a control. The microtiter plate was incubated for 24 h at 37 °C. Subsequently, about 10 μL of TTC solution with an initial concentration of 20 mg/mL was added to each well and then incubated for 3 h at 37 °C. The MIC value was considered in the wells without the red color formation.

Meanwhile, the MBC is the lowest concentration of antibacterial agent that fully kills the bacteria, where no bacterial growth is observed. The MBC was assessed by subculturing the well suspension from MIC results onto nutrient agar aseptically. Briefly, about 10 μL of bacterial suspension was dropped on the agar plate and incubate for 24 h at 37 °C. The lowest concentration that did not display any bacterial growth was considered as the MBC. All experiments were carried out in triplicate.

2.4.3. Antibiofilm Activities of Zinc Oxide Nanoparticles
Biofilm Inhibition Assay

The potential of ZnO NPs to inhibit initial cell attachment was determined using a biofilm inhibition assay following the method described by Famuyide et al. [31]. Briefly, the inhibitory activity was measured using a 96-well microtiter plate filled with 180 μL nutrient broth (Merck, Darmstadt, Germany). A 10 μL inoculum, with the $OD_{560} = 1.0$ of *Salmonella* spp., *E. coli*, and *S. aureus* was pipetted to the individual broth and incubated for 6 h at 37 °C without shaking. After incubation, a series of ZnO NPs concentration (final concentration of $0.5\times$, $2\times$, $4\times$, and $8\times$ MIC) were added into the wells and further incubated for 24 h at 37 °C in a static condition. After incubation, the cultures were gently discarded and rinsed three times with PBS to remove free-floating cells before air-dried in the laminar flow. The wells were stained with 100 μL 0.1% (*w/v*) crystal violet and after the incubation at room temperature for 15 min, the dye was discarded and washed with dH_2O repeatedly and dried at 60 °C for 30 min. Finally, the dye was destained with 95% ethanol and allowed to sit for 30 min, and the OD of the biofilm formed associated with crystal violet was determined at 590 nm. The untreated wells served as the control. The assay was conducted in triplicate, and the percentage of biofilm inhibition was calculated using the following formula:

$$Biofilm\ inhibition\ (\%) = \frac{(OD\ control - OD\ treatment)}{OD\ control} \times 100$$

Biofilm Eradication Assay

Similar to the biofilm inhibition assay, the bacterial cells were added into each well of a 96-well microtiter plate and incubated at 37 °C for 24 h (irreversible attachment phase) and 48 h (mature biofilm). The plates were then incubated in a static condition to enable the formation of a multilayer biofilm. After biofilm formation for the respective incubation periods, ZnO NPs were added in the wells at concentrations corresponding to $0.5\times$, $2\times$, $4\times$, and $8\times$ MIC values and were further incubated for 24 h. After incubation, the wells were washed with distilled water and stained with crystal violet according to the previously described procedure. The untreated wells served as the control. The assay was performed in triplicate, and the biofilm eradication was calculated using the following formula:

$$Biofilm\ eradiction\ (\%) = \frac{(OD\ control - OD\ treatment)}{OD\ control} \times 100$$

2.4.4. Time–Kill Assay of Zinc Oxide Nanoparticles

Time–kill assay was carried out following the procedure described by Mohamed et al. [32] with modification. A time-killing assessment was conducted to evaluate the bactericidal activity of ZnO NPs on poultry-associated foodborne pathogens for 24 h. An inoculum size of 3×10^9 CFU/mL of each bacterial cell was used for the time-killing assay treated with ZnO NPs. The time–kill test consisted of untreated bacteria as the control (without ZnO NPs addition) and bacteria-treated samples at series concentrations of ZnO NPs (final concentrations of $2\times$, $4\times$, and $8\times$ MIC). The bacterial suspension was then incubated at 37 °C on an orbital shaker at 150 rpm. Aliquots of 100 µL of bacterial suspension from each treatment group were removed at time intervals of 0, 4, 8, 12, and 24 h, and plated on nutrient agar after a 100-fold dilution in 0.85% NaCl for the determination of CFU/mL by the plate count technique. The agar plates were then incubated for 24 h at 37 °C prior to colony counting. The time–kill curve was plotted as $\log_{10}$ CFU/mL against time. The assay was conducted in triplicate.

2.5. Assessment of Cell Membrane Integrity (Trypan Blue Exclusion Assay)

The membrane integrity of bacterial cells was determined by trypan blue exclusion assay, according to Hossain et al. [33]. Briefly, the bacterial strains were exposed to ZnO NPs at the concentration of $8\times$ MIC following the time-killing assay procedure described earlier. About 100 µL of bacterial suspension was removed at 0 and 24 h and then mixed with 0.4% trypan blue solution at 1:1 ratio, mixed gently, and incubated for 10 min. After incubation, 20 µL of trypan blue-bacterial suspension was loaded on a microscope glass slide and air-dried, followed by viewing the live and dead cells using a phase-contrast microscope (Olympus CX21, Tokyo, Japan) at $100\times$ magnification.

2.6. Quantification of Reactive Oxygen Species (ROS)

All the bacterial strains were exposed to ZnO NPs (final concentration of $2\times$, $4\times$, and $8\times$ MIC) to evaluate the difference in intracellular ROS generation at different time intervals (0, 8, and 24 h). ROS quantification was carried out following the method previously outlined by Tiwari et al. [34] with some modifications. The bacterial pellet exposed to ZnO NPs was obtained by centrifugation at $10,000\times$ g for 10 min at 4 °C. About 500 µL of 2% nitro blue tetrazolium (NBT) (Merck, Darmstadt, Germany) solution was introduced to the bacterial cell, vortexed, and incubated for 1 h in darkness. Afterward, the mixture was centrifuged ($10,000\times$ g for 10 min) to remove the supernatant, followed by multiple washed with PBS and centrifuged. The pellet was then suspended in 2 M KOH for cell membrane disruption, followed by the addition of 50% dimethyl sulfoxide (DMSO) solution and incubation at room temperature for 10 min to dissolve the formazan crystal. A blue-colored mixture was observed, indicating the reduction of NBT by ROS. The mixture was centrifuged and the absorbance of the supernatant was measured at 620 nm.

2.7. Assay for the Membrane Leakage of Protein and Reducing Sugar

The effect of ZnO NPs on membrane leakage was determined by quantifying protein and reducing sugars from the intracellular cytosol of the cells after treated with ZnO NPs. As described earlier, each bacterial strain was treated with ZnO NPs at different concentrations. The bacterial suspension was withdrawn at time intervals (0, 8, and 24 h), and the supernatant was collected by centrifugation ($10,000\times$ g for 5 min). The obtained supernatant was stored at -20 °C until further use. This supernatant was used to quantify protein and reducing sugars. Protein was quantified using the Bradford assay method [35] while reducing sugars was determined using the dinitrosalicylic acid assay [36].

2.8. Morphological Analysis of Bacterial Cell by Scanning Electron Microscope (SEM)

A SEM analysis was carried out to investigate the effects of ZnO NPs on bacterial cell morphology. Briefly, each of the bacterial cells was treated with ZnO NPs at the concentration of $8 \times$ MIC and incubated for 24 h at 37 °C on an orbital shaker at 150 rpm. The untreated cells were used as a control. Bacteria cells were collected by centrifugation and processed before viewing under SEM (JSM-IT100, JEOL, Tokyo, Japan).

2.9. Data Analysis

All the antibacterial activity experiments were conducted in triplicates. Data obtained were analyzed by one-way analysis of variance (ANOVA) and mean comparisons were carried out using Tukey's test with $p < 0.05$ indicating significance. Data analysis was performed using GraphPad Prism software (Version 7.0).

3. Results and Discussion

3.1. Biosynthesized Zinc Oxide Nanoparticles Characterization

Biosynthesis of Zinc oxide nanoparticles (ZnO NPs) by using microorganisms has been vastly employed due to their eco-friendly and cost-effective method [6] that offers a highly sustainable economic alternative to the conventional synthesis method. In this present study, the biosynthesized ZnO NPs were prepared using a supernatant of *L. plantarum* TA4. We previously reported that strain TA4 could reduce Zn^{2+} to ZnO NPs both intra- and extracellularly. A whitish precipitate was observed in the CFS upon reacting with Zn^{2+}, indicating the metal ion reduction and formation of ZnO NPs. The resulted ZnO NPs were collected and dried to obtain a white powder form of ZnO NPs (Figure 1a). The formation of ZnO NPs was validated using UV-Vis spectroscopy which exhibits a characteristic surface plasmon band (SPR) center at 360 nm (Figure 1b). The obtained SPR peak was in the range of the typical ZnO NPs band, as reported in the literature [37]. Furthermore, the size and shape morphology of biosynthesized ZnO NPs were characterized by HR-TEM. Figure 1c shows the HR-TEM image of bulk ZnO and ZnO NPs. The ZnO NPs displayed an oval shape with various sizes, while bulk ZnO was presented as hexagonal shaped and large aggregated particles. The average particle size of ZnO NPs was 29.7 nm as determined by ImageJ software (Figure 1d).

We hypothesized that Zn^{2+} is one of the key mechanisms for antibacterial activity; therefore, we studied the dissolved Zn^{2+} released from ZnO NPs and bulk ZnO suspended in media similar to the antibacterial assay. As shown in Figure 1e, a significant amount of Zn^{2+} was released from ZnO NPs and bulk ZnO. Both ZnO NPs and bulk ZnO were observed to continuously release ions into the media suspension, and the number of ions released increased in parallel with the incubation time. In addition, the fractions of dissolved Zn^{2+} released from ZnO NPs were higher than those released from bulk ZnO. The finding indicated that a smaller size of ZnO NPs could accelerate the release of Zn^{2+}. Accumulated evidence has revealed that size has a significant impact on the rates of reactions. For instance, particles of smaller size have a larger surface area and demonstrate higher chemical reactivity; resulting in increased antibacterial efficacy as compared to their bulkier counterparts [38–40].

In addition, due to their small particle size, ZnO NPs have been widely used as a feed supplement in livestock and poultry industries, based on their capacity to improve the bioavailability and utilization rate of nutrients in animal bodies [6,41]. ZnO NPs also exhibited potential antibacterial activity for the control of bacterial diseases. The larger surface area of the NPs allows for more exposure to the bacterial cell surface, leading to better antibacterial activity than bulk ZnO. Although ZnO NPs are effective against a wide range of microorganisms [42], there is data paucity on their antibacterial effect on specific pathogens that are common in the livestock and poultry industry.

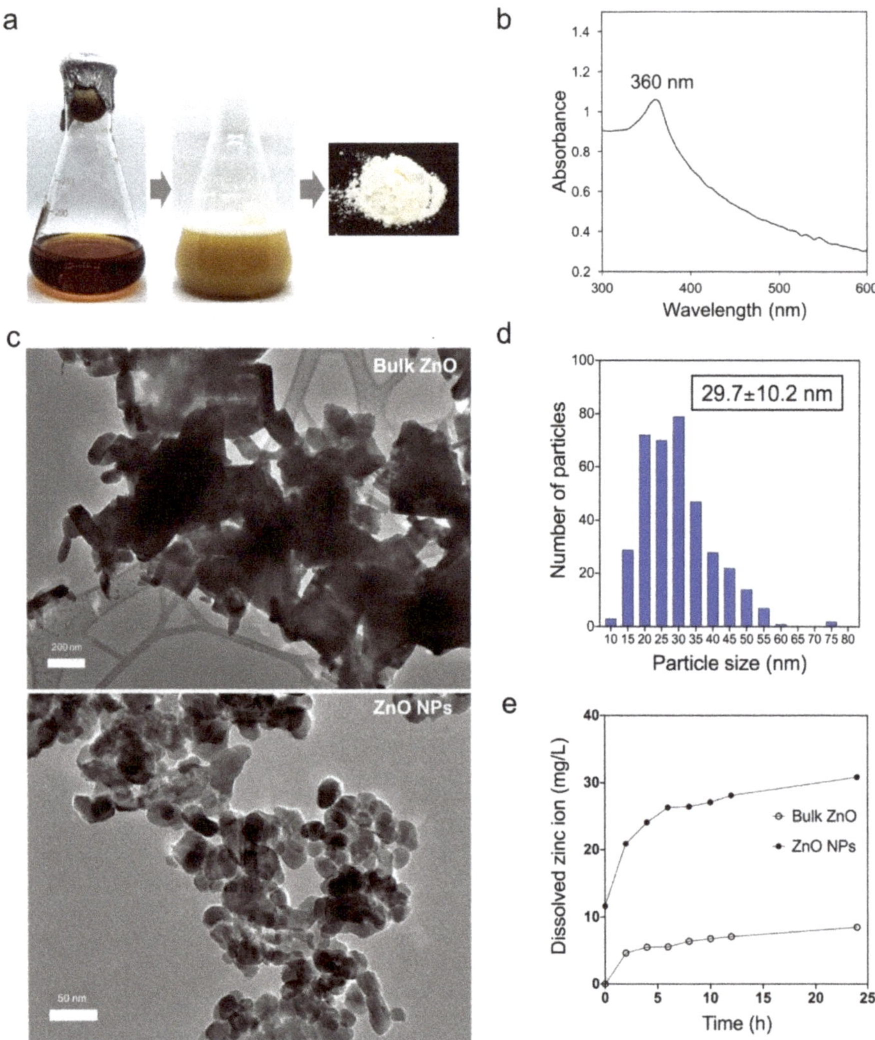

Figure 1. (**a**) Reduction of Zn^{2+} to ZnO NPs using *L. plantarum* TA4. (**b**) UV-Visual spectroscopy of biosynthesized ZnO NPs. (**c**) HR-TEM micrographs of bulk ZnO and biosynthesized ZnO NPs. (**d**) The size distribution of ZnO NPs (based on HR-TEM image). (**e**) Dissolution of Zn^{2+} from bulk ZnO and ZnO NPs in 0.85% NaCl suspension over incubation time assessed by ICP-OES. The concentration of Zn used in this study was 100 mg/L.

3.2. Agar Well Diffusion Assay

The antibacterial property of biosynthesized ZnO NPs was examined by agar well diffusion method against three types of poultry-associated foodborne pathogens such as *Salmonella* spp., *E. coli*, and *S. aureus*, isolated from broiler chicken. These three pathogens were chosen due to their prevalence in the gastrointestinal tract of poultry. In this study, a clear bacterial growth zone was observed in wells containing ZnO NPs, with different zones of inhibition diameters at different levels of concentration against the three pathogens. The inhibitory zones are presented in Table 1. A notable increase in diameter is observed with the increment of the ZnO NPs concentration. The study revealed that the inhibitory activity value of the ZnO NPs is higher against *S. aureus* than the other two pathogens.

Whereas the inhibition zones of *Salmonella* spp. and *E. coli* are more or less similar to each other (Figure 2). Bulk ZnO was used as a control to compare with the nano-sized ZnO antimicrobial activity. Notably, the well-containing bulk ZnO does not show any inhibitory activity against *Salmonella* spp. in contrast to ZnO NPs, which exhibit an increase in the inhibitory zone at concentrations of 4000 and 5000 µg/mL ($p < 0.05$). However, although no inhibition zone is observed for *E. coli* at lower ZnO concentration, inhibition zones are recorded at 4000 and 5000 µg/mL concentrations, which is significantly lower ($p > 0.05$) compared to ZnO NPs. Meanwhile, *S. aureus* is more susceptible to bulk ZnO but at a lower degree of activity than the ZnO NPs, where the inhibitory zone increases significantly ($p < 0.05$) with increasing concentration for each treatment group (Table 1).

Table 1. Antibacterial activity of bulk ZnO and zinc oxide nanoparticles (ZnO NPs) against *Salmonella* spp., *E. coli*, and *S. aureus* at various concentrations. The antibacterial activity was measured by the inhibition zone in millimeters (mm).

Concentration (µg/mL)	Zone of Inhibition (mm)					
	Salmonella **spp.**		*E. coli*		*S. aureus*	
	Bulk ZnO	**ZnO NPs**	**Bulk ZnO**	**ZnO NPs**	**Bulk ZnO**	**ZnO NPs**
1000	ND	8.00 ± 0.00 [d]	ND	8.00 ± 0.00 [c]	7.33 ± 0.58 [c,y]	11.33 ± 1.15 [c,x]
2000	ND	9.33 ± 0.58 [c]	ND	9.00 ± 1.00 [b]	7.67 ± 0.58 [c,y]	12.00 ± 1.00 [c,x]
3000	ND	10.67 ± 0.58 [b]	ND	10.33 ± 0.58 [a,b]	8.67 ± 0.58 [b,y]	15.00 ± 1.00 [b,x]
4000	ND	12.00 ± 0.00 [a]	7.00 ± 1.00 [a,y]	11.00 ± 1.00 [a,x]	9.67 ± 0.58 [a,y]	16.00 ± 1.00 [b,x]
5000	ND	12.33 ± 1.53 [a]	8.00 ± 0.00 [a,y]	12.00 ± 1.00 [a,x]	10.00 ± 1.00 [a,y]	19.67 ± 0.58 [a,x]

[a–d] Mean values with different superscripts in the same column corresponding to each type of bacteria are considered statistically different ($p < 0.05$). [x,y] Mean value with different superscripts in the same row corresponding to different concentrations of ZnO NPs are considered statistically different ($p < 0.05$). Values are mean $\pm$ SD error of three replicates. ND = No diameter for inhibition zone.

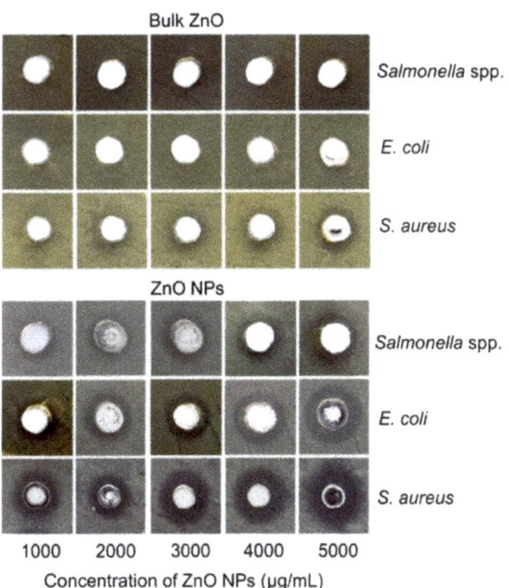

Figure 2. Antibacterial activity of zinc oxide nanoparticles (ZnO NPs) was investigated using the agar well diffusion method. Plates were incubated at 37 °C for 24 h, and images were taken at 24 h. Images are representative of three biological replicates.

The agar well diffusion method is commonly used to evaluate in vitro antimicrobial activity. Nonetheless, this method could not ascertain the exact mechanisms of the antibacterial capacity of ZnO NPs, as there was no direct interaction between the ZnO NPs and the bacterial cells. The inhibitory mechanisms are likely due to the free Zn^{2+} from ZnO NPs that diffuses through the agar and inhibits bacterial growth, causing the clear zone. It should be observed that Zn^{2+} is toxic to bacterial cells at higher concentrations. Furthermore, the agar well diffusion assay results corroborated with the results of the Zn^{2+} dissolution analysis (Figure 1e), which showed that Zn^{2+} release from ZnO NPs was higher than that of their bulkier counterparts, exerting their antibacterial potency.

3.3. Determination of Minimum Inhibitory Concentration (MIC) and Minimum Bactericidal Concentration (MBC) Value

The MIC assay was performed to determine the lowest concentration of ZnO NPs that inhibited bacterial growth, while MBC was conducted to determine the lowest concentration of ZnO NPs that killed 99.9% of the bacterial cells. Bacterial growth in the MIC assay was studied by visually inspecting the color changes from yellow to red following the addition of TTC in the culture. The changes to red formazan are directly proportional to the viable active bacterial cells. The MIC and MBC values are presented in Table 2. The MIC values of ZnO NPs against *Salmonella* spp., *E. coli*, and *S. aureus* are 80, 60, and 30 µg/mL, with the corresponding MBC values of 160, 140, and 100 µg/mL. The MIC results in Figure 3a illustrate that *S. aureus* is more vulnerable to ZnO NPs than the other tested bacteria, with a low concentration value for inhibiting bacterial growth. Meanwhile, the bulk ZnO exhibits high MIC and MBC values against the test organisms (Table 2). It should be observed that lower MIC and MBC values indicate greater antibacterial effectiveness.

Table 2. Minimum inhibitory concentration (MIC) and minimum bactericidal concentration (MBC) values of zinc oxide nanoparticles (ZnO NPs) and their bulk forms against *Salmonella* spp., *E. coli*, and *S. aureus*.

Bacteria	Bulk ZnO		ZnO NPs	
	MIC (µg/mL)	MBC (µg/mL)	MIC (µg/mL)	MBC (µg/mL)
Salmonella spp.	200	800	80	160
E. coli	200	1000	60	140
S. aureus	100	800	30	100

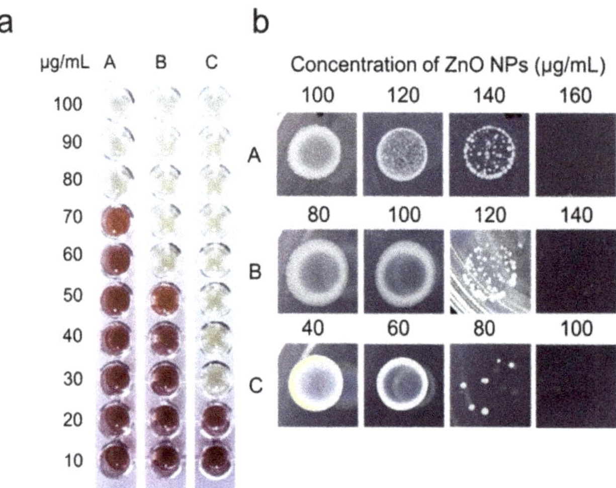

Figure 3. (a) Minimum inhibitory concentration (MIC) and (b) minimum bactericidal concentration (MBC) of biosynthesized ZnO NPs against (**A**) *Salmonella* spp., (**B**) *E. coli*, and (**C**) *S. aureus*.

As illustrated in Figure 3b, the reduction of the viable cells on the agar plate is observed with the increment of the ZnO NPs concentration within 24 h of incubation, indicating their bactericidal efficacy at a certain level of concentration, normally higher than the

MIC value. Similar to the agar well diffusion results (Figure 2), the growth of *S. aureus* is efficiently repressed at lower ZnO NPs concentrations compared to *Salmonella* spp. and *E. coli*. Our findings are consistent with other studies, suggesting that Gram-positive bacteria (*S. aureus*) are more vulnerable to NPs [43,44]. The efficiency of ZnO NPs as an antibacterial agent can be associated with the cell wall properties of the microorganisms. The cell wall of Gram-positive bacteria is composed of almost 60% teichoic acids [45], an anionic glycopolymer that served as an anion site for the metal cation from ZnO NPs, elucidating why *S. aureus* is more susceptible to ZnO NPs. On the other hand, Gram-negative bacteria have two cell membranes, the plasma membrane, and the outer membrane, whereas Gram-positive bacteria only have one. The primary function of the outer membrane is to act as a selective permeability barrier, shielding the cells from any harmful compounds [4]. In addition, *Salmonella* spp. and *E. coli* are capable of producing extracellular polymeric substances (EPS) [46] that protect them from adverse environmental conditions, contributing to their resistance to a certain concentration of ZnO NPs. Nonetheless, bulk ZnO had higher MIC and MBC values, suggesting their less effective antibacterial activity against the tested bacteria.

As described earlier, the antibacterial effects of ZnO NPs are mostly attributed to Zn^{2+} [6,38]. The results demonstrated that the smaller particle size of ZnO NPs enhances the antibacterial activity due to their greater surface area to volume ratio, which enhances their surface reactivity and the release of more ions [40]. Padmavathy and Vijayaraghavan [40] demonstrated that the bactericidal efficacies of ZnO NPs with 12 nm particle size are more effective than those with a larger size, which increased the surface reactivity of ZnO NPs. In their study, the specific surface area of ZnO NPs with a size of 12 nm yielded from the Brunauer–Emmett–Teller (BET) measurement analysis was 112 m^2g^{-1}, higher than the bulk ZnO (5.11 m^2g^{-1}). They suggested that ZnO NPs with a large surface area produced more oxygen species and ions on their surface, resulting in increased antibacterial activity [40]. On the contrary, the bulk ZnO showed less bactericidal activity due to their large particles, with a small surface area; thus, releasing fewer oxygen species and ions. This observation was consistent with the data obtained from the Zn^{2+} dissolution study (Figure 1e), i.e., more Zn^{2+} was dissolved from ZnO NPs than from bulk ZnO.

3.4. Antibiofilm Activity of Zinc Oxide Nanoparticles

Biofilm formation is an important factor employed by bacterial pathogens to invade host cells and enhance an on-going infectious process. Biofilm is defined as bacterial groups that clump together and firmly adhere to a surface surrounded by the extracellular polymeric substance (EPS) produced by the bacteria [47]. The primary function of biofilm is to protect the microorganisms from unfavorable environments, including antibacterial agents, thus resulting in resistance to antimicrobial agents and host defenses. A serious health problem might ensue when there is a failure to either prevent or eradicate microbial biofilm. In this study, the ZnO NPs ability to inhibit and eradicate the biofilm was investigated, and the results were evaluated by the formation of a thin layer of biofilms after staining with crystal violet.

As presented in Figure 4a, the ZnO NPs effectively inhibited biofilm formation in all the bacteria in a dose-dependent manner. The amount of biofilm formation decreased with the increasing concentration of ZnO NPs. Moreover, lower ZnO NPs concentrations resulted in a lower percentage of biofilm inhibition (below the MIC value for each pathogen). The antibiofilm result in this study revealed that ZnO NPs could reduce biofilm formation by affecting the growth of planktonic bacteria. These results are consistent with that of Bhattacharyya et al. [48], who found that ZnO NPs reduced *Streptococcus pneumonia* cell adhesion to the surface at sub-MIC doses. Likewise, Khan et al. [49] suggested that the generation of Zn^{2+} from ZnO NPs inhibits the enzymatic action of the DapE protein involved in peptidoglycan synthesis, thus resulting in biofilm formation failure at the early stage.

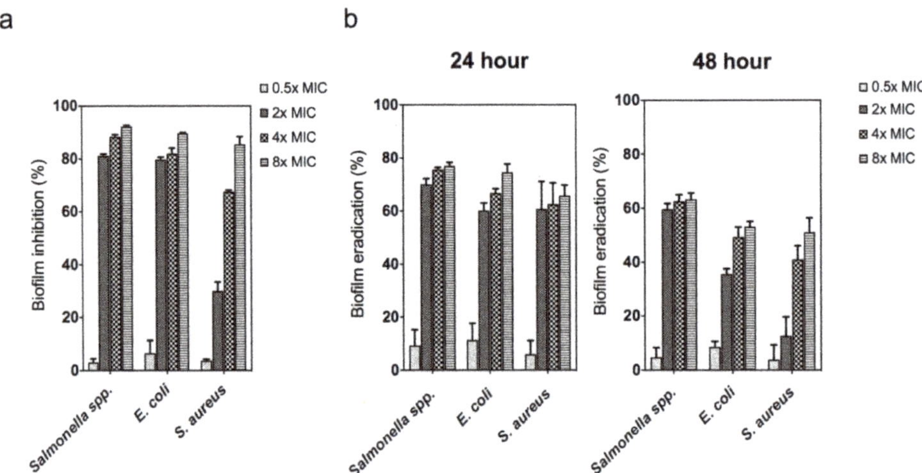

Figure 4. (**a**) Biofilm inhibitory following treatment with ZnO NPs 0.5×, 2×, 4×, and 8× minimum inhibitory concentration (MIC) against *Salmonella* spp., *E. coli*, and *S. aureus*. (**b**) Biofilm eradication at 24 h and 48 h following treatment with ZnO NPs 0.5×, 2×, 4×, and 8× MIC against *Salmonella* spp., *E. coli*, and *S. aureus*. Biofilm was assessed by crystal violet staining and measured at an absorbance of 590 nm. The data represent the mean ± SD of three replicates.

Generally, biofilm formation can be classified into three main stages, namely adhesion, cell colonization, and cell maturation. The production of biofilms is initiated during the adhesion stage, where the adherent cells aggregate on the cell's surface. Cell colonization is characterized by the formation of EPS and irreversible attachment to the surface of the host cells. Cell maturation is the stage where the biofilm reaches its maximum cell density [50,51]. In the present study, the eradication of 24 h (irreversible) and 48 h (matured) old biofilms was investigated. As shown in Figure 4b, the percentage of eradicated biofilms at 24 h ranged from 60 to 80% at all concentrations, whereas the eradication of the biofilms at 48 h ranged from 60 to 70%. This finding may be due to the production of multi-layer biofilm at the mature stage, requiring a higher concentration of ZnO NPs for its eradication. Chen et al. [52] demonstrated that *S. aureus* young biofilms (24 h) treated with antibiotics were easier to eradicate than older biofilms (72 h). This implies that matured biofilms have lower antibiotic susceptibility compared with young biofilms. Overall, both biofilms and planktonic cultures were susceptible to ZnO NPs, which makes antibacterial-based NPs a good candidate to prevent widespread biofilm formation.

3.5. Time–Kill Assay and Trypan Blue Exclusion Assay

The time–kill assay of ZnO NPs against the poultry-associated foodborne pathogens was assessed to determine the effect of ZnO NPs on bacterial viability and to ascertain the time required to meet the bactericidal activity threshold. The assay was carried out by suspending the bacterial cell in ZnO NPs-containing media. The suspension without the addition of ZnO NPs was used as control. Based on the results presented in Figure 5a, a rapid decrease in CFU among all bacterial cells is observed after 8 h exposure to ZnO NPs, where more than 5-fold CFU reduction is observed. Pronounced bactericidal activity is observed at higher ZnO NPs concentrations (4× and 8× MIC). The bacteria are completely killed after 24 h exposure to ZnO NPs. Interestingly, *E. coli* demonstrated a shorter time for the bactericidal effect within 12 h at 4× and 8× MIC concentrations.

In this study, it was also observed that the higher the concentration of ZnO NPs, the shorter the time required for the bactericidal effect on the tested bacteria. Xie et al. [53] demonstrated that ZnO NPs effectively killed *Campylobacter jejuni* cells in less than 3 h, even at low concentrations, indicating that the bacteria were highly susceptible to ZnO

NPs. Likewise, Hoseinzadeh et al. [54] reported that the killing time for *E. coli* and *S. aureus* at 2 × MIC was 6 and 3 h. Compared to other studies, our time–kill results took a bit longer to completely kill the bacteria, possibly attributable to the variation of NPs size, bacterial species, and the method used.

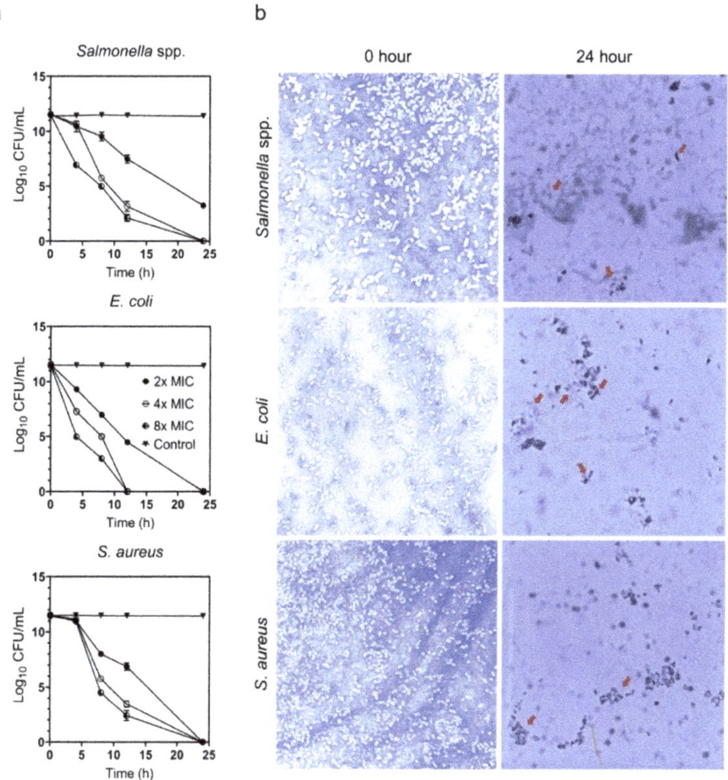

Figure 5. (**a**) Time–kill assay based on cell viability ($\log_{10}$ CFU/mL) on the poultry-associated foodborne pathogens, *Salmonella* spp., *E. coli*, and *S. aureus* at the concentrations of 2×, 4×, and 8× MIC at different time intervals (0, 4, 8, 12, and 24 h). Bars represent the standard deviation of the mean for three replicates. (**b**) Trypan blue dye exclusion assay of *Salmonella* spp., *E. coli*, and *S. aureus* at 0 h and 24 h treated with ZnO NPs at the concentrations of 8× MIC. Dead cells are indicated with red arrows. The 0 h selection were used as a control of live cells. The images were taken using a light microscope at 100× magnification.

A trypan blue exclusion assay was performed to verify the damage of the bacterial cell membrane. In this study, the killing effect of ZnO NPs is speculated to be associated with the damage of the cell membrane, which contributes to cell death. As a result, a trypan blue exclusion assay was performed to distinguish dead bacteria with damaged cell membranes from live bacteria with intact cells. Generally, the trypan dye penetrates the damaged cell membrane, causing the cells to appear blue under the microscope. Based on the findings obtained in Figure 5b, live bacteria displayed an intact cell membrane at 0 h, but after 24 h incubation, the bacteria appeared blue, indicating cell membrane damage. Therefore, our results suggested that ZnO NPs disrupted the membrane cell, leading to cell death.

3.6. Quantification of Reactive Oxygen Species (ROS) and Bacterial Cellular Leakage (Protein and Sugar)

The production of ROS was assessed in all the bacteria by treating them with ZnO NPs. Bacterial cells produce excessive ROS, including superoxide radical, when the ambient environment is unconducive, which means more enzymes and antioxidants need to be produced to scavenge the ROS [55]. However, the accumulation of ROS results in oxidative stress-induced cell death. Various antibacterial mechanisms of NPs have been proposed [6,38] (Figure 6), with the production of ROS being the most commonly proposed antibacterial mechanism. Metal and metal oxide NPs are well documented in their ability kill microorganisms through the generation of ROS. The chain of events leading to cell death following the production of ROS from ZnO NPs includes cell membrane disruption, attenuation of DNA replication, and dysfunctional production of adenosine triphosphate (ATP), leading to impaired bacterial metabolism [56].

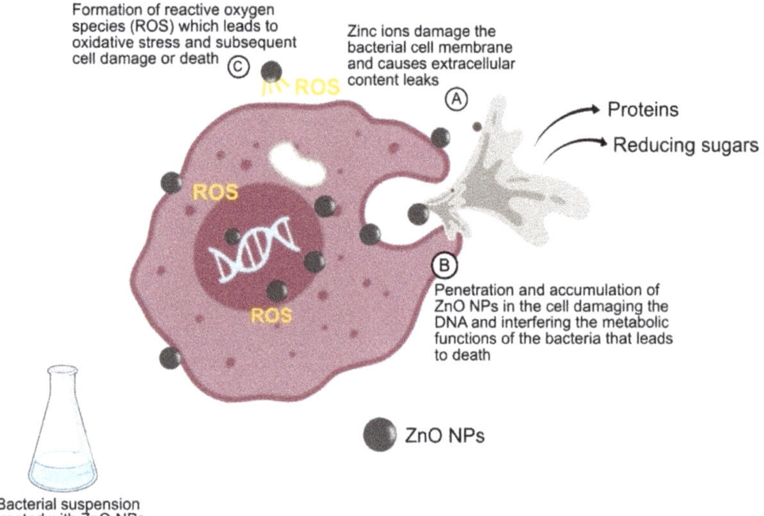

Figure 6. The proposed antibacterial actions of zinc oxide nanoparticles (ZnO NPs).

In this study, the ROS concentration formed in the bacterial cell was quantified using nitro blue tetrazolium (NBT). NBT is a pale-yellow water-soluble nitro-substituted aromatic tetrazolium compound that reacts with cellular ROS in the bacterial cell to form a formazan derivative with a dark purple coloration that can be monitored spectrophotometrically [57]. Therefore, the NBT reaction can be simultaneously used for an indirect reflection of the ROS-generating activity in the bacterial cell and to quantify the amount of ROS produced. In Figure 7a, a remarkable increase in ROS level was observed in all the bacteria over the incubation time. However, the level of ROS was depleted after 24 h in *E. coli* and *S. aureus* (Figure 7a), which could be explained by the loss of ROS in the suspension. The production of ROS peaked at 8 hours, and afterward, none of the bacterial cells generated ROS. Moreover, ROS production is likely to occur during the early stage of the interaction between ZnO NPs and the bacteria. Liao et al. [58] also reported that silver NPs induced excessive ROS production in multidrug-resistant *Pseudomonas aeruginosa* in a time- and concentration-dependent manner.

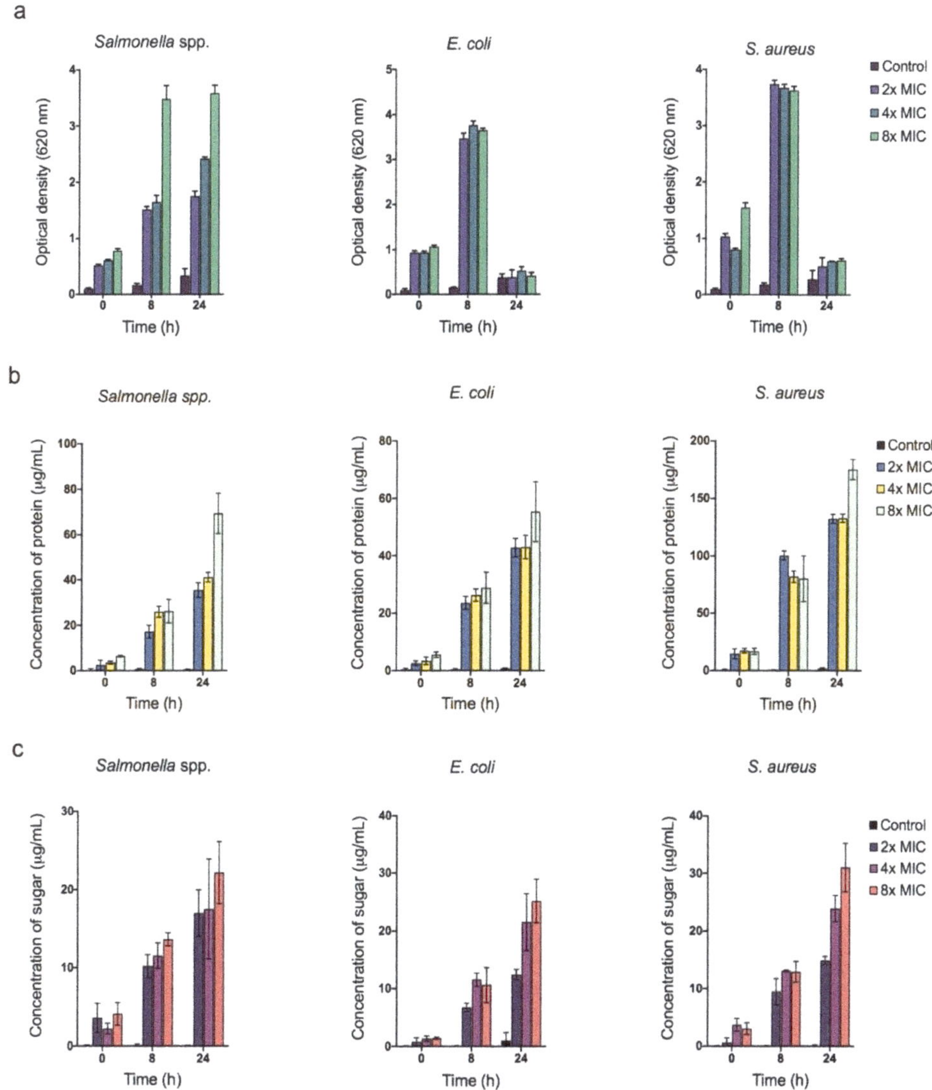

Figure 7. (**a**) Reactivity oxygen species (ROS) quantification. Cellular leakage analysis of (**b**) protein content and (**c**) reducing sugar from the cell suspensions of *Salmonella* spp., *E. coli*, and *S. aureus* bacteria treated with different concentrations of ZnO NPs at 0, 8, and 24 h of incubation time. Data were shown as the mean ± SD of three replicates.

Further studies revealed that cellular leakage is the possible mechanism of ZnO NPs bactericidal action. The release of cellular content such as protein and reducing sugars is considered to be a reliable indicator of cell membrane permeability and damage. As depicted in Figure 7b,c, both protein and reducing sugars content showed a similar increment pattern, where the leakage of cellular content increased with the increase in incubation time. A probable explanation may be that more cells were damaged as the incubation time increased, which lead to higher amounts of intracellular content leakage. Furthermore, the extent of cellular content leakage was more profound in ZnO NPs with the highest concentration in all the bacterial strains. This indicated that a higher concentration of ZnO NPs had the potential to kill more cells.

In conclusion, ROS exerts mechanical stress on the cell membrane, compromising the membrane integrity of bacteria, leading to cellular content leakage and cell death. A similar observation was observed by Kim et al. [59], based on the effects of silver NPs on *E. coli* and *S. aureus*. The authors reported a greater amount of protein leaked from the cell due to the generated ROS from silver NPs in bacterial cells leading to membrane permeability. Likewise, ZnO NPs have been shown to have a bactericidal effect on *Campylobacter jejuni*, a common poultry pathogen, by inducing oxidative stress through ROS in bacterial cells and disrupting the cell membrane [53]. These results were corroborated by Jiang et al. [60], who confirmed that the bactericidal activity of ZnO NPs was inhibited in the presence of radical scavenger, thus providing clear evidence that the antibacterial activity of ZnO NPs is ROS-mediated.

3.7. Bacterial Surface Morphology Study Using Scanning Electron Microscope (SEM)

The damage to the bacterial cell membrane was further investigated by SEM analysis. SEM micrographs in Figure 8 illustrate several morphological changes at 8 and 24 h cell exposure time to ZnO NPs, which differed from the control group. The cell membrane of all bacterial strains appears distorted and becomes severely damaged after 24 h. The micrographs portray holes on the membrane surface of ZnO NPs-treated cells, indicating bacterial membrane deformities upon exposure to ZnO NPs. The morphological changes of all treated bacteria are similar, where the cell wall becomes crumpled, with some becoming atrophic and rupturing, leading to subsequent lethal events. Our SEM results revealed a good correlation with the trypan blue and time-killing assays (Figure 5), where no viable cells were observed after 24 h. Likewise, the formation of holes (pits) on the cell membrane caused cellular contents leakage which corroborated our protein and reducing sugars quantification results. Similarly, Wu et al. [61] reported that silver NPs showed an effective antibacterial activity against *S. aureus* and *E. coli*, as observed by the formation of pits in the cell walls. The pit formations allow the NPs to enter the periplasm and destroy the cell membrane. It also enables the NPs to penetrate the cytoplasm, causing the interaction that resulted in the leakage of cellular components [61]. Significant changes in the cell morphology of bacteria were also observed in several studies [62–64], congruent to our findings.

The exact mechanisms for the antibacterial action mode of ZnO NPs are still debated; however, our findings suggested that the oxidative stress caused by ZnO NPs might be one of the reasons for bactericidal activity. Another possible mechanism is the attachment of metal NPs to the cell membrane, which exerts mechanical stress on the membrane surface [65]. The main composition of the bacterial cell membrane is proteins and lipids, which are also the sites of interaction of most metal NPs. Durán et al. [66] reported that silver ions react with the thiol groups of protein, causing the inactivation of membrane-bound enzymes and proteins. Moreover, Wong and Liu [67] proposed that the small particle size also plays a significant role in the antibacterial mechanism by providing a large surface area that allows them to attach to the cell membrane and penetrate the bacteria. Moreover, the presence of metal NPs (metal ions) in the cells interferes with the respiratory chain in the bacterial mitochondria, hampering energy production, leading to cell death [67,68].

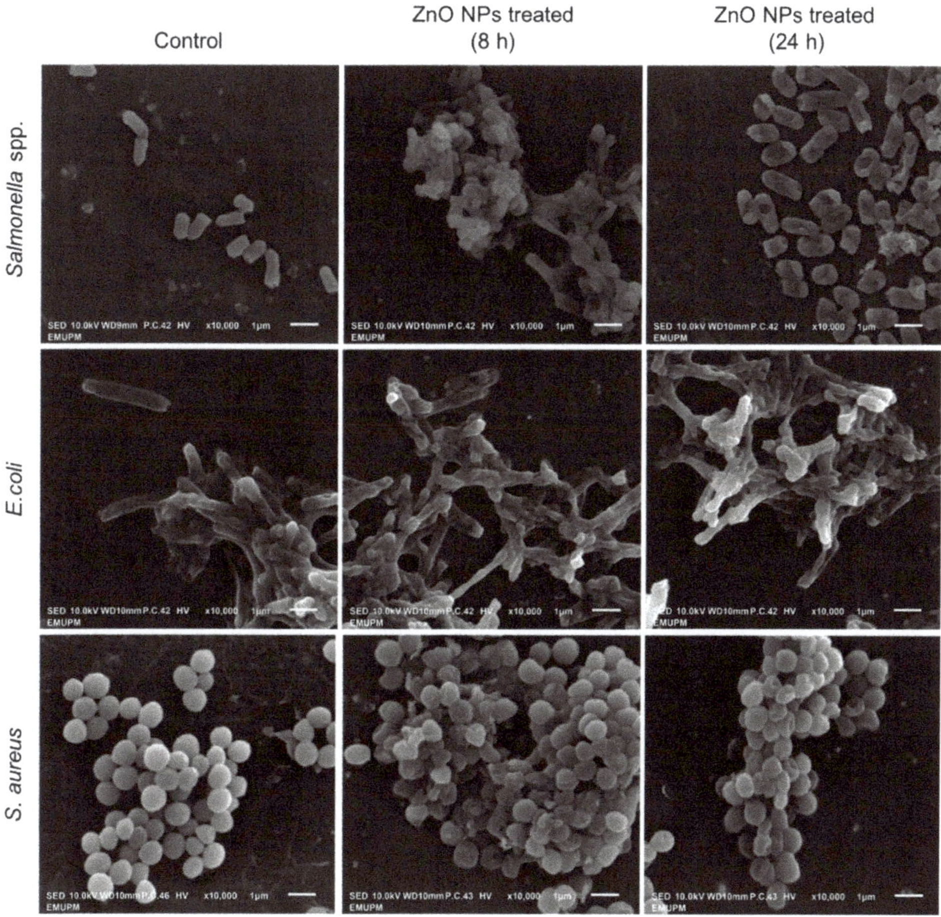

Figure 8. Visualization of zinc oxide nanoparticles (ZnO NPs) treated *Salmonella* spp., *E. coli*, and *S. aureus* at 8 and 24 h using a scanning electron microscope (SEM). The absence of ZnO NPs was used as control.

3.8. Overview of Risk Assessment for the Application of Zinc Oxide Nanoparticles in the Poultry Industry

In this present study, the ZnO NPs demonstrated a good in vitro antibacterial activity against Gram-positive and Gram-negative bacteria. However, the use of ZnO NPs as feed antibiotics in animals based on their destructive effects on the gut microbiota, especially on the beneficial microorganism, remains debatable. The intestine of poultry contains trillions of microorganisms that form a complex microbial community, coexisting with the host, and plays a critical role in the body's physiological function, such as nutrition digestion, gut mucosal morphology, and immune system development [69].

Several studies reported that the supplementation of ZnO NPs on animals induced microbial alterations in the gut. For example, Reda et al. [70] reported that the dietary supplement of ZnO NPs at 0.1 g/kg on Japanese quail led to a significant increase in all cecal microbiota counts, including *Lactobacillus* and *Salmonella*, except *E. coli* and *Enterococcus* spp. However, supplementing ZnO NPs at a higher dosage of 0.4 g/kg resulted in a significant reduction in the microbiota population, indicating that the level of ZnO NPs concentration plays an important role in regulating the intestinal microbiota. Likewise, Feng et al. [14] discovered that the ileal microbiota of hens decreased with the administration of increasing

ZnO NPs concentration, with a higher concentration of ZnO NPs at 100 mg/kg causing more microbiota reduction. They also found that ZnO NPs remarkably reduced the relative abundance of *Lactobacillus*. On the other hand, the reduction of *Lactobacillus* does not affect the growth performance of the hens [14]. Furthermore, bulk ZnO was traditionally used in higher doses (up to 3000 mg/kg feed) of pig nutrition to combat enterobacteria-caused post-weaning diarrhea. A recent study by Pei et al. [71] reported that the population of *E. coli* in the cecum and colon of pigs supplemented with 450 mg/kg and 3000 mg/kg of ZnO NPs and bulk ZnO was significantly decreased. Meanwhile, no decrease in the amount of *Lactobacillus* and *Bacillus bifidus* (Bifidobacteria) was reported in their research, implying that ZnO NPs reduced *E. coli* and retained the number of beneficial bacteria.

Lactobacillus is a predominant beneficial gut bacterial genus in the animal gut ecosystem that plays a significant role in the intestinal health and growth performance of poultry. However, it is easily impacted by various factors such as feed composition, feed additives, and antibiotic growth promoters. Nonetheless, previous studies were inconsistent, with some studies reporting no reduction of a beneficial bacteria population, notably *Lactobacillus*, vice versa. A large reduction of *Lactobacillus* in poultry's gut was thought to be due to their dominating population, which can be easily changed. Meanwhile, the increasing number of *Lactobacillus* is probably due to its ability to reproduce rapidly in the gut ecosystem. Even though ZnO NPs caused a decrease in the gut microbiota population, supplementing animals with ZnO NPs did not cause severe growth and gut health problems. Ali et al. [72] documented that supplementing ZnO NPs at 40 mg/kg in broiler enhanced gut health by improving intestinal microarchitecture (increasing the villus height and surface area) in all segments of the small intestine and increasing the immunomodulatory effect. Subsequently, long villi and greater surface area are correlated with greater absorption of available nutrients and better gut health. Furthermore, the higher bioavailability of ZnO NPs was assumed to be responsible for the increased villus microarchitecture, which contributed to a reduction in cell turnover rate in the villi and resulting in higher villus height [72,73]. Likewise, several studies also reported that supplementation of ZnO NPs improved the intestinal health of animals [20,70,74]. Nevertheless, more investigation is required to further elucidate the antibacterial effects of ZnO NPs and ensure its potential applications as a safe regulator of gut microbiota of poultry because the result may vary depending on the segment of the intestine, dose, shape, and size of ZnO NPs.

Although ZnO NPs are widely used in many bio-applications in animals, their toxicity risk in the body remains controversial because while some studies have found that ZnO NPs have therapeutic benefits, others reported them to be toxic to living organisms. Nonetheless, investigations have shown that dose [75], size, shape [76], and functionalized groups [77] of ZnO NPs are the key factors contributing to the toxic effects. In addition, chemically manufactured ZnO NPs were demonstrated less biocompatible due to the use of toxic chemicals for stabilizing agents that may reside in the final product of NPs, becoming one of the factors that leads to innate toxicity of NPs. ZnO NPs in this work were produced using the biological route without the inclusion of toxic chemicals, which may reduce the toxicity effects of NPs. Furthermore, administering higher concentrations or doses of ZnO NPs to the animals may also result in toxicity. For example, Wang et al. [78] found that supplementing mice with high doses of ZnO NPs at 5000 mg/kg caused toxicity by reducing body weight and increasing the relative weight of organs. Therefore, the use of ZnO NPs in animal feed should be limited to a specific minimum concentration to avoid this hazardous effect, which necessitates additional research. In addition, it is critical to assess ZnO NPs cytotoxicity in both cancer and normal cell lines because much of the current literature documented cytotoxicity effects on cancer cell lines [79] and claims that ZnO NPs are toxic to all cells might be deceptive, as it may not be hazardous to healthy cells.

Several studies reported that metal and metal oxide NPs have excellent antibacterial activity against multidrug-resistant bacteria [1,58,63,64]. However, one of the biggest issues with using NPs is the possibility for bacteria to develop metal-tolerance mechanisms, particularly in the gut ecosystem of animals and in the environment. To date, there has been a lack of research carried out on the emergence of metal-resistant bacteria. Nonetheless, some research has discovered that bacteria's tolerance to NPs may be due to electrostatic repulsion, biofilm adaptation to NPs, and ion efflux pumps in the bacterial system [80]. Furthermore, earlier research has only identified mechanisms of tolerance to NPs in in vitro exposure studies, although such mechanisms will differ in vivo, particularly in the intestinal environment of animals [81]. Therefore, additional research is required to emphasize the toxicology and possible development of metal-resistant bacteria, concerning the type of metal NPs, level of concentration, and the method used.

The widespread usage of ZnO NPs in various industries has raised concerns regarding their environmental impact. Excessive usage of ZnO NPs would inevitably release many NPs into the environment, resulting in heavy metal contamination. In the animal feed industry, ZnO NPs are introduced into poultry feed in a small amount due to their small particle size, which improves bioavailability and Zn absorption in the body. In this sense, no or low quantities of ZnO NPs will be discharged into the environment, reducing heavy metal contamination. However, a continuous and dynamic risk assessment is still required. To summarize, the safety and risk of the application of ZnO NPs in the animal industry, particularly poultry, should be assessed carefully before the direct implementation of ZnO NPs for any application.

4. Conclusions

In the present study, we reported the antibacterial activity of biologically synthesized ZnO NPs against poultry-associated foodborne pathogens, *Salmonella* spp., *E. coli*, and *S. aureus*. The ZnO NPs were synthesized by using *L. plantarum* TA4, followed by the analysis of the physicochemical characteristics. The biosynthesized ZnO NPs had spherical shapes with an average particle size of 29.7 nm. The Zn^{2+} dissolution study revealed that ZnO NPs dissolved more ions than their bulkier counterparts, implying that the release of high Zn^2 was responsible for the antibacterial activity. Based on the agar well diffusion assay, MIC and MBC, time–kill assay, and antibiofilm activity tests, the ZnO NPs exhibited effective antibacterial actions against poultry-associated foodborne pathogens, with *S. aureus* as the most susceptible to ZnO NPs. The ROS formation, cellular leakage, and SEM study revealed that the underlying antibacterial mechanisms of ZnO NPs include the generation of ROS, oxidative stress on the bacterial cell membrane leading to membrane damage, and cellular material leakage, which ultimately leads to cell death. In conclusion, the results of this in vitro antibacterial study demonstrated that biosynthesized ZnO NPs have great potential to be used as an alternative antibacterial agent (nanobiotic) in poultry production to control the gut burden of poultry-associated foodborne pathogens. Finally, further studies are needed to elucidate the in vivo antibacterial efficacy in poultry production.

Author Contributions: Conceptualization, H.M.Y., N.A.R., R.M., U.H.Z. and A.A.S.; data curation, H.M.Y., N.A.R., R.M., U.H.Z. and A.A.S.; funding acquisition N.A.R.; methodology, H.M.Y. and N.A.R.; writing—original draft preparation, H.M.Y.; writing—review and editing, H.M.Y., N.A.R., R.M., U.H.Z. and A.A.S. All authors have read and agreed to the published version of the manuscript.

Funding: This research received no external funding. The APC was funded by the Research Management Centre, Universiti Putra Malaysia.

Institutional Review Board Statement: Not applicable.

Informed Consent Statement: Not applicable.

Data Availability Statement: The data presented in this study are available on request from the corresponding author.

Conflicts of Interest: The authors declare no conflict of interest.

References

1. Ye, Q.; Chen, W.; Huang, H.; Tang, Y.; Wang, W.; Meng, F.; Wang, H.; Zheng, Y. Iron and zinc ions, potent weapons against multidrug-resistant bacteria. *Appl. Microbiol. Biotechnol.* **2020**, *104*, 5213–5227. [CrossRef]
2. Nie, L.; Deng, Y.; Zhang, Y.; Zhou, Q.; Shi, Q.; Zhong, S.; Sun, Y.; Yang, Z.; Sun, M.; Politis, C.; et al. Silver-doped biphasic calcium phosphate/alginate microclusters with antibacterial property and controlled doxorubicin delivery. *J. Appl. Polym. Sci.* **2020**, *138*, 50433. [CrossRef]
3. Fiedot, M.; Maliszewska, I.; Rac-Rumijowska, O.; Suchorska-Wózniak, P.; Lewínska, A.; Teterycz, H. The relationship between the mechanism of zinc oxide crystallization and its antimicrobial properties for the surface modification of surgical meshes. *Materials* **2017**, *10*, 353. [CrossRef] [PubMed]
4. Li, W.R.; Xie, X.B.; Shi, Q.S.; Zeng, H.Y.; Ou-Yang, Y.S.; Chen, Y. Ben Antibacterial activity and mechanism of silver nanoparticles on *Escherichia coli*. *Appl. Microbiol. Biotechnol.* **2010**, *85*, 1115–1122. [CrossRef]
5. Ahmadi, A.; Ahmadi, P.; Ehsani, A. Development of an active packaging system containing zinc oxide nanoparticles for the extension of chicken fillet shelf life. *Food Sci. Nutr.* **2020**, *8*, 5461–5473. [CrossRef]
6. Mohd Yusof, H.; Mohamad, R.; Zaidan, U.H.; Abdul Rahman, N. Microbial synthesis of zinc oxide nanoparticles and their potential application as an antimicrobial agent and a feed supplement in animal industry: A review. *J. Anim. Sci. Biotechnol.* **2019**, *10*, 1–22. [CrossRef]
7. Azad, S.K.; Shariatmadari, F.; Torshizi, M.A.K.; Chiba, L.I. Comparative Effect of Zinc Concentration and Sources on Growth Performance, Accumulation in Tissues, Tibia Status, Mineral Excretion and Immunity of Broiler Chickens. *Braz. J. Poult. Sci.* **2020**, *22*, 1245. [CrossRef]
8. National Research Council (NRC). *Nutrient Requirements of Poultry*, 9th ed.; National Academy Press: Washington, DC, USA, 1994.
9. Abedini, M.; Shariatmadari, F.; Torshizi, M.A.K.; Ahmadi, H. Effects of a dietary supplementation with zinc oxide nanoparticles, compared to zinc oxide and zinc methionine, on performance, egg quality, and zinc status of laying hens. *Livest. Sci.* **2017**, *203*, 30–36. [CrossRef]
10. Zhao, C.Y.; Tan, S.X.; Xiao, X.Y.; Qiu, X.S.; Pan, J.Q.; Tang, Z.X. Effects of dietary zinc oxide nanoparticles on growth performance and antioxidative status in broilers. *Biol. Trace Elem. Res.* **2014**, *160*, 361–367. [CrossRef] [PubMed]
11. Khah, M.M.; Ahmadi, F.; Amanlou, H. Influence of dietary different levels of zinc oxide nano particles on the yield and quality carcass of broiler chickens during starter stage. *Indian J. Anim. Sci.* **2015**, *85*, 287–290.
12. Yausheva, E.; Miroshnikov, S.; Sizova, E. Intestinal microbiome of broiler chickens after use of nanoparticles and metal salts. *Environ. Sci. Pollut. Res.* **2018**, *25*, 18109–18120. [CrossRef]
13. Feng, Y.; Gong, J.; Yu, H.; Jin, Y.; Zhu, J.; Han, Y. Identification of changes in the composition of ileal bacterial microbiota of broiler chickens infected with *Clostridium perfringens*. *Vet. Microbiol.* **2010**, *140*, 116–121. [CrossRef] [PubMed]
14. Feng, Y.; Min, L.; Zhang, W.; Liu, J.; Hou, Z.; Chu, M.; Li, L.; Shen, W.; Zhao, Y.; Zhang, H. Zinc oxide nanoparticles influence microflora in ileal digesta and correlate well with blood metabolites. *Front. Microbiol.* **2017**, *8*, 992. [CrossRef] [PubMed]
15. Xia, T.; Lai, W.; Han, M.; Han, M.; Ma, X.; Zhang, L. Dietary ZnO nanoparticles alters intestinal microbiota and inflammation response in weaned piglets. *Oncotarget* **2017**, *8*, 64878–64891. [CrossRef] [PubMed]
16. Mehdi, Y.; Létourneau-Montminy, M.-P.; Gaucher, M.-L.; Chorfi, Y.; Suresh, G.; Rouissi, T.; Brar, S.K.; Côté, C.; Ramirez, A.A.; Godbout, S. Use of antibiotics in broiler production: Global impacts and alternatives. *Anim. Nutr.* **2018**, *4*, 170–178. [CrossRef]
17. Thames, H.T.; Theradiyil Sukumaran, A. A Review of *Salmonella* and *Campylobacter* in Broiler Meat: Emerging Challenges and Food Safety Measures. *Foods* **2020**, *9*, 776. [CrossRef]
18. Heredia, N.; García, S. Animals as sources of food-borne pathogens: A review. *Anim. Nutr.* **2018**, *4*, 250–255. [CrossRef]
19. King, T.; Osmond-McLeod, M.J.; Du, L.L. Trends in Food Science & Technology Nanotechnology in the food sector and potential applications for the poultry industry. *Trends Food Sci. Technol.* **2018**, *72*, 62–73.
20. Wang, C.; Zhang, L.; Su, W.; Ying, Z.; He, J.; Zhang, L.; Zhong, X.; Wang, T. Zinc oxide nanoparticles as a substitute for zinc oxide or colistin sulfate: Effects on growth, serum enzymes, zinc deposition, intestinal morphology and epithelial barrier in weaned piglets. *PLoS ONE* **2017**, *12*, e0181136. [CrossRef]
21. Duffy, L.L.; Osmond-McLeod, M.J.; Judy, J.; King, T. Investigation into the antibacterial activity of silver, zinc oxide and copper oxide nanoparticles against poultry-relevant isolates of *Salmonella* and *Campylobacter*. *Food Control* **2018**, *92*, 293–300. [CrossRef]
22. Kadiyala, U.; Turali-Emre, E.S.; Bahng, J.H.; Kotov, N.A.; VanEpps, J.S. Unexpected insights into antibacterial activity of zinc oxide nanoparticles against methicillin resistant *Staphylococcus aureus* (MRSA). *Nanoscale* **2018**, *10*, 4927–4939. [CrossRef] [PubMed]
23. Ajitha, B.; Kumar Reddy, Y.A.; Reddy, P.S.; Jeon, H.J.; Ahn, C.W. Role of capping agents in controlling silver nanoparticles size, antibacterial activity and potential application as optical hydrogen peroxide sensor. *RSC Adv.* **2016**, *6*, 36171–36179. [CrossRef]
24. Mohd Yusof, H.; Abdul Rahman, N.; Mohamad, R.; Zaidan, U.H. Microbial Mediated Synthesis of Silver Nanoparticles by *Lactobacillus plantarum* TA4 and its Antibacterial and Antioxidant Activity. *Appl. Sci.* **2020**, *10*, 6973. [CrossRef]
25. Darvishi, E.; Kahrizi, D.; Arkan, E. Comparison of different properties of zinc oxide nanoparticles synthesized by the green (using *Juglans regia* L. leaf extract) and chemical methods. *J. Mol. Liq.* **2019**, *286*, 110831. [CrossRef]
26. Mohd Yusof, H.; Abdul Rahman, N.; Mohamad, R.; Zaidan, U.H.; Samsudin, A.A. Biosynthesis of zinc oxide nanoparticles by cell-biomass and supernatant of *Lactobacillus plantarum* TA4 and its antibacterial and biocompatibility properties. *Sci. Rep.* **2020**, *10*, 19996. [CrossRef]

27. Mohd Yusof, H.; Mohamad, R.; Zaidan, U.H.; Rahman, N.A. Sustainable microbial cell nanofactory for zinc oxide nanoparticles production by zinc-tolerant probiotic *Lactobacillus plantarum* strain TA4. *Microb. Cell Fact.* **2020**, *19*, 10. [CrossRef]

28. Haque, M.A.; Imamura, R.; Brown, G.A.; Krishnamurthi, V.R.; Niyonshuti, I.I.; Marcelle, T.; Mathurin, L.E.; Chen, J.; Wang, Y. An experiment-based model quantifying antimicrobial activity of silver nanoparticles on *Escherichia coli*. *RSC Adv.* **2017**, *7*, 56173–56182. [CrossRef]

29. CLSI. *Performance Standards for Antimicrobial Susceptibility Testing*; Twenty-Seventh Informational Supplement; CLSI Document M100-S27.; CLSI: Wayne, PA, USA, 2017; ISBN 1562387855.

30. Ashengroph, M.; Khaledi, A.; Bolbanabad, E.M. Extracellular biosynthesis of cadmium sulphide quantum dot using cell-free extract of *Pseudomonas chlororaphis* CHR05 and its antibacterial activity. *Process Biochem.* **2020**, *89*, 63–70. [CrossRef]

31. Famuyide, I.M.; Aro, A.O.; Fasina, F.O.; Eloff, J.N.; McGaw, L.J. Antibacterial and antibiofilm activity of acetone leaf extracts of nine under-investigated south African Eugenia and Syzygium (*Myrtaceae*) species and their selectivity indices. *BMC Complement. Altern. Med.* **2019**, *19*, 141. [CrossRef]

32. Mohamed, D.S.; El-Baky, R.M.A.; Sandle, T.; Mandour, S.A.; Ahmed, E.F. Antimicrobial activity of silver-treated bacteria against other multi-drug resistant pathogens in their environment. *Antibiotics* **2020**, *9*, 181. [CrossRef]

33. Hossain, M.M.; Polash, S.A.; Takikawa, M.; Shubhra, R.D.; Saha, T.; Islam, Z.; Hossain, S.; Hasan, M.A.; Takeoka, S.; Sarker, S.R. Investigation of the Antibacterial Activity and in vivo Cytotoxicity of Biogenic Silver Nanoparticles as Potent Therapeutics. *Front. Bioeng. Biotechnol.* **2019**, *7*, 239. [CrossRef]

34. Tiwari, V.; Mishra, N.; Gadani, K.; Solanki, P.S.; Shah, N.A.; Tiwari, M. Mechanism of Anti-bacterial Activity of Zinc Oxide Nanoparticle against Carbapenem-Resistant *Acinetobacter baumannii*. *Front. Microbiol.* **2018**, *9*, 1218. [CrossRef]

35. Bradford, M.M. A rapid and sensitive method for the quantitation of microgram quantities of protein utilizing the principle of protein-dye binding. *Anal. Biochem.* **1976**, *72*, 248–254. [CrossRef]

36. Miller, G.L. Use of Dinitrosalicylic Acid Reagent for Determination of Reducing Sugar. *Anal. Chem.* **1959**, *31*, 426–428. [CrossRef]

37. Ezealisiji, K.M.; Siwe-Noundou, X.; Maduelosi, B.; Nwachukwu, N.; Krause, R.W.M. Green synthesis of zinc oxide nanoparticles using *Solanum torvum* (L.) leaf extract and evaluation of the toxicological profile of the ZnO nanoparticles–hydrogel composite in Wistar albino rats. *Int. Nano Lett.* **2019**, *9*, 99–107. [CrossRef]

38. Sirelkhatim, A.; Mahmud, S.; Seeni, A.; Kaus, N.H.M.; Ann, L.C.; Bakhori, S.K.M.; Hasan, H.; Mohamad, D. Review on zinc oxide nanoparticles: Antibacterial activity and toxicity mechanism. *Nano-Micro Lett.* **2015**, *7*, 219–242. [CrossRef]

39. Hanif, M.; Lee, I.; Akter, J.; Islam, M.; Zahid, A.; Sapkota, K.; Hahn, J. Enhanced Photocatalytic and Antibacterial Performance of ZnO Nanoparticles Prepared by an Efficient Thermolysis Method. *Catalysts* **2019**, *9*, 608. [CrossRef]

40. Padmavathy, N.; Vijayaraghavan, R. Enhanced bioactivity of ZnO nanoparticles—An antimicrobial study. *Sci. Technol. Adv. Mater.* **2008**, *9*, 035004. [CrossRef] [PubMed]

41. Swain, P.S.; Rao, S.B.N.; Rajendran, D.; Dominic, G.; Selvaraju, S. Nano zinc, an alternative to conventional zinc as animal feed supplement: A review. *Anim. Nutr.* **2016**, *2*, 134–141. [CrossRef] [PubMed]

42. Hill, E.K.; Li, J. Current and future prospects for nanotechnology in animal production. *J. Anim. Sci. Biotechnol.* **2017**, *8*, 26. [CrossRef] [PubMed]

43. Umar, H.; Kavaz, D.; Rizaner, N. Biosynthesis of zinc oxide nanoparticles using Albizia lebbeck stem bark, and evaluation of its antimicrobial, antioxidant, and cytotoxic activities on human breast cancer cell lines. *Int. J. Nanomed.* **2018**, *14*, 87–100. [CrossRef]

44. Banerjee, A.; Das, D.; Andler, R.; Bandopadhyay, R. Green Synthesis of Silver Nanoparticles Using Exopolysaccharides Produced by *Bacillus anthracis* PFAB2 and Its Biocidal Property. *J. Polym. Environ.* **2021**, *3*, 02051.

45. Sobhanifar, S.; Worrall, L.J.; Gruninger, R.J.; Wasney, G.A.; Blaukopf, M.; Baumann, L.; Lameignere, E.; Solomonson, M.; Brown, E.D.; Withers, S.G.; et al. Structure and mechanism of *Staphylococcus aureus* TarM, the wall teichoic acid α-glycosyltransferase. *Proc. Natl. Acad. Sci. USA* **2015**, *112*, E576–E585. [CrossRef]

46. Garrett, T.R.; Bhakoo, M.; Zhang, Z. Bacterial adhesion and biofilms on surfaces. *Prog. Nat. Sci.* **2008**, *18*, 1049–1056. [CrossRef]

47. Rossi, D.A.; Melo, R.T.; Mendonça, E.P.; Monteiro, G.P. Biofilms of *Salmonella* and *Campylobacter* in the Poultry Industry. In *Poultry Science*; InTechOpen: London, UK, 2017; pp. 93–113.

48. Bhattacharyya, P.; Agarwal, B.; Goswami, M.; Maiti, D.; Baruah, S.; Tribedi, P. Zinc oxide nanoparticle inhibits the biofilm formation of *Streptococcus pneumoniae*. *Antonie Van Leeuwenhoek* **2018**, *111*, 89–99. [CrossRef]

49. Khan, S.T.; Ahamed, M.; Musarrat, J.; Al-Khedhairy, A.A. Anti-biofilm and antibacterial activities of zinc oxide nanoparticles against the oral opportunistic pathogens *Rothia dentocariosa* and *Rothia mucilaginosa*. *Eur. J. Oral Sci.* **2014**, *122*, 397–403. [CrossRef]

50. Shkodenko, L.; Kassirov, I.; Koshel, E. Metal Oxide Nanoparticles Against Bacterial Biofilms: Perspectives and Limitations. *Microorganisms* **2020**, *8*, 1545. [CrossRef] [PubMed]

51. Verderosa, A.D.; Totsika, M.; Fairfull-Smith, K.E. Bacterial Biofilm Eradication Agents: A Current Review. *Front. Chem.* **2019**, *7*, 824. [CrossRef]

52. Chen, X.; Thomsen, T.R.; Winkler, H.; Xu, Y. Influence of biofilm growth age, media, antibiotic concentration and exposure time on *Staphylococcus aureus* and *Pseudomonas aeruginosa* biofilm removal in vitro. *BMC Microbiol.* **2020**, *20*, 264. [CrossRef]

53. Xie, Y.; He, Y.; Irwin, P.L.; Jin, T.; Shi, X. Antibacterial Activity and Mechanism of Action of Zinc Oxide Nanoparticles against *Campylobacter jejuni*. *Appl. Environ. Microbiol.* **2011**, *77*, 2325–2331. [CrossRef]

54. Hoseinzadeh, E.; Alikhani, M.-Y.; Samarghandi, M.-R.; Shirzad-Siboni, M. Antimicrobial potential of synthesized zinc oxide nanoparticles against gram positive and gram negative bacteria. *Desalin. Water Treat.* **2014**, *52*, 4969–4976. [CrossRef]

55. Dong, Y.; Zhu, H.; Shen, Y.; Zhang, W.; Zhang, L. Antibacterial activity of silver nanoparticles of different particle size against *Vibrio Natriegens*. *PLoS ONE* **2019**, *14*, e0222322. [CrossRef]

56. Sharma, D.; Chaudhary, A. One Pot Synthesis of Gentamicin Conjugated Gold Nanoparticles as an Efficient Antibacterial Agent. *J. Clust. Sci.* **2020**, *20*, 1864. [CrossRef]

57. Griendling, K.K.; Touyz, R.M.; Zweier, J.L.; Dikalov, S.; Chilian, W.; Chen, Y.R.; Harrison, D.G.; Bhatnagar, A. Measurement of Reactive Oxygen Species, Reactive Nitrogen Species, and Redox-Dependent Signaling in the Cardiovascular System: A Scientific Statement from the American Heart Association. *Circ. Res.* **2016**, *119*, e39–e75. [CrossRef] [PubMed]

58. Liao, S.; Zhang, Y.; Pan, X.; Zhu, F.; Jiang, C.; Liu, Q.; Cheng, Z.; Dai, G.; Wu, G.; Wang, L.; et al. Antibacterial activity and mechanism of silver nanoparticles against multidrug-resistant *Pseudomonas aeruginosa*. *Int. J. Nanomed.* **2019**, *14*, 1469–1487. [CrossRef] [PubMed]

59. Kim, S.H.; Lee, H.S.; Ryu, D.S.; Choi, S.J.; Lee, D.S. Antibacterial activity of silver-nanoparticles against *Staphylococcus aureus* and *Escherichia coli*. *Korean J. Microbiol. Biotechnol.* **2011**, *39*, 77–85.

60. Jiang, Y.; Zhang, L.; Wen, D.; Ding, Y. Role of physical and chemical interactions in the antibacterial behavior of ZnO nanoparticles against *E. coli*. *Mater. Sci. Eng. C* **2016**, *69*, 1361–1366. [CrossRef] [PubMed]

61. Wu, C. Antimicrobial activity and the mechanism of silver nanoparticle thermosensitive gel. *Int. J. Nanomed.* **2011**, *6*, 2873–2877. [CrossRef] [PubMed]

62. Ravinayagam, V.; Rehman, S. Zeolitic imidazolate framework-8 (ZIF-8) doped TiZSM-5 and Mesoporous carbon for antibacterial characterization. *Saudi J. Biol. Sci.* **2020**, *27*, 1726–1736. [CrossRef]

63. Alavi, M.; Karimi, N.; Salimikia, I. phytosynthesis of zinc oxide nanoparticles and its antibacterial, antiquorum sensing, antimotility, and antioxidant capacities against multidrug resistant bacteria. *J. Ind. Eng. Chem.* **2019**, *72*, 457–473. [CrossRef]

64. Gad El-Rab, S.M.F.; Abo-Amer, A.E.; Asiri, A.M. Biogenic Synthesis of ZnO Nanoparticles and Its Potential Use as Antimicrobial Agent Against Multidrug-Resistant Pathogens. *Curr. Microbiol.* **2020**, *77*, 1767–1779. [CrossRef]

65. Durán, N.; Durán, M.; de Jesus, M.B.; Seabra, A.B.; Fávaro, W.J.; Nakazato, G. Silver nanoparticles: A new view on mechanistic aspects on antimicrobial activity. *Nanomed. Nanotechnol. Biol. Med.* **2016**, *12*, 789–799. [CrossRef]

66. Durán, N.; Marcato, P.D.; De Conti, R.; Alves, O.L.; Costa, F.T.M.; Brocchi, M. Potential use of silver nanoparticles on pathogenic bacteria, their toxicity and possible mechanisms of action. *J. Braz. Chem. Soc.* **2010**, *21*, 949–959. [CrossRef]

67. Wong, K.K.Y.; Liu, X. Silver nanoparticles—the real "silver bullet" in clinical medicine? *Medchemcomm* **2010**, *1*, 125. [CrossRef]

68. Holt, K.B.; Bard, A.J. Interaction of silver(I) ions with the respiratory chain of *Escherichia coli*: An electrochemical and scanning electrochemical microscopy study of the antimicrobial mechanism of micromolar Ag. *Biochemistry* **2005**, *44*, 13214–13223. [CrossRef] [PubMed]

69. Pan, D.; Yu, Z. Intestinal microbiome of poultry and its interaction with host and diet. *Gut Microbes* **2013**, *5*, 108–119. [CrossRef]

70. Reda, F.M.; El-Saadony, M.T.; El-Rayes, T.K.; Attia, A.I.; El-Sayed, S.A.A.; Ahmed, S.Y.A.; Madkour, M.; Alagawany, M. Use of biological nano zinc as a feed additive in quail nutrition: Biosynthesis, antimicrobial activity and its effect on growth, feed utilisation, blood metabolites and intestinal microbiota. *Ital. J. Anim. Sci.* **2021**, *20*, 324–335. [CrossRef]

71. Pei, X.; Xiao, Z.; Liu, L.; Wang, G.; Tao, W.; Wang, M.; Zou, J.; Leng, D. Effects of dietary zinc oxide nanoparticles supplementation on growth performance, zinc status, intestinal morphology, microflora population, and immune response in weaned pigs. *J. Sci. Food Agric.* **2019**, *99*, 1366–1374. [CrossRef] [PubMed]

72. Ali, S.; Masood, S.; Zaneb, H.; Faseh-ur-Rehman, H.; Masood, S.; Khan, M.-R.; Khan Tahir, S.; ur Rehman, H. Supplementation of zinc oxide nanoparticles has beneficial effects on intestinal morphology in broiler chicken. *Pak. Vet. J.* **2017**, *37*, 335.

73. Shao, Y.; Lei, Z.; Yuan, J.; Yang, Y.; Guo, Y.; Zhang, B. Effect of Zinc on Growth Performance, Gut Morphometry, and Cecal Microbial Community in Broilers Challenged with *Salmonella enterica* serovar typhimurium. *J. Microbiol.* **2014**, *52*, 1002–1011. [CrossRef]

74. Khajeh Bami, M.; Afsharmanesh, M.; Salarmoini, M.; Tavakoli, H. Effect of zinc oxide nanoparticles and *Bacillus coagulans* as probiotic on growth, histomorphology of intestine, and immune parameters in broiler chickens. *Comp. Clin. Path.* **2018**, *27*, 399–406. [CrossRef]

75. Deng, X.Y.; Luan, Q.X.; Chen, W.T.; Wang, Y.L.; Wu, M.H.; Zhang, H.J.; Jiao, Z. Nanosized zinc oxide particles induce neural stem cell apoptosis. *Nanotechnology* **2009**, *20*, 115101. [CrossRef] [PubMed]

76. Hanley, C.; Layne, J.; Punnoose, A.; Reddy, K.M.; Coombs, I.; Coombs, A.; Feris, K.; Wingett, D. Preferential killing of cancer cells and activated human T cells using ZnO nanoparticles. *Nanotechnology* **2008**, *19*, 10. [CrossRef] [PubMed]

77. Pujalte, I.; Passagne, I.; Brouillaud, B.; Treguer, M.; Durand, E.; Ohayon-Courtes, C.; L'Azou, B. Cytotoxicity and oxidative stress induced by different metallic nanoparticles on human kidney cells. *Part. Fibre Toxicol.* **2011**, *8*, 10. [CrossRef] [PubMed]

78. Wang, C.; Lu, J.; Zhou, L.; Li, J.; Xu, J.; Li, W.; Zhang, L.; Zhong, X.; Wang, T. Effects of long-term exposure to zinc oxide nanoparticles on development, zinc metabolism and biodistribution of minerals (Zn, Fe, Cu, Mn) in mice. *PLoS ONE* **2016**, *11*, e0164434. [CrossRef]

79. Bisht, G.; Rayamajhi, S. ZnO Nanoparticles: A Promising Anticancer Agent. *Nanobiomedicine* **2016**, *3*, 9. [CrossRef] [PubMed]

80. Niño-Martínez, N.; Salas Orozco, M.F.; Martínez-Castañón, G.-A.; Torres Méndez, F.; Ruiz, F. Molecular Mechanisms of Bacterial Resistance to Metal and Metal Oxide Nanoparticles. *Int. J. Mol. Sci.* **2019**, *20*, 2808. [CrossRef]

81. Barreto, M.S.R.; Andrade, C.T.; da Silva, L.C.R.P.; Cabral, L.M.; Flosi Paschoalin, V.M.; Del Aguila, E.M. In vitro physiological and antibacterial characterization of ZnO nanoparticle composites in simulated porcine gastric and enteric fluids. *BMC Vet. Res.* **2017**, *13*, 181. [CrossRef]

Review

Nanominerals: Fabrication Methods, Benefits and Hazards, and Their Applications in Ruminants with Special Reference to Selenium and Zinc Nanoparticles

Sameh A. Abdelnour [1], Mahmoud Alagawany [2,*], Nesrein M. Hashem [3,*], Mayada R. Farag [4], Etab S. Alghamdi [5], Faiz Ul Hassan [6], Rana M. Bilal [7], Shaaban S. Elnesr [8], Mahmoud A. O. Dawood [9], Sameer A. Nagadi [10], Hamada A. M. Elwan [11], Abeer G. ALmasoudi [12] and Youssef A. Attia [10,13,14,*]

[1] Animal Production Department, Faculty of Agriculture, Zagazig University, Zagazig 44511, Egypt; saelnour@zu.edu.eg

[2] Department of Poultry, Faculty of Agriculture, Zagazig University, Zagazig 44511, Egypt

[3] Department of Animal and Fish Production, Faculty of Agriculture (El-Shatby), Alexandria University, Alexandria 21545, Egypt

[4] Forensic Medicine and Toxicology Department, Faculty of Veterinary Medicine, Zagazig University, Zagazig 44511, Egypt; dr.mayadarf@gmail.com

[5] Department of Food and Nutrition, Faculty of Human Sciences and Design, King Abdulaziz University, Jeddah 21589, Saudi Arabia; asalghamdy2@kau.edu.sa

[6] Institute of Animal & Dairy Sciences, Faculty of Animal Husbandry, University of Agriculture, Faisalabad 38040, Pakistan; faizabg@gmail.com

[7] University College of Veterinary and Animal Sciences, The Islamia University of Bahawalpur, Bahawalpur 63100, Pakistan; bilaladam@gmail.com

[8] Poultry Production Department, Faculty of Agriculture, Fayoum University, Fayoum 63514, Egypt; ssn00@fayoum.edu.eg

[9] Department of Animal Production, Faculty of Agriculture, Kafrelsheikh University, Kafrelsheikh 33516, Egypt; mahmoud.dawood@agr.kfs.edu.eg

[10] Department of Agriculture, Faculty of Environmental Sciences, King Abdulaziz University, Jeddah 21589, Saudi Arabia; snagadi@kau.edu.sa

[11] Animal and Poultry Production Department, Faculty of Agriculture, Minia University, El-Minya 61519, Egypt; hamadaelwan83@mu.edu.eg

[12] Food Science Department, College of Science, Branch of the College at Turbah, Taif University, Taif 21944, Saudi Arabia; a.sleman@tu.edu.sa

[13] Department of Animal and Poultry Production, Faculty of Agriculture, Damanhour University, Damanhour 22516, Egypt

[14] The Strategic Center to Kingdom Vision Realization, King Abdulaziz University, Jeddah 21589, Saudi Arabia

* Correspondence: dr.mahmoud.alagwany@gmail.com (M.A.); nesreen.hashem@alexu.edu.eg (N.M.H.); yaattia@kau.edu.sa (Y.A.A.)

Citation: Abdelnour, S.A.; Alagawany, M.; Hashem, N.M.; Farag, M.R.; Alghamdi, E.S.; Hassan, F.U.; Bilal, R.M.; Elnesr, S.S.; Dawood, M.A.O.; Nagadi, S.A.; et al. Nanominerals: Fabrication Methods, Benefits and Hazards, and Their Applications in Ruminants with Special Reference to Selenium and Zinc Nanoparticles. *Animals* **2021**, *11*, 1916. https://doi.org/10.3390/ani11071916

Academic Editors: Massimo Trabalza-Marinucci and Javier Álvarez-Rodríguez

Received: 16 April 2021
Accepted: 22 June 2021
Published: 28 June 2021

Publisher's Note: MDPI stays neutral with regard to jurisdictional claims in published maps and institutional affiliations.

Simple Summary: Nanomaterials can contribute to the sustainability of the livestock sector through improving the quantitative and qualitative production of safe, healthy, and functional animal products. Given the diverse nanotechnology applications in the animal nutrition field, the administration of nanominerals can substantially enhance the bioavailability of respective minerals by increasing cellular uptake and avoiding mineral antagonism. Nanominerals are also helpful for improving reproductive performance and assisted reproductive technologies outcomes of animals. Despite the promising positive effects of nanominerals on animal performance (growth, feed utilization, nutrient bioavailability, antioxidant status, and immune response), there are various challenges related to nanominerals, including their metabolism and fate in the animal's body. Thus, the economic, legal, and ethical implications of nanomaterials must also be considered by the authority.

Abstract: Nanotechnology is one of the major advanced technologies applied in different fields, including agriculture, livestock, medicine, and food sectors. Nanomaterials can help maintain the sustainability of the livestock sector through improving quantitative and qualitative production of safe, healthy, and functional animal products. Given the diverse nanotechnology applications in the animal nutrition field, the use of nanomaterials opens the horizon of opportunities for enhancing feed utilization and efficiency in animal production. Nanotechnology facilitates the development of nano vehicles for nutrients (including trace minerals), allowing efficient delivery to improve digestion and absorption for better nutrient metabolism and physiology. Nanominerals are interesting alternatives for inorganic and organic minerals for animals that can substantially enhance the bioavailability and reduce pollution. Nanominerals promote antioxidant activity, and improve growth performance, reproductive performance, immune response, intestinal health, and the nutritional value of animal products. Nanominerals are also helpful for improving assisted reproductive technologies (ART) outcomes by enriching media for cryopreservation of spermatozoa, oocytes, and embryos with antioxidant nanominerals. Despite the promising positive effects of nanominerals on animal performance and health, there are various challenges related to nanominerals, including their metabolism and fate in the animal's body. Thus, the economic, legal, and ethical implications of nanomaterials must also be considered by the authority. This review highlights the benefits of including nanominerals (particularly nano-selenium and nano-zinc) in animal diets and/or cryopreservation media, focusing on modes of action, physiological effects, and the potential toxicity of their impact on human health.

Keywords: nanominerals; manufacturing; bioavailability; hazards; health; livestock

1. Introduction

Nanotechnology is somehow a novel scientific field, and it is widely applied in many parts of our life, such as in the therapeutic field, nutrition, disease diagnosis, chemical industries, and biological research [1–5]. Globally, nanotechnology is an emerging and promising technology that has a formidable prospect in terms of generally revolutionizing agriculture, especially in the livestock sector [6,7]. The definition of nanotechnology is expressed as the control and comprehension of matter structure at the nanoscale range within 1–100 nanometer in size, more than 1000 times smaller than the diameter of a human hair. The first introduction of the connotation of nanotechnology was by Nobel Laureate Richard Feynman (1959), in his talk entitled "There's plenty of room at the bottom". He is considered the father of nanotechnology. The process of converting molecules into a nanoscale is named nanotechnology. This conversion is implemented by various chemical, physical, biological (microorganism and plants), or mixed methods [8–11]. The main idea in these approaches (converting to nanoparticles) is to induce alterations in the parent material's fundamental chemical and physical nature. Although chemical and physical devices are further widespread for synthesizing nanoparticles, the low-dosage applications of those materials are often toxic. Moreover, they may be found in unstable status, thus critically limiting their biomedical enforcement, mainly in clinical scopes.

Moreover, this exerts chemicals into the environment [11]. Therefore, avoiding the limitation associated with the traditional synthesis methods (chemical and physical) to establish an environmentally benign procedure to create nanoparticles. Subsequently, investigators have utilized easy techniques to synthesize nanoparticles by reducing metal ions with the assistance of biological extracts, which contain flavanones, phenolics, terpenoids, amines, and alkaloids as sources of reductants [12,13]. Usually, biological approaches are safe to use and can be successfully used without other residual effects.

A novel approach in livestock production is applying nanominerals, especially selenium (Se) and zinc (Zn), which can serve as a platform to incorporate these elements into the body. This approach enables direct transportation of active compounds to target organs, avoiding their fast degradability and encouraging several health benefits [14,15]. So far, studies have shown that the application of nanominerals in the production, immunity, and reproduction of animals is promising [6,7], but adverse effects and toxicity are also reported [14,15]. Before it is implemented in the livestock industry, the application of nanomaterials must be evaluated. Shi et al. [16] stated that nano-Se addition Se content in blood and tissues was improved. The dietary complementation of nano-Se can be used more efficiently compared to inorganic or organic Se. Inclusion of different types of selenite (sodium selenite, selenized yeast, and elemental nano-Se) increased Se levels in whole blood, serum, and tissue. Shi et al. [17] noted that nano-Se supplementation in the basal diet had enhanced the fermentation of rumen and feed.

Many beneficial impacts of nano-Zn have been reported, such as production-promoting, enhancing animal reproductive efficiency, and antibacterial and immunomodulatory properties. Nano-Zn oxide treatment has been identified in cows, milk value has increased in clinical mastitis, milk yield has been suppressed by nano-Zn intake in dairy animals, and subclinical mastitis has been suppressed (reduced somatically counting) [6,15,18–26].

Researchers try to maximize the benefit from the application of nanotechnology by inserting nanominerals in animal nutrition and using their advantages to aim towards the better performance and health of animals and benefits for humans. Accordingly, this review aims to present the current knowledge related to this technology, starting with manufacturing procedures and potential applications in the ruminants' industry, in order to impact human nutrition and the hazards for different biological systems and environmental settings.

2. Preparation of Nanominerals

Nanoparticles can be mainly categorized into organic and inorganic materials, based on their chemical characteristics. In the livestock sector, the nutritional values of feed can be enhanced by using organic nanoparticle (such as proteins, fats, and sugar molecules) supplementation [1]. Nutrients, in the form of nanoparticles, can be encased as nanocapsules and transported through the gastrointestinal tract (GIT) into the bloodstream, and then into many body organs, such as the brain, liver, kidney, heart, stomach, intestine, and spleen, multiplying the bioavailability of the delivered nutrients [19,20]. These capsules are proposed to transport the nutrients without any effect on appearance or taste. Encapsulated nanomaterials are combined into fodder as liposomes, micelles, and in-feed bundle systems as recognition markers, biosensors, antimicrobials, and shelf-life extenders [21]. As for inorganic nanoparticles, minerals have been used widely as nanoparticles such as calcium, magnesium, silicon dioxide, and silver nanoparticles in water and animal-related [2,22–24]. There are many manufacturing methods used for nanomineral fabrication with different physicochemical properties [25,26].

The intrinsic properties of nanominerals are generally determined by their shape, size, crystalline structure, composition, and morphology [27]. The shapes of the nanoparticles are numerous, including spheres, cones, rolls, worms, rectangular discs, canes, and circular or elliptical discs. These shapes strongly influence the biological behavior of nanoparticles. All these cases come up in the first, second, or third dimensions, depending on the materials used and the preparation method. The thickness and viscosity of the material used to control the particle will either be with flat or sharp endings [19]. Many factors that affect nanoparticles' effectiveness, such as thickness and viscosity, are the base fluid viscosity, amount of nanoparticles, shape, type, diameter of particles, type, pressure, temperature, shear rate, and pH value [28,29]. Moreover, nanoparticles may show regions with several curvatures, texture concavity, and other properties that critically impact the adhesion strength that affects the effectiveness of the delivery and the efficacy of nanoparticles [19,20,27–29].

The notable characteristics of nanominerals are determined mainly by their shape, size, crystalline aspects, composition, morphology, and structure. Functional activities (catalytic, chemical, or biological impact) of nanominerals are strongly affected by their particles' sizes [27]. Nanominerals have large surface areas, allowing better interface with other organic and inorganic constituents.

With the increasing demand for nanomaterials, there is a great need to provide these materials. So, the creation of some sensitive and practical approaches to synthesize the desired nanoparticle is required. During the initial stages of nanoparticle synthesis, the main aim was to have a preferable hegemony over particle size, purity, morphology, quality, and amount [28]. As a result, different methods have been approved to synthesize nanoparticles, such as chemical, physical, and biological processes (Figure 1). In this section, we will describe the advantages and disadvantages of these techniques.

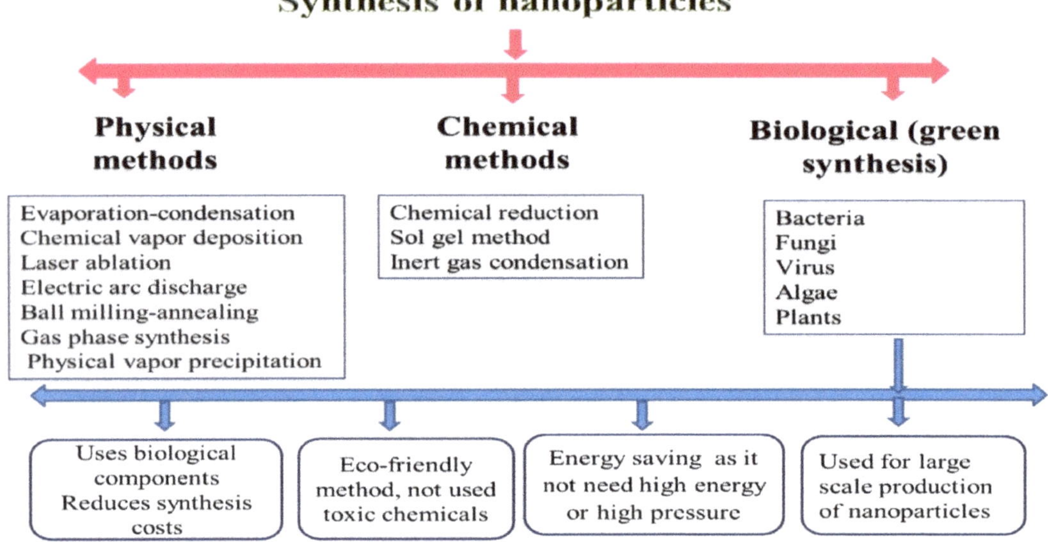

Figure 1. Preparation and synthesis of nanoparticles.

2.1. Physical Methods

According to many previous investigations, there is a broad framework of physical approaches for nanoparticle preparation. For instance, these approaches are similar to evaporation–condensation, which takes place by applying a boiling tube at atmospheric pressure [9]; ablation, which takes place by laser and evaporation–condensation together [29]; chemical vapor deposition; electric arc discharge; ball milling–annealing; gasphase synthesis methods [21]; physical vapor precipitation [30]; etc. The ball mill approach is used for grinding materials into a highly fine crumb of nanosize in livestock diets. The inclusion of high energy and ball milling (HEBM) methods is quite effective (1000 times) for synthesizing the nanominerals more than traditional ball mills [31]. HEBM is a simple, inexpensive, and efficient method for the preparation of powder in bulk amounts. This processing technique, also called mechanochemical synthesis, has already been used to prepare various materials such as amorphous metallic alloys, composites, and the modification of different classes of inorganic materials [32].

Generally, a longer milling period is required for HEBM to stimulate and complete the structural alterations. However, controlled milling temperature and atmosphere must be monitored when using the HEBM. The shortfall in the synthesis of the gas phase of nanomaterials is that it usually leads to the deposition of particles with larger sizes (from 10 to 200 nm) [21]. The advantages of physical methods for the synthesis of nanominerals include the absence of solvent contamination and the maximal recovery of nanoparticles [21].

2.2. Chemical Methods

The chemical method represents a direct approach for synthesizing and producing materials, including several steps in the gas or liquid phase. First, the atoms' formation can be achieved using chemical reactions under control. Thus, newly formed atoms can then undergo elementary nucleation followed by growth processes, leading to specific nanomaterials [33]. The use of chemical methods in the synthesis of nanoparticles is characterized by the extraction of nanomineral particles from modification, solvent, mass production, processing control, and stabilization of nanominerals particles from agglomeration, as well as the possibility of achieving effective and controlled bulk production compared with physical methods [34].

Chemical methods produce uniform and nanosized particles, but physical processes have an ample particle size range [35]. It is the most suitable for reducing the size of molecules [36]. Metal particle sizes mainly depend on the reduced capacity of reagents, where a powerful reducing reagent enhances the rate of rapid reaction and produces smaller nanometal particles [36]. However, there may be disadvantages, such as the possibility of toxicity due to hazardous chemicals during synthesis. Hence, there have been many attempts to use eco-friendly chemicals, fungal components, and plants in the production of nanomineral particles [37]. This method is called the green chemistry method. The nontoxic and eco-friendly substances used in this method include plant extracts, amino acids, starch, and glucose. Otherwise, microwave synthesis and nonchemical methods are substitutes to toxic chemical methods for the creation of nanomineral particles in a cost-effective manner and large scale [21].

In chemical methods, surfactants, such as polyvinyl pyrrolidone, cyclodextrin, quaternary ammonium salts, or polyvinyl alcohol, and stabilizing agents are required to inhibit the agglomeration of the metal particles [38]. Stabilizers keep the produced nanomineral particle aggregation in check, prevent the reduction of uncontrollable particle size, control particle size, and allow solubility of particles in different solvents [39]. Solvent molecules can further stabilize nanomineral particles [40]. Ligands, such as amines, phosphine, carbon monoxide, thiol, etc., can be used as stabilizers in the production of nanomineral particles through the coordination between the metal nanomineral particles and the ligand moiety [34].

2.3. Biosynthesis Methods

Because traditional methods are dangerous and consume more energy, there is a tendency to use "green synthesis" of nanomineral particles due to its eco-friendly, easy, and efficient nature [41] as well as the fact that it is less toxic [42]. Biosynthetic methods have emerged, using plant extracts or biological microorganisms as a simple substitute for chemical and physical processes [41]. Bacteria, viruses, algae, and plants are now used in the production of nanoparticles due to their biological advantages (low cost, nontoxic, and energy efficient) [35,42]. Biological methods were successfully used in the synthesis of different metal molecules such as gold, silver, selenium, cadmium, barium titanate, titanium, and palladium by using various plant materials [41,43], while the biosynthesis of ZnO nanoparticles was previously prepared by using *Parthenium hysterophorous* leaves [42]. Metal nanomineral particles have been synthesized with *Aloe vera*, *Avena sativa*, alfalfa, *Azadirachta indica*, *Sesbania drummondii*, lemongrass, the latex of *Jatropha curcas*, and papaya fruit extract [44]. The use of plant materials in nanomineral synthesis is advantageous and easier as this process is safe and straightforward. In addition, it follows one-step synthesis procedures, allowing effortless product recovery from the final solutions. Additionally, it then takes place as a one-pot synthesis and is eco-friendly, compatible with biomedical and pharmaceutical applications, economically viable, cost-effective, less time-consuming, and nontoxic, and does not need to sustain a specific culture [21]. Although there are many advantages to biological methods, there are limitations of these methods, such as maintaining the culture media, the culture condition, the difficulty in product recovery, and the period in the creation of the nanomineral particles [36].

Little is known about the potential synthetic methods used to produce Se nanoparticles. Anu et al. [45] reported the potential of using *Allium sativum* extracts in the green synthesis of Se nanoparticles. These extracts make Se nanoparticles spherical and crystalline with a size from 40 to 100 nm. The conventional chemically synthesized and green-fabricated Se nanoparticles were investigated to assess their cytotoxicity against Vero cells. The values of CC50 (cytotoxicity concentration 50%) indicate biologically synthesized Se nanoparticles show biocompatible features and decrease cytotoxicity compared with chemically synthesized Se nanoparticles.

3. Mechanism of Actions of Nanominerals

Nanoparticles, including nanominerals, can be used as functional units. Additionally, they can act as delivery means for materials associated with their surface or encapsulate inside. An animal study stated that the nanoparticle's action mechanisms are as follows [20]: (1) raise the available surface area to connect with biological support, (2) lengthen compound residence time in GIT, (3) efficiently deliver active components to target sites in the body, (4) minimize the effect of intestinal clearance mechanisms, (5) enable efficient uptake by cells, (6) induce cross epithelial lining fenestration, e.g., liver, and (7) permeate deeply into the tissues through fine capillaries.

Recently, nanominerals have been successfully used as feed additives to fulfill livestock and poultry from the minerals. These nanoparticles are expected to possess the features of a small dose rate, better bioavailability, and stable interaction with other compounds [1,2]. Because of their low-dose use, they can be used as promoters of growth as substitutes for antibiotics, which benefit from eliminating antibiotic residues in final products, reducing environmental pollution, and producing contamination-free animal products. Additionally, these nano-additives can be integrated with micelles or capsules of protein or any other natural feed ingredient [23].

Nanoparticles enter the GIT in many ways, including inhalation and ingestion pathways and smart or oral delivery into GIT (an oral path). The different processes (absorption, metabolism, distribution, or excretion) of nanoparticles in the body rely on their physico-chemical characteristics (solubility, size, and charge). For example, less than 300 nm can travel in the bloodstream, but particles smaller than 100 nm can enter into different organs and tissues [46].

The physiological activity of nanoparticles in the gastrointestinal tract of animals occurs through the ability of these particles for bioavailability and absorption. Nanoparticles, including nanominerals, have a larger surface-area-to-volume ratio, which provides a greater surface area for interaction with the mucosal surface, according to Corbo et al. [47]. The nanoparticulate dosage forms have shown the following advantages for gastrointestinal nanominerals delivery, owing to their smaller size: (1) easier transport through the GI tract, (2) increase in residence time of particles in the GI tract, (3) more uniform distribution and nanominerals release, (4) improved uptake into mucosal tissues and cells, and (5) specific accumulation to the site of disease, such as inflamed tissues. When a nanomineral is inserted into a biological medium such as blood or mucus, it is instantly covered with proteins adsorbed on its surface, which give it a specific "biological identity". This protein "crown" (corona) can condition the bio-distribution as well as the possible toxicity of the nanoparticle.

Nanoparticles are usually tinier than 100 nanometers, so they easily can pass through the stomach wall and diffuse into body cells quicker than common elements with larger particle sizes. Bunglavan et al. [22] observed that the particle size of minerals, as feed additives, in the nanoparticle form is believed to be smaller than 100 nanometers. Therefore, they can pass through the stomach wall and into body cells faster than ordinary ones with larger particle sizes.

4. Applications of Nanominerals in Ruminants

Microminerals can be useful for improving health and immunity, digestive system functions, microbiota homeostasis, metabolism, and reproductive performance in ruminants [48]. Additionally, they can be used for producing functional and safe animal products, maybe through eliminating the antibiotic use and increasing concentrations of trace minerals in animal products (meat and milk) required for better human health [1,6,7,14]. The health benefits and practical application of nanominerals in ruminants are illustrated in Figure 2. These effects will be displayed in detail in the following sections.

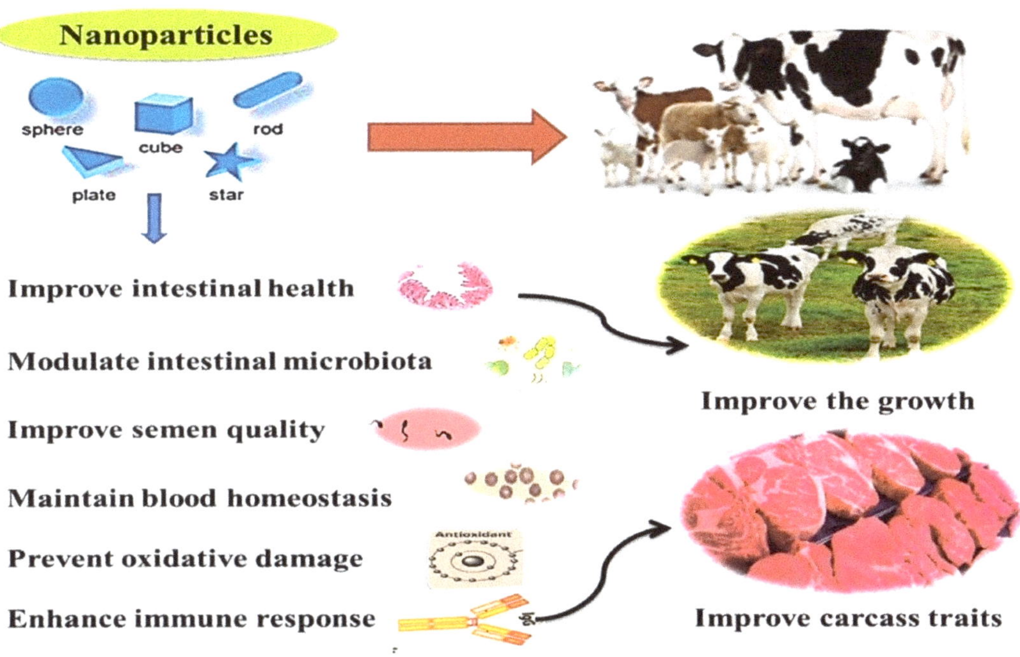

Figure 2. Health benefits and beneficial application of nano-minerals in ruminant.

4.1. Effects on Rumen Fermentation and Growth Performance

The effects of nanoparticles on growth performance, feed digestibility, and milk yield parameters in ruminants are illustrated in Table 1. Mostly dietary Zn is included in animal diets in two primary forms, i.e., organic (Zn-amino acids) and inorganic (such as ZnO and $ZnSO_4$) zinc. Recently, nano-ZnO nanoparticles (ZnNPs with size within 1–100 nm) are getting more attention for use in the mineral nutrition of livestock to address dietary requirements and to promote animal growth [30]. The inclusion of ZnNPs (100 and 200 mg/kg) has increased the volatile fatty acids, microbial crude protein, and degradation of organic matter at the 6th and 12th hours of incubation period under in vitro rumen fermentation conditions [49]. Similarly, improvement in the microbial biomass production and reduction in methane emanation were recorded with 20 mg of Zn as ZnNPs compared with other higher Zn levels (40 and 60 mg/kg dry matter, DM) during an in vitro fermentation study [50]. These positive effects are also seen in vivo with adult and/or growing animals; in ewes, the dietary supplementation of ZnNPs significantly increased the digestibility of DM, organic matter, nitrogen, and crude fiber-free extract compared with Zn larger particle and control ewes [51,52]. In growing animals, the inclusion of ZnNPs in the diets of lambs enhanced the digestibility and feeding value of the diet, as revealed by the higher feed utilization efficiency observed among the experimental dietary treatments [53]. Kojouri et al. [54] reported the beneficial impact of SeNP on the weight gain of lambs during the second and fourth weeks. Male goats that received 0.03 mg/kg of nano-Se (SeNPs) had higher final body weights and average daily gains than those that received either sodium selenite or Se-yeast [16]. A positive effect has also been found in response to dietary supplementation with 50 mg Zn/kg DM, in the form of ZnO or nano-ZnO, on dry matter digestibility in Holstein calves [26].

Overall, these positive effects on rumen fermentation and nutrient digestibility could be ascribed to the increased surface area to volume ratio, nanoscale size, rapid and specific movement, and catalytic effectiveness. These contribute to improving absorption bioavailability of nanominerals in the GIT [55,56]. The enhancing effects of nanominerals on growth can be related to their ability to beneficially alter the gut microarchitecture of animals [57] and improve rumen fermentation, specifically fiber digestion and redox homeostasis [58–60].

Table 1. Effects of nanominerals on growth performance, feed digestibility, and milk yield parameters in ruminants.

Element	Dose	Species	Major Effects
Nano-Se [17]	0, 0.3, 3 and 6 g/kg DM diet fed for 75 days	Sheep (Dorset sheep × Small Tail Han × Tan sheep)	Nano-Se at 3 g/kg DM: • Increased rumen fermentation and feed digestibility.
Nano-Se and SS [61]	1 mg/kg DM diet nano-Se and SS for 10 consecutive days	Sheep (Lori–Bakhtiari breed)	Nano-Se: • Exhibited better anti-oxidative effects than SS.
Nano-Se [62]	0.5 mg/kg DM diet nano-Se during gestation	Cashmere goat	Nano-Se: • Improved the development of hair follicles and promoted fetal growth.
Nano-Se and SY [63]	4 mg nano-Se and YS with 4 g Se-yeast	Sheep	Nano-Se: • Enhanced rumen fermentation and feed conversion efficiency as compared with YS
Nano-Se [54]	0.1 mg/kg DM diet for 60 days	Sheep (neonatal lambs)	Nano-Se: • Enhanced the body growth and antioxidant parameters
Nano-Se, SS, and SY [16]	0.3 mg/kg DM diet of nano-Se, SS and SY as compared to control (0.03mg/kg Se)	Taihang black goats	ADG was higher in Nano-Se and SY than SS or control group. Nano-Se: • Improved serum antioxidant enzymes (GSH-Px, SOD, and CAT) • Improved serum Se contents
Nano-Se and SS [64]	0.1 mg/kg live weight of nano-Se	Sheep (Makuei breed)	Nano-Se: • Enhanced weight gain • Reduced the oxidative stress as compared to SS
Nano-ZnO and ZnO [52]	30 or 40 mg/kg DM diet of nano-ZnO or ZnO for pre-partum and post-partum periods	Sheep (Khorasan-Kurdish breed)	Nano-ZnO: • Improved DMI, DMD, TAC in the rumen fluid • Increased leukocytes and milk Zn contents.
Nano-ZnO [65]		Iranian Angora goat	Nano-ZnO: • Exhibited no effect on DMI in goat kids

Table 1. *Cont.*

Element	Dose	Species	Major Effects
Nano-ZnO [49]	0, 50, 100, 200 or 400 mg/kg DM diet of nano-ZnO	In vitro ruminal fermentability	• Inclusion of 100 and 200 mg of nZnO/kg: Increased the OM fermentation and VFA content • Decreased the acetate-to-propionate ratio and ammonia-N
Nano-ZnO [26]	Cows exhibiting subclinical mastitis supplemented with 60 ppm inorganic zinc, zinc methionine, and nano-ZnO	Dairy cattle	Nano-ZnO: • Improved milk production Reduce SCC as compared to ZnO
Nano-ZnO [66]		In vitro	Nano-ZnO: • Increased the in vitro ruminal VFA contents without affecting number of protozoa.

NS = nano-Se; SS = sodium selenite; SY = selenium yeast, Nano-ZnO = nano Zinc Oxide; DM = dry matter; DMI = dry matter intake; DMD = dry matter digestibility; OM = organic matter; DWG = daily weight gain; ADG = average daily gain; FCR = feed conversion ratio TAC = total antioxidant capacity; VFA = volatile fatty acids; GSH-Px = glutathione peroxidase; SOD = superoxide dismutase, CAT = catalase.

4.2. Effects on Reproductive Performance

In recent years, the nanotechnology revolution has dominated all scientific fields, including farm animals' reproduction, and facilitating particular improvements in this domain while offering many innovative interventions. The positive effects of supplementing dietary nanomaterials on reproductive performance include improving ART outcomes and addressing technical issues regarding the application of different ART in animals. Moreover, the semen purification and preservation processes have been established utilizing various nanomaterials and methods to obtain semen doses with high sperm quality [67]. Semen's enhancement extender with antioxidant agents, such as antioxidant minerals, has been reported to upgrade the semen quality properties of cooling or post-thawing sperm cells, mainly if they are in nano-forms. Defective sperm cause failure at fertilization under both in vitro and in vivo conditions. So, maintaining perfect conditions during the storage of semen is a prerequisite for viability maintenance. During the freezing/cooling processes, sperm is preserved in synthetic extenders, which always need adjustment to maintain adequate semen quality traits. Accordingly, nanominerals have been utilized to modify semen extender properties, aiming to achieve antioxidant and antibacterial effects. Supplementation of bull semen extender with ZnNPs during cryopreservation has been detected to diminish lipid peroxidation and enhance the mitochondrial activity and functionality of sperm plasma membrane in a dose-dependent manner, without any deleterious effect on motility parameters [68]. Enrichment of semen extenders with SeNPs at a concentration of 1.0 mg/mL improved post-thawing sperm quality in Holstein bulls and, thus, the in vivo fertility rate by reducing apoptosis, lipid peroxidation, and sperm damage occurring by cryopreservation [69]. Furthermore, the addition of ZnNPs at 0.1 mg/mL to the ram semen extender enhanced total and progressive motility and the proportion of survival spermatozoa and reduced oxidative markers [70]. Similarly, the dilution of semen with a zinc nano complex resulted in a higher activity of post-thawed sperm plasma membrane integrity with a better mitochondrial activity in a dose-dependent manner. Additionally, ZnNPs exhibited no adverse effects on semen motility and raised antioxidant status. Furthermore, nanoparticles of Zn may experience beneficial actions for enhancing bovine gametes quality without affecting pregnancy rate [48]. It is likely that selenium (Se NPs) nanoparticles were employed in various research as a ROS scavenger to protect against the oxidative damage in cells' sperm.

The affirmative result of ZnNPs and SeNPs administrations on semen traits could be owing to the possible role of these minerals as a co-factor in the activities of several antioxidant enzymes and the protective functions against reactive oxygen species [71,72]. Both Zn and Se have been recognized as favorable for the stability and viability of spermatozoa through avoiding protein degradation and inhabited enzymes, which leads to damaged DNA of spermatozoa [67,73]. Furthermore, they maintain the stability of lysosomes, ribosomes, DNA, and RNA, which supports the survival and normal functions of sperm cells [74]. In ART related to oocytes and embryos culture, the addition of appropriate levels of nanominerals to the culture media (in vitro maturation, in vitro fertilization, and embryo culture media) can improve the developmental competence of oocytes during in vitro maturation, as well as the fertilization rate, the cleavage rate and the quality of embryos [75]. However, compared to ART related to males, studies on the effects of nanominerals on the reproductive performance of female ruminants are too limited and require further exploring.

It is well documented that minerals with potent antioxidant capacity, such as Se and Zn, are crucial trace elements in maintaining the animal's reproductive physiology. These microminerals play an indispensable role in spermatogenesis, sperm viability, sperm cell membrane integrity, and in maintaining the chromatin structure of sperm nuclei [73–77]. Abaspour et al. [69] found that oral administration of Zn oxide nanoparticles (ZnNPs) at a level of 80 ppm to rams significantly enhanced sperm motility and viability rates, semen volume, sperm concentration, and the functionality of sperm membrane by 20.96, 24.03, 33.9, 11.86, and 24.4%, respectively. Moreover, it also significantly reduced sperm morphological abnormalities by 28.3%, compared with the non-supplemented group. Similarly, ZnNPs had a beneficial effect on the qualitative properties of sperm, leading to a noteworthy enhancement in some antioxidant parameters of Moghani ram seminal plasma in the non-breeding season [70]. Studies have shown positive effects of mineral administrations during different reproductive windows in females [78,79]. Many reproductive events in females, such as pregnancy and lactation, may increase the nutritional requirements of dams and change their metabolism [78]. In this respect, SeNPs supplementations to late pregnant goats significantly increased Se content in the whole blood and serum compared to selenomethionine and sodium selenite [79]. However, selenomethionine was more efficient in transferring Se into kids through the placenta and colostrum. These results refer to the role of a chemical or physical variety of Se supplementation on different physiological and biological processes.

Overall, the available literature highlights the positive roles of nanominerals on ART outcomes and on animals' reproductive performance when supplemented with nanominerals. However, it is essential to note that the studies on in vivo models, either in males or females, are limited in drawing a complete overview for the effects of such additivies on reproductive performance and in knowing the dynamic mode of actions. The impacts of nano-Se and -Zn on the ruminant's reproduction are found in Table 2.

Table 2. Effect of nanominerals on the ruminant's reproduction.

Element	Dose	Species	Major Effects
SeNPs [70]	1.0 μg/mL	Bull	• Improved post-thawing kinematic and morphologic sperm quality • Decreased apoptotic and necrotic sperm cells • Improved seminal plasma & antioxidant status • Increasing in vivo fertility rate

Table 2. Cont.

Element	Dose	Species	Major Effects
ZnNPs [68]	$10^{-6}, 10^{-2}$ molar/mL	Bull	• Increasing levels of Zn-nano-complex • Improved plasma membrane functionality and mitochondrial activity • No deleterious effect on motility parameters
ZnNPs [48]	$10^{-6}, 10^{-2}$ M	Bull	• Promoted Plasma membrane integrity • Increased live spermatozoa with active mitochondria
ZnNPs [74]	50 mg/kg or 100 mg/kg	Rams	• Improved epididymal semen quality, and seminal plasma
SeNPs [13]	0.3 mg/kg	Bucks	• Reduced sperm abnormality rate, abnormalities in the mitochondria of the midpiece of spermatozoa • Enhanced the testis Se content
ZnONPs [74]	80 ppm level	Arabic Ram	• Increased the functionality of sperm membrane
CuONPs and ZnONPs Abdel-Halim et al., 2018 [75]	0, 0.4, 0.7, 1.0 or 1.5 µg/mL	In vitro maturation (IVM) of bovine oocytes	• 7 and 1.0 µg/mL of CuONPs or ZnONPs decreased DNA damage and increased glutathione concentrations in oocytes and cumulus, blastocyst rates • 1.5 µg/mL of CuONPs or ZnO-NPs had detrimental effects on the developmental competence of bovin oocytes
SeNPs, Sodium Selenite, and L-Selenomethionine [79]	0.6 mg/head/day	Late preganat goats	• Se NPs increased total Se content of the whole blood and serum • L-Selenomethionine increased placental and colostral transfer of Se into kids
SeNPs [80]	1 µg/mL	Camel	• Improved the progressive motility, vitality and ultrastructural morphology • Decreased apoptosis of frozen semen
ZnONPs [81]	50 µg/mL	Camel	• Improved sperm membrane integrity
SeNPs, [82]	2 µg/mL	Ram	• Increased sperm motility
SeNPs [83]	0.5 and 1 µg/mL	Ram	• Positive effects were observed on motility, acrosome protection and preservation of sperm membrane integrity
SeNPs [82,83]	0.5, 1 µg/mL	Ram	• SeNPs (0.5 and 1 µg/mL) improved sperm motility, viability index, and membrane integrity

SeNPs = selenium nanoparticles, ZnNPs = zinc nanoparticles, ZnONPs = zinc oxide nanoparticles, and CuONPs = copper oxide nanoparticles.

4.3. Effects on Antioxidant Status and Health

Nanominerals may promote antioxidant activity by inhibiting the free radical's production because of the increased surface area, leading to a higher number of active sites for scavenging an increased number of free radicals [78]. Sheep fed a basal diet containing ZnNPs exhibited a better antioxidant level [47,79]. Supplementation of SeNPs in newborn lambs promoted superoxide dismutase (SOD) levels with a concurrent reduction in thiobarbituric acid–reactive substances (TBARS) values [51].

The inclusion of Se NPs (0.6 mg Se head/day) in Khalkhali goat diets during the late stage of pregnancy significantly increased Se level in the blood (584.15µg/L) and serum (351.62 µg/L) of goats at kidding, compared with control goats (123.74 µg/L and 66.94 µg/L, respectively) [79]. Moreover, it was found that introducing SeNPs to goats during the late stage of pregnancy significantly increased iron levels in the blood and serum of kids or goats and colostrum [80]. A better iron homeostasis capability was observed after the addition of SeNPs compared with other Se sources [81]. This may be due to the distinguished physicochemical features of SeNPs such as small size, large surface area, enhanced absorption via epithelial cells, and other functional properties.

The increase in blood antioxidant minerals is usually associated with particular improvements in the antioxidant status of animals and, thus, the health status of animals. Male goats offered SeNPs increased serum Se and superoxide SOD, catalase (CAT), and glutathione peroxidase (GSH-Px) compared with provided Se yeast and sodium selenite [16]. These antioxidant enzymes play the primary role of removing oxidative stress agents, such as malondialdehyde (MDA) and nitric oxide. Parallel studies by Zhan et al. [56] and Zhang et al. [80] showed that SeNPs exhibited an excellent bioavailability due to their high catalytic efficiency, low toxicity, and absorbing solid ability. These specific properties and the different absorption patterns may emphasize how nano-Se is more bioavailable than organic or inorganic Se [16].

Nanominerals also have protective effects against some physiological disorders; SeNPs showed a defensive impact on the cardiac cells from ischemia [78]. Additionally, due to the antibacterial activity of antioxidant nanominerals, some nanominerals, such as ZnO NPs, could be helpful in the prevention and curation of some bacterial-borne diseases such as sub-clinical mastitis in cows [26]. The impacts of nano-Se and nano-Zn on serum antioxidant parameters, immune response, and serum/milk Se contents in ruminants are summarized in Table 3.

Table 3. Effects of nanominerals on serum antioxidant parameters, immune response and serum/milk composition in ruminants.

Element	Dose	Species	Major Effects
Nano-Se and SS [61]	Nano-Se and SS for 63 days	Sheep	Similar GSH-Px content in both sources of Se
Nano-Se, SS, and Se-Met [84]	0.6 mg/head/d for 4 weeks before parturition	Pregnant goats	Nano-Se: • Increased serum Se level Se-Met: • Improved Se transfer efficacy of placenta and colostrum into kid
Nano-Se [85]	0, 1 and 2 mg/kg DM diet	Sheep (male Moghani lambs)	2 mg/kg DM nano-Se: • Improved the expression of liver GSH-Px and selenoprotein W1

Table 3. *Cont.*

Element	Dose	Species	Major Effects
Nano-Se and SS [86]	0.30 mg/kg of DM for one month	Dairy cows	Nano-Se: • Improved milk Se and serum GSH-Px contents
Nano-Se and SS [87]	0.055 mg/kg BW for three months	Sheep (Lambs)	Nano-Se: • Increased Se contents in plasma, erythrocytes, platelets, and GSH-Px activity
Nano-Se [88]	5 mg/kg BW/day	Wumeng semi-fine wool sheep	Nano-Se: • Induce Se poisoning • Reduced the immune and antioxidant parameters
Nano-Zn, ZnO, Zn-Met [89]	28 mg/kg DM diet	Sheep	Nano-ZnO: • Increased the Zn bioavailability in rumen and blood • Enhaned serum IgG • Decreasing BUN contents
Nano-ZnO and ZnO [49]	Nano-ZnO supplemented at 30 or 40 mg/kg DM for pre-partum and post-partum periods	Sheep (Khorasan-Kurdish breed)	Nano-ZnO: • Improved the TAC in the rumen fluid • Improved milk Zn contents

NS = nano-Se; SS = sodium selenite; SY = Selenium Yeast; GSH-Px = glutathione peroxidases; DM = dry matter; IgG = immunoglobulin G; BUN = blood urea nitrogen; TAC = total antioxidant capacity.

5. Impact on the Environment and Toxicity of Nanominerals

Anthropogenic processes manufacture incidental nanomaterials as side-products [90], while engineered nanomaterials with unique properties are intentionally developed [91]. There is a review of air, water, and soil exposure from different paths. Researchers have focused on different aspects of nanomaterials, such as their innovative applications for the removal of ions and chemical molecules and contaminants, such as adsorbents, ion exchangers, and disinfectants in water and air [92,93], as well as the evaluation of their related adverse effects on human health, ecology, and the environment [94].

A specific emphasis of the topic was on papers that can cope with their health hazards and the consequences for both indoor and outdoor applications for rules and legislation [95].

Several investigations were implemented over the last few years, recommending the potential role of nanominerals such as Se and Zn in different animal nutrition and pathways [96–98]. Consequently, it was observed that Zn is mainly excreted from the body (because of less availability) to the environment, causing environmental contamination [73,99]. In this light, nano-Zn, as an alternative to the traditional sources of Zn, can be a great substitute in the livestock sector [71]. Therefore, nano-Zn may be used in livestock feed at lower levels to reach better findings than other Zn resources and to indirectly avoid environmental pollution. The nanoform of supplements augments the surface area that can enhance absorption and, thereby, utilize minerals, resulting in reduced dietary supplements and, ultimately, reduced feed cost and more sustainability [67]. However, the increase in nanotechnology and extensive usage of nanoparticles in everyday human life has led to worries concerning their plausible hazard impacts on live organisms and human health. The adverse impacts of nanoparticles on many cellular and molecular modifications have been well-considered, while the potential experimental toxicity needs further investigation in the studies on lab animals. Additionally, further research is essential in the future to comprehend the influence of nanominerals and their mechanisms and sites of absorption, transcript expression analysis of distribution, and mode of action [6,15].

Studies should be carried out on cellular and molecular modifications within animals to verify the effectiveness of nanominerals compared with conventional sources of minerals [2]. Furthermore, exploration should be focused on finding the ideal levels of nano-Zn in diets that can provide a better performance and reduce the hazardous impacts on the environment.

Due to the fast development of nanotechnology and future bulk manufacture of nanomaterials, there comes the need to understand, identify, and counteract any adverse health effects of these materials that may take place during manufacture, usage, or by accident [99–101].

The conceivable toxicological influence in both nonruminants and ruminants, along with the toxic levels, needs to be researched before they can be used in feeds. There should be systematic and comprehensive studies in order to understand the harmful effects of various doses of nanomaterials because the toxicological investigations provide various results in animal models [99].

The prime target for the use of nanomineral in animal and poultry feed is for it to be easy to absorb with no toxicity [101]. The study of the toxicity of nanominerals helps the producers and researchers in animal production and poultry to make safer decisions about the use of nanotechnologies and also increases the community's consciousness towards nanotechnology-based applications in livestock production systems [74,102].

Selenium (Se) has been one of the crucial and necessary nutritional trace minerals for the physiological activities in the human body due to the high recovery potential of glutathione peroxidase and selenoenzymes [103,104]. However, the widespread use of Se-NPs in nanoelectronics and medication has increased the risk of their environmental contamination, which might affect living organisms, though it is useful to understand the assessment of the toxicity of Se-NPs to the biological ecosystem [105].

For instance, the nano-Se toxicity is lower than that of selenomethionine, and its toxicity is currently the lowest of all Se supplements. The toxicity of nano-Se is three times lower than that of organic Se, and seven times lower than that of inorganic Se [106].

The available literature indicates that nanoparticles of copper (CuNP) are more toxic than copper's iconic form [20,107–109]. However, the explanation of the result is difficult to be made given that they indicate that CuNP safeguards proteins and DNA more effectively against nitrates and oxidation processes than Cu salts. This may be due to the clumping of nanoparticles onto agglomerates, making them less available to the body, thus reducing the damage level generated by the body. Some nanoparticles are presently described as toxic to animal and human health or the environment.

Orally feeding lambs caused severe observed renal damage (75% of animals) and mild liver toxicity (degeneration and edema in hepatocytes) when offered ZnNPs (20 mg/kg BW for 25 days) [110].

Moreover, long-term exposure to ZnNPs (250 mg/kg BW) induced liver damage in rats, which might be due to the accumulation of zinc [111]. Nanoparticles can join inside the nucleus of cells, which is the main concern that toxic nanoparticles may be able to modulate several physiobiological events of cells that may lead to cell death [112]. Higher coated ZnNPs (100 mg/kg) could damage the intestinal tract in growing pigs [113]. Studies in mice revealed that Se-NPs cause abnormal body weight, disturbances in blood biochemistry, and degeneration in hepatic and lung cells [114–116]. The degree of toxicity severity of nanoparticles might be attributed to mineral (organic or metallic) sources, size, shapes, dosage, synthesis methods, age of the animal, and exposure period. Lesser NPs (3–6 nm) are more easily cleared out from the kidneys compared to bigger NPs (around 30 nm), which remain in the hepatic cell [117]. Additionally, the greater size of NPs manages to stay longer in the kidney tissues due to the slower excretion machineries of glomerular filtration. This long-term retention can lead to tissue toxicity.

It was reported that the chemically synthesized ZnO NPs could be one of the possible causes of the innate toxicity of NPs, due to the chemical reaction conditions in the conventional method [118]. Accordingly, our previous work indicated that the SeNPs mainly synthesized by the biological technique at diet levels of 25 or 50 mg/kg enhanced the heat tolerance of growing rabbits [119].

Based on the literature, the toxicity of nanominerals in ruminants has not been investigated; instead, the majority of chronic or sub-chronic toxicity reports were explored using mice as an animal model [114–116]. In general, the applications of ZnNPs or SeNPs in animal feeding should be bordered to the precise lowest levels to avoid their negative impacts. Likewise, for better safety of NPs synthesis, the use of a microbe mediated combination approach should be considered due to its biocompatibility and the manageable shape and size of NPs, which can be realized via the optimization procedure. The best toxicological valuations have applied a cost-effective in vitro scheme, in that only precise biological paths can be confirmed under particular situations. However, the toxicologic assessments in an in vivo scheme are complex, costly, and incorporate many challenges, especially in ruminants, which have a special characterization of a microbial ecosystem [112]. Many investigations from epidemiological and experimental trials reveal that epigenetic modifications may be employed to detect toxicity produced by NPs and, more notably, to expect their potentiality of toxicity in preclinical assessment. This proposes that epigenetic modifications can be valuable indicators of NPs toxicity and can be plausible translational biomarkers for detecting unfavorable impacts of silica NPs in animals. After exposing mice cells to NPs, the damage of global DNA methylation was detected in the cells [120]. Moreover, they exhibited a substantial reduction in global cytosine DNA methylation in white blood cells from workers subjected to silica NPs [121]. Indeed, the actual molecular machinery and targets for NPs on the cellular organelles and cell membrane structure are yet to be discovered.

From the human perspective, it is recognized that alveoli contain nanoparticle deposits, and mediated mechanisms clear them through typical macrophages. A proportion of particles can translocate. This may be related to physicochemical features; however, nothing can be said regarding whether chronic exposure leads to sufficient particle accumulation to trigger disease or not [122–124]. Several studies reported a strong connection between high morbidity and ultrafine particles (UFP) exposure, particularly for the elderly. Moreover, recent studies indicated that there are impaction variations during particle levels and exposures for a short time as an essential cardiac activity agent.

Furthermore, specific nanoparticles may negatively induce inflammation and oxidative stress. Other materials only show toxicity at the nanoscale, impair kidney cell growth, and negatively affect cell growth and turnover [95–99]. According to this previous apprehension, the Toxic Substances Control Act (TSCA) was also reputable for evaluating the risk of many types of nanomaterials and to offer authority to the Environmental Protection Agency (EPA) to regulate them. Moreover, legislation administering the employ of NPs is bordered around the world in the twenty-first century, nanotechnology is one of the fast-developing areas of exploration. Therefore, a toxicity estimation of NPs is presently a demanded investigation field, the focus being NPs interaction with biological and ecological systems.

It is urgently needed that exposure to nanomaterials be considered and assessed, given that children are significantly more prone than adults when it comes to hazardous chemicals due to their larger relative body surface area. Additionally, it is essential to note that some nanoproducts are intended for use by specific subgroups, such as children and the elderly [114].

6. Recent Applications of Nanominerals for Human Benefits

Nanotechnology has expanded rapidly in animal health applications. Still, it is noteworthy that it is at the cutting edge of many potential human health benefits, including the production of functional food, disease diagnosis and treatment, and drug delivery [95]. Nanotechnology is shown to have beneficial applications in the human food chain, mainly through increasing the bioavailability and providing adequate amounts of essential nutrients, minerals, and vitamins in animal products consumed by humans [54,99,100]. Furthermore, the consumers' demand for foods and their awareness has been elevated as consumers seek safe and high-quality foods with beneficial health characteristics, high sensory quality, and prolonged shelf life [110].

Improving animal production products brings many advantages to humans. The use of nano-Zn and nano-Se has been found to enhance both the quantity and quality of animal products [1]. From the human perspective, improving animal products' contents of microelements is of particular importance. The bioavailability of minerals that originate from these sources is higher than that of minerals arising from plant sources. Considering Zn and Se, human requirements for Zn and Se are 8–15 and 55–400 µg/day, respectively, relatively higher than those provided by any nutritional source. In beef meat, the levels of Zn and Se in fresh tissue of steers average between 2.14 mg/100 g and 0.42–1.30 µg/g, respectively. Recently, nano-Zn and Se can be used as innovative and novel vehicles to improve the mineral contents of organic meats via the fortification of beef cattle diets with Se and Zn nanoelements is a candidate strategy [125]. The work on the possibility of improving the composition and quality of milk with the use of nanominerals was hardly ever produced. However, Rajendran et al. [126] used nano-zinc in feeding dairy cows and found that the use of this nanomineral reduces the number of somatic cells in cow's milk with subclinical mastitis. Although a significant focus has been to remove potentially harmful contaminants from milk, there has also been some interest in mixing nanoparticle supplements directly into cow's milk for human consumption [127]. The addition of combined nanopowdered oyster shells into milk to enrich the calcium content from 100 to 120 mg/mL for growing children and postmenopausal women did not negatively alter the sensory or physicochemical qualities [128].

Many studies proved the potential of supplementing nanominerals to increase mineral contents in animal products; however, most of these studies were carried out on poultry, meat and eggs [22,107,108]. More pieces of research are needed to evaluate the potential of nanominerals to change the quality and nutritional value of red meat and milk, since such food is produced by ruminants, which have a different pattern of food digestion and metabolism. However, the impact of nanominerals on health and the environment needs further investigation [129–132].

7. Conclusions

The global livestock sector faces the continuous pressure of ever-increasing raw material prices, including prices of minerals, which necessitates other potential sources of minerals with better bioavailability, efficacy, and lower antagonism. Nanotechnology offers potential advantages of using nanominerals in livestock nutrition, in addition to their potential impacts on human and animal health. The use of nanominerals in livestock nutrition has shown promising results in enhancing the performance, nutrient bioavailability, and immunity status of animals, and in improving the quality and composition of animal products while reducing environment-related hazards. Nevertheless, more research efforts are still needed to validate nanomaterials' effectiveness, efficacy, and safety, in order to avoid any adverse effects on the livestock, environment, and humans.

Author Contributions: All authors listed have made a substantial, direct, and intellectual contribution to the work; they all approved it for publication. All authors have read and agreed to the published version of the manuscript.

Funding: The review article received no funds.

Institutional Review Board Statement: Ethical review and approval were waived for this study as a review study which depends only on the published articles so far and the material and methods of this manuscript did not contain any humans or animals subjects.

Informed Consent Statement: Not applicable.

Data Availability Statement: The datasets used in this study are available in text and cited in the reference section.

Conflicts of Interest: The authors declare no conflict of interest and the research was conducted in the absence of any commercial or financial relationships that could be construed as a potential conflict of interest.

References

1. Abd El-Hack, M.; Alagawany, M.; Farag, M.; Arif, M.; Emam, M.; Dhama, K.; Sarwar, M.; Sayab, M. Nutritional and pharmaceutical applications of nanotechnology: Trends and advances. *Int. J. Pharmacol.* **2017**, *13*, 340–350. [CrossRef]
2. Hashem, N.M.; El-Desoky, N.; Hosny, N.S.; Shehata, M.G. Gastrointestinal microflora homeostasis, immunity and growth performance of rabbits supplemented with innovative non-encapsulated or encapsulated synbiotic. *Multidiscip. Digit. Publ. Inst. Proc.* **2020**, *73*, 5.
3. El-Kholy, M.S.; Ibrahim, Z.-G.; El-Mekkawy, M.M.; Alagawany, M. Influence of in ovo administration of some water-soluble vitamins on hatchability traits, growth, carcass traits and blood chemistry of Japanese quails. *Ann. Anim. Sci.* **2019**, *19*, 97–111. [CrossRef]
4. Hassanein, E.M.; Hashem, N.M.; El-Azrak, K.E.-D.M.; Gonzalez-Bulnes, A.; Hassan, G.A.; Salem, M.H. Efficiency of GnRH–loaded chitosan nanoparticles for inducing LH secretion and fertile ovulations in protocols for artificial insemination in rabbit does. *Animals* **2021**, *11*, 440. [CrossRef] [PubMed]
5. Hashem, N.M.; Sallam, S.M. Reproductive performance of goats treated with free gonadorelin or nanoconjugated gonadorelin at estrus. *Domest. Anim. Endocrinol.* **2020**, *71*, 106390. [CrossRef]
6. Huang, S.; Wang, L.; Liu, L.; Hou, Y.; Li, L. Nanotechnology in agriculture, livestock, and aquaculture in China. A review. *Agron. Sustain. Dev.* **2015**, *35*, 369–400. [CrossRef]
7. Osama, E.; El-Sheikh Sawsan, M.A.; Khairy, M.H.; Galal Azza, A.A. Nanoparticles and their Potential Applications in Veterinary Medicine. *J. Adv. Vet. Res.* **2020**, *10*, 268–273.
8. Mohanpuria, P.; Rana, N.K.; Yadav, S.K. Biosynthesis of nanoparticles: Technological concepts and future applications. *J. Nanopart. Res.* **2008**, *10*, 507–517. [CrossRef]
9. Ingale, A.; Chaudhari, A. Biogenic Synthesis of Nanoparticles and Potential Applications: An Eco-Friendly Approach. *J. Nanomed. Nanotechnol.* **2013**, *4*. [CrossRef]
10. Ghashghaei, S.; Emtiazi, G. The methods of nanoparticle synthesis using bacteria as biological nanofactories, their mechanisms and major applications. *Curr. Bionanotechnol. (Discontinued)* **2015**, *1*, 1–15. [CrossRef]
11. Singh, A.; Prasad, S.M. Nanotechnology and its role in agro-ecosystem: A strategic perspective. *Int. J. Environ. Sci. Technol.* **2017**, *14*, 2277–2300. [CrossRef]
12. Tripathy, A.; Raichur, A.M.; Chandrasekaran, N.; Prathna, T.C.; Mukherjee, A. Process variables in biomimetic synthesis of silver nanoparticles by aqueous extract of Azadirachta indica (Neem) leaves. *J. Nanopart. Res.* **2010**, *12*, 237–246. [CrossRef]
13. Reda, F.M.; El-Saadony, M.T.; Elnesr, S.S.; Alagawany, M.; Tufarelli, V. Effect of dietary supplementation of biological curcumin nanoparticles on growth and carcass traits, antioxidant status, immunity and caecal microbiota of Japanese quails. *Animals* **2020**, *10*, 754. [CrossRef]
14. Fesseha, H.; Degu, T.; Getachew, Y. Nanotechnology and its application in animal production: A review. *Vet. Med. Open J.* **2020**, *52*, 43–50. [CrossRef]
15. Reddy, P.R.K.; Yasaswini, D.; Reddy, P.P.R.; Zeineldin, M.; Adegbeye, M.J.; Hyder, I. Applications, challenges, and strategies in the use of nanoparticles as feed additives in equine nutrition. *Vet. World* **2020**, *13*, 1685–1696. [CrossRef] [PubMed]
16. Shi, L.; Xun, W.; Yue, W.; Zhang, C.; Ren, Y.; Shi, L.; Wang, Q.; Yang, R.; Lei, F. Effect of sodium selenite, Se-yeast and nano-elemental selenium on growth performance, Se concentration and antioxidant status in growing male goats. *Small Rumin. Res.* **2011**, *96*, 49–52. [CrossRef]
17. Shi, L.; Xun, W.; Yue, W.; Zhang, C.; Ren, Y.; Liu, Q.; Wang, Q.; Shi, L. Effect of elemental nano-selenium on feed digestibility, rumen fermentation, and purine derivatives in sheep. *Anim. Feed Sci. Technol.* **2011**, *163*, 136–142. [CrossRef]
18. Swain, P.S.; Rao, S.B.N.; Rajendran, D.; Dominic, G.; Selvaraju, S. Nano zinc, an alternative to conventional zinc as animal feed supplement: A review. *Anim. Nutr.* **2016**, *2*, 134–141. [CrossRef] [PubMed]
19. Champion, J.A.; Katare, Y.K.; Mitragotri, S. Making polymeric micro- and nanoparticles of complex shapes. *Proc. Natl. Acad. Sci. USA* **2007**, *104*, 11901–11904. [CrossRef] [PubMed]

20. Iravani, S.; Korbekandi, H.; Mirmohammadi, S.V.; Zolfaghari, B. Synthesis of silver nanoparticles: Chemical, physical and biological methods. *Res. Pharm. Sci.* **2014**, *9*, 385–406.
21. Chen, Z.; Meng, H.; Xing, G.; Chen, C.; Zhao, Y.; Jia, G.; Wang, T.; Yuan, H.; Ye, C.; Zhao, F. Acute toxicological effects of copper nanoparticles in vivo. *Toxicol. Lett.* **2006**, *163*, 109–120. [CrossRef] [PubMed]
22. Bunglavan, S.J.; Garg, A.; Dass, R.; Shrivastava, S. Use of nanoparticles as feed additives to improve digestion and absorption in livestock. *Livest. Res. Int.* **2014**, *2*, 36–47.
23. Al-Beitawi, N.A.; Shaker, M.M.; El-Shuraydeh, K.N.; Bláha, J. Effect of nanoclay minerals on growth performance, internal organs and blood biochemistry of broiler chickens compared to vaccines and antibiotics. *J. Appl. Anim. Res.* **2017**, *45*, 543–549. [CrossRef]
24. Raje, K.; Ojha, S.; Mishra, A.; Munde, V.; Chaudhary, S.K. Impact of supplementation of mineral nano particles on growth performance and health status of animals: A review. *J. Entomol. Zool. Stud.* **2018**, *6*, 1690–1694.
25. Stoimenov, P.K.; Klinger, R.L.; Marchin, G.L.; Klabunde, K.J. Metal oxide nanoparticles as bactericidal agents. *Langmuir* **2002**, *18*, 6679–6686. [CrossRef]
26. Rajendran, D.; Kumar, G.; Ramakrishnan, S.; Shibi, T. Enhancing the milk production and immunity in Holstein Friesian crossbred cow by supplementing novel nano zinc oxide. *Res. J. Biotechnol.* **2013**, *8*, 11–17.
27. Rosi, N.L.; Mirkin, C.A. Nanostructures in biodiagnostics. *Chem. Rev.* **2005**, *105*, 1547–1562. [CrossRef]
28. Allers, W.; Hahn, C.; Löhndorf, M.; Lukas, S.; Pan, S.; Schwarz, U.D.; Wiesendanger, R. Nanomechanical investigations and modifications of thin films based on scanning force methods. *Nanotechnology* **1996**, *7*, 346–350. [CrossRef]
29. Cardenas, G.; Meléndrez, M.; Cruzat, C.; Díaz, J. Synthesis of tin nanoparticles by physical vapour deposition technique (vd). *Acta Microsci.* **2007**, *1*, 1–2.
30. Swain, P.S.; Rajendran, D.; Rao, S.B.N.; Dominic, G. Preparation and effects of nano mineral particle feeding in livestock: A review. *Vet. World* **2015**, *8*, 888–891. [CrossRef] [PubMed]
31. Lane, R.; Craig, B.; Babcock, W. Materials engineering with nature's building blocks. *AMPTIAC Newslett. Spring* **2002**, *6*, 31–37.
32. Luz, M.S.D.; Tofanello, F.P.; Esposto, M.S.; Campos, A.D.; Ferreira, B.; Santos, C.A.M.D. Synthesis of SnSb2Te4 microplatelets by high-energy ball milling. *Mater. Res.* **2015**, *18*, 953–956. [CrossRef]
33. Yu, C.-H.; Tam, K.; Tsang, E.S.C. Chapter 5 Chemical Methods for Preparation of Nanoparticles in Solution. In *Handbook of Metal Physics*; Blackman, J.A., Ed.; Elsevier: Amsterdam, The Netherlands, 2008; Volume 5, pp. 113–141. [CrossRef]
34. Khan, I.; Saeed, K.; Khan, I. Nanoparticles: Properties, applications and toxicities. *Arab. J. Chem.* **2019**, *12*, 908–931. [CrossRef]
35. Oremland, R.S.; Herbel, M.J.; Blum, J.S.; Langley, S.; Beveridge, T.J.; Ajayan, P.M.; Sutto, T.; Ellis, A.V.; Curran, S. Structural and spectral features of selenium nanospheres produced by se-respiring bacteria. *Appl. Environ. Microbiol.* **2004**, *70*, 52–60. [CrossRef] [PubMed]
36. Phan, H.T.; Haes, A.J. What Does Nanoparticle Stability Mean? *J. Phys. Chem. C* **2019**, *123*, 16495–16507. [CrossRef] [PubMed]
37. Szczepanowicz, K.; Stefańska, J.; Socha, R.; Warszyński, P. Preparation of silver nanoparticles via chemical reduction and their antimicrobial activity. *Fizykochem. Probl. Miner.* **2010**, *45*, 85–98.
38. Yang, J.; Deivaraj, T.C.; Too, H.-P.; Lee, J.Y. Acetate stabilization of metal nanoparticles and its role in the preparation of metal nanoparticles in ethylene glycol. *Langmuir* **2004**, *20*, 4241–4245. [CrossRef]
39. Zhou, Y. Recent Advances in Ionic Liquids for Synthesis of Inorganic Nanomaterials. *Curr. Nanosci.* **2005**, *1*, 35–42. [CrossRef]
40. Mary, E.; Inbathamizh, L. Green synthesis and characterization of nano silver using leaf extract of morinda pubescens. *Asian J. Pharm. Clin. Res.* **2012**, *5*, 159–162.
41. Narayanan, K.B.; Sakthivel, N. Phytosynthesis of gold nanoparticles using leaf extract of Coleus amboinicus Lour. *Mater. Charact.* **2010**, *61*, 1232–1238. [CrossRef]
42. Sri, S.K.; Prasad, T.N.V.K.V.; Panner, S.P.; Hussain, O.M. Synthesis, characterization and evaluation of effect of phytogenic zinc nanoparticles on soil exo-enzymes. *Appl. Nanosci.* **2014**, *4*, 819–827. [CrossRef]
43. Philip, D. Green synthesis of gold and silver nanoparticles using Hibiscus rosa sinensis. *Phys. E Low-Dimens. Syst. Nanostruct.* **2010**, *42*, 1417–1424. [CrossRef]
44. Shankar, S.S.; Rai, A.; Ahmad, A.; Sastry, M. Rapid synthesis of Au, Ag, and bimetallic Au core-Ag shell nanoparticles using Neem (*Azadirachta indica*) leaf broth. *J. Colloid Interface Sci.* **2004**, *275*, 496–502. [CrossRef]
45. Anu, K.; Singaravelu, G.; Murugan, K.; Benelli, G. Green-synthesis of selenium nanoparticles using garlic cloves (*Allium sativum*): Biophysical characterization and cytotoxicity on vero cells. *J. Clust. Sci.* **2017**, *28*, 551–563. [CrossRef]
46. Hett, A. *Nanotechnology: Small Matter, Many Unknowns*; Swiss Reinsurance Co.: Zurich, Switzerland, 2004.
47. Corbo, C.; Molinaro, R.; Parodi, A.; Toledano Furman, N.E.; Salvatore, F.; Tasciotti, E. The impact of nanoparticle protein corona on cytotoxicity, immunotoxicity and target drug delivery. *Nanomedicine* **2016**, *11*, 81–100. [CrossRef] [PubMed]
48. Jahanbin, R.; Yazdanshenas, P.; Rahimi, M.; Hajarizadeh, A.; Tvrda, E.; Nazari, S.A.; Mohammadi-Sangcheshmeh, A.; Ghanem, N. In vivo and in vitro evaluation of bull semen processed with zinc (zn) nanoparticles. *Biol. Trace Elem. Res.* **2021**, *199*, 126–135. [CrossRef]
49. Chen, J.; Wei, W.; ZhiSheng, W. Effect of nano-zinc oxide supplementation on rumen fermentation in vitro. *Chin. J. Anim. Nutr.* **2011**, *23*, 1415–1421.
50. Riyazi, H.; Rezaei, J.; Rouzbehan, Y. Effects of supplementary nano-ZnO on in vitro ruminal fermentation, methane release, antioxidants, and microbial biomass. *Turk. J. Vet. Anim. Sci.* **2019**, *43*, 737–746. [CrossRef]

51. Mohamed, M.Y.; Ibrahim, E.M.M.; Abd El-Mola, A.M. Effect of selenium yeast and/or vitamin E supplemented rations on some physiological responses of post-lambing ossimi ewes under two different housing systems. *Egypt. J. Nutr. Feeds* **2017**, *20*, 361–378. [CrossRef]

52. Hosseini-Vardanjani, S.F.; Rezaei, J.; Karimi-Dehkordi, S.; Rouzbehan, Y. Effect of feeding nano-ZnO on performance, rumen fermentation, leukocytes, antioxidant capacity, blood serum enzymes and minerals of ewes. *Small Rumin. Res.* **2020**, *191*, 106170. [CrossRef]

53. Mohamed, A.; Mohamed, M.; Abd-El-Ghany, F.; Mahgoup, A.; Ibrahim, K. Influence of some trace minerals in form of nanoparticles as feed additives on lambs performance. *J. Anim. Poult. Prod.* **2015**, *6*, 693–703. [CrossRef]

54. Kojouri, G.; Arbabi, F.; Mohebbi, A. The effects of selenium nanoparticles (SeNPs) on oxidant and antioxidant activities and neonatal lamb weight gain pattern. *Comp. Clin. Pathol.* **2020**, *29*, 369–374. [CrossRef]

55. Seven, P.T.; Seven, I.; Baykalir, B.G.; Mutlu, S.I.; Salem, A.Z.M. Nanotechnology and nano-propolis in animal production and health: An overview. *Ital. J. Anim. Sci.* **2018**, *17*, 921–930. [CrossRef]

56. Zhan, X.; Wang, M.; Zhao, R.; Li, W.; Xu, Z. Effects of different selenium source on selenium distribution, loin quality and antioxidant status in finishing pigs. *Anim. Feed Sci. Technol.* **2007**, *132*, 202–211. [CrossRef]

57. Adegbeye, M.J.; Elghandour, M.M.M.Y.; Barbabosa-Pliego, A.; Monroy, J.C.; Mellado, M.; Ravi, K.R.P.; Salem, A.Z.M. Nanoparticles in equine nutrition: Mechanism of action and application as feed additives. *J. Equine Vet. Sci.* **2019**, *78*, 29–37. [CrossRef] [PubMed]

58. Murthy, N. *Investigation of Biochemical and Histopathological Modifiations Induced by Selenium Toxicity on Albino Mice*; LAP Lambert Academic Publishing: Saarbrucken, Germany, 2013.

59. Zhai, X.; Zhang, C.; Zhao, G.; Stoll, S.; Ren, F.; Leng, X. Antioxidant capacities of the selenium nanoparticles stabilized by chitosan. *J. Nanobiotechnol.* **2017**, *15*, 4. [CrossRef]

60. Mohamed, A.H.; Mohamed, M.Y.; Ibrahim, K.; Abd El Ghany, F.T.F.; Mahgoup, A.A.S. Impact of nano-zinc oxide supplementation on productive performance and some biochemical parameters of ewes and offspring. *Egypt. J. Sheep Goats Sci.* **2017**, *12*, 1–16. [CrossRef]

61. Sadeghian, S.; Kojouri, G.A.; Mohebbi, A. Nanoparticles of selenium as species with stronger physiological effects in sheep in comparison with sodium selenite. *Biol. Trace. Elem. Res.* **2012**, *146*, 302–308. [CrossRef]

62. Wu, X.; Yao, J.; Yang, Z. Improved fetal hair follicle development by maternal supplement of selenium at nano size (Nano-Se). *Livest. Sci.* **2011**, *142*, 270–275. [CrossRef]

63. Xun, W.; Shi, L.; Yue, W.; Zhang, C.; Ren, Y.; Liu, Q. Effect of high-dose nano-selenium and selenium–yeast on feed digestibility, rumen fermentation, and purine derivatives in sheep. *Biol. Trace Elem. Res.* **2012**, *150*, 130–136. [CrossRef]

64. Yaghmaie, P.; Ramin, A.; Asri-Rezaei, S.Z. Evaluation of glutathion peroxidase activity, trace minerals and weight gain following administration of selenium compounds in lambs. *Vet. Res. Forum* **2017**, *8*, 133–137.

65. Zaboli, K.; Aliarabi, H.; Bahari, A.; Abbasalipourkabir, R. Role of dietary nano-zinc oxide on growth performance and blood levels of mineral: A study on Iranian Angora (Markhoz) goat kids. *J. Pharm. Health Sci.* **2013**, *2*, 19–26.

66. Kumar, R.; Umar, A.; Kumar, G.; Nalwa, H.S. Antimicrobial Properties of ZnO Nanomaterials: A Review. *Ceram. Int.* **2017**, *43*, 3940–3961. [CrossRef]

67. Hashem, N.M.; Gonzalez-Bulnes, A. State-of-the-art and prospective of nanotechnologies for smart reproductive management of farm animals. *Animals* **2020**, *10*, 840. [CrossRef]

68. Jahanbin, R.; Yazdanshenas, P.; Amin, A.; Mohammadi, S.; Varnaseri, H.; Chamani, M.; Nazaran, M.; Bakhtiyarizadeh, M. Effect of zinc nano-complex on bull semen quality after freeze-thawing process. *J. Anim. Prod.* **2015**, *17*, 371–380.

69. Khalil, W.A.; El-Harairy, M.A.; Zeidan, A.E.B.; Hassan, M.A.E. Impact of selenium nanoparticles in semen extender on bull sperm quality after cryopreservation. *Theriogenology* **2019**, *126*, 121–127. [CrossRef]

70. Heidari, J.; Seifdavati, J.; Mohebodini, H.; Sharifi, R.S.; Benemar, H.A. Effect of nano zinc oxide on post-thaw variables and oxidative status of moghani ram semen. *Kafkas Univ. Vet. Fak. Derg.* **2019**, *25*, 71–76. [CrossRef]

71. Rahman, H.U.; Qureshi, M.S.; Khan, R.U. Influence of dietary zinc on semen traits and seminal plasma antioxidant enzymes and trace minerals of beetal bucks. *Reprod. Domest. Anim.* **2014**, *49*, 1004–1007. [CrossRef]

72. Hosny, N.S.; Hashem, N.M.; Morsy, A.S.; Abo-Elezz, Z.R. Effects of organic selenium on the physiological response, blood metabolites, redox status, semen quality, and fertility of rabbit bucks kept under natural heat stress conditions. *Front. Vet. Sci.* **2020**, *7*, 290. [CrossRef] [PubMed]

73. Hernández-Sierra, J.F.; Ruiz, F.; Pena, D.C.C.; Martínez-Gutiérrez, F.; Martínez, A.E.; Guillén, A.; Tapia-Pérez, H.; Castañón, G.M. The antimicrobial sensitivity of Streptococcus mutans to nanoparticles of silver, zinc oxide, and gold. *Nanomedicine* **2008**, *4*, 237–240. [CrossRef] [PubMed]

74. Abaspour, A.M.H.; Mamoei, M.; Tabatabaei, V.S.; Zareei, M.; Dadashpour, D.N. The effect of oral administration of zinc oxide nanoparticles on quantitative and qualitative properties of arabic ram sperm and some antioxidant parameters of seminal plasma in the non-breeding season. *Arch. Razi. Inst.* **2018**, *73*, 121–129. [CrossRef]

75. Abdel-Halim, B.R.; Moselhy, W.A.; Helmy, N.A. Developmental competence of bovine oocytes with increasing concentrations of nano-copper and nano-zinc particles during in vitro maturation. *Asian Pac. J. Reprod.* **2018**, *7*, 161–166. [CrossRef]

76. Lewis-Jones, D.I.; Aird, I.A.; Biljan, M.M.; Kingsland, C.R. Effects of sperm activity on zinc and fructose concentrations in seminal plasma. *Hum. Reprod.* **1996**, *11*, 2465–2467. [CrossRef]

77. Afifi, M.; Almaghrabi, O.A.; Kadasa, N.M. Ameliorative effect of zinc oxide nanoparticles on antioxidants and sperm characteristics in streptozotocin-induced diabetic rat testes. *BioMed Res. Int.* **2015**, *2015*, e153573. [CrossRef] [PubMed]

78. Konvicna, J.; Vargova, M.; Paulikova, I.; Kovac, G.; Kostecka, Z. Oxidative stress and antioxidant status in dairy cows during Cousins RJ, Liuzzi, J.P. Trace metal absorption and transport. prepartal and postpartal periods. *Acta Vet. Brno* **2015**, *84*, 133–140. [CrossRef]

79. Kachuee, R.; Abdi-Benemar, H.; Mansoori, Y.; Sánchez-Aparicio, P.; Seifdavati, J.; Elghandour, M.M.M.Y.; Guillén, R.J.; Salem, A.Z.M. Effects of sodium selenite, l-selenomethionine, and selenium nanoparticles during late pregnancy on selenium, zinc, copper, and iron concentrations in khalkhali goats and their kids. *Biol. Trace. Elem. Res.* **2019**, *191*, 389–402. [CrossRef] [PubMed]

80. Zhang, J.; Wang, X.; Xu, T. Elemental selenium at nano size (Nano-Se) as a potential chemopreventive agent with reduced risk of selenium toxicity: Comparison with se-methylselenocysteine in mice. *Toxicol. Sci.* **2008**, *101*, 22–31. [CrossRef]

81. Shahin, M.A.; Khalil, W.A.; Saadeldin, I.M.; Swelum, A.A.; Mostafa, A.; El-Harairy, M.A. Comparison between the effects of adding vitamins, trace elements, and nanoparticles to shotor extender on the cryopreservation of dromedary camel epididymal spermatozoa. *Animals* **2020**, *10*, 78. [CrossRef]

82. Hozyen, H.F.; El Shamy, A.A. Screening of genotoxicity and oxidative stress effect of selenium nanoparticles on ram spermatozoa. *Curr. Sci. Int.* **2019**, *7*, 799–807.

83. Hozyen, H.F.; El Shamy, A.A.; Farghali, A.A. In vitro supplementation of nano selenium minimizes freeze-thaw induced damage to ram spermatozoa. *Inter. J. Vet. Sci.* **2019**, *8*, 249–254.

84. Bąkowski, M.; Kiczorowska, B.; Samolińska, W.; Klebaniuk, R.; Lipiec, A. Silver and zinc nanoparticles in animal nutrition—A review. *Ann. Anim. Sci.* **2018**, *18*, 879–898. [CrossRef]

85. Amini, S.M.; Pirhajati, M.V. Selenium nanoparticles role in organ systems functionality and disorder. *Nanomed. Res. J.* **2018**, *3*, 117–124. [CrossRef]

86. Soumya, R.S.; Vineetha, V.P.; Raj, P.S.; Raghu, K.G. Beneficial properties of selenium incorporated guar gum nanoparticles against ischemia/reperfusion in cardiomyoblasts (H9c2). *Metallomics* **2014**, *6*, 2134–2147. [CrossRef]

87. Azarmi, F.; Kumar, P.; Mulheron, M. The exposure to coarse, fine and ultrafine particle emissions from concrete mixing, drilling and cutting activities. *J. Hazard. Mater.* **2014**, *279*, 268–279. [CrossRef] [PubMed]

88. Peralta-Videa, J.R.; Zhao, L.; Lopez-Moreno, M.L.; de la Rosa, G.; Hong, J.; Gardea-Torresdey, J.L. Nanomaterials and the environment: A review for the biennium 2008–2010. *J. Hazard. Mater.* **2011**, *186*, 1–15. [CrossRef] [PubMed]

89. Novoselec, J.; Šperanda, M.; Klir, Ž.; Mioč, B.; Steiner, Z.; Antunović, Z. Blood biochemical indicators and concentration of thyroid hormones in heavily pregnant and lactating ewes depending on selenium supplementation. *Acta Vet. Brno* **2017**, *86*, 353–363. [CrossRef]

90. Ghaderzadeh, S.; Aghjegheshlagh, M.F.; Nikbin, S.; Navidshad, B. Correlation effects of nano selenium and conjugated linoleic acid on the performance, lipid metabolism and immune system of male moghani lambs. *Iran. J. Appl. Anim. Sci.* **2019**, *9*, 443–451.

91. Han, L.; Kun, P.; Tong, F.; Clive, J.; Phillips, C.; Tengyun, G. Nano selenium supplementation increases selenoprotein (sel) gene expression profiles and milk selenium concentration in lactating dairy cows. *Biol. Trace Elem. Res.* **2021**, *199*, 113–119. [CrossRef]

92. Saadi, A.; Dalir-Naghadeh, B.; Asri-Rezaei, S.; Anassori, E. Platelet selenium indices as useful diagnostic surrogate for assessment of selenium status in lambs: An experimental comparative study on the efficacy of sodium selenite vs. selenium nanoparticles. *Biol. Trace Elem. Res.* **2020**, *194*, 401–409. [CrossRef]

93. Li, Y.; He, J.; Shen, X. Effects of nano-selenium poisoning on immune function in the wumeng semi-fine wool sheep. *Biol. Trace Elem. Res.* **2020**, *1*, 6. [CrossRef]

94. Alijani, K.; Rezaei, J.; Rouzbehan, Y. Effect of nano-ZnO, compared to ZnO and Zn-methionine, on performance, nutrient status, rumen fermentation, blood enzymes, ferric reducing antioxidant power and immunoglobulin G in sheep. *Anim. Feed Sci. Technol.* **2020**, *267*, 114532. [CrossRef]

95. Khin, M.M.; Nair, A.S.; Babu, V.J.; Murugan, R.; Ramakrishna, S. A review on nanomaterials for environmental remediation. *Energy Environ. Sci.* **2012**, *5*, 8075–8109. [CrossRef]

96. Pradeep, T. Noble metal nanoparticles for water purification: A critical review. *Thin Solid Films* **2009**, *517*, 6441–6478. [CrossRef]

97. Grieger, K.D.; Linkov, I.; Hansen, S.F.; Baun, A. Environmental risk analysis for nanomaterials: Review and evaluation of frameworks. *Nanotoxicology* **2012**, *6*, 196–212. [CrossRef]

98. Prashant, K.; Arun, K.; Teresa, F.; Godwin, A. Nanomaterials and the Environment. *J. Nanomater.* **2014**, *528606*. [CrossRef]

99. Korish, M.A.; Attia, Y.A. Evaluation of Heavy Metal Content in Feed, Litter, Meat, Meat Products, Liver, and Table Eggs of Chickens. *Animals* **2020**, *10*, 727. [CrossRef] [PubMed]

100. Attia, Y.A.; Abd Al-Hamid, A.E.; Zeweil, H.S.; Qota, E.M.; Bovera, F.; Monastra, G.; Sahledom, M.D. Effect of dietary amounts of inorganic and organic zinc on productive and physiological traits of White Pekin ducks. *Animal* **2013**, *7*, 895–900. [CrossRef] [PubMed]

101. Attia, Y.A.; Abdalah, A.A.; Zeweil, H.S.; Bovera, F.; Tag El-Din, A.A.; Araft, M.A. Effect of inorganic or organic selenium supplementation on productive performance, egg quality, and some physiological traits of dual-purpose breeding hens. *Czech J. Anim. Sci.* **2010**, *55*, 505–519. [CrossRef]

102. Attia, Y.A.; Addeo, N.F.; Al-Hamid, A.A.-H.E.A.; Bovera, F. Effects of phytase supplementation to diets with or without zinc addition on growth performance and zinc utilization of white pekin ducks. *Animals* **2019**, *9*, 280. [CrossRef] [PubMed]

103. Sertova, N.M. Contribution of Nanotechnology in Animal and Human Health Care. *Adv. Mater. Lett.* **2020**, *11*, 1–7. Available online: https://www.vbripress.com/aml/articles/details/1545 (accessed on 24 January 2021). [CrossRef]

104. Chung, I.-M.; Rajakumar, G.; Gomathi, T.; Park, S.-K.; Kim, S.-H.; Thiruvengadam, M. Nanotechnology for human food: Advances and perspective. *Front. Life Sci.* **2017**, *10*, 63–72. [CrossRef]

105. Faisal, S.; Yusuf, H.F.; Zafar, Y.; Majeed, S.; Leng, X.; Zhao, S.; Saif, I.; Malik, K.; Li, X. A review on nanoparticles as boon for biogas producers—nano fuels and biosensing monitoring. *Appl. Sci.* **2019**, *9*, 59. [CrossRef]

106. Shafi, B.U.D.; Kumar, R.; Jadhav, S.E.; Kar, J. Effect of zinc nanoparticles on milk yield, milk composition and somatic cell count in early-lactating barbari does. *Biol. Trace Elem. Res.* **2020**, *196*, 96–102. [CrossRef] [PubMed]

107. Menon, S.; Devi, K.S.S.; Agarwal, H.; Shanmugam, V.K. Efficacy of biogenic selenium nanoparticles from an extract of ginger towards evaluation on antimicrobial and antioxidant activities. *Colloid Interface Sci. Commun.* **2019**, *29*, 1–8. [CrossRef]

108. Alagawany, M.; Elnesr, S.S.; Farag, M.R.; Tiwari, R.; Yatoo, M.I.; Karthik, K.; Dhama, K. Nutritional significance of amino acids, vitamins and minerals as nutraceuticals in poultry production and health—A comprehensive review. *Vet. Q.* **2021**, *41*, 1–29. [CrossRef]

109. Chaudhary, S.; Mehta, S.K. Selenium nanomaterials: Applications in electronics, catalysis and sensors. *J. Nanosci. Nanotechnol.* **2014**, *14*, 1658–1674. [CrossRef] [PubMed]

110. Konkol, D.; Wojnarowski, K. The use of nanominerals in animal nutrition as a way to improve the composition and quality of animal products. *J. Chem.* **2018**, *2018*, 1–7. [CrossRef]

111. Wang, C.; Cheng, K.; Zhou, L.; He, J.; Zheng, X.; Zhang, L.; Zhong, X.; Wang, T. Evaluation of Long-Term Toxicity of Oral Zinc Oxide Nanoparticles and Zinc Sulfate in Mice. *Biol Trace Elem. Res.* **2017**, *178*, 276–282. [CrossRef] [PubMed]

112. Pogribna, M.; Hammons, G. Epigenetic Effects of Nanomaterials and Nanoparticles. *J. Nanobiotechnol.* **2021**, *19*, 2. [CrossRef]

113. Bai, M.M.; Liu, H.N.; Xu, K.; Wen, C.Y.; Yu, R.; Deng, J.P.; Yin, Y.L. Use of coated nano zinc oxide as an additive to improve the zinc excretion and intestinal morphology of growing pigs. *J. Anim. Sci.* **2019**, *97*, 1772–1783. [CrossRef]

114. Jia, X.; Li, N.; Chen, J.A. Subchronic toxicity study of elemental Nano-Se in Sprague-Dawley rats. *Life Sci.* **2005**, *76*, 1989–2003. [CrossRef] [PubMed]

115. Benko, I.; Nagy, G.; Tanczos, B.; Ungvari, E.; Sztrik, A.; Eszenyi, P.; Prokisch, J.; Banfalvi, G. Subacute toxicity of nano-selenium compared to other selenium species in mice. *Environ. Toxicol. Chem.* **2012**, *31*, 2812–2820. [CrossRef] [PubMed]

116. He, Y.; Chen, S.; Liu, Z.; Cheng, C.; Li, H.; Wang, M. Toxicity of selenium nanoparticles in male Sprague-Dawley rats at supranutritional and nonlethal levels. *Life Sci.* **2014**, *115*, 44–51. [CrossRef] [PubMed]

117. Saman, S.; Moradhaseli, S.; Shokouhia, A.; Ghorbani, M. Histopathological effects of ZnO nanoparticles on liver and heart tissues in Wistar rats. *Adv. Biores.* **2013**, *4*, 83–88.

118. Yusof, M.H.; Mohamad, R.; Zaidan, U.H.; Abdul Rahman, N.A. Microbial synthesis of zinc oxide nanoparticles and their potential application as an antimicrobial agent and a feed supplement in animal industry: A review. *J. Anim. Sci. Biotechnol.* **2019**, *9*, 10–57.

119. Sheiha, A.M.; Abdelnour, S.A.; El-Hack, A.; Mohamed, E.; Khafaga, A.F.; Metwally, K.A.; Ajarem, J.S.; Maodaa, S.N.; Allam, A.A.; El-Saadony, M.T. Effects of Dietary Biological or Chemical-Synthesized Nano-Selenium Supplementation on Growing Rabbits Exposed to Thermal Stress. *Animals* **2020**, *4*, 430. [CrossRef]

120. Sooklert, K.; Nilyai, S.; Rojanathanes, R.; Jindatip, D.; Sae-Liang, N.; Kitkumthorn, N.; Mutirangura, A.; Sereemaspun, A. N-acetylcysteine reverses the decrease of DNA methylation status caused engineered gold, silicon, and chitosan nanoparticles. *Int. J. Nanomed.* **2019**, *14*, 4573–4587. [CrossRef]

121. Liou, S.H.; Wu, W.T.; Liao, H.Y.; Chen, C.Y.; Tsai, C.Y.; Jung, W.T.; Lee, H.L. Global DNA methylation and oxidative stress biomarkers in workers exposed to metal oxide nanoparticles. *J. Hazard. Mater.* **2017**, *331*, 329–335. [CrossRef]

122. Abedini, M.; Shariatmadari, F.; Karimi, T.M.A.; Ahmadi, H. Effects of zinc oxide nanoparticles on the egg quality, immune response, zinc retention, and blood parameters of laying hens in the late phase of production. *J. Anim. Physiol. Anim. Nutr.* **2018**, *102*, 736–745. [CrossRef]

123. Jose, N.; Elangovan, A.V.; Awachat, V.B.; Shet, D.; Ghosh, J.; David, C.G. Response of in ovo administration of zinc on egg hatchability and immune response of commercial broiler chicken. *J. Anim. Physiol. Anim. Nutr.* **2018**, *102*, 591–595. [CrossRef]

124. Bribiesca, R.E.; Casas, L.; Cruz, M.R.; Romero, P.A. Supplementing selenium and zinc nanoparticles in ruminants for improving their bioavailability meat. In *Nutrient Delivery*; Academic Press: New York, NY, USA, 2017; pp. 713–747. [CrossRef]

125. Hassan, H.A.; Arafat, A.R.; Farroh, K.Y.; Bahnas, M.S.; El-Wardany, I.; Elnesr, S.S. Effect of in ovo copper injection on body weight, immune response, blood biochemistry and carcass traits of broiler chicks at 35 days of age. *Anim. Biotechnol.* **2021**, 1–8. [CrossRef]

126. Howard, V. General toxicity of N.M. WHO Workshop on Nanotechnology and Human Health: Scientific Evidence and Risk Governance. In Proceedings of the WHO Expert Meeting, Bonn, Germany, 10–11 December 2012.

127. Hill, E.K.; Li, J. Current and future prospects for nanotechnologyin animal production. *J. Ani. Sci. Biotechnol.* **2017**, *8*, 26. [CrossRef] [PubMed]

128. Lee, Y.K.; Ahn, S.I.; Chang, Y.H.; Kwak, H.S. Physicochemical and sensory properties of milk supplemented with dispersible nanopowdered oyster shell during storage. *J. Dairy Sci.* **2015**, *98*, 5841–5849. [CrossRef]

129. Prabhu, B.M.; Ali, S.F.; Murdock, R.C.; Hussain, S.M.; Srivatsan, M. Copper nanoparticles exert size and concentration dependent toxicity on somatosensory neurons of rat. *Nanotoxicology* **2010**, *4*, 150–160. [CrossRef]
130. Peng, D.; Zhang, J.; Liu, Q.; Taylor, E.W. Size effect of elemental selenium nanoparticles (Nano-Se) at supranutritional levels on selenium accumulation and glutathione S-transferase activity. *J. Inorg. Biochem.* **2007**, *101*, 1457–1463. [CrossRef] [PubMed]
131. Liao, H.-Y.; Chung, Y.-T.; Lai, C.-H.; Wang, S.-L.; Chiang, H.-C.; Li, L.-A.; Tsou, T.-C.; Li, W.-F.; Lee, H.-L.; Wu, W.-T. Six-month follow-up study of health markers of nanomaterials among workers handling engineered nanomaterials. *Nanotoxicology* **2014**, *8* (Suppl. 1), 100–110. [CrossRef] [PubMed]
132. Najafzadeh, H.; Ghoreishi, S.; Mohammadian, B.; Rahimi, E.; Afzalzadeh, M.; Kazemivarnamkhasti, M.; Ganjealidarani, H. Serum biochemical and histopathological changes in liver and kidney in lambs after zinc oxide nanoparticles administration. *Vet. World* **2013**, *6*, 534. [CrossRef]

Article

Silver and Copper Nanoparticles Inhibit Biofilm Formation by Mastitis Pathogens

Agata Lange [1], Agnieszka Grzenia [1], Mateusz Wierzbicki [1], Barbara Strojny-Cieslak [1], Aleksandra Kalińska [2], Marcin Gołębiewski [2], Daniel Radzikowski [2], Ewa Sawosz [1] and Sławomir Jaworski [1,*]

[1] Department of Nanobiotechnology, Institute of Biology, Warsaw University of Life Sciences, 02-786 Warsaw, Poland; agata_lange@sggw.edu.pl (A.L.); agnieszka.grzenia22@gmail.com (A.G.); mateusz_wierzbicki@sggw.edu.pl (M.W.); barbara_strojny@sggw.edu.pl (B.S.-C.); ewa_sawosz_chwalibog@sggw.edu.pl (E.S.)

[2] Department of Animal Breeding, Institute of Animal Sciences, Warsaw University of Life Sciences, 02-786 Warsaw, Poland; aleksandra_kalinska1@sggw.edu.pl (A.K.); marcin_golebiewski@sggw.edu.pl (M.G.); daniel_radzikowski@sggw.edu.pl (D.R.)

* Correspondence: slawomir_jaworski@sggw.edu.pl; Tel.: +48-225-936-675

Simple Summary: Bovine mastitis is a common disease in cows. It is caused by many pathogen species, which can form three-dimensional structures composed of bacterial cells, known as biofilms. These structures are almost impermeable to antimicrobials, making treatment difficult. We looked at the influence of metal nanometre-scale particles on biofilm formation by several pathogen species. We analysed the properties of these nanoparticles, determined the concentration needed to inhibit the growth of pathogens and to damage their membranes, and finally, checked how nanoparticles influence biofilm formation. We show that metal nanoparticles (silver and copper nanoparticles and their mixture) limit the formation of biofilm very effectively. These results mean that nanoparticles can be used to cure cattle suffering from mastitis, which will lead to higher milk production and less financial loss.

Abstract: Bovine mastitis is a common bovine disease, frequently affecting whole herds of cattle. It is often caused by resistant microbes that can create a biofilm structure. The rapidly developing scientific discipline known as nanobiotechnology may help treat this illness, thanks to the extraordinary properties of nanoparticles. The aim of the study was to investigate the inhibition of biofilms created by mastitis pathogens after treatment with silver and copper nanoparticles, both individually and in combination. We defined the physicochemical properties and minimal inhibitory concentration of the nanoparticles and observed their interaction with the cell membrane, as well as the extent of biofilm reduction. The results show that the silver–copper complex was the most active of all nanomaterials tested (biofilm was reduced by nearly 100% at a concentration of 200 ppm for each microorganism species tested). However, silver nanoparticles were also effective individually (biofilm was also reduced by nearly 100% at a concentration of 200 ppm, but at concentrations of 50 and 100 ppm, the extent of reduction was lower than for the complex). Nanoparticles can be used in new alternative therapies to treat bovine mastitis.

Keywords: biofilm; mastitis pathogens; bovine mastitis; silver nanoparticles; copper nanoparticles; silver-copper complex

Citation: Lange, A.; Grzenia, A.; Wierzbicki, M.; Strojny-Cieslak, B.; Kalińska, A.; Gołębiewski, M.; Radzikowski, D.; Sawosz, E.; Jaworski, S. Silver and Copper Nanoparticles Inhibit Biofilm Formation by Mastitis Pathogens. *Animals* **2021**, *11*, 1884. https://doi.org/10.3390/ani11071884

Academic Editors: Antonio Gonzalez-Bulnes and Nesreen Hashem

Received: 25 May 2021
Accepted: 21 June 2021
Published: 24 June 2021

Publisher's Note: MDPI stays neutral with regard to jurisdictional claims in published maps and institutional affiliations.

1. Introduction

1.1. Biofilms

A biofilm is a multicellular structure with specific composition, formed by microorganisms. In addition to bacterial cells, biofilm contains water and extracellular polymeric substance (EPS), which consists mainly of polysaccharides, proteins, nucleic acids, and surfactants [1]. It allows microorganisms to adhere particularly strongly to both biotic and abiotic surfaces, especially when exposed to unfavourable environmental factors, and hence biofilm forming is recognised as protection against damage [2]. The exopolysaccharide matrix may also be impermeable to antimicrobials and inhibit their penetration into the biofilm [3].

Maturation of the biofilm includes steps such as initial reversible and irreversible attachments, maturation, and dispersion. The last step is critical, allowing the biofilm to embed in new areas through detachment of planktonic forms, which initiate new biofilm structures in other locations [4].

The phenotype structure of a biofilm, regardless of its shape, is always similar. Small cells with limited multiplication potential are located inside the bacterial community, whereas outward-facing areas are occupied by metabolically active cells. This is due to the reduced availability of oxygen and nutrients in the centre of the structure [5]. Bacterial heterogeneity may promote the development of various resistant features, which, as a result, may encompass the whole community [3]. Remarkably, bacteria without resistant features initially become less sensitive to antimicrobials when grown in a biofilm structure [6]. "Quorum sensing" is the main means of communication in bacterial clusters, where every cell produces chemical compounds (such as bacteriocins), and certain genes are expressed. This phenomenon is considered a resistant mechanism [7]. The larger the population of microorganisms, the more self-inducing compounds are secreted. Changes in gene expression occur, and eventually the whole population is altered [8]. The prevalence of resistant bacteria in biofilms is one of the reasons bacterial infections are difficult to treat. The mechanisms of medications may also be ineffective against bacteria embedded in a biofilm [9]. Initially, infections are usually easily treatable because of the sensitivity of the planktonic forms that cause them. The difficulty arises when infections become chronic. They become much tougher to combat because by then they form advanced biofilm structures [10].

1.2. Bovine Mastitis

Bovine mastitis, caused by many bacterial species, is one of the most common bovine illnesses, affecting whole herds of cattle [10]. This affects the amount of milk produced, which in turn leads to financial losses in the dairy industry due to low productivity [11].

Mastitis pathogens are characterised as either contagious (spreading through the milking process) or environmental microorganisms. The contagious group includes *Staphylococcus aureus* and *Streptococcus agalactiae*, whereas the main environmental pathogens are *Escherichia coli*, *Streptococcus dysgalactiae*, and other streptococci [12]. While more than 130 bacterial species can cause the illness, *S. aureus* is one of the most common causes of chronic mastitis, because it can form a biofilm structure [13]. The biofilm-forming process allows bacteria such as staphylococci to colonise the internal part of the udder, and in particular, the biofilm structure with a polymeric matrix allows microbials to survive antimicrobial treatments [14]. Likewise, *E. faecalis*, which also creates a biofilm structure, has additional inherent resistance to certain antimicrobials, making it even more challenging to eradicate [15].

The rise in the number of resistant bacteria in the dairy industry is related to frequent use of antibiotics on herds of cattle. These pathogens are not thought to be dangerous to humans if only pasteurised milk is consumed, but they may still constitute a health hazard to the human population, especially since an increasing number of people consume raw milk [16]. Furthermore, the detrimental influence of mastitis on cattle health and milk production emphasises the urgent need for effective strategies to prevent and control the development of the disease [11].

1.3. Nanoparticles

Nanoparticles are widely used in biology and medicine due to their ability to freely penetrate the organisms' barriers [17]. The most popular medical nanomaterial is silver nanoparticles (AgNPs), which have been used as a component of antiseptics since ancient times because of their remarkable antibacterial properties and relatively low toxicity [18]. Nanosilver is thought to be effective against many of the resistant pathogens, and may become an alternative to antibiotics in the future. Nanoparticles interact with bacterial cells in several ways, including by disturbing cell layers and generating reactive oxygen species that damage internal structures [19]. Silver ions interact with the outer membrane (or wall), depriving bacteria of protection against harmful external factors [20]. They are also able to intercalate into nucleic acids, as well as to disturb ribosomes, which may lead to the inhibition of basic life processes in a cell, such as transcription and translation [20]. Nanoparticles with a diameter greater than 10 nm interact with the cell wall and membrane, causing its disintegration and cell death [21].

Copper nanoparticles (CuNPs), characterised by a high surface to volume ratio, also have great potential as antimicrobials due to easy surface functionalisation with other compounds to amplify the primary antibacterial effect [22]. According to a previous study [23], a combination of copper nanoparticles with carbon nanotubes inhibits the growth of the bacterial biofilm of *Methylobacterium* spp. and is not toxic to human fibroblasts. Similar effects have been observed for highly resistant *Pseudomonas aeruginosa* treated with copper nanoparticles [24]. Several nanomaterials exhibit tremendous antibacterial properties and inhibit biofilm formation on various surfaces for an extended period [9]. However, although nanoparticles are known to damage cells and to penetrate and disrupt biofilms, their precise mechanism of action is not fully understood [2].

1.4. Objective

The aim of the study was to investigate the ability of silver and copper nanoparticles both together and separately to inhibit biofilm formation produced by mastitis pathogens.

2. Materials and Methods

2.1. Nanoparticles

AgNPs and CuNPs were obtained from aXonnite (Nano-Tech, Warsaw, Poland). Silver-copper (Ag-Cu) complexes were prepared by mixing 50 ppm of the hydrocolloids of each of the nanoparticles in a 1:1 ratio, and the obtained mixture was sonicated for 45 min at room temperature. Prior to use in experiments, each compound was subjected to ultrasonic treatment for 30 min.

2.2. Physicochemical Analysis

Physicochemical analysis was conducted at room temperature (25 °C). The dynamic light scattering method was used for size distribution, and laser Doppler electrophoresis was used for zeta potential analysis with the Zetasizer Nano-ZS ZEN 3600 (Malvern, Malvern Town, UK). For morphology analysis, transmission electron microscopy with a voltage of 80 keV was used. 10 μL of nanoparticles was applied to copper grids (Mesh Cu Grids, Agar Scientific, Stansted, UK) and dried. Samples were observed under a microscope (JEM-1220, JEOL, Tokyo, Japan).

2.3. Bacterial Strains

Bacterial strains *Streptococcus agalactiae* (ATCC-31475), *Streptococcus dysgalactiae* (ATCC-12388), *Enterococcus faecalis* (ATCC-47077), *Staphylococcus aureus* (ATCC-27821), *Salmonella Enteritidis* (ATCC-BAA-1734), *Escherichia coli* (ATCC-12814), and *Enterobacter cloacae* (ATCC-35030), as well as yeast *Candida albicans* (ATCC-24433), were purchased from LGC Standards (Lomianki, Poland). The Mueller-Hinton (MH) broth used for growth and maintenance of the bacterial cultures was supplied by Biomaxima (cat. PS15, Lublin, Poland), whereas the yeast nitrogen base (YNB) for yeast was supplied by Merck Millipore (cat. 51483, Darmstadt, Germany).

2.4. Microbial Cultures

Each microbial strain was stored as a suspension in 20% (v/v) glycerol at -20 °C. Prior to experiments, the glycerol was removed and the microbial cells were washed with distilled water. Then microbial cultures were grown in media with an optimal availability of nutrients: MH broth for bacteria and YNB for yeast. Then the microbial cultures were kept in a bacterial incubator (NUAire, Plymouth, MN, USA) under standard conditions (37 °C).

2.5. The Minimal Inhibitory Concentration (MIC) Test

The first step of the MIC test was the preparation of microbial cell dilutions with an optical density (OD) of 0.1, which is equivalent to 10^6 cells per millilitre. Optical density was measured at a wavelength of 660 nm. For this purpose, 100 µL of the overnight microbial culture was added to 20 mL of liquid medium (MH for bacteria and YNB for yeast). The resulting suspension was then diluted again, yielding the final concentration of 2×10^4 cells per mL. Serial dilution was carried out in a 96-well plate in the presence of a blank control (medium without cells or nanoparticles) and a growth control group (inoculum without nanoparticles). After 24 h of incubation at 30 °C, a reading was taken using a microplate reader (Tecan M200 Infinite, Monachium, Germany; absorbance at 600 nm). This allowed us to select the appropriate concentrations to continue the research. The concentrations selected were 50 ppm, 100 ppm, and 200 ppm, for silver, copper, and silver-copper complex, respectively.

2.6. Membrane Integrity

To evaluate cell membrane integrity, the lactate dehydrogenase (LDH) activity was examined using a Cytotoxicity Detection Kit (In Vitro Toxicology Assay Kit, based on lactic dehydrogenase, LDH, Sigma-Aldrich, Hamburg, Germany). 100 µL of bacteria and yeast cells (1×10^6 CFU/mL) were cultured in liquid medium (MH for bacteria and YNB for yeast) on 96-well plates, with the addition of nanoparticles in the concentrations identified as the minimal inhibitory concentration for microbial species used (3.125, 6.25, 12.5, 25 ppm). After 24 h of incubation, 100 µL of the LDH assay mixture was added to each well. The plates were kept in the dark and incubated for 30 min at room temperature (25 °C). The absorbance was recorded at 490 nm on an ELISA reader (Infinite M200, Tecan, Männedorf, Switzerland). LDH leakage was expressed as a percentage of the test sample (reduced by the value of the blank) in relation to the control sample (also reduced by the value of the blank), where a blank probe was the medium without cells, and the control sample was inoculum treated with 100 µL of Triton X-100 (Sigma-Aldrich, Hamburg, Germany).

2.7. Biofilm Formation

Hydrocolloids of the nanoparticles at the selected concentrations (50, 100, and 200 ppm) were affixed to wells in a 96-well plate, and the prepared plate was left under the laminar flow cabinet for 24 h until completely dry. After this, 100 µL of microbial culture (1.5×10^8 CFU/mL) was added to each well, and the plate was incubated again for 24 h at 37 °C in a microbiological incubator (NUAire, Plymouth, MN, USA).

Planktonic cells were removed carefully by pipetting the liquid culture from the plate, leaving only the attached biofilm, which was fixed for 5 min with a 2.5% glutaraldehyde solution to inhibit further growth. The fixative was then removed and the wells were washed three times with sterile phosphate-buffered saline (PBS; Sigma-Aldrich, Darmstadt, Germany). To determine the exact quantity of biofilm, cells were stained with 100 µL of 0.25% crystal violet dye and washed gently three times with sterile PBS to remove any additional unbound dye. Plates were dried overnight, and crystal violet was then extracted using a 1:1 acetone:ethanol solution. The biofilm formation level was determined by measuring absorbance at a wavelength of 570 nm (microplate reader Tecan M200 Infinite, Monachium, Germany), which was related to the amount of dye attached to cells compared to controls (cultured in uncoated wells).

2.8. Data Analysis

The results were analysed by one-way analysis of variance (ANOVA) with Statgraphics Plus 4.1 (StatPoint Technologies Inc., Warrenton, VA, USA). All data were compiled with ANOVA (conforming to the assumptions), and differences were assumed to be statistically significant at $p \leq 0.05$.

3. Results

3.1. Physicochemical Analysis

AgNPs had the smallest average size and a spherical structure (Figure 1). The Ag-Cu complex showed a mean value of hydrodynamic diameter among tested samples. CuNPs had the biggest average diameter of over 300 nm, although size distribution suggests that two fractions were present, one with nanoparticles smaller than 100 nm, and the other much bigger. However, transmission electron microscopy (TEM) showed that small particles had agglomerated into large structures according to size distribution.

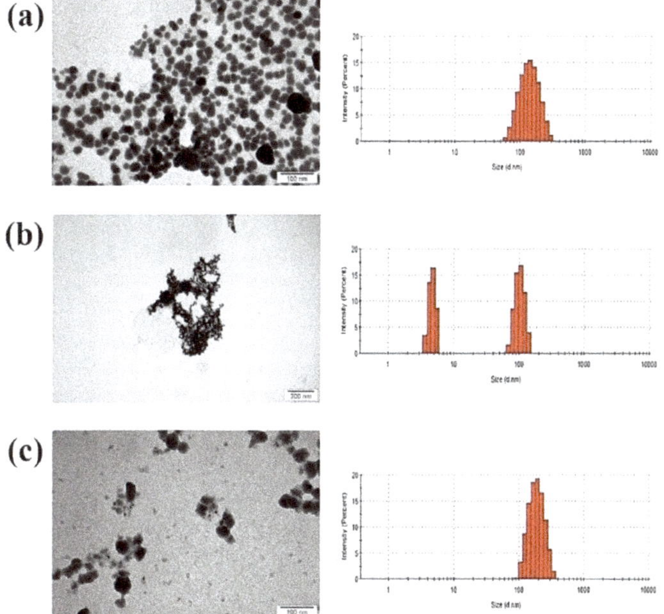

Figure 1. TEM images and size distribution (hydrodynamic diameter) of nanoparticles: (**a**) silver nanoparticles; (**b**) copper nanoparticles; (**c**) silver–copper complex.

Zeta potential values also confirm the tendency of CuNPs to agglomerate, as the value was close to zero (Table 1). Nevertheless, all samples analysed had a negative zeta potential not exceeding ±30 mV and no colloidal stability.

Table 1. Physicochemical parameters (average hydrodynamic diameter, zeta potential, and structure) of nanoparticles used (Ag, silver nanoparticles; Cu, copper nanoparticles; Ag-Cu, silver–copper complex).

Nanomaterial	Average Hydrodynamic Diameter (nm)	Zeta Potential (mV)	Structure
Ag	154.1	−26.7	spherical
Cu	345.6	−0.463	spherical
Ag-Cu	174.2	−9.09	spherical

3.2. Minimal Inhibitory Concentration

Minimal inhibitory concentration (MIC) was determined using serial dilution based on the concentration at which 50% of the bacterial growth was inhibited, which is considered one of the basic parameters for evaluating the effectiveness of tested substances [25]. Both AgNPs and the Ag-Cu complex could inhibit microbial growth at a relatively low concentration (Table 2). Optical density measurements showed that the minimal inhibitory concentration reduced microbial growth by approximately half (Figure 2). In the case of CuNPs, a concentration four times higher (25 ppm) was needed to limit the growth of most species. *Salmonella* spp. and *E. coli* were more sensitive, and the inhibitory concentration of CuNPs for these species was twice that of AgNPs and Ag-Cu complex (12.5 ppm for CuNPs and 6.25 ppm for the others).

Table 2. Values of minimal inhibitory concentration (ppm) of the nanoparticles used (Ag, silver nanoparticles; Cu, copper nanoparticles; Ag-Cu, silver–copper complex) for each microorganism strain.

Nanomaterial	*S. agalactiae*	*S. dysagalactiae*	*Salmonella* spp.	*E. faecalis*	*E. cloacae*	*C. albicans*	*E. coli*	*S. aureus*
Ag	3.125	6.25	6.25	6.25	6.25	6.25	6.25	12.5
Cu	25	25	12.5	25	25	25	12.5	25
Ag-Cu	3.125	6.25	6.25	6.25	6.25	6.25	6.25	12.5

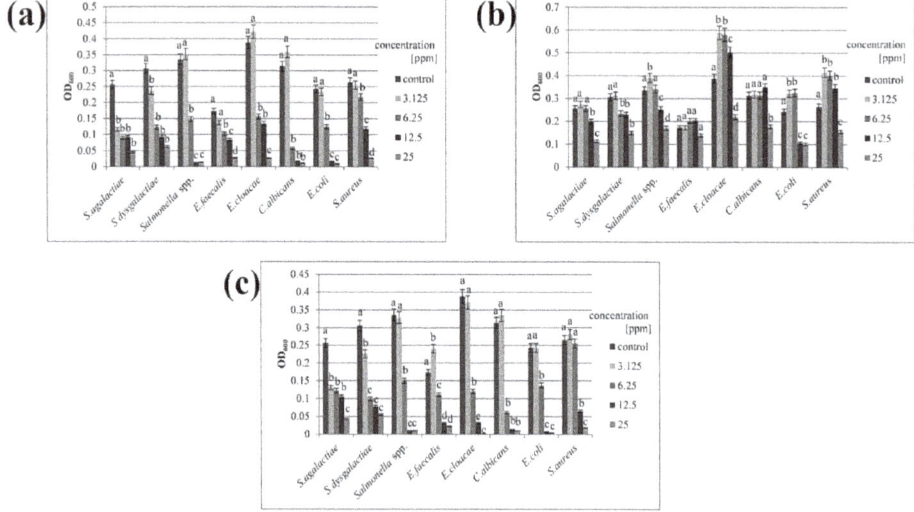

Figure 2. Optical density of microbial growth. Microbes were treated with (**a**) silver nanoparticles; (**b**) copper nanoparticles; (**c**) silver–copper complex. All samples were measured in triplicate and the results were averaged. Columns labelled a–d indicate statistically significant differences between groups within a species, and error bars show standard deviation.

3.3. Membrane Integrity

In all samples where the cell membrane was disturbed, the LDH release was observed in the culture medium. All three types of nanoparticles disrupted the cell membrane, but the greatest effect was observed in the presence of the nanoparticle complex. Results from samples with AgNPs and the complex were slightly similar, while CuNPs caused a weaker disruption of cell membranes. However, all nanomaterials caused disturbance in a dose-dependent manner (Figure 3).

3.4. Biofilm Formation

The application of crystal violet dye allowed us to determine the number of microorganisms able to attach to the biotic or abiotic surfaces. The greatest biofilm reduction occurred after treatment with AgNPs and the Ag-Cu complex at a concentration of 200 ppm (Figure 4). This effect was reproducible for all species, including both Gram-positive and Gram-negative bacteria and the yeast *Candida albicans*. No inhibition of biofilm formation was observed when using CuNPs at concentrations of 50 or 100 ppm. Copper nanoparticles at a concentration of 200 ppm did not eliminate the biofilm entirely in any species tested, while silver nanoparticles and the Ag-Cu complex both did. Interestingly, out of the three nanoparticle types, the Ag-Cu complex was most effective at inhibiting biofilm formation, even at the lowest concentration. The results were variable depending on the microorganism species, but in all cases, biofilm reduction was improved with a higher dose concentration.

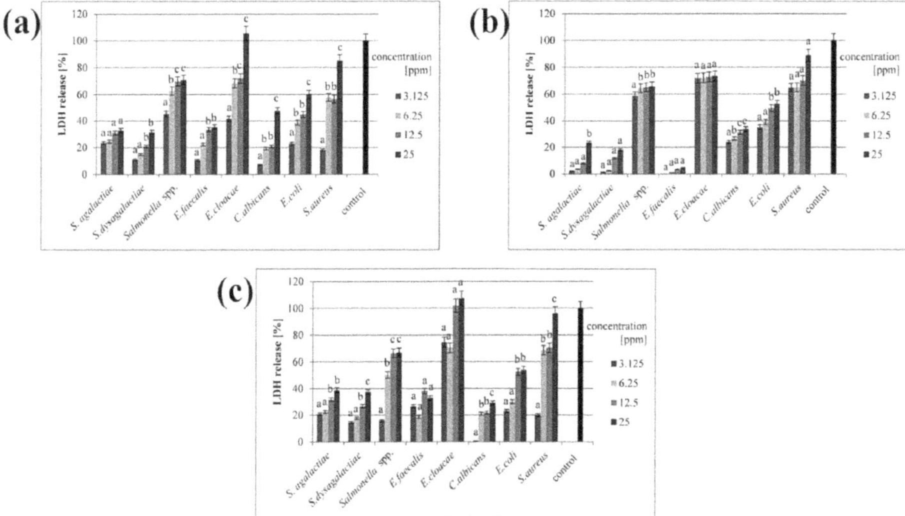

Figure 3. Percentage of LDH release after treatment with (**a**) silver nanoparticles; (**b**) copper nanoparticles; (**c**) silver–copper complex. All samples were measured in triplicate and the results were averaged. Columns labelled a–d indicate statistically significant differences between groups within a species, and error bars show standard deviation.

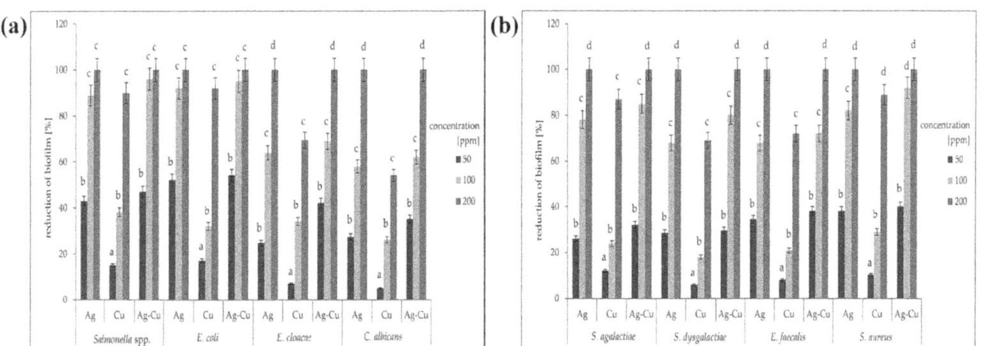

Figure 4. Percentage reduction of the biofilm for each of the test microorganisms treated with different concentrations (ppm) of nanoparticles: (**a**) Gram-negative bacteria and yeast; (**b**) Gram-positive bacteria. All samples were measured in triplicate and the results were averaged. Columns labelled a–d indicate statistically significant differences between all groups within a species, and error bars show standard deviation.

4. Discussion

4.1. Antibacterial Properties

We examined the influence of widely used metal nanoparticles on biofilm formation by mastitis pathogens. The analysis shows the high potential of nanomaterials, used both separately and in combination, to treat bovine mastitis, which is currently challenging. Overall, the Ag-Cu complex was more effective than either individual nanoparticle at inhibiting biofilm formation by mastitis pathogens.

Both AgNPs and CuNPs have great potential as antibacterials for inhibiting microbial growth [26]. The combination of AgNPs and CuNPs results in a synergy of effects, even though the activity of separate nanoparticles (Ag or Cu) is distinct. This means that the Ag-Cu complex is more promising than either nanoparticle used alone [27]. A similar effect has been observed in an earlier study on mastitis pathogens, where AgNPs and CuNPs caused a high degree of disruption to microbial viability [28]. We also found that the Ag-Cu complex caused the greatest biofilm reduction (Figure 4). In all samples, even the lowest concentration of the complex (50 ppm) disrupted biofilm growth more than either individual component. In all samples, higher nanomaterial concentration caused greater biofilm reduction. Interestingly, this was true not only for every nanomaterial but also for every microorganism species (Figure 4).

The antibacterial properties of AgNPs and CuNPs are associated with various factors, including their stability in hydrocolloids. Stable nanoparticle hydrocolloids tend not to agglomerate; their surface area is not limited, resulting in stronger antibacterial properties [29]. This was observed in our studies, where AgNPs and the Ag-Cu complex were more stable than CuNPs, and also had better antibacterial properties (Table 1). CuNPs had a zeta potential value of -0.463 mV, whereas AgNPs reached -26.7 mV, which is close to the limit value of colloidal stability (±30 mV). The value of the Ag-Cu complex was closer to the limit than that of CuNPs (-9.09 mV vs. -0.463 mV).

In toxicological studies, certain molecular characteristics must be considered (e.g., shape or size), since they determine the impact of a nanomaterial on the in vitro model [30]. It is assumed that small nanoparticles penetrate deeper into cell structures, but agglomerates, which reach a much greater average size, may have weaker interactions with cells [31]. Therefore, the low toxicity of CuNPs might result from their tendency to form large agglomerates, while the other two nanomaterials did not agglomerate and had a smaller diameter (Figure 1). Although the CuNPs hydrocolloid was composed of two fractions, the average diameter was 345.6 nm, indicating that there were more agglomerates than small nanoparticles (smaller than 100 nm). The toxicity of the nanoparticles is probably determined by their size, not their shape, since all the nanoparticles used had a similar shape (Figure 1). TEM analysis of nanoparticle shape yielded similar results to those reported by Paszkiewicz et al. in 2016 [32], where the shape was also found to be spherical.

According to one of the latest reports, a combination of copper and silver nanoparticles can be used in antibacterial therapies, although the main mode of action is related not to bacterial species, but to cell type and growth, or to nanoparticle uptake ability [33]. Our results confirm that cellular response does not depend on species, which is illustrated in the MIC analysis, where there were no clear distinctions in the reduction of viability of Gram-positive bacteria, Gram-negative bacteria, and yeasts (Table 2, Figure 2). However, in terms of the reduction of biofilm formation, two Gram-negative species (*Salmonella* spp. and *E. coli*) seemed to the most sensitive to the nanoparticles. For other species, biofilm reduction was lower, but for the yeast *C. albicans*, the reduction was similar to that for *E. cloacae* (Figure 4). This is probably the result of the high capacity of both species for developing bacterial resistance [34,35]. The greater sensitivity of Gram-negative bacteria can be explained by their cell wall structure. Their cell walls are much thinner than those of Gram-positive bacteria, despite the presence of an external layer [36].

4.2. Possible Mechanisms

The antibacterial properties of nanomaterials result mainly from the generation of free radicals, which disturb the cell wall, membrane, or organelles of bacterial cells [37]. However, there are other mechanisms of microbial cell inactivation. Some of the commonly reported ones include the disruption of intracellular ATP, damage to DNA structures, and damage to other organelles [29]. It is known that metal nanoparticles kill major Gram-positive and Gram-negative pathogens, and that they penetrate and eradicate biofilms; however, the precise mechanism is not fully understood [2]. The interaction of metal nanoparticles with bacterial cells is very complicated due to the enormous number of characteristics that nanoparticles exhibit and to the fact that their mechanism of interaction is still poorly understood. There are many plausible hypotheses for the interactions between nanoparticles and biofilms. These interactions take place on several levels, including disturbing the cell layer and producing reactive oxygen species that damage internal structures [19]. Silver ions, which are generated from silver nanoparticles, bind to the negatively-charged layer, causing cell perforation and cell death [38]. The accumulation of metal nanoparticles around bacterial cells and in biofilm networks has been visualised in a previous study [39]; the effect was dependent on nanoparticle type. Larger nanoparticles (more than 10 nm in diameter) interact with the cell wall or membrane [21]. In our research, the nanoparticles and their complex agglomerates had a diameter over 10 nm (AgNPs: 154.1 nm; CuNPs: 345.6 nm; Ag-Cu complex: 174.2 nm); therefore, they attacked the internal parts of cell. Furthermore, even if nanoparticles or their agglomerates are too large to penetrate through the entire biofilm, they interact with planktonic cells, which still reduces biofilm formation. Dispersion is a critical step in biofilm formation because under natural conditions it allows cells to spread to new areas [4]. By attacking planktonic cells, nanoparticles prevent this spread.

In our research, LDH release was dose-dependent in all samples (Figure 3), which supports our hypothesis about the interaction of nanomaterials with the outer part of microbial cells. The mechanism of this interaction is well-known, but biofilms show entirely different phenotypes from planktonic forms [40]. The penetration of antimicrobials and their impact on microbial cells in biofilms is hampered, mainly due to the presence of exopolysaccharide (EPS), which presumably binds directly to antimicrobial agents [40]. The presence of EPS and the complex structure of biofilms contribute to the acquisition of resistance, and thus make treatment of infections more difficult [9]. This is affected by the location of cells in the biofilm, where metabolically active cells are located in the other parts of the structure [5], and this microbial heterogeneity makes it possible for resistance characteristics to spread throughout the entire biofilm [3]. It is believed that nanomaterials may damage signalling molecules, leading to the inhibition of gene expression pathways required to develop and modify the biofilm structure. This can cause the biofilm to lose its resistant traits [41].

The positive effects of metal nanoparticles on the inhibition of biofilms made up of certain bacteria species have been observed in research by Gurunathan et al. [42], who suggest that AgNPs may constitute an adjuvant for curing bacterial infections. The same was demonstrated by Martinez-Gutierrez et al. [43], who found that AgNPs not only inhibited biofilm formation, but also induced cell death.

Thanks to their high antibacterial potential, nanoparticles are considered one of the most promising agents for preventing bovine mastitis [44–46]. However, despite their excellent properties, a great number of aspects must be considered before use, such as their influence on mammalian tissues and on whole organisms [47]. Nevertheless, the proposed solution for treating mastitis infections might alleviate serious hazards for animals, entrepreneurs, and the human population [10,11].

5. Conclusions

This research (and its possible follow-up studies) proposes a promising treatment for mastitis, an illness caused to a large extent by biofilm formation by microorganisms. The presented results show that metal nanoparticles are able to disrupt the biofilm. Particularly noteworthy is the combination of AgNPs and CuNPs, which yielded the best results. These results are important, especially since the threat of mastitis may be more serious than it seems at first glance.

Author Contributions: Conceptualisation, S.J., M.G., and E.S.; methodology, A.L., A.G., D.R., and A.K.; validation, B.S.-C. and M.W.; formal analysis, A.G. and A.L.; investigation, A.L. and S.J.; writing—original draft preparation, A.L. and S.J.; writing—review and editing, B.S.-C.; visualisation, M.W.; supervision, M.G and S.J.; project administration, M.G.; funding acquisition, M.G. All authors have read and agreed to the published version of the manuscript.

Funding: This study was funded by National Centre for Research and Development (LIDER/6/0070/L-7/15/NCBR/2016).

Institutional Review Board Statement: Not applicable.

Data Availability Statement: The data presented in this study are available on request from the corresponding author.

Conflicts of Interest: The authors declare no conflict of interest.

References

1.   Czyzewska-Dors, E.; Dors, A.; Pomorska-Mol, M. Właściwości biofilmu bakteryjnego warunkujące oporność na antybiotyki oraz metody jego zwalczania. *Życie Weter.* **2018**, *93*, 765–771. (In Polish)
2.   Liu, Y.; Shi, L.; Su, L.; van der Mei, H.C.; Jutte, P.C.; Ren, Y.; Busscher, H.J. Nanotechnology-based antimicrobials and delivery systems for biofilm-infection control. *Chem. Soc. Rev.* **2019**, *48*, 428–446. [CrossRef]
3.   Mah, T.-F.C.; O'Toole, G.A. Mechanisms of biofilm resistance to antimicrobial agents. *Trends Microbiol.* **2001**, *9*, 34–39. [CrossRef]
4.   Rabin, N.; Zheng, Y.; Opoku-Temeng, C.; Du, Y.; Bonsu, E.; Sintim, H.O. Biofilm formation mechanisms and targets for developing antibiofilm agents. *Future Med. Chem.* **2015**, *7*, 493–512. [CrossRef]
5.   Bryers, J.D. Medical biofilms. *Biotechnol. Bioeng.* **2008**, *100*, 1–18. [CrossRef]
6.   Stewart, P.S. Mechanisms of antibiotic resistance in bacterial biofilms. *Int. J. Med. Microbiol.* **2002**, *292*, 107–113. [CrossRef]
7.   Li, Y.-H.; Tian, X. Quorum Sensing and Bacterial Social Interactions in Biofilms. *Sensors* **2012**, *12*, 2519–2538. [CrossRef]
8.   Matejczyk, M.; Suchowierska, M. Charakterystyka zjawiska quorum sensing i jego znaczenie w aspekcie formowania i funkcjonowania biofilmu w inżynierii środowiska, budownictwie, medycynie oraz gospodarstwie domowym. *Bud. I Inżynieria Sr.* **2008**, *2*, 71–75. (In Polish)
9.   Venkatesan, N.; Perumal, G.; Doble, M. Bacterial resistance in biofilm-associated bacteria. *Future Microbiol.* **2015**, *10*, 1743–1750. [CrossRef] [PubMed]
10.   Melchior, M.B.; Vaarkamp, H.; Fink-Gremmels, J. Biofilms: A role in recurrent mastitis infections? *Vet. J.* **2006**, *171*, 398–407. [CrossRef] [PubMed]
11.   Hossain, M. Bovine Mastitis and Its Therapeutic Strategy Doing Antibiotic Sensitivity Test. *Austin J. Vet. Sci. Anim. Husb.* **2017**, *4*, 1030. [CrossRef]
12.   Dufour, S.; Labrie, J.; Jacques, M. The Mastitis Pathogens Culture Collection. *Microbiol. Resour. Announc.* **2019**, *8*, e00133-19. [CrossRef]
13.   Raza, A.; Muhammad, G.; Sharif, S.; Atta, A. Biofilm Producing *Staphylococcus aureus* and Bovine Mastitis: A Review. *Mol. Microbiol. Res.* **2013**, *3*. [CrossRef]
14.   Oliveira, M.; Bexiga, R.; Nunes, S.F.; Carneiro, C.; Cavaco, L.M.; Bernardo, F.; Vilela, C.L. Biofilm-forming ability profiling of *Staphylococcus aureus* and *Staphylococcus epidermidis* mastitis isolates. *Vet. Microbiol.* **2006**, *118*, 133–140. [CrossRef] [PubMed]
15.   Gomes, F.; Saavedra, M.J.; Henriques, M. Bovine mastitis disease/pathogenicity: Evidence of the potential role of microbial biofilms. *Pathog. Dis.* **2016**, *74*, 1–7. [CrossRef] [PubMed]
16.   Oliver, S.P.; Murinda, S.E. Antimicrobial Resistance of Mastitis Pathogens. *Vet. Clin. N. Am. Food Anim. Pract.* **2012**, *28*, 165–185. [CrossRef]
17.   Malina, D.; Sobczak-Kupiec, A.; Wzorek, Z. Nanobiotechnologia—Dziś i jutro. *Chemik* **2011**, *65*, 1027–1034. (In Polish)
18.   Franci, G.; Falanga, A.; Galdiero, S.; Palomba, L.; Rai, M.; Morelli, G.; Galdiero, M. Silver Nanoparticles as Potential Antibacterial Agents. *Molecules* **2015**, *20*, 8856–8874. [CrossRef]
19.   Fulaz, S.; Vitale, S.; Quinn, L.; Casey, E. Nanoparticle–Biofilm Interactions: The Role of the EPS Matrix. *Trends Microbiol.* **2019**, *27*, 915–926. [CrossRef]
20.   Siddiqi, K.S.; Husen, A.; Rao, R.A.K. A review on biosynthesis of silver nanoparticles and their biocidal properties. *J. Nanobiotechnol.* **2018**, *16*, 14. [CrossRef]
21.   Salas Orozco, M.F.; Niño-Martínez, N.; Martínez-Castañón, G.A.; Méndez, F.T.; Ruiz, F. Molecular mechanisms of bacterial resistance to metal and metal oxide nanoparticles. *Int. J. Mol. Sci.* **2019**, *20*, 2808.

22. Ingle, A.P.; Duran, N.; Rai, M. Bioactivity, mechanism of action, and cytotoxicity of copper-based nanoparticles: A review. *Appl. Microbiol. Biotechnol.* **2014**, *98*, 1001–1009. [CrossRef] [PubMed]

23. Seo, Y.; Hwang, J.; Lee, E.; Kim, Y.J.; Lee, K.; Park, C.; Choi, Y.; Jeon, H.; Choi, J. Engineering copper nanoparticles synthesized on the surface of carbon nanotubes for anti-microbial and anti-biofilm applications. *Nanoscale* **2018**, *10*, 15529–15544. [CrossRef] [PubMed]

24. LewisOscar, F.; MubarakAli, D.; Nithya, C.; Priyanka, R.; Gopinath, V.; Alharbi, N.S.; Thajuddin, N. One pot synthesis and anti-biofilm potential of copper nanoparticles (CuNPs) against clinical strains of *Pseudomonas aeruginosa*. *Biofouling* **2015**, *31*, 379–391. [CrossRef] [PubMed]

25. Prost, M.E.; Prost, R. Basic parameters of evaluation of the effectiveness of antibiotic therapy. *OphthaTherapy Ther. Ophthalmol.* **2017**, *4*, 233–236. [CrossRef]

26. Ruparelia, J.P.; Chatterjee, A.K.; Duttagupta, S.P.; Mukherji, S. Strain specificity in antimicrobial activity of silver and copper nanoparticles. *Acta Biomater.* **2008**, *4*, 707–716. [CrossRef]

27. Fan, X.; Yahia, L.; Sacher, E. Antimicrobial Properties of the Ag, Cu Nanoparticle System. *Biology* **2021**, *10*, 137. [CrossRef]

28. Kalińska, A.; Jaworski, S.; Wierzbicki, M.; Gołębiewski, M. Silver and copper nanoparticles—An alternative in future mastitis treatment and prevention? *Int. J. Mol. Sci.* **2019**, *20*, 1672. [CrossRef]

29. Khodashenas, B. The Influential Factors on Antibacterial Behaviour of Copper and Silver Nanoparticles. *Indian Chem. Eng.* **2016**, *58*, 224–239. [CrossRef]

30. Powers, K.W.; Palazuelos, M.; Moudgil, B.M.; Roberts, S.M. Characterization of the size, shape, and state of dispersion of nanoparticles for toxicological studies. *Nanotoxicology* **2007**, *1*, 42–51. [CrossRef]

31. Montes-Burgos, I.; Walczyk, D.; Hole, P.; Smith, J.; Lynch, I.; Dawson, K. Characterisation of nanoparticle size and state prior to nanotoxicological studies. *J. Nanopart. Res.* **2010**, *12*, 47–53. [CrossRef]

32. Paszkiewicz, M.; Gołąbiewska, A.; Rajski, Ł.; Kowal, E.; Sajdak, A.; Zaleska-Medynska, A. Synthesis and Characterization of Monometallic (Ag, Cu) and Bimetallic Ag-Cu Particles for Antibacterial and Antifungal Applications. *J. Nanomater.* **2016**, *2016*, 1–11. [CrossRef]

33. Valdez-Salas, B.; Beltrán-Partida, E.; Zlatev, R.; Stoytcheva, M.; Gonzalez-Mendoza, D.; Salvador-Carlos, J.; Moreno-Ulloa, A.; Cheng, N. Structure-activity relationship of diameter controlled Ag@Cu nanoparticles in broad-spectrum antibacterial mechanism. *Mater. Sci. Eng. C* **2021**, *119*, 111501. [CrossRef]

34. Gulati, M.; Nobile, C.J. Candida albicans biofilms: Development, regulation, and molecular mechanisms. *Microbes Infect.* **2016**, *18*, 310–321. [CrossRef] [PubMed]

35. Annavajhala, M.K.; Gomez-Simmonds, A.; Uhlemann, A.-C. Multidrug-Resistant *Enterobacter cloacae* Complex Emerging as a Global, Diversifying Threat. *Front. Microbiol.* **2019**, *10*, 44. [CrossRef]

36. Al-Sharqi, A.; Apun, K.; Vincent, M.; Kanakaraju, D.; Bilung, L.M. Enhancement of the antibacterial efficiency of silver nanoparticles against gram-positive and gram-negative bacteria using blue laser light. *Int. J. Photoenergy* **2019**, *2019*. [CrossRef]

37. Miller, K.P.; Wang, L.; Benicewicz, B.C.; Decho, A.W. Inorganic nanoparticles engineered to attack bacteria. *Chem. Soc. Rev.* **2015**, *44*, 7787–7807. [CrossRef] [PubMed]

38. Liu, C.; Guo, J.; Yan, X.; Tang, Y.; Mazumder, A.; Wu, S.; Liang, Y. Antimicrobial nanomaterials against biofilms: An alternative strategy. *Environ. Rev.* **2017**, *25*, 225–244. [CrossRef]

39. Sawosz, E.; Chwalibog, A.; Szeliga, J.; Sawosz, F.; Grodzik, M.; Rupiewicz, M.; Niemiec, T.; Kacprzyk, K. Visualization of gold and platinum nanoparticles interacting with *Salmonella* Enteritidis and *Listeria monocytogenes*. *Int. J. Nanomed.* **2010**, *5*, 631–637. [CrossRef]

40. Donlan, R.M. Biofilms: Microbial Life on Surfaces. *Emerg. Infect. Dis.* **2002**, *8*, 881–890. [CrossRef]

41. Kołwzan, B. Analiza zjawiska biofilmu—Warunki jego powstawania i funkcjonowania. *Ochr. Sr.* **2011**, *33*, 3–14. (In Polish)

42. Gurunathan, S.; Han, J.W.; Kwon, D.N.; Kim, J.H. Enhanced antibacterial and anti-biofilm activities of silver nanoparticles against Gram-negative and Gram-positive bacteria. *Nanoscale Res. Lett.* **2014**, *9*. [CrossRef]

43. Martinez-Gutierrez, F.; Boegli, L.; Agostinho, A.; Sánchez, E.M.; Bach, H.; Ruiz, F.; James, G. Anti-biofilm activity of silver nanoparticles against different microorganisms. *Biofouling* **2013**, *29*, 651–660. [CrossRef] [PubMed]

44. Algharib, S.A.; Dawood, A.; Xie, S. Nanoparticles for treatment of bovine *Staphylococcus aureus* mastitis. *Drug Deliv.* **2020**, *27*, 292–308. [CrossRef] [PubMed]

45. Dehkordi, S.H.; Hosseinpour, F.; Kahrizangi, A.E. An in vitro evaluation of antibacterial effect of silver nanoparticles on *Staphylococcus aureus* isolated from bovine subclinical mastitis. *Afr. J. Biotechnol.* **2011**, *10*, 10795–10797. [CrossRef]

46. Radzikowski, D.; Kalińska, A.; Ostaszewska, U.; Gołębiewski, M. Alternative solutions to antibiotics in mastitis treatment for dairy cows—A review. *Anim. Sci. Pap. Rep.* **2020**, *38*, 117–133.

47. Orzechowska, A.; Szymanska, R. Nanotechnologia w zastosowaniach biologicznych—Wprowadzenie. *Wszechświat* **2016**, *117*, 60–69. (In Polish)

 *animals* 

Review

Modulation of Bovine Endometrial Cell Receptors and Signaling Pathways as a Nanotherapeutic Exploration against Dairy Cow Postpartum Endometritis

Ayodele Olaolu Oladejo [1,2], Yajuan Li [1], Xiaohu Wu [1], Bereket Habte Imam [1], Jie Yang [1], Xiaoyu Ma [1], Zuoting Yan [1,*] and Shengyi Wang [1,*]

[1] Key Laboratory of Veterinary Pharmaceutical Development of Ministry of Agriculture, Lanzhou Institute of Husbandry and Pharmaceutical Sciences of Chinese Academy of Agricultural Science, Lanzhou 730050, China; aygenesis2succeed@yahoo.com (A.O.O.); 18719826933@163.com (Y.L.); wuxiaohu01@caas.cn (X.W.); Bekihi14@gmail.com (B.H.I.); 15535384771@163.com (J.Y.); 18903865563@163.com (X.M.)

[2] Department of Animal Health Technology, Oyo State College of Agriculture and Technology, Igboora 201103, Nigeria

[*] Correspondence: yanzuoting@caas.cn (Z.Y.); wangshengyi@caas.cn (S.W.)

Citation: Oladejo, A.O.; Li, Y.; Wu, X.; Imam, B.H.; Yang, J.; Ma, X.; Yan, Z.; Wang, S. Modulation of Bovine Endometrial Cell Receptors and Signaling Pathways as a Nanotherapeutic Exploration against Dairy Cow Postpartum Endometritis. *Animals* **2021**, *11*, 1516. https://doi.org/10.3390/ani11061516

Academic Editors: Antonio Gonzalez-Bulnes and Nesreen Hashem

Received: 18 March 2021
Accepted: 20 May 2021
Published: 23 May 2021

Publisher's Note: MDPI stays neutral with regard to jurisdictional claims in published maps and institutional affiliations.

Simple Summary: The provision of updated information on the molecular pathogenesis of bovine endometritis with host-pathogen interactions and the possibility of exploring the cellular sensors mechanism in a nanotechnology-based drug delivery system against persistent endometritis were reported in this review. The mechanism of Gram-negative bacteria and their ligands has been vividly explored, with the paucity of research detail on Gram-positive bacteria in bovine endometritis. The function of cell receptors, biomolecules proteins, and sensors were reportedly essential in transferring signals into cell signaling pathways to induce immuno-inflammatory responses by elevating pro-inflammatory cytokines. Therefore, understanding endometrial cellular components and signaling mechanisms across pathogenesis are essential for nanotherapeutic exploration against bovine endometritis. The nanotherapeutic discovery that could inhibit infectious signals at the various cell receptors and signal transduction levels, interfering with transcription factors activation and pro-inflammatory cytokines and gene expression, significantly halts endometritis.

Abstract: In order to control and prevent bovine endometritis, there is a need to understand the molecular pathogenesis of the infectious disease. Bovine endometrium is usually invaded by a massive mobilization of microorganisms, especially bacteria, during postpartum dairy cows. Several reports have implicated the Gram-negative bacteria in the pathogenesis of bovine endometritis, with information dearth on the potentials of Gram-positive bacteria and their endotoxins. The invasive bacteria and their ligands pass through cellular receptors such as TLRs, NLRs, and biomolecular proteins of cells activate the specific receptors, which spontaneously stimulates cellular signaling pathways like MAPK, NF-kB and sequentially triggers upregulation of pro-inflammatory cytokines. The cascade of inflammatory induction involves a dual signaling pathway; the transcription factor NF-κB is released from its inhibitory molecule and can bind to various inflammatory genes promoter. The MAPK pathways are concomitantly activated, leading to specific phosphorylation of the NF-κB. The provision of detailed information on the molecular pathomechanism of bovine endometritis with the interaction between host endometrial cells and invasive bacteria in this review would widen the gap of exploring the potential of receptors and signal transduction pathways in nanotechnology-based drug delivery system. The nanotherapeutic discovery of endometrial cell receptors, signal transduction pathway, and cell biomolecules inhibitors could be developed for strategic inhibition of infectious signals at the various cell receptors and signal transduction levels, interfering on transcription factors activation and pro-inflammatory cytokines and genes expression, which may significantly protect endometrium against postpartum microbial invasion.

Keywords: endometritis; cell receptors; signaling pathways; PAMP; biomolecules; cytokines; nanotherapy; dairy cow; nanosystem

1. Introduction

Researchers have reported several reports on reducing dairy cattle reproductive efficiency because of postpartum development of uterine disease or infection in dairy cows [1–4]. Several diagnostic, treatment and prevention such as purulent vulval discharges, return to oestrus after breeding, presence of microorganism from microbiological examination, infiltration of polymorphonuclear cells, hormonal variation, antibiotic and anti-inflammatory treatment regimen, cleanliness of cattle environment, avoidance of coitus with infected animals, fomite sterilization and vaccinationhave been adopted to eliminate this postpartum uterine infection, which includes rectal palpation of the genital tract to detect uterine abnormalities [3,5–10]. Postpartum endometritis of dairy cows usually occurs, causing economic losses due to high infectivity percentages, which has been a significant concern to dairy scientists, particularly at the onset of this millennium [3,11]. Postpartum uterine disease reflects disturbance on the normal physiological characteristics during postpartum in the dairy cow uterus, leading to impaired uterine involution. Monitoring the uterine involution, endometrial restoration, resumption of ovarian cyclicity, and pathogenic bacteria control are the major postpartum activities to ensure continuous reproductive ability [6,12]. Postpartum massive pathogens invasion of endometrium leads to the incapability of the uterine immune system to combat pathogens. The inflammatory response to postpartum bacterial infection during subclinical endometritis is correlated with the secretion of pro-inflammatory mediators that cause cell proliferation by either reacting on the oocyte, forming the embryo, endometrium, or impairing the hypothalamic-pituitary-gonadal axis [4]. The anatomical position and vascularization of the endometrium are physiological arrangements that make it prone to override by bacteria invasion. Endometritis occurs as a result of its inadequacy to disrupt pathogen growth and multiplication in the uterine epithelium. *Escherichia coli* are the most abundant endometrial pathogenic bacteria interacting with other bacteria isolated from cows with uterine disease [13,14].

In addition to bacteria, the in vitro infection of bovine epithelial cells with bovine herpesvirus IV [BoHV-4] has been reported suggesting the endometrium could be susceptible to viral infection [12]. However, the molecular mechanism implicated in the pathogenesis of endometritis has not been deeply investigated. The innate immune system recognizes pathogenic bacteria that invade the endometrial tissue by host cell PRRs to bacteria PAMPs [14]. PAMPs like LPS, a major component of cell walls unique to most gram-negative bacteria, stimulate immune system cells and induce a strong inflammatory response [15]. The inflammatory intracellular signaling pathway cascades induced by the interactions between PAMPs and PRRs contribute to the transcription and release of crucial pro-inflammatory mediators, including IL-β, TNFα, and chemokines responsible for complement activation and acute phase protein response that results in successful immune and inflammatory responses to eradicate the pathogens [16,17]. However, chronic uterine inflammation attributable to insufficient uterine pathogen eradication, excessive inflammatory signals, and anti-inflammatory pathway deficiencies is likely to affect fertilization and pregnancy [18], ultimately leading to subclinical endometritis. It was reported that early postpartum endometrial inflammation is a normal physiological process aimed at repairing the damaged endometrial uterine lining, though consequently bacteria [7] and their ligands invade the lumen, resulting in disruption of the typical anatomical architecture of the uterine mucosa, leading to endometritis. The quest to provide an in-depth review highlighting the understanding of the mechanism of dairy cow endometritis pathophysiology through synergy existing between bacterial PAMPs, activation of PRRs of the endometrial cells, cellular signaling pathways activation and phosphorylation co-exist, leading to production and regulation of inflammatory cytokines, which are mediators to the onset of bovine postpartum endometritis. Therefore, there are critical needs to evaluate the pathogenesis cascade of endometritis and endometrial cell sensors and biomolecules. We reviewed the current knowledge and research article about cell molecular characterization that may be affecting the endometrium being prone to microbial invasion during postpartum and a possible pointer tool for nanodrug development or immunotherapeutic index in curbing the persistence of dairy cow endometritis.

2. Function of Bacterial Pampsin Bovine Endometritis

Bacterial identification in mammalian tissues relies on the perception of their PAMP on innate immune system cells. PAMP recognition of PRRs triggers a cascade of events that cause host protection mechanisms to deter and resist the initial infection and prompt immune response. Endogenous molecules and PAMPs may directly tie to the pathogenesis of endometrial diseases [19–22]. The Gram-positive bacteria cell wall comprises a thick layer of peptidoglycan mixed with teichoic acid and extracellular proteins. Some PAMPs produced by Gram-positive bacteria are Lipoteichoic Acid [LTA], lipoproteins, peptidoglycan, glycosyl phosphatidyl, hemagglutinin protein, and phospholipomannan [20–22]. The Gram-negative bacteria cell wall comprises a small, elastic protein peptidoglycan layer coated in a PAMP additional layer [23]. The Gram-negative bacteria produce PAMP capable of inducing inflammatory reaction activating specific PRR and phosphorylation of relevant cellular signaling pathways [24,25], stimulating the immune response. PAMPs are subject to evolutionary change due to selective pressure from host defense mechanisms. This discovery may eventually expand the range of ligands to a given PRR and the cellular signals it can elicit [25].

Most unique PAMPs are identified by various PRRs to synergistically cause inflammatory responses and guide the adaptive immune responses [22]. Some ligands expressed their function either on the cell wall or cytoplasm and other subcellular organelles of the host cell. Evaluating the pathogenesis of bacteria causing dairy cow endometritis needs to review the role of the entire Gram-negative bacteria endotoxins and virulence factors in penetrating the cell and activating cellular function. *Escherichia coli* produces cell-wall bacterial components such as LPS [26], cholesterol-dependent cytotoxin and pyolysin, *Truperella pyogenes*, and *Fusobacterium necrophorum* produce growth factor F and Leukotoxin; whereas *Provetella* species develops a phagocytose-inhibiting agent [27]. The most abundant Gram-negative PAMP that can induce inflammatory and immune responses of invaded endometrial cells is LPS, a major component of Gram-negative bacteria external membrane [28]. LPS from *E. coli* is the principal Gram-negative bacteria PAMP that has been evaluated in endometritis pathomechanisms [11,26]. However, the molecular mechanisms of other Gram-negative bacteria to induce endometritis remain untapped.

Hexa-acetylated lipid A types present in *E.coli* and *Salmonella enterica serovar Typhimurium* are effective cell receptor activators [27]. As shown in experimental models of *Salmonella canoni*, *Legionella pneumophila*, and *E.coli* infections [29,30], the flagellin component of specific gram-negative cell walls is potent PAMPS. Still, their role in bovine endometritis pathogenesis remains unknown. *E. coli* flagella [H serogroup], Shiga-toxin [a heat-stable and labile toxin], and lipopolysaccharide [O serogroup LPS] [31,32], fimbriae, endotoxigenic of other Gram-negative bacteria cell wall proteins and adhesins are putative ligands for various PRRs.

At the time of this review's compilation, studies on the molecular and cellular evaluation of dairy cow endometrium inflammatory response after calving due to invasion of Gram-positive bacteria and their ligand endotoxins were absent. The PAMPs, bacterial load, and virulence factors of the postpartum endometrial pathogens determined whether infection of the postpartum uterus evolves toward inflammatory conditions like clinical or subclinical endometritis [25,32]. Certain Gram-positive bacteria, including *Corynebacterium*, *Nocardia*, *Bacillus*, *Listeria*, *Staphylococcus*, *Clostridium*, *Enterococcus*, *Streptococcus*, and *Mycobacterium species* are some of the pathogens implicated in postpartum endometritis pathogenesis but not specifically characterized ding uterine infection [21,33]. They produced PAMPs such as lipopeptides, peptidoglycan, glycolipids, LTA [34], though it remains elucidated whether the upregulation of those specific PRRs can induce immuno-inflammatory responses.

Triacylated lipoprotein from a Gram-positive bacterium has been shown to stimulate an inflammatory response by activating a specific PRR with subsequent phosphorylation and activation of the intracellular signaling pathway [35,36]. The apparent dearth of studies on the impact of Gram-positive bacteria and their ligands on the pathogenesis of postpartum endometritis in dairy cows, as it was reported [37] suggesting that these pathogens and their PAMPs may have a long-term retarding effect on the current method used in the evaluation of postpartum uterine immune suppression related to pro-inflammatory cytokines and chemokines activation induced by uterine exposure to the influx of bacteria and their ligands. Hence, the concern for prompt research actions in postpartum dairy cows. The mechanism through which the bacteria and PAMPs penetrate the various cell receptors to induce cellular and molecular changes needs to be in the pathogenesis of bovine endometritis as it is not thoroughly analyzed up today.

3. Endometrial Cell Receptors and Signalling Biomolecules

The bovine endometrium epithelial and stromal cells have been reported to have multiple PRRs, which normally identify PAMPs that contribute to the stimulation of cell signaling pathways to enhanced production of numerous inflammatory mediators correlated with endometrial inflammatory disease pathogens is [37,38]. The review of patterns by which different endometrial cell receptors recognize various signals from microorganisms and their ligands in responsive actions leading to postpartum dairy cow endometritis. Several pathogen receptors have been detected in endometrial cells involved in immuno-inflammatory reaction complexes [39]. Each of the endometrial PRRs had an affinity for different bacteria PAMPs and cellular locations [40].

TLRs are cell membrane-bound proteins with an additional transmembrane domain implicated in ligand recognition, whether on the extracellular surface, within endosomes, and cytoplasmic domain engaged in signal transduction [38,41]. NLRs and RLRs are cytosolic helicase receptors able to detect PAMPs' cellular invasion [21]. TLRs and NLRs are major PRRs identified in endometrial cells, synergize and complement each other intracellularly [42,43], but the mechanism through which their synergism results in the activation and phosphorylation of intracellular signaling pathways remain unknown. Bovine endometrial revealed in healthy non-gravid cows, the uterus ipsilateral and contralateral to the corpus luteum horn have expressed TLR1 to TLR10 [43,44]; TLR2, TLR3, TLR4, and TLR6 are expressed ante- and postpartum in caruncular and inter-caruncular areas of the endometrium, and whereas TLR9 has a greater expression in caruncular areas [44]. Also, endometrial epithelial cells in culture expressed TLR1, TLR7and TLR9 [45], and stromal cells expressed TLR1, TLR4, TLR6, TLR7, TLR9, and TLR10 [45,46]. TLRs are divided into two subgroups regarding their cellular localization and PAMP detection, in which TLR1, TLR2, TLR4, TLR5, TLR6, and TLR10 are expressed exclusively on the cell surface and stimulated by lipids, lipoproteins, and proteins of pathogen membrane components [46] and TLR3, TLR7, Ref. [47] and TLR9 are expressed on intra-cellular organelle membranes such as endosomes (Figure 1) [26,47–49]. The negative feedback mechanism of TLR regulation results in aberrant signaling by improper identification of self-proteins as foreign antigens, thus, resulting in excessive inflammation, systemic infectious diseases, and autoimmune diseases [43,50–52].

TLR signaling pathways contain signaling biomolecules and adaptor proteins, including MyD88, Mal/TIRAP, TRIF, TRAM, and SARM proteins, cell-specific and species-specific [39,49]. Deep cellular and molecular knowledge of TLRs and their biomolecule proteins function may lead to their role as therapeutic targets for medical application. TLRs are studied predominantly in uterine infection and endometritis, but their physiological role in other female reproductive events needs further investigation [50]. TLR2 detects invasion of bacterial and mycobacterial triacylated lipopeptides, mycoplasma diacylated lipopeptides, peptidoglycan [PGN], and LTA from gram-positive bacteria [34]. The bovine endometrial epithelial and stromal cells culture stimulated by diacylated and triacylated bacterial lipopeptides activates TLR2 leading to increased production of inflammatory cytokines infiltrating leukocytes. Still, the signaling pathway through which the expression occurred remains unsolved [37,51]. The TLR1–TLR2, TLR2–TLR6 heterodimeric interactions within adaptor protein, and hydrophobic interactions are responsible for increased immuno-inflammatory response [52]. TLR1–TLR10 formed dimers from the same subfamily [53], while TLR4–TLR6 heterodimer is formed in reaction to endogenous ligands established to foster inflammatory responses [51]. The intracellular immuno-signaling within subcellular organelles regulates the extend of TLR 2, 4, or 6 in immune cells, evoke innate and cellular immune responses and promote lysosomal degradation [53,54]. TLR4 signaling complex is a complete cell receptor that enhanced signal transfer within the cell and its micromolecules to elicit an inflammatory response via the transcription of cytokine genes [54,55] and the most examined receptors in bovine endometrium [40]. The TLR4 signaling pathway contributes to the development of different pro-inflammatory cytokines and chemokines linked to inflammatory responses. Immunological assays of TLRs will be vital for improving our understanding of early inflammatory events regulating immunological response due to endometritis [55]. The NLR family is the second group of endometrial cytoplasmic receptors that recognize microbial products and pathogen-associated molecular patterns. These proteins shared a common domain in organization with an NH2-terminal protein-protein interaction domain, NOD domain, and COOH-terminal leucine-rich repeats (LRR). NLRs are an active category of intracellular receptors, a key component of the hosts' innate immune systems, which includes NOD1, NOD2, NALP, LRR, and PYD protein-containing domains 3 [56] (Figure 2).

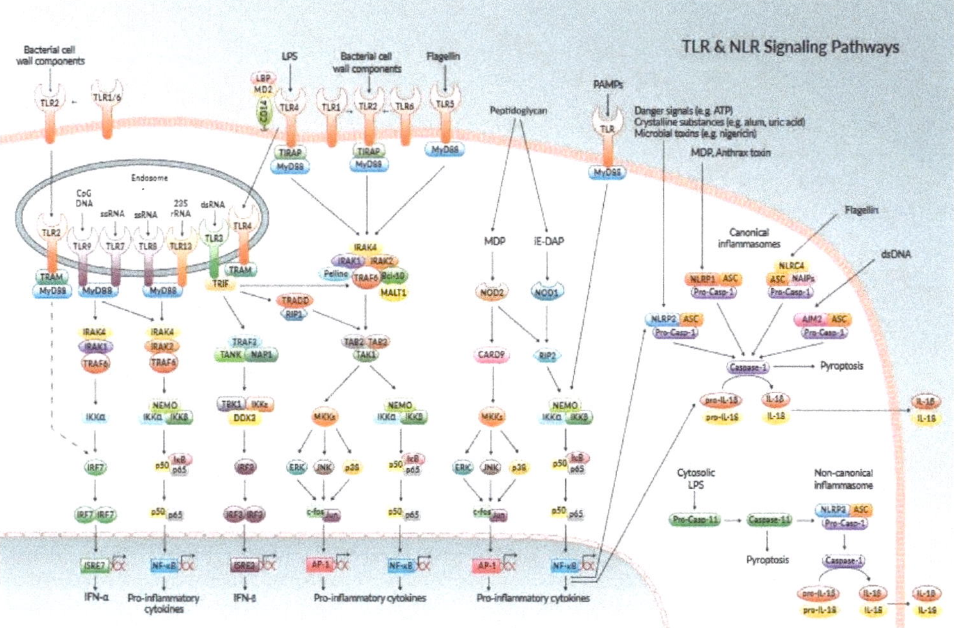

Figure 1. The mechanism of pathogen molecular recognition by Toll-like receptors triggering cellular signaling pathways activation (26,48). The mechanism of PAMP stimulating the cell receptor with their major transduction by biomolecules, which induced various isoforms of the signaling pathway and led to upregulation of pro-inflammatory cytokines and chemokines. The delivery of nanodrugs in suppressing inflammatory response in any part of the signaling pathways is essential in nanotherapeutic development against endometritis.

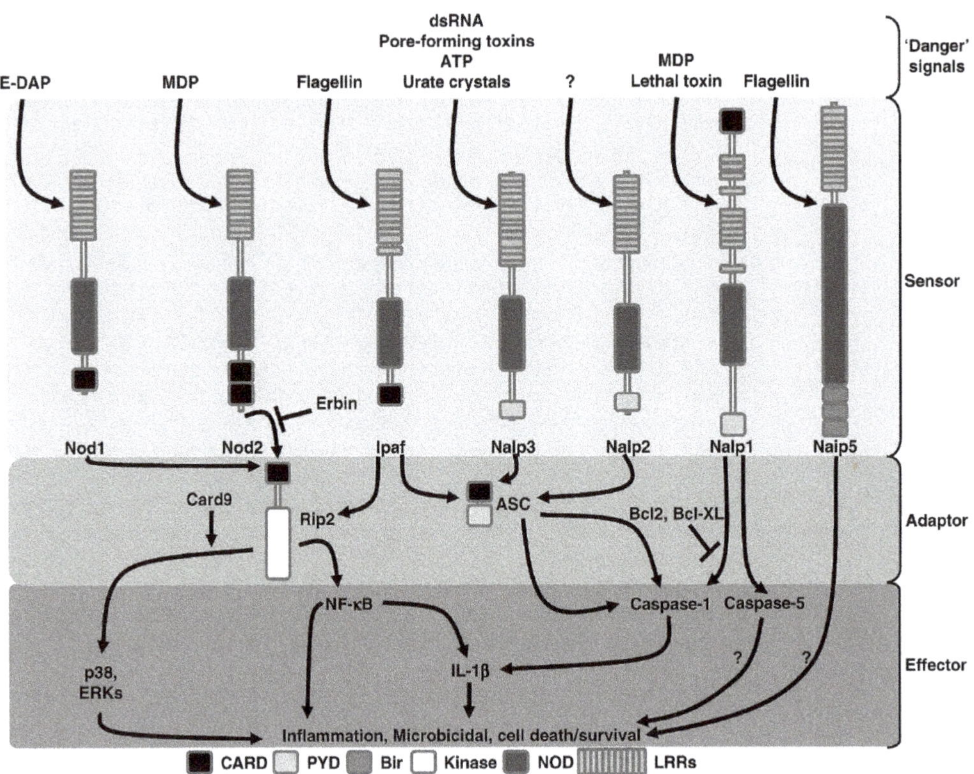

Figure 2. Intracellular Debugging of the NLR Signaling Pathways showing the cellular interconnectivity between PAMPs, cellular sensors, adaptors, and effectors to activate intracellular signaling cascade and stimulation of inflammation ultimate. The intracellular connections between its biomolecules are the basis for nanomedical and nanotechnology discovery against lingering subclinical endometritis.

This review analyzed NLRs expression in dairy cow endometrial cells both in the normal and diseased state. NLRs protein expression in endometrial cells exposed to PAMPs, receptor–receptor connectivity, interaction with various cellular signaling enhanced the pathophysiology of the endometrial inflammatory response, and subsequent prolonged effect leads to subclinical endometritis. The NLR family recognizes PAMPs and biomolecules in the cytosol. NOD1 (also named as NLRC1) and NOD2 [NLRC2] are most characterized by NLR family members [56,57], which sense bacterial molecules derived from the synthesis and degradation of peptidoglycan [PGN], but its reactivity has not been explored using other bacteria ligands, especially pathogenic bacteria causing postpartum bovine endometritis. NOD1 and NOD2 identify intracellular PAMPs, and NLRP3 reacts to multiple stimuli to form a multi-protein complex.NOD1 is active in the intracellular recognition of different pathogenic bacteria, including entero-invasive *E coli*. The activation of NOD1 and NOD2 results in a signaling cascade, triggering an inflammatory response resulting in increased cytokine production [58,59]. The mechanisms by which it also delivers PGN into the cytosol to gain access to NOD1 and NOD2 are unclear. NODs and NLRP3 are essential to cell cytoplasmic-to-cytoplasmic interactions receptors, which recognize endogenous molecules and microbial molecules [60]. The molecular analysis of NLRs is defined by a centrally located NOD that induces oligomerization, a C-terminal LRR that mediates ligand with TLRs, and an N-terminal CARD responsible for the initiation of signaling [56]. If NOD1 and NOD2 are specific PAMP receptors or sense shifts in host factors resulting from the existence of microbial molecules in the cytosol is currently unclear [61]. NOD1 and NOD2 stimulation results primarily in the activation of pro-inflammatory gene expression. Other NLR proteins are involved in the activation of caspases [62]. NOD1 recognizes gD-glutamyl-meso-diaminopimelic acid [iE-DAP] present in all Gram-negative PGN structures andvarious Gram-positive bacteria, such as *Bacillus subtilis* and *L. cytogenes*. At the same time, NOD2 detects muramyl dipeptide [MDP], the largest component of the PGN motif often found in all Gram-negative and Gram-positive bacteria.

Meanwhile, the evaluation of the pathophysiological function of PGN in subclinical endometritis has not been elucidated. A rare and essential antibacterial role in intestinal cells, such as control of antimicrobial peptides, was documented in NOD1 and NOD2 receptors [62,63]. Schroder & Tschopp 2010 [64] stated that activation of NLRs contributes to developing a multi-protein inflammasome complex, characteristically attached to caspase-1, which activates pro-inflammatory cytokine production.The molecular role of caspase-1 in the progressing onset of endometrial inflammatory response remains unevaluated. NLRs and their inflammasome are innate immune receptors that gained increased interest over the past years and are considered main intracellular pathogen sensors, and danger signals played an important role in infection and immunity [65]. It usually stimulates bovine endometrial inflammasome through molecular activities of caspase-4 dependent cascade called non-canonical inflammasome signaling pathway [65,66]. NLRP1 [NALP1] requires a protein complex dimeric reaction with NOD2 to mediate caspase-1 activation in response to MDP recognition [62,67]. NLRP2 was characterized as a maternal-specific gene in oocytes and granulosa cells, and its loss in zygotes triggered early embryonic death [68,69]. NLRP2 activation has been described in the central nervous system immune cells [70] and stimulated pro-inflammatory caspase 1, which results in pro-inflammatory cytokine and chemokine production. It has also been implicated in untoward anti-fetal responses through suppression of NF–κBsignaling pathway in linkage with the subcortical maternal complex to fertility issues were reported by Mahadevan et al. 2017 [71]. NLRP 2 activity was reported to be essential in reproductive potential and infertility [71], with no available record as at the moment of compiling this review on bovine endometritis. The NLRP3 inflammasome comprises NLRP3, ASC, and caspase-1, which are critical regulators of multiple inflammatory diseases [72].

IPAF belong to the NLR protein family and contains 22 members [73] and reacted to the exposure of endogenous ligands such as PAMPs and DAMPs to seldomly stimulates inflammatory responses in which the dysregulation in the function of NLRP3 was associated with the pathogenesis of several inflammatory diseases was reported [74,75]. Caspase1 was implicated in controlling the release of the inflammasome substrate, which is a signal produced to amplify the release of PAMPs by activating caspase-1 in neighbor cells [73,75]. The molecular mechanism and pathways through which the inflammasome complex could be activated in the endometrial remain undocumented. Several inflammasome families were identified in different body tissues and can recognize potentially harmful signals or PAMPs through their respective signaled cell receptors [57]. The NLRP3 inflammasome was reportedly formed by various PAMPs exposure of the inflicted cells with various structures. NLRP3 is majorly required for caspase-1 activation in response to bacteria ligands such as LPS, dsRNA, PGN, and LTAs when stimulated together with extracellular ATP [65].

As mentioned above, these NLRP3 functions could be explored to understand bovine endometritis pathogenesis with exposure to the PAMPs. The ionic flux, mitochondrial dysfunction, reactive oxygen species [ROS] generation, and lysosomal damage are molecular or cellular events shown to activate the NLRP3 inflammasome [76,77].

There is a need to investigate their roles in the pathogenesis of bovine endometritis. Nakahira et al. 2014 [78] reported dysfunctional mitochondria to enhance the production of mtROS to activate inflammatory activation of NLRP3 in response to LPS and ATP, which implies that NLRP3 senses cellular stress and culminate in activation of inflammasome remain to be fully elucidated, so the mitochondrial DNA [mtDNA] released into both NLRP3 and mtROS-dependent cytosols, mtDNA also interacts with both NLRP3 and AIM2, which are essential for the activation of the inflammasome during infectious phases [70,75]. Therefore, thiscould be an insight for a probable function of mtDNA of the endometrial cells' immuno-inflammatory responses upon cell exposure to inflicting pathogens and their ligands, which at present remain uninvestigated. The urgent quest of molecular research into NLRP3 inflammatory activation of mitochondrial dysfunction, mtROS, and mtDNA could be pharmacological helpful [79]. The mechanism through which mitochondrion activates NLRP3 in the cellular pathway of bovine endometrium during prolonged inflammatory reaction needs further verification; probably it could give an insight to immunotherapeutic against incessant dairy cow endometritis.NLRP5 was expressed in mice autoantigen-specific oocyte linked with autoimmune premature ovarian failure in an infectious reproductive tract [80]. The intracellular NLRP5 in oocyte mitochondria and nucleoli near the nuclear pores show cytoplasmic and nuclear functions [79]. Decrease or absence of NLRP5 in women of advanced reproductive age result in impaired fertility and mediated mitochondrion function in mouse's oocytes and embryo and localization of this protein in the reproductive cells, potentiating increase ROS production, depressed the cell morphology and physiology [81], but the molecular role of this inflammasome still call for concern in the molecular pathophysiology of bovine endometritis with addition to the NLRP6 inflammasomes. Likewise, the NLRP6 inflammasome was reported to promote the production and repair of intestinal epithelial cells of IL-18 as a response to inflammatory-induced intestinal wounds [82,83]. The NLRP6 is expressed in macrophages to reduce intestinal inflammation caused by bacteria to induce inflammatory cytokines expression [84,85]. Transcripts of NLRP7 were found in various human tissues, including endometrium, placenta, hematopoietic cells, all oocyte levels, and preimplantation embryos [86,87]. NLRP7 acts in chromatin reprogramming and DNA methylation across germline and premature quiescent before inflammatory response, leading to high cytokine production levels due to chorionic gonadotropin hormone related to abortion [88]. A network of biomolecules and sensors controls activation of NLRs inflammasomes in pathological conditions within NF–κBand MAPK signaling during oxidative stress and inflammatory stimuli [72]. The identified NLRs pathomechanism functions are poorly understood in innate immunity and the transcriptional factors regulation. The precise molecular mechanisms of some NLRs in bovine endometrial cells are unknown. Several reports have it that reception of PAMPs signals by cell receptors may be transported onto the cell signaling pathway depending on the immuno-inflammatory potency of the cells and their environment.

4. Evaluation of MAPKSignaling Transduction Pathways in Endometritis Pathogenesis

Mitogen-activated protein kinase (MAPK) pathways are implicated in several cellular processes, including proliferation, differentiation, apoptosis, cell survival, cell motility, metabolism, stress response, and inflammation. The conscious reflection of existing knowledge about cellular mechanisms adopted by pathogens and their PAMPs to target the host's cellular MAPK signaling pathways; specialized cell receptors and highjack the immune response in dairy cow endometrium, in the manner of promoting enabling parasite maintenance in the host to induce clinical/ subclinical endometritis [52,89–91]. The MAPKs are signaling cascades that include various extracellular stimuli when the inflammatory response is initiated, including the development of pro-inflammatory cytokines and their substrates [90,91]. Seven MAPK families, ERK1/2, ERK3/4, ERK5, ERK7/8, NLK, C-JUN, and p38 groups, have been documented in mammalian cells [92,93]. There are two subgroups; the classical MAPKs of ERK1/2, p38, JNK, ERK5, and the atypical MAPKs of ERK3, ERK4, ERK7, and NLK [94], and are independent of or interacting with each other. Three well-known MAPK pathways are the ERK1/2, JNK1/2/3, p38 MAPK, and their isoforms, which are grouped based on their activation motif, structure, and function [95,96]. ERK1/2 is induced in response to growth factors, hormones, and pro-inflammatory stimuli, whereas JNK1/2/3 and p38 MAPKs are activated by cell environmental stresses and resultant inflammatory processes [96–98]. The atypical ERK remains elusive in cellular activities and disease reactions. ERK3 and ERK4 are predominant in the gastrointestinal tract and colon, respectively [99]. The elevated phosphorylation of MAPK proteins (ERK1/2, p38, and JNK) in LPS-induced endometritis with upregulated expression of pro-inflammatory cytokines and chemokines was reported [92]. MAPK phosphatases such as MKP-M targets JNK primarily, while others have a broader range, such as MKP-1 acting on most MAPKs [100]. ERK1/2 is the most studied in physiological tissues and cells, inhibiting the MAP2 K, MEK, phosphorylating the threonine and tyrosine residues in the TEY domain ERK1/2 [101]. The MAPKs share a similar organizational structure, but extracellular stress factors mainly regulate p38MAPK and JNK. ERK is preferably a target for mitogenic stimuli. The JNKs were characterized first because of their activation in response to various extracellular pressures and phosphorylate N—terminal transactivation domain transcription factors and cell regulatory in response to protein synthesis suppression [98–100]. The JNK1 and 2 are ubiquitous; JNK3 is confined to the brain. JNKs regulation is complex and influenced by many MKKs, with several MAPKKs activated, which are the same as p38, the TAK1, MEKK1/4, and ASK1. They can, however, activate JNK to phosphorylate MEK4 or 7 through the MAPKKs, which are MAPKKs specific to JNK [92,97,99] p38MAPK comprises four protein isoforms termed α, β, γ, and δ [98,102]. The p38δ is expressed in lung, kidney, testis, pancreas, other reproductive organs, and small intestine, p38γ is expressed in skeletal muscle and, p38α, and p38β are ubiquitously expressed [103,104]. Stimulating cells with LPS evidenced the expression of alpha-isoform of p38MAPK involved in the synthesis of pro-inflammatory cytokines. The action of cellular receptor and sequence leading to transduction of cellular signaling remains unverified. The p38β shared 50%, 63%, and 57% structural homology with p38α, p38γ, and p38δMAPK respectively.

All p38MAPK isoforms reportedly shared activation cascade with not exclusively upstream kinases targeting common molecular substrate to validate the transcripts. There are also differences among the isoforms concerning their mode of activation, regulation and inhibition, substrate specificity, and molecular cell function, which underlie differences in the expression pattern of p38MAPK isoforms in different tissue and organs of the body implicated in different infectious disease occurrence [90]. It was reported that p38αMAPK specifically induces the synthesis of proteases, which are essential substrates in the inflammatory process. Hence, there is a need to explore the mechanism of the protease's biosynthesis as it functions in regulating molecular and cellular processes [105]. However, the molecular target of the pyridinyl imidazole family of compounds suppressing inflammatory cytokine biosynthesis was p38αMAPK [106]. Induction of several other inflammatory molecules such as COX2 and inducible nitric oxide synthase [iNOS] was implicated in the p38 MAPK pathway [107,108]. MAPK cell signaling cascades have been reported to transduce endotoxin impulse in the cell's nucleus, leading to activation of several cell nuclear factors to stimulate genetic and inflammatory changes.

5. Component of Nuclear Factor Kappa Beta [NF–κB] Activation in Endometrium Inflammatory Response

NF–κB transcripts were reported in many inflammatory diseases due to their ability to activate through specialized assigned cell receptors reactivity for bacteria and their ligands invasion to modulate the expression of upstream pro-inflammatory cytokines and chemokines [109,110]. NF–kB signaling nuclear activation stimulates inflammatory reactions in endometritis induced by LPS result in phosphorylation and release of specific NF–kB transcriptional factor [75]. NF–kB regulates genes involved in various immune response processes, including mobilization of innate immune cells, inflammation, maturation of dendritic cells, and lymphocytes' stimulation. In this way, the NF–kB feature can be combined with other signaling pathways and signaling regulated by the prototypical IkB member proteasomal degradation [IkBα] stimulate the production and release of inflammatory cytokines, resulting in prolonged inflammatory responses. NF–kB canonical and atypical [non-canonical] pathways regulate proteolysis of IkB [an NF–kB inhibitor], and IkB related proteins by the mechanisms which control the transcription of inflammatory genes at the same time [111–113]. Signal transductions are triggered by the attachment of bacteria ligands to the cell leading to the activation of the two catalytic subunits [IKK1 and IKK2] and the NEMO regulatory subunit IKK complex [75,114]. Activated IKK phosphorylates IkBα, initiating its lysine-48-linked polyubiquitination and proteasomal degradation primarily via the action of IKK2, enabling associated NF–kB subunits to translocate into the nucleus. The NF–kB cloning consists of NF-kB1 [p50 and its ancestor p105], NF-kB2 [p52 and its ancestor p100], RelA [p65], RelC, and RelB, all of which are distinguished by a possibility of an N-terminal Rel homology stratum [RHD] mandatory for homo and hetero-dimerization in sequence-specific DNA boundaries. Non-canonical NF–kB channels, including p105 and p100, also have the crucial immuno-inflammatory potential of the transcriptional signaling pathways [75,112,115]. A subset of TNF family members and triggers of NF–kB activating kinase activated the p100-mediated pathway [NIK] and IKK1 in sequence. IKK1 [IKKα] phosphorylates p100, initiating its poly-ubiquitination and proteolysis through the proteasome to create p52, which translocated distinctly into the nucleus RelB, activating the alternate NF–kB signaling pathway [111,116]. The triggered NF–κBsubunit p65 dissociated from its intracellular protein IκB-α and translocated from the cytoplasm to the nucleus, where specific target genes are transcribed [117].

IKKα and IKKβ are intracellular protein kinases with high sequence relation, often influenced IKB protein phosphorylation, and serves as convergence points for some transduction pathways leading to NF–KB activation. IKKβ is necessary for the rapid activation of NF–kB by pro-inflammatory signaling cascades. Stimulation of cells contributes to phosphorylation of IKKβ, provides a signal identification for reactive enzymes that marks a rapid proteasomal degradation of IkBs. For an effective immune response, prompt activation of NF–kB is needed, but this response cannot last indefinitely and must be correctly terminated to prevent tissue damage. IKKα/NEMO, a 48 kDa regulatory subunit at its N-terminus with a kinase-binding domain and ubiquitous domain at its C-end [37,118]. IKKγ/NEMO serves as a scaffolding or adapter feature in several signal pathways, and the pathophysiological effects of its defects were seen in reproductive failure in mice [116]. IkBs degradation leading NF–kB dimers activated inside the nucleus to stimulate transcription of the target gene [117].

The IKKβ-dependent pathway is essential for natural immunity activation. The IKKα dependent pathway played an important role in adaptive immunity regulation and lymphoid organogenesis of domestic animals. IKKα attenuated signals by the IKKβ-dependent pathway. IKKβ is most critical for the rapid degradation of NF–kB bound IkBs; IKKα enhanced p100 synthesis, contributing to p52 activation: RelB dimers [117,119]. IKKβ can also engage in a negative feedback loop, downregulating the signaling pathways contributing to its activation. The IKKβ-dependent pathway, known as the classical NF–kB pathway to process p100 and the activation of p52, while RelB is known as an alternative NF–kB pathway [119]. IKKγ is essential for binding, catalytic subunits to upstream activation in the classical NF–κBpathway. IKKα phosphorylation is not essential for most pro-inflammatory response triggers to activate the classical IKK complex or NF–kB activation [120,121]. The molecular activation of NF–κBand its component biomolecules stimulate increased pro-inflammatory cytokines and other genetic factors in the cell nucleus.

6. Inflammatory Cytokines and Chemokines in Bovine Endometritis

Shreds of evidence exist that cytokines (including chemokines) are essential in a pregnancy-dependent manner by the bovine endometrium [43,66,122–124]. The cytokines and chemokines are key mediators of innate immunity, necessary for the clearance of this infection and resolution of inflammation. They are also necessary for the postpartum endometrial involution and antimicrobial peptides [AMPs] [44–47]. Cytokine reportedly functioned in cell apoptosis in which they exhibit enormous stimulatory, synergetic, and redundancy interaction among cells. Chemokines were thought to be surfactant cytokines with various biological events in controlling leukocyte trafficking, immune response, cell migration, and growth factors [16,110,123]. The uterus is an immunocompetent site usually poised at defeating microbial infections as soon as they are established during postpartum. Still, its clearance capacity may wear off, causing prolonged or excessive inflammatory response [125,126]. Lee, 2020 [127] reported that AP-1 and IRF-3 in the cell nucleus are considered pivotal factors in regulating inflammation by producing pro-inflammatory mediators and cytokines. The bovine endometrium was reported to express a wide range of immune factors, including interleukin [IL] 1β-2, -6, -8, and -10, and tumor necrosis factors. They also involved cytokines in triggering an immune reaction, acute inflammatory events, and chronic inflammatory changes or changes in uterine flush from cows with endometritis [43,126]. Several scientists have asserted that pro-inflammatory cytokines are necessary for the maturation and disintegration of the follicle, ovulatory process, and corpus luteum development. This evidence might offer an insight into their roles in the pathomechanism of infertility because of subclinical endometritis [40,43,47,126]. Sometimes the pro-inflammatory cytokines are generated and triggered by the association of macrophages with the endotoxin microorganism. Often, they perform important roles in inflammatory disease pathophysiology, including bovine endometritis. The TNF-α is the earliest endogenous mediator of inflammatory responses. IL-1 was the host's central and secretive mediator of the inflammatory immune response to infections when the endometrial inflammatory response increased [128,129].

IL1 upregulated expression in cows with clinical endometritis seven days postpartum and maintained IL1α and IL1β in cows to 21 days postpartum. Il-1β plays a crucial role in the local and systemic stages of inflammatory reaction. Its early upregulation enhanced and increased endogenous activation of other inflammatory mediators and contributor to cytological endometrial pathophysiology, causing significant tissue damage and septic observations [130]. IL6 is mainly important in inflammatory acute-section reactions. After seven days of parturition, IL6 levels were reliably elevated in apparently healthy cows, cows with clinical endometritis, the cow with dystocia [131], and cervicovaginal mucus [132]. The inflammatory endometrial response elevation of IL-6 is normal to ensure uterine involution [133]. Chemokines improve trophoblast cell migration inhibition and invasion during placental treatment. LPS-stimulated bovine endometritis induced an elevated concentration of chemokines [36].

CXCL8 and CXCL5 function immunologically in uterine tissue infection's physiological and pathological processes and endometrial cell inflammation [134]. CXCL1 and CXCL6 aid in neutrophil activation, which is associated with mechanisms of inflammation, which apoptosis. In human and canine endometriosis pathogenesis, it was observed that CCL5, CXCL5, and GRO1 chemokines play a role. However, there is a paucity of research information on the pathomechanism of receptors and signal pathways through which activation or upregulation of chemokines occur. Effective immunotherapy needs to be developed to curb incessant cytokine/chemokine-related pathology in subclinical endometritis [135]. The definite connection between cytokine levels and endometrial infectious pathogenesis requires uncompromising research to provide a way out for immunomodulatory treatment and prevention against subclinical endometritis [136].

7. Application of Nanotherapeutic in Reproductive Diseases

Due to recent developments in terms of excessive resistance and abuse of antibiotics, the effect in the reproductive tracts of production animals has signaled the discovery of the concept of nanomedicine in livestock development [137]. In the area of infection control of veterinary medicine, nanotechnology has a promising role in preventing or treating infections [137–139]. Nanoparticles may present a feasible alternative to antibiotics and help bar pathogens from entering the endometrium at postpartum. More focus has been placed on organic nanoparticles' use in reproductive medicine due to their biocompatibility and biodegradability [140]. Organic nanoparticles can either encapsulate the drug inside or integrate the drug on the nanoparticle's surface [141]. A major advantage with nanotherapy is the ability to perform surface modification, conjugating the nanoparticle to targeting peptides or antibodies [142]. There have been many studies about the antimicrobial activity of polymeric materials. Nanoscale systems could contribute greatly to improving innovative therapies for infectious reproduction diseases because of their tunable size and increased suspendibility and surface tailorability, which enhances interactions with biological systems at the molecular level [139,143]. For example, studies have shown that silver ions can make structural changes in the cell membrane [144]. Silver has a high affinity for negatively charged side groups on biological molecules such as sulfhydryl, carboxyl, phosphate, and other charged groups distributed throughout microbial cells. Silver ions inhibit several enzymatic activities by reacting with electron donor groups, especially sulfhydryl groups [145,146]. Silver ions induce the inactivation of critical physiological functions such as cell wall synthesis, membrane transport, nucleic acid (RNA and DNA) synthesis, translation, protein folding and function, and electron transport [144,146]. Kim et al. [147] found that silver nanoparticles could inhibit the growth of hemorrhagic enteritis-inciting *E. coli* O157:H7 and yeast isolated from a case of bovine mastitis. Iron oxide nanoparticles have previously been used as a thermal ablation treatment [148], and it has been suggested that they could form the future basis for hyperthermia-related endometriosis treatment. Cerbu et al. 2021 [149] sought to control the release of tilmicosin by using hydrogenated castor oil–solid lipid nanoparticle carriers in mastitis.

Poly (ε-caprolactone) (PCL) lipid-core nanocapsules (LNC) are a promising drug/bioactive compound carrier with a high potential for biomedical applications due to their biodegradability and biocompatibility characteristics. LNC is a vesicular carrier comprising a core structured by a dispersion of solid lipids (sorbitan monostearate) and liquid lipids (caprylic/capric triglyceride), creating a PCL-surrounded polymeric wall [150].

In 2013, Tu'uhevaha et al. [151] reported that EGFR-targeted EnGeneIC Delivery Vehicles (EDVs) loaded with doxorubicin significantly inhibited trophoblastic tumor cell growth in vivo and in vitro and induced significant cell death ex vivo, potentially mediated by increasing apoptosis and decreasing proliferation. EDVs may be a novel nanoparticle treatment for ectopic pregnancy and other disorders of trophoblast growth to militate against Ectopic pregnancy.

Polymeric biodegradable form of Engineered Nanoparticles (pbENPs) has been proposed as effective platforms for the protection and controlled release of reproductive hormones, including steroid or gonadotropic hormones [152]. A previous study of chitosan nanoparticles on hCG (Human Chorionic Gonadotrophin) hormone increased dairy cattle ovulation induction [153]. There was the dissolution of endometrial tumor cell in endometrial cancer in a polymeric nanoparticle (NP) delivery system, which improves efficacy and safety of the combinatorial strategy [154] and also the development of a targeted drug delivery system for the uterus utilizing an immunoliposome platform targeting the oxytocin receptor leading to targeted liposomal drug delivery to the myometrium is reduced dose and reduced toxicity to both mother and fetus [155]. Likewise, Hassanein et al. 2021 [156] reported the fabrication of Gonadotropin-releasing hormone (GnRH)–loaded–Chitosan Nanoparticles could allow a reduction in the conventional intramuscular GnRH dose used for AI in rabbits to half without affecting fertility. The combinatorial effect, poly(lactic-co-glycolic) acid nanoparticles loaded with epigallocatechin gallate(EGCG) and doxycycline (Dox) in a single vehicle appears to be promising for treating endometriosis. The significant decreases in endometrial glands and microvessel density in the Dox-EGCG NP-treated group compared to the groups treated with Dox NPs and EGCG NPS confirmed the increased efficacy of Dox-EGCG NPs compared to the single drug-loaded nanoparticles [157]. Thus far, the literature examining nanoparticles and nanotherapy in subclinical endometritis has been sparse, calling for research concerns.

8. Conclusions and Perspectives

The purpose is to treat, control and prevent lingering subclinical endometritis, which has been a significant bottleneck to the development of the dairy industry. The roles of Gram-positive bacteria and their ligands potentials in the pathomechanism of bovine endometritis need further exploration. The cell receptor and their biomolecular component were the primary cell wall receptors point of this review and found crucial research loopholes in the viability of cell receptors that may be distinct for multifaceted research development. Several TLRs have been reported; TLR4 is the most evaluated with little or no record on the molecular activities of other TLRs in the pathogenesis of endometritis. The possibility of heterodimerization within TLRs was reported between TLR2/4/6 exhaust much interaction between the cell receptors, which may be a good clue in drug discovery. TLRs cytosolic adaptor proteins such as MyD88, IRAK1/2/4, TRAF6, TIRAP, TRAM, and SARM, are essential in the inactivation and potentiation of TLRs during stimulation by bacteria PAMPs. NLRs are specialized intracellular or cytoplasmic receptors, enhanced the innate immune function of bovine endometrium in early postpartum, but their function was compromised due to massive invasion of microorganism. The NLRP inflammasomes also serve as the hallmark of inflammation in immune cells, and there are several kinds of it; only NLRP3 inflammasome was bit explored in the molecular pathogenesis of bovine endometritis.

Other NLRs need to be investigated [52,84]. The mode of activation, regulation, and inhibition, substrate specificity, and molecular cell function underlies the difference in the expression pattern of ERK, p38MAPK, and C-JNK isoforms in bovine endometrium need further molecular and immunological characterization [158]. The MAPKinhibitors are promising therapeutic agents in inflammatory diseases in which bacterial products and pro-inflammatory cytokines play a critical role in their pathogenesis. The neglect of some NF–kB protein kinases and their signaling subunits in evaluating and regulating serial cascades in the molecular pathogenesis of dairy cow postpartum endometritis call for concern [159]. The therapeutic invention of NF–kB and IKK inhibitors for treating inflammation and cancers, some of which are under clinical trials (Figure 3). The cellular physiological role of IKK-induced p105 needs to be established in the proteolysis of NF–kB activation in the pathogenesis of bovine endometritis [117].

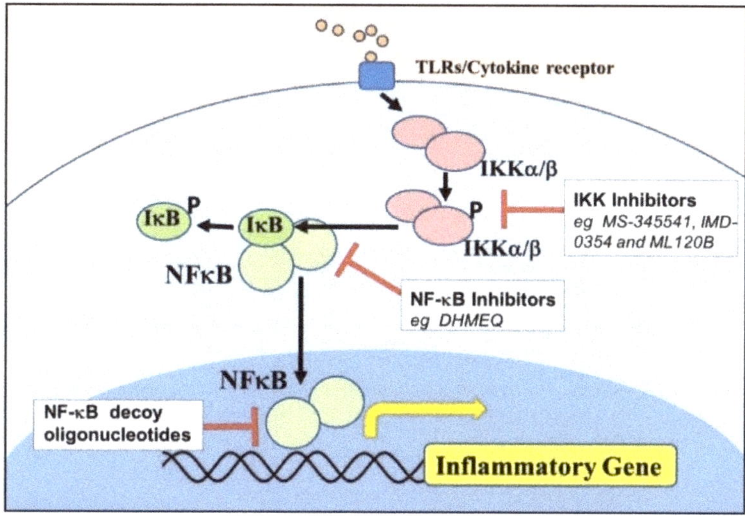

Figure 3. Schematic diagram narrating the potential cascade of NF–κBand IKKαβ inhibitors to protect inflammatory gene.Table 88. Myeloid Differentiation factor 88.

Numerous studies have demonstrated that cell biomolecules are good delivery platforms that increase the uptake of antigens and adjuvant, leading to better immune responses [160–162]. Understanding the molecular and cellular mechanism underlying the pathogenesis of postpartum endometritis may help in immunotherapies or nanotherapeutic discoveries for preventing and regulatingthe menace of dairy cow uterine disease. With the strong demand to develop alternative therapeutic options to address unrealized therapeutic needs, novel nanotechnology-based platforms have recently provided an important baseline for cell health and protection.The development of nano-based systems has provided protection strategies for incorporated agents, such as biomolecules—nucleic acids, peptides, and proteins which are generally quickly degraded when administered in vivo.

The development of nano-based systems based on cell particles has also been described as platforms for targeting and delivering therapeutic agents and nanodevices and analytical systems for theragnostics. The range of applications of nanosystems can include drug delivery, cancer, gene therapy, and imaging and cell tracking through biomarkers and biosensors [163], thereby allowing for the use of prophylactic measures to avoid the progress of the disease or to greater efficacy of therapies due to an earlier treatment [162]. Therapeutic agents can be embedded, encapsulated, or even adsorbed or conjugated onto the nanosystems, which can be modified and associated with other cell biomolecules to achieve an optimized release profile [164,165]. Future in-depth research into the series of cell signaling, sensors, effectors, and biomolecules of the bovine endometrium that activate the pathomechanisms of infection is essential in developing an alternative nanotechnology-based therapy against endometritis in dairy cows, as it was recently found to cure some diseases. During the writing of the manuscript, more than 308 published articles were reviewed. In the 144 articles relevant to the molecular exploration of the pathogenesis of bovine endometritis, nanotherapeutic discoveries were evaluated. The articles published in the last and these decades were selected for the write-up. The journals selected were science citation index journals with good impact factors. Most articles have wide views by different researchers, which means most the article is within the researchers' reach.

Author Contributions: Conceptualization, A.O.O. and Z.Y.; methodology, A.O.O., Y.L., S.W., B.H.I.; investigation, X.M., A.O.O., J.Y.; resources, A.O.O., B.H.I., Z.Y.; writing—original draft preparation, A.O.O.; writing—review and editing, Z.Y., S.W., X.W.; supervision, Z.Y.; project administration, Z.Y., A.O.O., S.W.; funding acquisition, Z.Y. All authors have read and agreed to the published version of the manuscript.

Funding: The research was financed by the National Key Research and Development Program of China (2017YFD0502200) and the Science and Technology Innovation Project of CAAS (CAAS-ASTIP-2014LIHPS-03).

Institutional Review Board Statement: Not applicable.

Informed Consent Statement: Not applicable.

Data Availability Statement: Not applicable.

Acknowledgments: The authors acknowledge that the funders of this researcher work for the provision of financial needs.

Conflicts of Interest: The authors declare they have no competing interests.

Abbreviations

PAMP	Pathogen Associated Membrane Protein
LPS	Lipopolysaccharide
IL-β	Interleukin Beta
TNFα	Tissue Necrosis Factor-alpha
TLR	Toll-like Receptor
RLR	Rig-like receptor
NLR	NOD Like Receptor
MyD88	Myeloid Differentiation factor 88
Mal	MyD88 associated ligand

TIRAP	TIR domain-containing adaptor protein
TRAF	tumor necrosis factor receptor (TNFR)-associated factor (TRAF)
TRIF	TIR-domain-containing adapter-inducing interferon-β
TRAM	TRIF-related adaptor molecule
SARM	Sterile alpha and TIR motif-containing
NOD	Nucleotide-binding Oligomerization Domains
NLRP	NOD Like Receptor Proteins
MAPK	mitogen-activated protein kinase
ERK	extracellular signal-regulated kinases
JNK	c-Jun N-terminal kinases

References

1. de Boer, M.W.; LeBlanc, S.J.; Dubuc, J.; Meier, S.; Heuwieser, W.; Arlt, S.; Gilbert, R.O.; McDougall, S. Invited review: Systematic review of diagnostic tests for reproductive-tract infection and inflammationin dairy cows. *J. Dairy Sci.* **2014**, *97*, 3983–3999. [CrossRef] [PubMed]
2. Gautam, G.; Nakao, T.; Yusuf, M.; Koike, K. Prevalence of endometritis during the postpartum period and its impact on subsequent reproductive performance in two Japanese dairy herds. *Anim. Reprod. Sci.* **2009**, *116*, 175–187. [CrossRef] [PubMed]
3. Gautam, G.; Nakao, T.; Koike, K.; Long, S.T.; Yusuf, M.; Ranasinghe, R.M.; Hayashi, A. Spontaneous recovery or persistence of postpartum endometritis and risk factors for its persistence in Holstein cows. *Theriogenology* **2010**, *73*, 168–179. [CrossRef] [PubMed]
4. Plöntzke, J.; Madoz, L.W.; De la Sota, R.L.; Heuwieser, W.; Drillich, M. Prevalence of clinical endometritis and its impact on reproductive performance in grazing dairy cattle in Argentina. *Reprod. Domest. Anim.* **2011**, *46*, 520–526. [CrossRef] [PubMed]
5. Leblanc, S.J.; Duffield, T.F.; Leslie, K.E.; Bateman, K.G.; Keefe, G.P.; Walton, J.S.; Johnson, W.H. Defining and Diagnosing Postpartum Clinical Endometritis and its Impact on Reproductive Performance in Dairy Cows. *J. Dairy Sci.* **2002**, *85*, 2223–2236. [CrossRef]
6. Okawa, H.; Fujikura, A.; Wijayagunawardane, M.M.P.; Vos, P.L.A.M.; Taniguchi, M.; Takagi, M. Effect of diagnosis and treatment of clinical endometritis based on vaginal discharge score grading system in postpartum Holstein cows. *J. Vet. Med. Sci.* **2017**, *79*, 1545–1551. [CrossRef]
7. Kumar, D.; Satish, S.; Purohit, G.N. A Discussion on Risk Factors, Therapeutic Approach of Endometritis and Metritis in Cattle. *Int. J.Curr. Microbiol. Appl. Sci.* **2019**, *8*, 403–421. [CrossRef]
8. Barański, W.; Podhalicz-Dziegielewska, M.; Zduńczyk, S.; Janowski, T. The diagnosis and prevalence of subclinical endometritis in cows evaluated by different cytologic thresholds. *Theriogenology* **2012**, *78*, 1939–1947. [CrossRef]
9. Dubuc, J.; Duffield, T.F.; Leslie, K.E.; Walton, J.S.; LeBlanc, S.J. Definitions and diagnosis of postpartum endometritis in dairy cows. *J. Dairy Sci.* **2010**, *93*, 5225–5233. [CrossRef]
10. Gilbert, R.O.; Shin, S.T.; Guard, C.L.; Erb, H.N.; Frajblat, M. Prevalence of endometritis and its effects on reproductive performance of dairy cows. *Theriogenology* **2005**, *64*, 1879–1888. [CrossRef] [PubMed]
11. Sheldon, I.M.; Rycroft, A.N.; Dogan, B.; Craven, M.; Bromfield, J.J.; Chandler, A.; Simpson, K.W. Specific strains of *Escherichia coli* are pathogenic for the endometrium of cattle and cause pelvic inflammatory disease in cattle and mice. *PLoS ONE* **2010**, *5*, e9192. [CrossRef]
12. Donofrio, G.; Herath, S.; Sartori, C.; Cavirani, S.; Flammini, C.F.; Sheldon, I.M. Bovine herpesvirus 4 is tropic for bovine endometrial cells and modulates endocrine function. *Reproduction* **2007**, *134*, 183–197. [CrossRef] [PubMed]
13. Sheldon, I.M.; Owens, S.E. Postpartum uterine infection and endometritis in dairy cattle. *Anim. Reprod.* **2017**, *14*, 622–629. [CrossRef]
14. Cronin, J.G.; Turner, M.L.; Goetze, L.; Bryant, C.E.; Sheldon, I.M. Toll-Like Receptor 4 and MYD88Dependent Signaling Mechanisms of the Innate Immune System are Essential for the Response to Lipopolysaccharide by Epithelial and Stromal Cells of the Bovine Endometrium. *Biol. Reprod.* **2012**, *86*, 1–9. [CrossRef]
15. Chandler, E.C.; Ernst, R.K. Bacterial lipids: Powerful modifiers of the innate immune response. *F1000Research* **2017**, *6*, 1–11. [CrossRef] [PubMed]
16. Koh, Y.Q.; Mitchell, M.D.; Almughlliq, F.B.; Vaswani, K.; Peiris, H.N. Regulation of inflammatory mediator expression in bovine endometrial cells: Effects of lipopolysaccharide, interleukin1beta, and tumor necrosis factor-alpha. *Physiol. Rep.* **2018**, *6*, e13676. [CrossRef]
17. Harju, K.; Ojaniemi, M.; Rounioja, S.; Glumoff, V.; Paananen, R.; Vuolteenaho, R.; Hallman, M. Expression of toll-like receptor four and endotoxin responsiveness in mice during the perinatal period. *Pediatr. Res.* **2005**, *57*, 644–648. [CrossRef]
18. Oguejiofor, C.F.; Cheng, Z.; Abudureyimu, A.; Fouladi-Nashta, A.A.; Wathes, D.C. Global transcriptomic profiling of bovine endometrial immune response in vitro. I. Effect of lipopolysaccharide on innate immunity. *Biol. Reprod.* **2015**, *93*, 100. [CrossRef] [PubMed]
19. Ryu, Y.H.; Baik, J.E.; Yang, J.S.; Kang, S.S.; Im, J.; Yun, C.H.; Kim, D.W.; Lee, K.; Chung, D.K.; Ju, H.R.; et al. Differential Immunostimulatory effects of Gram-positive bacteria due to their lipoteichoic acids. *Int. Immunopharmacol.* **2009**, *9*, 127–133. [CrossRef]

20. Müller-Anstett, M.A.; Müller, P.; Albrecht, T.; Nega, M.; Wagener, J.; Gao, Q.; Kaesler, S.; Schaller, M.; Biedermann, T.; Götz, F. Staphylococcal peptidoglycan co-localizes with Nod2 and TLR2 and activates innate immune response via both receptors in primary murine keratinocytes. *PLoS ONE* **2010**, *5*, e13153. [CrossRef] [PubMed]

21. Moranta, D.; Regueiro, V.; March, C.; Llobet, E.; Margareto, J.; Larrarte, E.; Garmendia, J.; Bengoechea, J.A. Klebsiella pneumoniae capsule polysaccharide impedes the expression of beta-defensins by airway epithelial cells. *Infect. Immun.* **2010**, *78*, 1135–1146. [CrossRef] [PubMed]

22. Bicalho, M.L.S.; Machado, V.S.; Oikonomou, G.; Gilbert, R.O.; Bicalho, R.C. Association between virulence factors of *Escherichia coli*, *Fusobacterium necrophorum*, and Arcano-bacterium pyogenes and uterine diseases of dairy cows. *Vet. Microbiol.* **2012**, *157*, 125–131. [CrossRef]

23. de Oliveira-Nascimento, L.; Massari, P.; Wetzler, L.M. The role of TLR2 in infection and immunity. *Front. Immunol.* **2012**, *3*, 79. [CrossRef]

24. Seibert, S.A.; Mex, P.; Köhler, A.; Kauf- mann, S.H.; Mittrücker, H.W. TLR2-, TLR4- and Myd88- independent acquired humoral and cellular immunity against *Salmonella enterica* Serovar typhimurium. *Immunol. Lett.* **2010**, *127*, 126–134. [CrossRef]

25. Kawai, T.; Akira, S. The roles of TLRs, RLRs, and NLRs in pathogen recognition. *Int. Immunol.* **2009**, *21*, 317–337. [CrossRef]

26. Akira, S.; Uematsu, S.; Takeuchi, O. Pathogen recognition and innate immunity. *Cell* **2006**, *124*, 783–801. [CrossRef] [PubMed]

27. Heil, F.; Hemmi, H.; Hochrein, H. Species-specific recognition of single-stranded RNA via Toll-like receptor 7 and 8. *Science* **2004**, *303*, 1526–1529. [CrossRef]

28. Lanzarotto, F.; Akbar, A.; Ghosh, S. Does innate immune response defect underlies inflammatory bowel disease in the Asian population? *Postgrad. Med. J.* **2005**, *81*, 483–485. [CrossRef]

29. Williams, E.J.; Sibley, K.; Miller, A.N.; Lane, E.A.; Fishwick, J.; Nash, D.M. The effect of Escherichia coli lipopolysaccharide and tumor necrosis factor-alpha on ovarian function. *Am. J. Reprod. Immunol.* **2008**, *60*, 462–473. [CrossRef] [PubMed]

30. Wira, C.R.; Grant-Tschudy, K.S.; Crane-Godreau, M.A. Epithelial cells in the female reproductive tract: A central role as sentinels of immune protection. *Am. J.Reprod. Immunol.* **2005**, *53*, 65–76. [CrossRef]

31. Lane, M.C.; Alteri, C.J.; Smith, S.N.; Mobley, H.L. Expression of flagella coincides with uropathogenic *Escherichia coli* ascension to the upper urinary tract. *Proc. Natl. Acad. Sci. USA* **2007**, *104*, 16669–16674. [CrossRef]

32. Leendertse, M.; Willems, R.J.; Giebelen, I.A.; van den Pangaart, P.S.; Wiersinga, W.J.; de Vos, A.F.; Florquin, S.; Bonten, M.J.; van der Poll, T. TLR2- dependent MyD88 signaling contributes to early host defense in murine Enterococcus faecium peritonitis. *J. Immunol.* **2008**, *180*, 4865–4874. [CrossRef]

33. Maldonado, R.F.; Sá-Correia, I.; Valvano, M.A. Lipopolysaccharide modification in gram-negative bacteria during chronic infection. *FEMS Microbiol. Rev.* **2016**, *40*, 480–493. [CrossRef]

34. De Tejada, G.M.; Heinbockel, L.; Ferrer-Espada, R.; Heine, H.; Alexander, C.; Bárcena-Varela, S.; Goldmann, T.; Correa, W.; Wiesmüller, K.H.; Gisch, N.; et al. Lipoproteins/ peptides are sepsis-inducing toxins from bacteria that can be neutralized by synthetic anti-endotoxin peptides. *Sci Rep.* **2015**, *5*, 14292. [CrossRef] [PubMed]

35. Shimizu, T.; Kida, Y.; Kuwano, K. A triacylated lipoprotein from Mycoplasma genitalium activates NF-κB through Toll-like receptor 1 (TLR1) and TLR2. *Infect. Immun.* **2008**, *76*, 3672–3678. [CrossRef]

36. Whelehan, C.J.; Meade, K.G.; Eckersall, P.D.; Young, F.J.; O'Farrelly, C. Experimental Staphylococcus aureus infection of the mammary gland induces region-specific changes in innate immune gene expression. *Vet. Immunol. Immunopathol.* **2011**, *140*, 181–189. [CrossRef]

37. Swangchan-Uthai, T.; Lavender, C.R.; Cheng, Z.; Fouladi-Nashta, A.A.; Wathes, D.C. Time course of defense mechanisms in bovine endometrium in response to lipopolysaccharide. *Biol. Reprod.* **2012**, *87*, 135. [CrossRef]

38. Chau, T.A.; McCully, M.L.; Brintnell, W.; An, G.; Kasper, K.J.; Vines, E.D. Toll-like receptor 2 ligands on the staphylococcal cell wall downregulate super antigen-induced T cell activation and prevent toxic shock syndrome. *Nat. Med.* **2009**, *15*, 641–648. [CrossRef] [PubMed]

39. Kumar, H.; Kawai, T.; Akira, S. Pathogen recognition by the innate immune system. *Int. Rev. Immunol.* **2011**, *30*, 16–34. [CrossRef] [PubMed]

40. Kharayat, N.S.; Sharma, G.C.; Kumar, G.R.; Bisht, D.; Chaudhary, G.; Singh, S.K.; Krishnaswamy, N. Differential expression of endometrial toll-like receptors (TLRs) and antimicrobial peptides (AMPs) in the buffalo (Bubalusbubalis) with endometritis. *Vet. Res. Commun.* **2019**, *43*, 261–269. [CrossRef]

41. Mogensen, T.H. Pathogen Recognition and Inflammatory Signaling in Innate Immune Defenses. *Clin. Microbiol. Rev.* **2009**, *22*, 240–273. [CrossRef]

42. Herath, S.; Fischer, D.P.; Werling, D.; Williams, E.J.; Lilly, S.T.; Dobson, H.; Bryant, C.E.; Sheldon, I.M. Expression and function of Toll-like receptor 4 in the endometrial cells of the uterus. *Endocrinology* **2006**, *147*, 562–570. [CrossRef] [PubMed]

43. Davies, D.; Meade, K.G.; Herath, S.; Eckersall, P.D.; Gonzalez, D.; White, J.O.; Conlan, R.S.; O'Farrelly, C.; Sheldon, I.M. Toll-like receptor and antimicrobial peptide expression in the bovine endometrium. *Reprod. Biol. Endocrinol.* **2008**, *6*, 53. [CrossRef]

44. Ritter, N. Peri-Partal Expression of Endometrial Toll-like Receptors and B-Defensins in Cattle. Ph.D. Thesis, University of Veterinary Medicine Hannover, Hanover, Germany, 2007.

45. Olmos-ortiz, A.; Flores-espinosa, P.; Mancilla-herrera, I.; Vega-Sánchez, R.; Díaz, L.; Zaga-Clavellina, V. Innate Immune Cells and Toll-like Receptor-Dependent Responses at the Maternal-Fetal Interface. *Int. J. Mol. Sci.* **2019**, *20*, 3654. [CrossRef]

46. Jacca, S.; Franceschi, V.; Colagiorgi, A.; Sheldon, M.; Donofrio, G. Bovine Endometrial Stromal Cells Support Tumor Necrosis Factor Alpha-Induced Bovine Herpesvirus Type 4 Enhanced Replication1. *Biol. Reprod.* **2013**, *88*. [CrossRef]

47. Matsumoto, M.; Kikkawa, S.; Kohase, M.; Miyake, K.; Seya, M. Establishment of a monoclonal antibody against human Toll-like receptor three blocks double-stranded RNA-mediated signaling. *Biochem. Biophys. Res. Commun.* **2002**, *293*, 1364–1369. [CrossRef]
48. Beutler, B. Inferences, questions, and possibilities in Toll-like receptor signaling. *Nature* **2004**, *430*, 257–263. [CrossRef]
49. Verstak, B.; Nagpal, K.; Bottomley, S.P.; Golenbock, D.T.; Hertzog, P.J.; Mansell, A. MyD88 adapter like (Mal)/TIRAP interaction with TRAF6 is critical for TLR2- and TLR4-mediated NF-kappaB pro-inflammatory responses. *J. Biol. Chem.* **2009**, *284*, 24192–24203. [CrossRef] [PubMed]
50. Young, O.M.; Tang, Z.; Niven-Fairchild, T.; Tadesse, S.; Krikun, G.; Norwitz, E.R.; Mor, G.; Abrahams, V.M.; Guller, S. Toll-like receptor-mediated responses by placental Hofbauer cells (HBCs): A potential pro-inflammatory role for fetal M2 macrophages. *Am. J. Reprod. Immunol.* **2015**, *73*, 22–35. [CrossRef] [PubMed]
51. Sheldon, I.M.; Cronin, J.G.; Bromfield, J.J. Tolerance and Innate Immunity Shape the Development of Postpartum Uterine Disease and the Impact of Endometritis in Dairy Cattle. *Ann. Rev. Anim. Biosci.* **2019**, *7*, 361–384. [CrossRef] [PubMed]
52. Turner, M.L.; Cronin, J.C.; Healey, G.D.; Sheldon, I.M. Epithelial and stromal cells of bovine endometrium has roles in innate immunity and initiate inflammatory responses bacterial lipopeptides in vitro via Toll-like receptors TLR2, TLR1, and TLR6. *Endocrinology* **2014**, *155*, 1453–1465. [CrossRef] [PubMed]
53. Hasan, U.; Chaffois, C.; Gaillard, C.; Saulnier, V.; Merck, E.; Tancredi, S.; Guiet, C.; Brière, F.; Vlach, J.; Lebecque, S.; et al. Human TLR10 is a functional receptor, expressed by B cells and plasmacytoid dendritic cells, which activates gene transcription through MyD88. *J. Immunol.* **2005**, *174*, 2942–2950. [CrossRef] [PubMed]
54. Stewart, C.R.; Stuart, L.M.; Wilkinson, K.; van Gils, J.M.; Deng, J.; Halle, A.; Rayner, K.J.; Boyer, L.; Zhong, R.; Frazier, W.A.; et al. CD36 ligands promote sterile inflammation through assembly of a toll-like receptor 4 and 6 heterodimers. *Nat. Immunol.* **2010**, *11*, 155–175. [CrossRef] [PubMed]
55. Botos, I.; Segal, D.M.; Davies, D.R. The structural biology of toll-like receptors. *Structure* **2011**, *19*, 447–459. [CrossRef]
56. Sirard, J.; Cecile, V.; RodrigueDessein, M.C. Nod-Like Receptors: Cytosolic Watchdogs forImmunity against Pathogens. *PLoS Pathog* **2007**, *3*, e152. [CrossRef]
57. Inohara, N.; Chamaillard, N.; McDonald, C. Nuñez G NOD-LRR proteins: Role in host-microbial interactions and inflammatory disease. *Annu. Rev. Biochem.* **2005**, *74*, 355–364. [CrossRef]
58. Chamaillard, M.M.; Hashimoto, Y.; Horie, J.; Masumoto, S.; Qiu, L.; Saab, Y.; Ogura, A.; Kawasaki, K.; Fukase, S.; Kusumoto, M.A.; et al. An essential role forNOD1 in host recognition of bacterial peptidoglycan containing diaminopimelic acid. *Nat. Immunol.* **2003**, *4*, 702–707. [CrossRef] [PubMed]
59. Uehara, A.; Yang, S.; Fujimoto, Y.; Fukase, K.; Kusumoto, S.; Shibata, K.; Sugawara, S.; Takada, H. Muramyldipeptide and diaminopimelic acid-containing desuramylpeptides in combination with chemically synthesized Toll-like receptor agonists synergistically induced production of interleukin-8 in a NOD2- and NOD1- dependent manner, respectively, in human monocytic cells in culture. *Cell Microbiol.* **2005**, *7*, 53–61.
60. Brodsky, I.E.; Monack, D. NLR-mediated control of inflammasome assembly in the host response against bacterial pathogens. *Semin. Immunol.* **2009**, *21*, 199–207. [CrossRef]
61. Kanneganti, T.D.; Lamkanfi, M.; Núñez, G. Intracellular NOD-like Receptors in Host Defenseand Disease. *Immunity* **2007**, *27*, 549–559. [CrossRef]
62. Takeuchi, O.; Akira, S. Pattern recognition receptors and inflammation. *Cell* **2010**, *140*, 805–820. [CrossRef] [PubMed]
63. Boughan, P.K.; Argent, R.H.; Body-Malapel, M.; Park, J.H.; Ewings, K.E.; Bowie, A.G.; Ong, S.J.; Cook, S.J.; Sorensen, O.E.; Manzo, B.A. Nucleotide-binding oligomerization domain-1 and epidermal growth factor receptor: Critical regulators of beta-defensins during Helicobacter pylori infection. *J. Biol. Chem.* **2006**, *281*, 11637–11648. [CrossRef] [PubMed]
64. Schroder, K.; Tschopp, J. The inflammasomes. *Cell* **2010**, *140*, 821–832. [CrossRef]
65. Man, S.M.; Kanneganti, T.D. Regulation of inflammasome activation. *Immunol Rev.* **2015**, *265*, 6–21. [CrossRef]
66. Yaribeygi, H.; Katsiki, N.; Butler, A.E.; Sahebkar, A. Effects of antidiabetic drugs on NLRP3inflammasome activity, with a focus on diabetic kidneys. *Drug Discov. Today* **2018**, *24*, 256–262. [CrossRef]
67. Kobayashi, K.S.; Chamaillard, M.; Ogura, Y.; Henegariu, O.; Inohara, N.; Nunez, G.; Flavell, R.A. Nod2-dependent regulation of innate and adaptive immunity in the intestinal tract. *Science* **2005**, *307*, 731–734. [CrossRef] [PubMed]
68. Hu, X.; Li, D.; Wang, J.; Guo, J.; Li, Y.; Cao, Y.; Fu, Y. Melatonin inhibits endoplasmic reticulum stress-associated TXNIP/NLRP3 inflammasome activation in lipopolysaccharide-induced endometritis in mice. *Int. Immunopharmacol.* **2018**, *64*, 101–109. [CrossRef]
69. Vizlin-Hodzic, D.; Zhai, Q.; Illes, S.; Södersten, K.; Truve, K.; Parris, T.Z.; Sobhan, P.K.; Salmela, S.; Kosalai, S.T.; Kanduri, C.; et al. Early-onset of inflammation during ontogeny of bipolar disorder: The NLRP2 inflammasome gene distinctly differentiates between patients and healthy controls in the transition between iPS cell and neural stem cell stages. *Transl. Psychiatry* **2017**, *7*, e1010. [CrossRef]
70. Minkiewicz, J.; de Rivero Vaccari, J.P.; Keane, R.W. Human astrocytes express a novel NLRP2 inflammasome. *Glia* **2013**, *61*, 1113–1121. [CrossRef]
71. Mahadevan, S.; Sathappan, V.; Utama, B.; Lorenzo, I.; Kaskar, K.; Van Den Veyver, I.B. Maternally expressed NLRP2 links the subcortical maternal complex (SCMC) to fertility, embryogenesis, and epigenetic reprogramming. *Sci. Rep.* **2017**, *7*, 1–17. [CrossRef]
72. Place, D.E.; Kanneganti, T.D. Recent advances in inflammasome biology. *Curr.Opin. Immunol.* **2018**, *50*, 32–38. [CrossRef]
73. Sharma, D.; Kanneganti, T.D. The cell biology of inflammasomes: Mechanisms of inflammasome activation and regulation. *J. Cell Biol.* **2016**, *213*, 617–629. [CrossRef]

74. Hoseini, Z.; Sepahvand, F.; Rashidi, B.; Sahebkar, A.; Masoudifar, A.; Mirzaei, H. NLRP3 inflammasome: It's regulation and involvement in atherosclerosis. *J. Cell Physiol.* **2018**, *233*, 2116–2132. [CrossRef] [PubMed]

75. Yang, X.D.; Huang, B.; Li, M.; Lamb, A.; Kelleher, N.L.; Chen, L.F. Negative regulation of NF-kappaB action by Set9-mediated lysine methylation of the RelA subunit. *EMBO J.* **2009**, *28*, 1055–1066. [CrossRef]

76. Lamkanfi, M.; Dixit, V.M. Inflammasomes and their roles in health and disease. *Annu. Rev. Cell Dev. Biol.* **2012**, *28*, 137–161. [CrossRef]

77. Kelley, N.; Jeltema, D.; Duan, Y.; He, Y. The NLRP3 inflammasome: An overview of mechanisms of activation and regulation. *Int. J. Mol. Sci.* **2019**, *20*, 3328. [CrossRef]

78. Nakahira, K.; Haspel, J.A.; Rathinam, V.A.; Lee, S.J.; Dolinay, T.; Lam, H.C.; Englert, J.A.; Rabinovitch, M.; Cernadas, M.; Kim, H.P.; et al. Autophagy proteins regulate innate immune responses by inhibiting the release of mitochondrial DNA mediated by the NALP3 inflammasome. *Nat. Immunol.* **2011**, *12*, 222–230. [CrossRef]

79. Shimada, K.; Crother, T.R.; Karlin, J.; Dagvadorj, J.; Chiba, N.; Chen, S.; Ramanujan, V.K.; Wolf, A.J.; Vergnes, L.; Ojcius, D.M.; et al. Oxidized mitochondrial DNA activates the NLRP3 inflammasome during apoptosis. *Immunity* **2012**, *36*, 401–414. [CrossRef]

80. Fernandes, A.C.C.; Davoodi, S.; Kaur, M.; Veira, D.; Melo, L.E.H.; Cerri, R.L.A. Effect of repeated intravenous lipopolysaccharide infusions on systemic inflammatory response and endometrium gene expression in Holstein heifers. *J. Dairy Sci.* **2019**, *102*, 3531–3543. [CrossRef] [PubMed]

81. Igosheva, N.; Abramov, A.Y.; Poston, L.; Eckert, J.J.; Fleming, T.P.; Duchen, M.R.; McConnell, J. Maternal diet-induced obesity alters mitochondrial activity and redox status in mouse oocytes and zygotes. *PLoS ONE* **2010**, *5*, e10074. [CrossRef]

82. Kim, B.; Kan, R.; Anguish, L.; Nelson, L.M.; Coonrod, S.A. Potential role for MATER in cytoplasmiclattice formation in murine oocytes. *PLoS ONE* **2010**, *5*, e12587. [CrossRef]

83. Levy, M.; Thaiss, C.A.; Zeevi, D.; Dohnalova, L.; Zilberman-Schapira, G.; Mahdi, J.A.; David, E.; Savidor, A.; Korem, T.; Herzig, Y.; et al. Microbiota modulated metabolites shape the intestinal microenvironments by regulating NLRP6 inflammasome signaling. *Cell* **2015**, *163*, 1428–1443. [CrossRef]

84. Chen, G.Y.; Liu, M.; Wang, F.; Bertin, J.; Nunez, G. A functional role for Nlrp6 in intestinalinflammation and tumorigenesis. *J. Immunol.* **2011**, *186*, 7187–7194. [CrossRef]

85. Anand, P.K.; Malireddi, R.K.; Lukens, J.R.; Vogel, P.; Bertin, J.; Lamkanfi, M.; Kanneganti, T.D. NLRP6Negatively regulates innate immunity and host defense against bacterial pathogens. *Nature* **2012**, *488*, 389–393. [CrossRef] [PubMed]

86. Murdoch, S.; Djuric, U.; Mazhar, B.; Seoud, M.; Khan, R.; Kuick, R.; Bagga, R.; Kircheisen, R.; Ao, A.; Ratti, B.; et al. Mutations in NALP7 cause recurrent hydatidiform moles and reproductive wastage in humans. *Nat. Genet.* **2006**, *38*, 300–302. [CrossRef] [PubMed]

87. Akoury, E.; Zhang, L.; Ao, A.; Slim, R. NLRP7 and KHDC3L, the two maternal-effect proteins responsible for recurrent hydatidiform moles, co-localize to the oocyte cytoskeleton. *Hum. Reprod.* **2015**, *30*, 159–169. [CrossRef]

88. Wang, C.M.; Dixon, P.H.; Decordova, S.; Hodges, M.D.; Sebire, N.J.; Ozalp, S.; Fallahian, M.; Sensi, A.; Ashrafi, F.; Repiska, V.; et al. Identification of 13 novel NLRP7mutations in 20 families with recurrent hydatidiform mole; missense mutations cluster in the leucine-rich region. *J. Med. Genet.* **2009**, *46*, 569–575. [CrossRef]

89. Loyi, T.; Kumar, H.; Nandi, S.; Patra, M.K. Expression of pathogen recognition receptors and pro-inflammatory cytokine transcripts in clinical and sub-clinical endometritis cows. *Anim. Biotechnol.* **2015**, *26*, 194–200. [CrossRef]

90. Miller, B.A.; Brewer, A.; Nanni, P.; Lim, J.J.; Callanan, J.J.; Grossmann, J.; Chapwanya, A. Characterization of circulating plasma proteins in dairy cows with cytological endometritis. *J. Proteom.* **2019**, *205*, 103421. [CrossRef] [PubMed]

91. Peti, W.; Page, R. Molecular basis of MAP kinase regulation. *Protein Sci.* **2013**, *22*, 1698–1710. [CrossRef]

92. Fang, L.; Wu, H.M.; Ding, P.S.; Liu, R.Y. TLR2 mediates phagocytosis and autophagy through JNKsignaling pathway in Staphylococcus aureus-stimulated RAW264.7 cells. *Cell. Signal.* **2014**, *26*, 806–814. [CrossRef]

93. Siljamaki, E.; Raiko, L.; Toriseva, M. p38delta mitogen-activated protein kinase regulates the expression of tight junction protein ZO-1 in differentiating human epidermal keratinocytes. *Arch. Dermatol. Res.* **2014**, *306*, 131–141. [CrossRef] [PubMed]

94. Coulombe, P.; Meloche, S. Atypical mitogen-activated protein kinases: Structure, regulation, andfunctions. *Biochim. Biophys. ActaMol. Cell Res.* **2007**, *1773*, 1376–1387. [CrossRef] [PubMed]

95. Gaestel, M.; Kotlyarov, A.; Kracht, M. Targeting innate immunity protein kinase signaling ininflammation. *Nat. Rev. Drug Discov.* **2009**, *8*, 480–499. [CrossRef] [PubMed]

96. Zlobin, A.; Bloodworth, J.C.; Osipo, C. Mitogen-Activated Protein Kinase (MAPK) Signaling BT. In *Predictive Biomarkers in Oncology: Applications in Precision Medicine*; Badve, S., Kumar, G.L., Eds.; Springer: Cham, Switzerland, 2019. [CrossRef]

97. Thalhamer, T.; McGrath, M.A.; Harnett, M.M. MAPKs and their relevance to arthritis and inflammation. *Rheumatology* **2008**, *47*, 409–414. [CrossRef] [PubMed]

98. Escós, A.; Risco, A.; Alsina-Beauchamp, D.; Cuenda, A. p38 γ and p38 δ Mitogen Activated ProteinKinases (MAPKs), New Stars in the MAPK Galaxy. *Front. Cell DevBiol.* **2016**, *4*, 31. [CrossRef]

99. Ittner, A.; Block, H.; Reichel, C.A.; Varjosalo, M.; Gehart, H.; Sumara, G.; Gstaiger, M.; Krombach, F.; Zarbock, A.; Ricci, R. Regulation of PTEN activity by p38delta-PKD1 signaling in neutrophils confer inflammatory responses in the lung. *J. Exp. Med.* **2012**, *209*, 2229–2246. [CrossRef]

100. Kourea, H.P.; Nikolaou, M.; Tzelepi, V. Expression of phosphorylated Akt, mTOR, and MAPK intype I endometrial carcinoma: Clinical significance. *Anticancer. Res.* **2015**, *35*, 2321–2331.

101. Knight, T.; Irving, J.A.E. Ras/Raf/MEK/ERK pathway activation in childhood acute lymphoblasticleukemia and its therapeutic targeting. *Front. Oncol.* **2014**, *4*, 160. [CrossRef]

102. Remy, G.; Risco, A.M.; Inesta-Vaquera, F.A.; González-Terán, B.; Sabio, G.; Davis, R.J.; Cuenda, A. Differential activation of p38MAPK isoformsby MKK6 and MKK3. *Cell. Signal.* **2010**, *22*, 660–667. [CrossRef]

103. Slattery, M.L.; Mullany, L.E.; Sakoda, L.C.; Wolff, R.K.; Samowitz, W.S.; Herrick, J.S. The MAPK signaling pathway in colorectal cancer: Dysregulated genes and their association with microRNAs. *Cancer Inform.* **2018**, *17*. [CrossRef] [PubMed]

104. Court, N.W.; dos Remedios, C.G.; Cordell, J.; Bogoyevitch, M.A. Cardiac Expression and Subcellular Localization of the p38 Mitogen-activated Protein Kinase Member, Stress-activated Protein Kinase-3(SAPK3). *J. Mol. Cell. Cardiol.* **2002**, *34*, 413–426. [CrossRef]

105. Del Reino, P.; Alsina-Beauchamp, D.; Escós, A.; Cerezo-Guisado, M.I.; Risco, A.; Aparicio, N. Prooncogenic role of alternative p38 mitogen-activated protein kinases p38gamma and p38delta, linking inflammation and cancer in colitis-associated colon cancer. *Cancer Res.* **2014**, *74*, 6150–6160. [CrossRef]

106. Kumar, S.; Boehm, J.; Lee, J. p38 MAP kinases: Key signaling molecules as therapeutic targets for inflammatory diseases. *Nat. Rev. Drug Discov.* **2003**, *2*, 717–726. [CrossRef]

107. Yoon, W.J.; Lee, N.H.; Hyun, C.G. Limonene suppresses lipopolysaccharide-induced production of nitric oxide, prostaglandin E2, and pro-inflammatory cytokines in RAW 264.7 macrophages. *J. Oleo Sci.* **2010**, *59*, 415–421. [CrossRef] [PubMed]

108. Ventura, J.J.; Tenbaum, S.; Perdiguero, E. p38α MAPkinase is essential in lung stem and progenitor cell proliferation and differentiation. *Nat. Genet.* **2007**, *39*, 750–758. [CrossRef] [PubMed]

109. Gonzalez-Ramos, R.; Van Langendonckt, A.; Defriere, S.; Lousse, J.C.; Colette, S.; Devoto, L. Involvement of the nuclear factor-kB pathway in the pathogenesis of endometriosis. *Fertil. Steril.* **2010**, *94*, 1985–1994. [CrossRef]

110. Karin, M.; Greten, F.R. NF-κB: Linking inflammation and immunity to cancer development and progression. *Nat. Rev. Immunol.* **2005**, *5*, 749–759. [CrossRef]

111. Shen, Y.; Liu, B.; Mao, W.; Gao, R.; Feng, S.; Qian, Y.; Cao, J. PGE2 downregulates LPS-induced inflammatory responses via the TLR4-NF-κB signaling pathway in bovine endometrial epithelial cells. *Prostaglandins Leukot. Essent. Fat. Acids* **2018**, *129*, 25–31. [CrossRef]

112. Huang, B.; Yang, X.D.; Lamb, A.; Chen, L.F. Posttranslational modifications of NF- kappaB: Another Layer of regulation for NF-kB signaling pathway. *Cell. Signal.* **2010**, *22*, 1282–1290. [CrossRef]

113. Bhatt, D.; Ghosh, S. Regulation of the NF-κB-mediated transcription of inflammatory genes. *Front. Immunol.* **2014**, *5*, 71. [CrossRef]

114. Hara, H.; Bakal, C.; Wada, T.; Bouchard, D.; Rottapel, R.; Saito, T.; Penninger, J.M. The Molecular Adapter Carma1 Controls Entry of IκB Kinase into the Central Immune Synapse. *J. Exp. Med.* **2004**, *200*, 1167–1177. [CrossRef]

115. Yang, X.D.; Tajkhorshid, E.; Chen, L.F. Functional interplay between acetylation and methylation of the RelA subunit of NF-kappaB. *Mol. Cell. Biol.* **2010**, *30*, 2170–2180. [CrossRef]

116. Ghosh, S.; Karin, M. Missing pieces in the NF-kB puzzle. *Cell* **2002**, *109*, S81–S96. [CrossRef]

117. Hoesel, B.; Schmid, J.A. The complexity of NF-κB signaling in inflammation and cancer. *Mol. Cancer* **2013**, *12*, 1–15. [CrossRef]

118. Marinho, H.S.; Real, C.; Cyrne, L.; Soares, H.; Antunes, F. Hydrogen peroxide sensing, signaling and regulation of transcription factors. *Redox Biol.* **2014**, *2*, 535–562. [CrossRef] [PubMed]

119. Moreno, R.; Sobotzik, J.M.; Schultz, C.; Schmitz, M.L. Specification of the NF-κB transcriptional response by p65 phosphorylation and TNF-induced nuclear translocation of IKKε. *Nucleic Acids Res.* **2010**, *38*, 6029–6044. [CrossRef] [PubMed]

120. Heissmeyer, V.; Krappmann, D.; Hatada, E.N.; Scheidereit, C. Shared pathways of IκB kinase- inducedSCF(βTrCP)-mediated ubiquitination and degradation for the NF-κB precursor p105 and IκBα. *Mol. Cell. Biol.* **2001**, *21*, 1024–1035.

121. Häcker, H.; Karin, M. Regulation and Function of IKK and IKK-Related Kinases. *Science's STKE* **2006**, *13*. [CrossRef]

122. Gabler, C.; Fischer, C.; Drillich, M.; Einspanier, R.; Heuwieser, W. Time-dependent mRNA expression of selected pro-inflammatory factors in the endometrium of primiparous cows postpartum. *Reprod. Biol. Endocrinol.* **2010**, *8*, 152. [CrossRef]

123. Almeria, S.; Serrano, B.; Yaniz, J.L.; Darwich, L.; Lopez-Gatius, F. Cytokine gene expression profiles in peripheral blood mononuclear cells from Neospora caninum naturally infected dams throughout gestation. *Vet. Parasitol.* **2012**, *183*, 237–243. [CrossRef] [PubMed]

124. Oliveira, L.J.; Barreto, R.S.N.; Perecin, F.; Mansouri-Attia, N.; Pereira, F.T.V.; Meirelles, F.V. Modulation of Maternal Immune System During Pregnancy in the Cow. *Reprod. Domest. Anim.* **2012**, *47*, 384–393. [CrossRef] [PubMed]

125. Nasu, K.; Narahara, H. Pattern recognition via the toll-like receptor system in the human female genital tract. *Mediat. Inflamm.* **2010**, *2010*. [CrossRef] [PubMed]

126. Baravalle, M.E.; Stassi, A.F.; Velazquez, M.M.L.; Belotti, E.M.; Rodríguez, F.M.; Ortega, H.H.; Salvetti, N.R. Altered expression of pro-inflammatory cytokines in ovarian follicles of cows with the cystic ovarian disease. *J. Comp. Pathol.* **2015**, *153*, 116–130. [CrossRef] [PubMed]

127. Lee, H.S. Orostachys japonicus extract inhibits the lipopolysaccharide-induced pro-inflammatory factors by suppression of transcription factors. *Food Sci. Nutr.* **2020**, *8*, 1812–1817. [CrossRef]

128. Fengyang, L.; Yunhe, F.; Bo, L.; Zhicheng, L.; Depeng, L.; Dejie, L.; Zhengtao, Y. Steviosidesuppressed inflammatory cytokine secretion by downregulation of NF-kB and MAPK signaling pathways in LPS-Stimulated RAW264.7 cells. *Inflammation* **2012**, *35*, 1669–1675. [CrossRef] [PubMed]

129. Cui, L.; Wang, Y.; Wang, H.; Dong, J.; Li, Z.; Li, J. Different effects of cortisol on pro-inflammatory gene expressions in LPS-, heat-killed E. coli-, or live E. coli-stimulated bovine endometrial epithelial cells. *BMC Vet. Res.* **2020**, *16*, 1–8. [CrossRef]

130. Kelly, P.; Meade, K.G.; O'Farrelly, C. Non-canonical inflammasome-mediated IL-1β production by primary endometrial epithelial and stromal fibroblast cells is NLRP3 and caspase-4 dependent. *Front. Immunol.* **2019**, *10*, 1–17. [CrossRef]

131. Healy, L.L.; Cronin, J.G.; Sheldon, I.M. Endometrial cells sense and react to tissue damage during infection of the bovine endometrium via interleukin 1. *Sci. Rep.* **2014**, *4*, 7060. [CrossRef]

132. Healy, L.L.; Cronin, J.G.; Sheldon, I.M. Polarized epithelial cells secrete interleukin 6 apically in the bovine endometrium. *Biol. Reprod.* **2015**, *92*, 151-1. [CrossRef]

133. Brodzki, P.; Kostro, K.; Brodzki, A.; Wawron, W.; Marczuk, J.; Kurek, L. Inflammatory cytokines and acutephase proteins concentrations in the peripheral blood and uterus of cows that developed endometritis during early postpartum. *Theriogenology* **2015**, *84*, 11–18. [CrossRef]

134. Zhang, H.; Wu, Z.; Yang, Y.; Ping, M.; Shaukat, A.; Yang, J.; Guo, Y.F.; Zhang, T.; Zhu, X.Y.; Qiu, J.X.; et al. Catalpol ameliorates LPS-induced endometritis by inhibiting inflammation and TLR4/NF-κB signaling. *J. Zhejiang Univ. Sci. B* **2019**, *20*, 816–827. [CrossRef]

135. Clark, I.A.; Vissel, B. The meteorology of cytokine storms, and the clinical usefulness of this knowledge. *Semin. Immunopathol.* **2017**, *39*, 505–516. [CrossRef] [PubMed]

136. Oh, Y.C.; Yun, H.J.; Cho, W.K.; Gu, M.J.; Jin, Y.M. Inhibitory effects of palmultang on inflammatory mediator production related to the suppression of NF-κB and MAPK pathways and induction of HO-1 expression in macrophages. *Int. J. Mol. Sci.* **2014**, *15*, 8443–8457. [CrossRef] [PubMed]

137. Hashem, N.M.; Gonzalez-Bulnes, A. State-of-the-art and prospective of nanotechnologies for smart reproductive management of farm animals. *Animals* **2020**, *10*, 840. [CrossRef] [PubMed]

138. Esmaeili, F.; Hosseini-Nasr, M.; Rad-Malekshahi, M.; Samadi, N.; Atyabi, F.; Dinarvand, R. Preparation and antibacterial activity evaluation of rifampicin-loaded poly lactide-co-glycolide nanoparticles. *Nanomed. Nanotechnol. Biol. Med.* **2007**, *3*, 161–167. [CrossRef]

139. Hill, E.K.; Li, J. Current and future prospects for nanotechnology in animal production. *J. Anim. Sci. Biotechnol.* **2017**, *8*, 1–13. [CrossRef] [PubMed]

140. Underwood, C.; van Eps, A.W. Nanomedicine and veterinary science: The reality and the practicality. *Vet. J.* **2012**, *193*, 12–23. [CrossRef]

141. Kim, B.Y.S.; Rutka, J.T.; Chan, W.C.W. Nanomedicine. *N. Engl. J. Med.* **2010**, *363*, 2434–2443. [CrossRef]

142. Pritchard, N.; Kaitu'u-Lino, T.; Harris, L.; Tong, S.; Hannan, N. Nanoparticles in pregnancy: The next frontier in reproductive therapeutics. *Hum. Reprod. Update* **2020**, *27*, 280–304. [CrossRef]

143. Jung, W.K.; Koo, H.C.; Kim, K.W.; Shin, S.; Kim, S.H.; Park, Y.H. Antibacterial activity and mechanism of action of the silver ion in Staphylococcus aureus and Escherichia coli. *Appl. Environ. Microbiol.* **2008**, *74*, 2171–2178. [CrossRef]

144. Keelan, J.A.; Leong, J.W.; Ho, D.; Iyer, K.S. Therapeutic and safety considerations of nanoparticle-mediated drug delivery in pregnancy. *Nanomedicine* **2015**, *10*, 2229–2247. [CrossRef]

145. Morones, J.R.; Elechiguerra, J.L.; Camacho, A.; Holt, K.; Kouri, J.B.; Ramirez, J.T. The bactericidal effect of silver nanoparticles. *Nanotechnology* **2005**, *16*, 2346–2353. [CrossRef]

146. Rai, M.; Yadav, A.; Gade, A. Silver nanoparticles as a new generation of antimicrobials. *Biotechnol. Adv.* **2001**, *27*, 76–83. [CrossRef]

147. Kim, J.C.; Shin, H.C.; Cha, S.W.; Koh, W.S.; Chung, M.K.; Han, S.S. Evaluation of developmental toxicity in rats exposed to the environmental estrogen bisphenol A during pregnancy. *Life Sci.* **2001**, *69*, 2611–2625. [CrossRef]

148. Carvalho, S.G.; Araujo, V.H.S.; dos Santos, A.M.; Duarte, J.L.; Silvestre, A.L.P.; FonsecSantos, B.; Villanova, J.C.O.; Gremião, M.P.D.; Chorilli, M. Advances and challenges in nanocarriers and nanomedicines for veterinary application. *Int. J. Pharm.* **2020**, *580*, 119214. [CrossRef] [PubMed]

149. Cerbu, C.; Kah, M.; White, J.C.; Astete, C.E.; Sabliov, C.M. Fate of Biodegradable Engineered Nanoparticles Used in Veterinary Medicine as Delivery Systems from a One Health Perspective. *Molecules* **2021**, *26*, 523. [CrossRef]

150. Greatti, V.R.; Oda, F.; Sorrechia, R.; Kapp, B.R.; Seraphim, C.M.; Weckwerth, A.C.V.B.; Chorilli, M.; Da Silva, P.B.; Eloy, J.O.; Kogan, M.J.; et al. Poly-ε-caprolactone Nanoparticles Loaded with 4-Nerolidylcatechol (4-NC) for Growth Inhibition of Microsporum canis. *Antibiotics* **2020**, *9*, 894. [CrossRef]

151. Tu'uhevaha, J.; Kaitu'u-Lino, T.J.; Pattison, S.; Ye, L.; Tuohey, L.; Sluka, P.; MacDiarmid, J.; Brahmbhatt, H.; Johns, T.; Horne, A.W.; et al. Targeted nanoparticle delivery of doxorubicin into placental tissues to treat ectopic pregnancies. *Endocrinology* **2013**, *154*, 911–919. [CrossRef]

152. Youssef, F.S.; El-Banna, H.A.; Elzorba, H.Y.; Galal, A.M. Application of some nanoparticles in the field of veterinary medicine. *Int. J. Vet. Sci. Med.* **2019**, *7*, 78–93. [CrossRef] [PubMed]

153. Pamungkas, F.A.; Sianturi, R.S.G.; Wina, E.; Kusumaningrum, D.A. Chitosan nanoparticle of hCG (Human Chorionic Gonadotrophin) hormone in increasing induction of dairy cattle ovulation. *J. Ilmu Ternak Dan Vet.* **2016**, *21*, 34. [CrossRef]

154. Ebeid, K.; Meng, X.; Thiel, K.W.; Do, A.V.; Geary, S.M.; Morris, A.S.; Pham, E.L.; Wongrakpanich, A.; Chhonker, Y.S.; Murry, D.J.; et al. Synthetically lethal nanoparticles for treatment of endometrial cancer. *Nat. Nanotechnol.* **2018**, *13*, 72–81. [CrossRef]

155. Paul, J.W.; Hua, S.; Ilicic, M.; Tolosa, J.M.; Butler, T.; Robertson, S.; Smith, R. Drug delivery to the human and mouse uterus using immunoliposomes targeted to the oxytocin receptor. *Am. J. Obstet. Gynecol.* **2017**, *216*, 283.e1–283.e14. [CrossRef]

156. Hassanein, E.M.; Hashem, N.M.; El-Azrak, K.E.D.M.; Gonzalez-Bulnes, A.; Hassan, G.A.; Salem, M.H. Efficiency of GnRH–loaded chitosan nanoparticles for inducing LH secretion and fertile ovulations in protocols for artificial insemination in rabbits does. *Animals* **2021**, *11*, 440. [CrossRef]
157. Singh, A.K.; Chakravarty, B.; Chaudhury, K. Nanoparticle-assisted combinatorial therapy for effective treatment of endometriosis. *J. Biomed. Nanotechnol.* **2015**, *11*, 789–804. [CrossRef] [PubMed]
158. Zhang, P.; Dixon, M.; Zucchelli, M.; Hambiliki, F.; Levkov, L.; Hovatta, O.; Kere, J. Expression analysis of the NLRP gene family suggests a role in human preimplantation development. *PLoS ONE* **2008**, *3*, e2755. [CrossRef]
159. Zhang, T.; Zhao, G.; Zhu, X.; Jiang, K.; Wu, H.; Deng, G.; Qiu, C. Sodium selenite induces apoptosis ROS-mediated NF-κB signaling and activation of the Bax-caspase-9-caspase-3 axis in 4T1 cells. *J. Cell. Physiol.* **2019**, *234*, 2511–2522. [CrossRef]
160. Diwan, M.; Tafaghodi, M.; Samuel, J. Enhancement of immune responses by co-delivery of a CpG oligodeoxynucleotide and tetanus toxoid in biodegradable nanospheres. *J. Control. Release* **2002**, *85*, 247–262. [CrossRef]
161. Schlosser, E.; Mueller, M.; Fischer, S.; Basta, S.; Busch, D.H.; Gander, B.; Groettrup, M. TLR ligands and antigen need to be co-encapsulated into the same biodegradable microsphere for the generation of potent cytotoxic T lymphocyte responses. *Vaccine* **2008**, *26*, 1626–1637. [CrossRef] [PubMed]
162. Florindo, H.F.; Pandit, S.; Goncalves, L.M.; Videira, M.; Alpar, O.; Almeida, A.J. Antibody and cytokine-associated immune responses to S. equi antigens entrapped in PLA nanospheres. *Biomaterials* **2009**, *30*, 5161–5169. [CrossRef]
163. Rawat, M.; Singh, D.; Saraf, S.; Saraf, S. Nanocarriers: Promising Vehicle for Bioactive Drugs. *Biol. Pharm. Bull.* **2006**, *29*, 1790–1798. [CrossRef] [PubMed]
164. Riehemann, K.; Schneider, S.W.; Luger, T.A.; Godin, B.; Ferrari, M.; Fuchs, H. Nanomedicine–challenge and perspectives. *Angew. Chem. Int. Engl.* **2009**, *48*, 872–897. [CrossRef] [PubMed]
165. Mahapatro, A.; Singh, D.K. Biodegradable nanoparticles are an excellent vehicle for site-directed in-vivo delivery of drugs and vaccines. *J. Nanobiotechnol.* **2011**, *9*, 55. [CrossRef]

*animals*

MDPI

Article

Efficiency of GnRH–Loaded Chitosan Nanoparticles for Inducing LH Secretion and Fertile Ovulations in Protocols for Artificial Insemination in Rabbit Does

Eman M. Hassanein [1], Nesrein M. Hashem [1,*], Kheir El-Din M. El-Azrak [1], Antonio Gonzalez-Bulnes [2,3,*], Gamal A. Hassan [1] and Mohamed H. Salem [1]

[1] Animal and Fish Production Department, Faculty of Agriculture, Alexandria University, Alexandria 21545, Egypt; eman.mostafa1177@yahoo.com (E.M.H.); kheir_elazrak@yahoo.com (K.E.-D.M.E.-A.); gamaleldeen.hassan@alexu.edu.eg (G.A.H.); mhsalem15@yahoo.com (M.H.S.)

[2] Departamento de Reproduccion Animal, INIA, Avda. Puerta de Hierro s/n., 28040 Madrid, Spain

[3] Departamento de Produccion y Sanidad Animal, Facultad de Veterinaria, Universidad Cardenal Herrera-CEU, CEU Universities, C/ Tirant lo Blanc, 7, 46115 Alfara del Patriarca, Valencia, Spain

* Correspondence: nesreen.hashem@alexu.edu.eg (N.M.H.); antonio.gonzalezbulnes@uchceu.es (A.G.-B.)

Citation: Hassanein, E.M.; Hashem, N.M.; El-Azrak, K.E.-D.M.; Gonzalez-Bulnes, A.; Hassan, G.A.; Salem, M.H. Efficiency of GnRH–Loaded Chitosan Nanoparticles for Inducing LH Secretion and Fertile Ovulations in Protocols for Artificial Insemination in Rabbit Does. *Animals* 2021, 11, 440. https://doi.org/10.3390/ani11020440

Academic Editors: Juan Vicente and Bermejo Delgado

Received: 11 January 2021
Accepted: 2 February 2021
Published: 8 February 2021

Publisher's Note: MDPI stays neutral with regard to jurisdictional claims in published maps and institutional affiliations.

Simple Summary: Nano-drug delivery systems can be employed for improving ovulation induction prior to artificial insemination (AI) in rabbits. In this study, different routes of administration and different doses of GnRH–loaded chitosan nanoparticles (GnRH–ChNPs) were assessed for inducing ovulation in rabbits, proving their usefulness to reduce the GnRH dose and animal handling and improving AI outcomes. The use of GnRH–ChNPs allows for the reduction of the conventional intramuscular GnRH dose to half without compromising fertility. However, the addition of GnRH–ChNPs to semen extenders, although successfully inducing ovulation, has negative impacts on fertility. Thus, more studies are needed to allow further adjustments.

Abstract: Gonadotropin-releasing hormone (GnRH)–loaded chitosan nanoparticles (GnRH–ChNPs) were used at different doses and routes of administration to induce ovulation in rabbits as an attempt to improve artificial insemination (AI) procedures and outcomes. In this study, the characteristics (size, polydispersity, loading efficiency, and zeta-potential) of GnRH–ChNPs and the GnRH release pattern were determined in an in vitro study. A first in vivo study assessed the pituitary and ovarian response to different GnRH–ChNPs doses and routes of administration (two i.m. doses, Group HM = 0.4 µg and Group QM = 0.2 µg, and two intravaginal doses, Group HV = 4 µg and Group QV = 2 µg) against a control group (C) receiving bare GnRH (0.8 µg). The HM, QM, and HV treatments induced an earlier LH-surge (90 min) than that observed in group C (120 min), whilst the QV treatment failed to induce such LH surge. The number of ovulation points was similar among treatments, except for the QV treatment (no ovulation points). A second in vivo study was consequently developed to determine the hormonal (progesterone, P_4, and estradiol, E_2) profile and pregnancy outcomes of both HM and HV treatments against group C. The treatment HM, but not the treatment HV, showed adequate P_4 and E_2 concentrations, conception and parturition rates, litter size, litter weight, and viability rate at birth. Overall, the use of GnRH–ChNPs allows for a reduction in the conventional intramuscular GnRH dose to half without compromising fertility. However, the addition of GnRH–ChNPs to semen extenders, although successfully inducing ovulation, has negative impacts on fertility. Thus, more studies are needed to explore this point and allow further adjustments.

Keywords: artificial-insemination; GnRH; nanotechnology; ovulation; rabbit

---

1. Introduction

The occurrence of external stimulation during mating (neurohormonal reflex) is required to evoke the gonadotropin-releasing hormone (GnRH)-dependent luteinizing hormone (LH) surge, and consequently ovulation, in non-spontaneously ovulating species like the rabbit [1]. This biological event plays a crucial limiting role for the successful application of assisted reproductive techniques such as artificial insemination (AI) because, in the absence of the male, ovulation needs to be induced by artificial hormonal stimulation. Hence, the induction of ovulation in routine AI protocols is usually achieved by the administration of either GnRH or LH or other gonadotrophins with LH activity (i.e., equine chorionic gonadotrophin, eCG, or human chorionic gonadotrophin, hCG). The GnRH analogues (such as gonadorelin, lecirelin, triptorelin, or buserelin) are the most recommended compounds because they have additional biological advantages and the possibility of repeated treatments without developing specific antibodies [2].

Currently, GnRH is included in semen extenders to be directly administered to the doe via the seminal dose, which improves animal management and welfare [2]. Certainly, the successful implementation of the intravaginal route in protocols for ovulation induction may minimize animal distress and staff workload when compared to traditional protocols via intramuscular doses. However, the enzymatic proteolytic activity of seminal plasma and vaginal fluids and the biological barriers for mucosal permeation may limit the efficiency of such protocols.

In addition, GnRH has a short half-life time in blood circulation, 2–4 min, because it is rapidly degraded by peptidases and cleared by glomerular filtration [3], which limits both its biological activity and sustained action. Consequently, a first approach for improving the efficiency of the treatment is to increase the dose or the activity of the GnRH analogue. However, doses more than ten to fifteen-fold higher are needed (about 8 and 16 µg buserelin/seminal dose), which constitutes a potential health risk for workers and significantly increases the cost-efficiency of the protocols [4].

A promising alternative may be based on the use of nano-drug delivery systems. Nano-drug delivery systems have been optimized to extend the half-life time of the compound to improve its passage across endothelial or epithelial barriers into blood or lymph circulation and to sustain its delivery to the target sites. Consequently, overcoming these biological barriers can improve cellular uptake and therefore lower doses of drugs/hormones may be used [5–9]. Hence, nano-drug systems for hormonal treatments (e.g., GnRH for ovulation induction) may benefit many biotechnological assisted reproductive techniques in the livestock field (e.g., AI [5]). However, there is a scarcity of data regarding both hormone bioactivity and animal performance when GnRH is used in nano-formula, although previous results are promising. To date, chitosan-dextran sulfate GnRH (buserelin acetate) nanoparticles have been added to seminal doses of rabbits [10], allowing the GnRH dose to be reduced to half the conventional dose without affecting the fertility of the does. Moreover, the results of a trial performed with the administration of chitosan-tripolyphosphate (TPP)-conjugated GnRH nanoparticles in goats indicate that the GnRH dose may be even reduced three- to four-fold without affecting fertility and prolificacy [11].

The practical application of such protocols makes necessary extensive studies on technical and physiological features that may limit the efficiency and final outcomes of treatments involving GnRH-loaded nanoparticles (fabrication, route of administration, bioavailability and hormonal balance). Therefore, the present study aimed to set the basis and evaluate the feasibility and efficiency of GnRH nanoparticles for inducing ovulation in rabbits. Three consecutive experiments (first an in vitro approach and afterward, two in vivo trials) were performed to (1) manufacture and evaluate the physicochemical properties and release patterns of GnRH-loaded chitosan nanoparticles, and (2) determine the ovarian response, hormonal balance, and reproductive performance of does after ovulation induction with different doses and administration routes (intramuscular vs. intravaginal) of such GnRH nanoparticles.

2. Materials and Methods

2.1. Manufacturing of Gonadotropin-Releasing Hormone (GnRH)-Loaded Chitosan Nanoparticles

Chitosan (chitosan extra pure; with a degree of deacetylation >85%; molecular weight = 300–350 KD; Alpha Chemika, Mubia, India) and sodium tripolyphosphate (TPP; Thermo Fisher GmbH, Kandel, Germany) were used in the fabrication of a nano-carrier polymer following the steps of the ionic gelation method [10–12]. Briefly, chitosan (0.1%, wt/vol) was vigorously stirred in an aqueous acidic solution (1%, wt/vol) to obtain polymeric chitosan cations. An aqueous solution of TPP (0.1 g/dL) was also prepared. Afterward, the chitosan-TPP nanoparticles were prepared by slowly dropping the TPP solution into a chitosan solution (chitosan to TPP weight ratio was 2:1) under constant magnetic stirring (800 rpm) for 2 h at room temperature. The pH of the chitosan-TPP solution was adjusted to 5.5 and then stored in the refrigerator before exposure to any analysis or application. GnRH–loaded chitosan nanoparticles (GnRH–ChNPs) were prepared by dripping-wise the hormone (GnRH; Receptal®, MSD, Intervet International GmbH, Unterschleißheim, Germany) solution to the fabricated ChNPs in a ratio of 1:1. The mixture was adjusted to a pH of 6.5 and after that stirred gently at 800 rpm for 60 min at room temperature and incubated overnight to allow the hormonal adsorption on the surface of the nanoparticles.

2.2. In Vitro Assessment of the Physicochemical Properties and Release Patterns of GnRH-Loaded Nanoparticles

Particle size, zeta potential, and size distribution (polydispersity index, PdI) of the free ChNPs and GnRH–ChNPs were measured using a Zetasizer instrument (Malvern Instruments, Malvern, UK), which is based on dynamic light scattering (DLS) techniques. The samples were analyzed in triplicate and the average value ($\pm$SD) of each parameter was calculated. The functional groups of chitosan (Ch), ChNPs, and GnRH–ChNPs were identified by Fourier transform infrared spectrophotometer (FTIR, a Perkins Elmer 1600, USA) in the range from 4000 to 400/cm using potassium bromide discs (KBr) (5 mg of particles, 100 mg KBr pellets). Morphological characteristics of the free ChNPs and GnRH–ChNPs were examined using transmission electron microscope (TEM; JEOL JEM-1400, 120 kV, Peabody, MA, USA). Freshly fabricated ChNPs and GnRH–ChNPs solutions were diluted with deionized water with adjusted pH close to neutral. A one-drop sample was placed on a carbon-coated film 300 mesh copper grid and was left for 10 min until air-dried. The sample was stained with 1 M uranyl acetate solution for 1.5 min at 7 °C and any excess uranyl acetate was removed with filter paper before viewing on the TEM at magnification $\times$ 20,000.

The hormonal loading efficiency (LE%) was determined by the separation of the nanoparticles from the aqueous medium by centrifugation at 1200$\times$ g at 4 °C for 20 min. The amount of free hormone in the supernatant of GnRH–ChNPs solution was measured using UV spectrophotometry at a wavelength of 280 nm (Optizen Pop, Mecasys Co. Ltd., Daejeon, Korea) and using the supernatant of ChNPs solution as a blank [13]. Afterward, LE% was calculated as ([initial GnRH concentration − free GnRH concentration]/initial GnRH concentration $\times$ 100).

The assessment of release patterns was carried out for 48 h to plot the kinetic release profile of GnRH from ChNPs according to the method described by Dounighi and co-workers [13]. In brief, a pre-weighed sample of GnRH-ChNPs was divided into three aliquots and dissolved in equal volumes of phosphate buffer solution (PBS; pH 7.4). The samples were then incubated at 37 °C in a shaker adjusted at 600 rpm for 48 h. At allocated time intervals (0.5, 1, 2, 4, 6, 10, 12, 24, 48 h), the samples were removed and centrifuged at 14,000 rpm and 4 °C for 20 min. The concentration of free GnRH released into the supernatant was evaluated by spectrophotometry at the 280 nm wavelength (Optizen pop, Mecasys Co. Ltd. Daejeon, Korea). The amount of GnRH was calculated using non-loaded nanoparticles as a blank and then LE% was determined as previously described.

2.3. In vivo Assessment of the Reproductive Response of Does Treated with GnRH–Loaded Nanoparticles

2.3.1. Animals and Ethic Statement

A total of 62 female rabbits were used. The experiments were carried out at the Laboratory of Rabbit Physiology Research, Faculty of Agriculture, Alexandria University, Egypt (31°20′N, 30°E). The does were individually housed in standard wire cage batteries (40 × 50 × 35 cm) at the same rabbitry (10.97 ± 0.36 h daylight length, 19.85 ± 1.03 °C temperature and 74.92 ± 2.543% relative humidity). Does were fed ad-libitum with a pelleted diet, containing 17.73% crude protein and 68.96% TDN) and had free access to water. All animals were handled according to the principles of animal care published by the European Union Directive 2010/63/UE on the protection of animals used for research and, in agreement, the experimental procedures used were previously assessed and approved by the INIA Committee of Ethics in Animal Research (report CEEA2014/087).

2.3.2. Trial 1. Patterns of LH Secretion, Ovulatory Efficiency, and Early Fertility

Twenty adult nulliparous rabbit does (5–5.5 months-old and 2.82 ± 0.21 kg of body-weight) were treated for estrus synchronization with 25 IU of eCG (Gonaser®, Hipra, Girona, Spain) [14] and 48 h later, does were subjected to one of the following five ovulation induction treatments: a control conventional i.m. protocol (0.8 µg of GnRH in distilled water; C-treatment) and four protocols using GnRH–ChNPs with half i.m. dose (0.4 µg. GnRH–ChNPs; HM-treatment), quarter i.m. dose (0.2 µg i.m. GnRH–ChNPs; QM-treatment), half vaginal dose (4 µg vs. GnRH–ChNPs; HV-treatment), and quarter vaginal dose (2 µg vs. GnRH–ChNPs; QV-treatment). Females in C, HM, and HV treatments were immediately inseminated artificially with 0.3 mL diluted semen (30 × 10⁶ sperm/insemination). Females in HV and QV treatments were inseminated, at the same time, using the same seminal dose, but supplemented with the corresponding GnRH–ChNPs. Semen samples used for AI were collected from five fertile rabbit bucks by using an artificial vagina and a teaser doe. The quality of semen was evaluated, and only semen samples fulfilling quality criteria were used for AI [15]. The doses of GnRH–ChNPs were based on the findings of previous studies, where the dose of 0.8 µg GnRH is the conventional dose in intramuscular treatment and dose of 8 µg GnRH is the conventional dose in intravaginal treatment [4,11]. The schematic diagram of the experiment is presented in Figure 1.

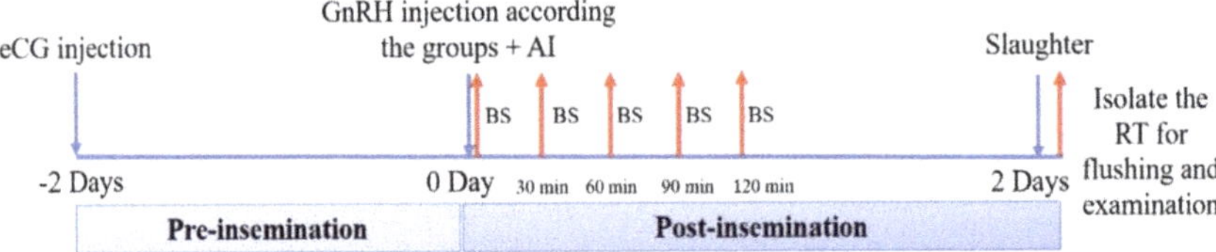

Figure 1. Experimental planning of trial 1, aiming to determine the effect of gonadotrophin-releasing hormone (GnRH) in different doses and routes of administration on LH secretion (by means of successive blood samples; BS) and presence of oocytes/embryos in the reproductive tract, RT, at day 2 after artificial insemination (AI).

The assessment of the patterns of LH secretion was performed on blood samples collected from the marginal ear vein of each doe with non-heparinized tubes at 0 (time of insemination) 30, 60, 90, and 120 min post-insemination. The samples were centrifuged at 700× g for 20 min to obtain serum, which was stored at −20 °C until analysis. Then, the serum LH concentration was determined using an enzyme immunoassay commercial kit (Cusabio Biotech, Hubei, China). The assay sensitivity was 0.35 mIU/mL and intra- and inter-assay coefficients of variation were 8% and 10%, respectively.

Afterward, all females were sacrificed 48 h post-insemination and the reproductive tracts were immediately obtained. The ovaries were dissected for determining the number of total ovarian follicles (clear visible follicles) and the number of ovulatory points (ovulated follicles), as previously described [14]. The oviducts were flushed with 15 mL of phosphate buffer solution (PBS) containing 0.2% of bovine serum albumin (BSA, Sigma-Aldrich, Madrid, Spain) to assess the number of embryos and their development stage and the number of non-fertilized oocytes with a binocular microscope (Olympus Optical Co. Ltd., Tokyo, Japan).

2.3.3. Trial 2. Fertility and Productive Traits

Forty-two adult nulliparous rabbit does (5–5.5 months-old and 2.41 ± 0.21 kg of body-weight) were treated for ovulation induction and AI as described in trial 1 but, according to the results of such experiment, only three treatments were used (C, HM, and HV; $n = 14$/treatment). The schematic diagram of the experiment is presented in Figure 2.

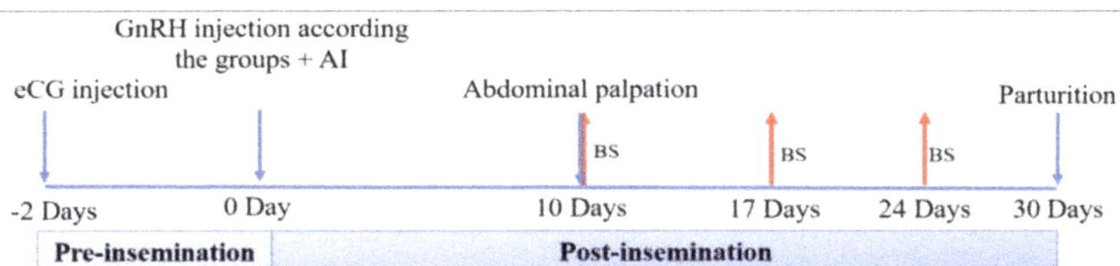

Figure 2. Experimental planning of trial 2, aiming to determine the effect of gonadotrophin-releasing hormone (GnRH) in different routes of administration on hormonal patterns (by means of successive blood samples; BS) and fertility and reproductive traits after artificial insemination (AI).

Pregnancy diagnosis was performed by abdominal palpation at Day 10 after AI. Blood samples were collected at days 10, 17, and 24 of pregnancy, as previously described, to determine the concentrations of progesterone (P_4) and estradiol (E_2) by using commercial enzyme immunoassay kits (Monobind Inc. Lake Forest, California, CA, USA.). The assay sensitivities were 0.105 ng/mL for P_4 and 8.2 pg/mL for E_2, while intra- and inter-assay coefficients of variation were 8.4% and 7.6% for P_4 and 8.6% and <5.6% for E_2.

Fertility and productive traits were recorded in terms of conception rate ([number of pregnant does on Day 10/total number of inseminated does] × 100), parturition rate ([number of delivering does/total number of inseminated does] × 100), abortion rate ([number of aborted does/number of pregnant does] × 100), litter size ([number of kits at birth/number of delivering does]) and viability rate ([number of live kits at birth/litter size at birth] × 100).

2.4. Statistical Analysis

Total numbers of visible and ovulated follicles, total numbers of oocytes, and embryos were subjected to square root transformation to approximate normal distribution before subjecting to ANOVA. Data observed once a time (results of the first trial 1) including weights of different reproductive organs, numbers of ovarian follicles, ovulation points, oocytes, and embryos were analyzed by a generalized linear model (GLM) of SAS [16]. The same procedure was used for litter size weight and viability variables in the second trial. Repeated measurements including concentrations of serum LH, P_4, and E_2 were analyzed using the MIXED procedure of SAS. The statistical model included the fixed effect of treatment, time of sampling/data collection, and the interactions as well as the random effect of an individual female were considered. Categorical data, expressed as percentages (conception, parturition, and abortion rates) were analyzed using a chi-square test (PROC FREQ). All results were presented as least square mean ± standard error (± S.E.M.). The statistically significant differences were accepted from $p < 0.05$).

3. Results

3.1. In Vitro Assessment of the Physicochemical Properties and Release Patterns of GnRH–Loaded Nanoparticles

Physicochemical characteristics and loading efficiency of GnRH–ChNPs are shown in Table 1. The loading efficiency of GnRH by ChNPs was 90% and the average size, PdI and zeta potential of ChNPs and GnRH–ChNPs were 95.19 ± 1.9 nm vs. 212 ± 2.69 nm, 0.165 vs. 0.295, and +34.0 vs. +8.0 mV, respectively. The FTIR showed that the Ch spectra exerted peaks at 3458.3 cm^{-1}, 2924.3 cm^{-1}, 1645.5 cm^{-1}, 1419.1 cm^{-1}, 1314.8 cm^{-1}, 1572.05 cm^{-1}, and 1077 cm^{-1}, which belong to the following functional groups: hydrogen-bond, O–H; C–H bond in pyranose rings; C=O in NHCOCH$_3$; C–H in CH$_2$OH; C–N stretching vibration of type II amine; N–H bond; and C–O–C in glucosidic-linkages. Addition of TPP to chitosan solution resulted in the disappearance of C–H and N–H functional groups, indicating binding of these groups with phosphate groups in TPP. Addition of GnRH to ChNPs resulted in shifts for the remaining functional groups identified in Ch spectra (Figure 3).

Table 1. Particle size, polydispersity index (PdI), zeta potential, and loading efficiency of chitosan nanoparticles (ChNPs) and GnRH-loaded chitosan nanoparticles (GnRH–ChNPs).

Nanoparticles.	Particle Size (nm)	PdI	Zeta Potential (mV)	Loading Efficiency (%)
ChNPs	95.19 ± 1.90	0.165	+34.0	-
GnRH–ChNPs	212 ± 2.69	0.295	+8.0	90

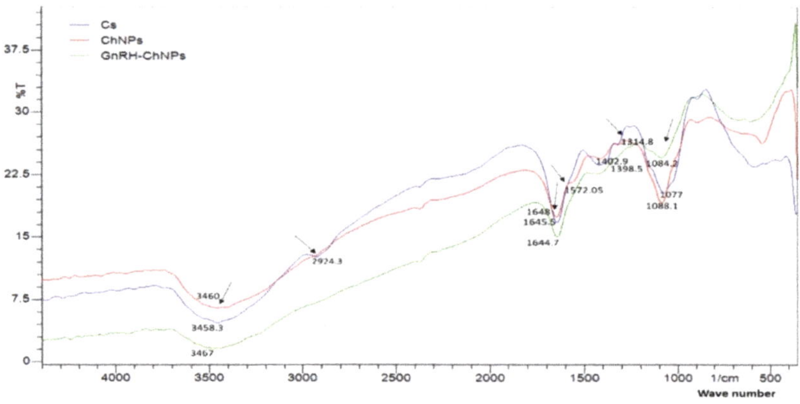

Figure 3. Fourier transform infrared (FTIR) spectra of chitosan (Ch), chitosan-TPP nanoparticles (ChNPs).

The images of ChNPs under the transmission electron microscope showed spherical nanoparticles with smooth surfaces and a diameter of about 100 nm and, after GnRH loading, the diameter of GnRH–ChNPs reached 200 nm with high agglomerating appearance (Figure 4).

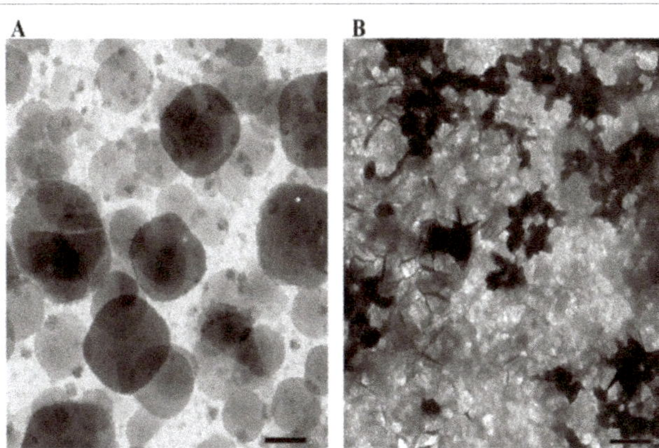

Figure 4. Transmission electron microscope image of (**A**) chitosan-TPP nanoparticles and (**B**) GnRH-loaded chitosan nanoparticles at a magnification × 20,000 and 25 °C.

The release profile of GnRH–ChNPs showed that the initial surge release of GnRH was about 50% at the first 10 h of incubation time, followed by a slow release of residual concentration during the subsequent 20 h. The release of the almost loaded hormone (about 99%) from ChNPs was after 30 h and remained steady up to 20 h later (Figure 5).

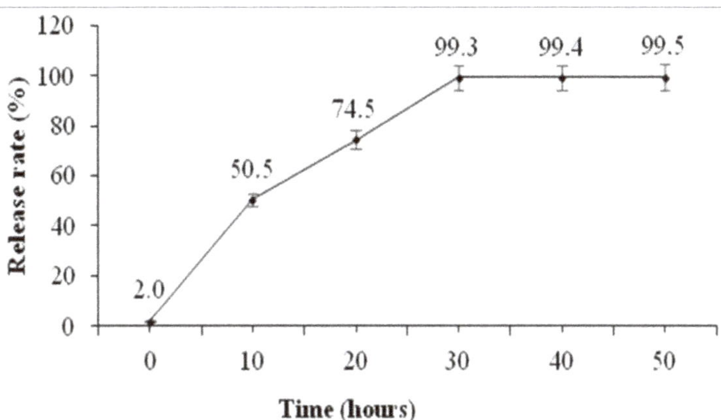

Figure 5. Mean (± S.E.M.) percentage of over-time release of gonadotropin-releasing hormone (GnRH) from GnRH-loaded chitosan nanoparticles.

3.2. In Vivo Assessment of Patterns of LH Secretion, Ovulatory Efficiency, and Early Fertility in Does Treated with GnRH-Loaded Nanoparticles

Figure 6 depicts the effect of different doses and routes of GnRH–ChNPs on serum LH concentrations at 0, 30, 60, 90, and 120 min. There were no significant effects on overall mean LH concentrations but on timing of the preovulatory LH surge. The groups HM, QM, and HV showed an earlier LH surge (90 min vs. 120 min post-insemination in the treatment C; $p < 0.05$), but the group QV failed to induce LH surge with LH concentrations remaining stable.

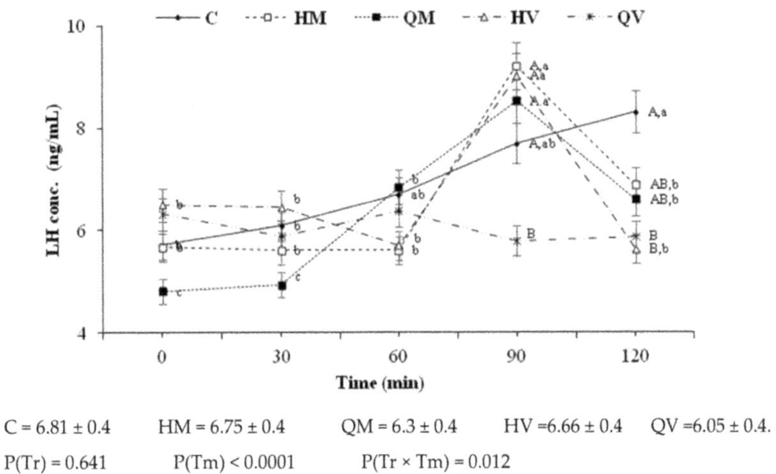

C = 6.81 ± 0.4   HM = 6.75 ± 0.4   QM = 6.3 ± 0.4   HV =6.66 ± 0.4   QV =6.05 ± 0.4.

P(Tr) = 0.641   P(Tm) < 0.0001   P(Tr × Tm) = 0.012

Figure 6. Changes over time in mean (±S.E.M.) serum LH concentrations in rabbit does receiving two i.m. doses (HM = 0.4 µg and QM = 0.2 µg) or two intravaginal doses (HV = 4 µg and QV = 2 µg) of GnRH–ChNPs or bare GnRH (C = 0.8 µg). Overall values are included below the figure. P(Tr) accounts for treatment effects, P(Tm) for time effects, and P(Tr × Tm) for their interactions. Different uppercase and lowercase superscript letters denote significant differences ($p < 0.05$) between or within treatments, respectively.

All groups (HM, QM, HV, and QV) significantly increased the number of total ovarian follicles when compared to group C ($p < 0.05$; Table 2). The number of total ovarian follicles was significantly lower in group HM, intermediate in groups QM and QV, and higher in group HV ($p < 0.05$). There were no significant differences in the number of ovulation points among groups, except for group QV, which showed no ovulation points.

Table 2. Ovarian structure and developmental stages of embryos collected 48 h post-insemination from rabbit does that received two intramuscular doses (HM = 0.4 µg and QM = 0.2 µg) or two intravaginal doses (HV = 4 µg and QV = 2 µg) of GnRH–ChNPs compared to control rabbit does that received bare GnRH (0.8 µg).

Variables	Groups (n = 4 Does/Group)					p-Value
	C	HM	QM	HV	QV	
	Ovarian structure (mean ± S.E.M.)					
Total ovarian follicles	15.5 ± 1.0 [c]	33.0 ± 3.6 [b]	36.5 ± 0.9 [ab]	40.2 ± 0.2 [a]	36.5 ± 2.0 [ab]	0.0001
Ovulation points	6.3 ± 0.9 [a]	9.3 ± 1.2 [a]	6.0±1.2 [a]	8.8 ± 1.2 [a]	0.00 ± 0.00 [b]	0.001
Total embryos	4.0 ± 0.7 [b]	6.2 ± 0.9 [a]	5.2±1.4 [ab]	8.0 ± 0.9 [a]	0.00 ± 0.00 [c]	0.001
Non-fertilized ova	0.25 ± 0.3	0.00 ± 0.0	0.00 ± 0.0	0.00 ± 0.0	0.00 ± 0.00	0.438
	Embryo developmental stages (%)					
Morula	50 [b](8/16)	48 [b](12/25)	76.2 [a](16/21)	25 [c](8/32)	-	0.004
Early blastocyst	25(4/16)	40(10/25)	23.8(5/21)	50(16/32)	-	0.175
Blastocyst	0(0/16)	12(3/25)	0(0/21)	0(0/32)	-	0.361
Expanded blastocyst	0 [b](0/16)	0 [b](0/16)	0 [b](0/21)	15.6 [a](5/32)	-	0.009
Hatching blastocyst	0 [b](0/16)	0 [b](0/16)	0 [b](0/21)	9.4 [a](3/32)	-	0.044

[a, b, c] Within rows, means with different superscripts differ at $p < 0.05$.

The number of recovered embryos was higher in groups HM and HV than in group C ($p < 0.05$), while group QM showed intermediate values and no embryos were collected at group QV. The development stages of embryos recovered 48 h post-insemination from does in groups C, HM, and QM were morula and early blastocyst stages. In contrast, embryo development was advanced in group HV since 15.6 and 9.4% of the embryos were expanded blastocyst and hatching blastocyst, respectively.

3.3. In Vivo Assessment of Fertility and Productive Traits in Does Treated with GnRH-Loaded Nanoparticles

The assessment in groups C, HM, and HV of the serum progesterone and estradiol concentrations (P_4 and E_2, respectively) at 10, 17, and 24 days of pregnancy (Figure 7) showed higher P_4 concentrations in group HM than in group HV ($p < 0.05$), with intermediate values in group C. The interaction of treatment-by-time showed no differences among groups at day 10 but, from days 17 to 24, P_4 concentrations decreased in group HV and increased in group HM ($p < 0.05$). The E_2 concentrations were also affected by a treatment-by-time interaction ($p < 0.05$). There were no effects on days 10 and 17, but the E_2 concentrations at day 24 were higher in group HM than in groups HV and C ($p < 0.05$). In fact, there were no changes over time in these groups whilst E_2 concentrations increased in group HM at day 24 ($p < 0.05$).

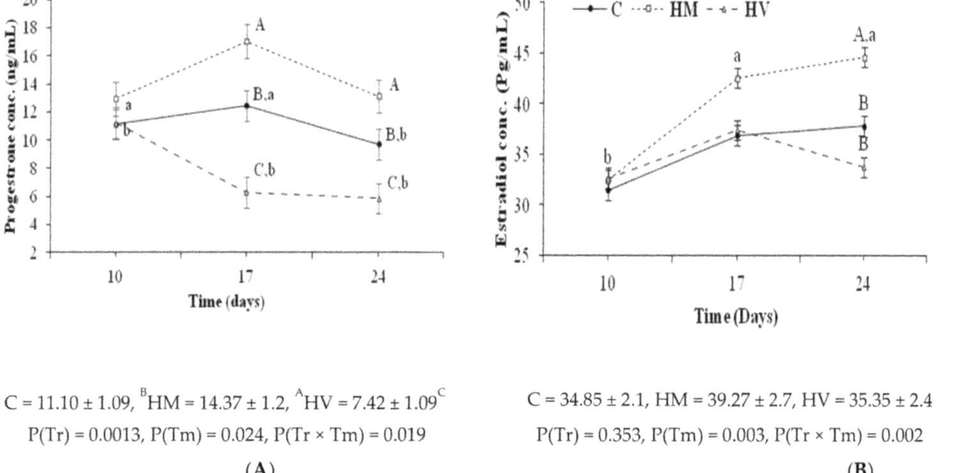

C = 11.10 ± 1.09, [B]HM = 14.37 ± 1.2, [A]HV = 7.42 ± 1.09[C]   C = 34.85 ± 2.1, HM = 39.27 ± 2.7, HV = 35.35 ± 2.4

P(Tr) = 0.0013, P(Tm) = 0.024, P(Tr × Tm) = 0.019   P(Tr) = 0.353, P(Tm) = 0.003, P(Tr × Tm) = 0.002

(**A**)   (**B**)

Figure 7. Changes over time in mean (±S.E.M.) serum progesterone (**A**) and E2 concentrations (**B**) in rabbit does receiving i.m. doses (HM = 0.4 µg) or intravaginal doses (HV = 4 µg) of GnRH–ChNPs or bare GnRH (C = 0.8 µg). Overall values are included below the figure. P(Tr) accounts for treatment effects, P(Tm) for time effects, and P(Tr × Tm) for their interactions. Different uppercase and lowercase superscripts denote significant differences ($p < 0.05$) between or within treatments, respectively.

Groups C and HM showed significantly higher conception and parturition rates than group HV and, without differences among groups in the abortion rate, also showed higher litter size, number of living litters, viability rate, and total litter body-weight at birth ($p < 0.05$; Table 3). In contrast, group HV showed a higher number of dead litter and mean body-weight at birth than groups C and HM.

Table 3. Productive traits of rabbit does receiving intramuscular (HM = 0.4 µg) or intravaginal doses (HV = 4 µg) of GnRH–ChNPs or bare GnRH (C = 0.8 µg).

Variables	Treatments (*n* = 14 Does/Group)			*p*-Value
	C	HM	HV	
Fertility evaluation parameters[1]				
Conception rate (%)	78.5 [a] (11/14)	71.5 [a] (10/14)	50 [b] (7/14)	0.019
Parturition rate (%)	78.5 [a] (11/14)	64.2 [a] (9/14)	35.7 [b] (5/14)	0.050
Abortion rate (%)	0.0 (0/11)	10.0 (1/10)	28.5 (2/7)	0.136
Pregnancy outcomes parameters (mean ± S.E.M.)[2]				
Litter size at birth	6.08 ± 0.70 [a]	5.47 ± 0.59 [a]	2.50 ± 0.54 [b]	0.001
No. of live litter	6.08 ± 0.70 [a]	5.47 ± 0.59 [a]	1.92 ± 0.63 [b]	0.001
No. of dead litter	0.00 ± 0.00 [b]	0.00 ± 0.00 [b]	0.54 ± 0.29 [a]	0.034
Viability rate at birth (%)	100 [a]	100 [a]	76.67 [b]	0.017
Litter weight at birth (g)	60.5 ± 3.4 [b]	66.0 ± 4.3 [ab]	76.0 ± 6.8 [a]	0.036
Total litter weight at birth (g)	344.2 ± 29.8 [a]	347.33 ± 39.2 [a]	177.3 ± 33.9 [b]	0.003

[a, b] Within rows, means with different superscripts differ at $p < 0.05$. [1] Conception rate %, (no. of pregnant does on Day 10/total no. of inseminated does × 100); parturition rate %, (no. of delivered does/no. of inseminated does × 100); and abortion rate %, (no. of aborted does/no. of pregnant doe × 100). [2] Litter size at birth (no. of kids at birth / no. of kindling does) and viability rate at birth % (no. of live kids at birth/litter size at birth × 100).

4. Discussion

The present study aimed to set the basis for implementing a nano-drug delivery system (i.e., GnRH-loading chitosan nanoparticles; GnRH–ChNPs) for improving procedures and yields of artificial insemination (AI) in rabbits by reducing hormone dose and facilitating animal handling during AI.

Achievement of these aims depends to a far extent on the physicochemical properties of the formula. The size of NPs is one of the most important determinants for passing barriers of mucosal tissues of and for the intracellular uptake [17]. The size range of most of the nanoparticles applied for the drug delivery system is between 50–250 nm. This size allows the particles to pass efficiently through different barriers and cell pores, and thus improves the cellular uptake [18,19]. The biological efficiency of NPs greatly depends on the potential of the nano-carrier to conjugate with loaded molecules. Increasing loading efficiency of active molecules within carriers improves its lifetime in the bloodstream due to the protection from enzymatic degradation. In the present study, chitosan (Ch):TPP with a 2:1 ratio resulted in 90% loading efficiency for GnRH (Table 1). In context, Hashem and Sallam [11] obtained 91.2% LE when the same Ch:TPP ratio was used for GnRH (gonadorelin) encapsulation. On the other hand, Rather et al. [19] obtained 69% LE for LH–RH when using the same Ch:TPP ratio in female fish.; Casares-Crespo et al. [10] also obtained only 43% LE when they used a Ch:dextran ratio of 4:1 for GnRH encapsulation. These results confirm the appropriateness of selected preparation conditions and the formula (Ch concentration 1 mg/mL and Ch:TPP ratio 2:1) to conjugate most of the loaded hormone used in the present study.

In the present study, the addition of GnRH to ChNPs increased the particle size from 95.19 ± 1.9 nm to 212 ± 2.69 nm and decreased the zeta potential value from +34.0 to +8.0 mV (Table 1). Similar results have been obtained in previous studies by Kumari et al. [20], who found that the addition of trypsin increased the size of ChNPs from 147 nm to ≈220 nm. On the other hand, Hashem and Sallam [11] obtained a lower GnRH–ChNPs size after loading the 93.91 ± 0.85 nm compound to non-loaded ChNPs of 125.9 ± 3.06 nm.

The variation in the size of nanoparticles after loading (increase or decrease) could be related to many factors including the type of loaded molecules, type of carrier molecules, the preparation conditions, and the surface charge (zeta potential). Kumari et al. [20] noted that the increased diameter of ChNPs after trypsin loading might be due to the size and the molecular weight of the enzyme during the loading process and adsorption on the

ChNP surface. However, in our study, the increase in GnRH–ChNP size could be ascribed to the occurrence of aggregation between GnRH–ChNPs. This suggestion is confirmed by the decreased values of zeta potential (Table 1) and image taken by transmission electronic microscope (TEM), which clearly showed the aggregation between GnRH–ChNPs (Figure 3B.). Zeta potential refers to the value of surface charge and is considered as an indicator of particle stability. Regardless of the surface charge (+ or −), high values of zeta potential (>30 mV) refer to high stability and repulsion of the particles in the solution because the high value of the electrical double-layer thickness prevents the aggregation between the particles. Thus, a low value of zeta potential accelerates the process of particle aggregation [21,22]. However, the increase in zeta potential values improves the stability of nanoparticles, this effect may not be suitable for cell viability. High zeta potential value, particularly the positive charge, stimulates cellular uptake, which affects cell survival [18,23].

The spectra of functional groups of Ch, ChNPs, and GnRH–ChNPs presented in Figure 4 were recorded in the region between 4000 cm^{-1} to 400 cm^{-1}. The Ch spectra showed peaks at 3458.3 cm^{-1}, 2924.3 cm^{-1}, 1645.5 cm^{-1}, 1419.1 cm^{-1}, 1314.8 cm^{-1}, 1572.05 cm^{-1}, and 1077 cm^{-1}, which belong to the following functional groups: hydrogen-bond, O–H; C–H bond in pyranose rings; C=O in $NHCOCH_3$; C–H in CH_2OH; C–N stretching vibration of type II amine; N–H band; and C–O–C in glucosidic-linkages. The addition of TPP to chitosan solution (ChNPs) resulted in the disappearance of C–H and N–H functional groups, indicating binding of these groups with phosphate groups in TPP. The addition of GnRH to ChNPs resulted in shifts for the remaining functional groups identified in the Ch spectra, which was confirmed in previous studies [19,24,25].

The results of the in vitro release study showed an initial surge release of GnRH from GnRH–ChNPs of around 50% during the first 10 h of incubation in phosphate buffer solution (PBS, pH = 7.4). This might be due to the weak bonds between adsorbed GnRH and the surface of ChNPs [26]. The other 50% of conjugated GnRH was released slowly through the incubation time (from 10 h to 30 h), which is due to the degradation of the GnRH–ChNPs' surface and the release of encapsulated GnRH [13]. In general, nanoparticles entrap, adsorb, bind, or encapsulate the active agents in their structure, depending on the preparation method [27]. The main use of the nano-formula innovated in the present study was the induction of ovulation in rabbit does. The preparation method used in our study seems to be suitable for this purpose, since the initial release of GnRH is required to stimulate the LH surge secretion, while the slow release maintains the GnRH concentration over a long period and this sustained the surge of gonadotrophins [28].

The assessment of the potential of the developed GnRH–ChNP formula for inducing the LH-surge and ovulation revealed the relevance of the HM, QM, and HV treatments. In these three treatments, a reduction from 25% to 50% in the conventional GnRH dose was achieved without negative effects on LH surge or ovulation rate. In contrast, the QV treatment failed to induce LH surge or ovulation. These results indicate the ability of the nano-formula to protect GnRH from enzymatic degradation [20], improving the cellular uptake and thus greater bioavailability.

It is interesting to note that HV treatment advanced the pre-implanted embryo development, as 25% of the collected embryos at 48 h post-treatment were beyond the blastocyst stage whilst morula and early blastocyst were the common stages in the other groups (C, HM, and QM). In the present study, the loading efficiency of ChNPs for GnRH reached 90%. Thus, intravaginal administration of GnRH–ChNPs may result in the sustained and high release of GnRH into the reproductive tract. This suggestion is highly acceptable since it is known that the mucus layer of the reproductive tract has a surface negative charge, whereas the GnRH–ChNPs developed in our study had a positive charge (+8.0 mV). This difference in charge might increase the adhesion between the GnRH–ChNPs and the mucosal layer of the reproductive tract, allowing sustained GnRH release [11,29]. The sustained release of GnRH for a longer time may extend the ovulation time, leading to asynchrony of ova shed time. The variation in ovulation time might have resulted in the occurrence of fertilization for a longer period, and consequently a variation in embryo developmental stages. In context, it is worthy to note that GnRH is synthesized by preimplantation embryos in many species including rabbits [30]. Furthermore, GnRH has a pivotal role in the pre-implantation division and implantation of the embryo. Nam et al. [31] found that pre-implantation embryonic development could be enhanced by incubation with increasing concentration of the GnRH agonist, while GnRH antagonist had a deleterious effect. Thus, intravaginal deposition of GnRH may facilitate a GnRH binding with its receptors on preimplantation embryos, affecting their developmental stage.

Analysis of progesterone (P_4) and estradiol (E_2) throughout pregnancy revealed a clear reduction in P_4 concentrations in group HV compared to groups C and HM. In rabbits, follicular and placental E_2 is the main luteotropic factor supporting CL maintenance [15,32]. This explains the obvious increase in serum P_4 concentration in group HM, as E_2 concentration was the highest among groups. On the other hand, the concentrations of E_2 showed a similar trend in groups C and HV, which indicates the implication of other factors in the maintenance of CL and P_4 concentration. In rabbits, the pre-implantation embryo can synthesize chorionic gonadotrophins, which play a role in the maternal embryo interface [33]. The failure of this process leads to luteolysis and early embryonic loss [34]. In the present study, a proportion of the embryos, about 25%, of group HV had advanced developmental stages, while the rest of the embryos were at 48 h at the typical developmental stage. The lack of synchrony between advanced developed embryos and normally developed embryos might drive to a disturbance in the maternal-embryonic cross talking, leading to luteolysis, P_4 reduction, and increased implantation failure and embryonic loss. Such hormonal and embryonic development disturbances may explain the lowest conception and parturition rates and litter size of group HV.

5. Conclusions

The fabrication of GnRH–ChNPs allows for a reduction in the conventional intramuscular GnRH dose used for AI in rabbits to half without affecting fertility. Conversely, the addition of GnRH–ChNPs to semen extenders, although successfully inducing ovulation, has negative impacts on fertility. Thus, more studies are needed to explore this point and allow further adjustments.

Author Contributions: Conceptualization, N.M.H. and E.M.H.; Methodology, E.M.H., N.M.H., and K.E.-D.M.E.-A.; Formal analysis and data curation, N.M.H. and E.M.H.; Writing—original draft preparation, E.M.H.; Writing—review and editing, N.M.H. and A.G.-B.; Supervision, N.M.H., K.E.-D.M.E.-A., G.A.H., and M.H.S. All authors have read and agreed to the published version of the manuscript.

Funding: This research received no external funding.

Institutional Review Board Statement: The study was conducted according to the principles of animal care published by the European Union Directive 2010/63/UE on the protection of animals used for research and, in agreement, the experimental procedures used were previously assessed and approved by the INIA Committee of Ethics in Animal Research (report CEEA2014/087).

Conflicts of Interest: The authors declare no conflict of interest.

References

1. Boiti, C. Underlying physiological mechanisms controlling the reproductive axis of rabbit does. In Proceedings of the Proc.: 8th World Rabbit Congress, Puebla, Mexico, 7–10 September 2004; pp. 7–10.
2. Quintela, L.A.; Peña, A.I.; Vega, M.D.; Gullón, J.; Prieto, C.; Barrio, M.; Becerra, J.J.; García-Herradón, P.J. Reproductive Performance of Rabbit Does Artificially Inseminated via Intravaginal Administration of [des-Gly 10, d-Ala6]-LHRH Ethylamide as Ovulation Inductor. *Reprod. Domest. Anim.* **2009**, *44*, 829–833. [CrossRef] [PubMed]
3. Kumar, P.; Sharma, A. Gonadotropin-releasing hormone analogs: Understanding advantages and limitations. *J. Hum. Reprod. Sci.* **2014**, *7*, 170. [CrossRef]
4. Quintela, L.A.; Peña, A.I.; Vega, M.D.; Gullón, J.; Prieto, M.C.; Barrio, M.; Becerra, J.J.; Maseda, F.; García-Herradón, P.J. Ovulation induction in rabbit does submitted to artificial insemination by adding buserelin to the seminal dose. *Reprod. Nutr. Dev.* **2004**, *44*, 79–88. [CrossRef] [PubMed]

5. Hashem, N.M.; Gonzalez-Bulnes, A. State-of-the-Art and Prospective of Nanotechnologies for Smart Reproductive Management of Farm Animals. *Anim.* **2020**, *10*, 840. [CrossRef]

6. Fakruddin, M.; Hossain, Z.; Afroz, H. Prospects and applications of nanobiotechnology: A medical perspective. *J. Nanobiotechnology* **2012**, *10*, 31. [CrossRef]

7. Rather, M.A.; Sharma, R.; Gupta, S.; Ferosekhan, S.; Ramya, V.L.; Jadhao, S.B. Chitosan-Nanoconjugated Hormone Nanoparticles for Sustained Surge of Gonadotropins and Enhanced Reproductive Output in Female Fish. *PLoS One* **2013**, *8*, e57094. [CrossRef]

8. Farokhzad, O.C.; Langer, R. Impact of Nanotechnology on Drug Delivery. *ACS Nano* **2008**, *3*, 16–20. [CrossRef]

9. Esquivel, R.; Juárez, J.; Almada, M.; Ibarra, J.; Valdez, M.A. Synthesis and Characterization of New Thiolated Chitosan Nanoparticles Obtained by Ionic Gelation Method. *Int. J. Polym. Sci.* **2015**, *2015*, 1–18. [CrossRef]

10. Casares-Crespo, L.; Fernández-Serrano, P.; Viudes-De-Castro, M.P. Protection of GnRH analogue by chitosan-dextran sulfate nanoparticles for intravaginal application in rabbit artificial insemination. *Theriogenology* **2018**, *116*, 49–52. [CrossRef] [PubMed]

11. Hashem, N.M.; Sallam, S. Reproductive performance of goats treated with free gonadorelin or nanoconjugated gonadorelin at estrus. *Domest. Anim. Endocrinol.* **2020**, *71*, 106390. [CrossRef]

12. Hassanein, E.M.; Hashem, N.M.; El-Azrak, K.M.; Hassan, G.A.; Salem, M.H. Effect of GnRH–loaded chitosan-TPP nanoparticles on ovulation induction and embryo recovery rate using artificial insemination in. In Proceedings of the 10th International Poultry Conference, Sharm El-Sheikh, Egypt, 26–29 November 2018.

13. Dounighi, N.M.; Eskandari, R.; Avadi; Zolfagharian, H.; Sadeghi, A.M.M.; Rezayat, M. Preparation and in vitro characterization of chitosan nanoparticles containing Mesobuthus eupeus scorpion venom as an antigen delivery system. *J. Venom. Anim. Toxins Incl. Trop. Dis.* **2012**, *18*, 44–52. [CrossRef]

14. Hashem, N.M.; Aboul-Ezz, Z.R. Effects of a single administration of different gonadotropins on day 7 post-insemination on pregnancy outcomes of rabbit does. *Theriogenology* **2018**, *105*, 1–6. [CrossRef]

15. Hashem, N.M.; Abo-Elsoud, M.; El-Din, A.N.; Kamel, K.; Hassan, G. Prolonged exposure of dietary phytoestrogens on semen characteristics and reproductive performance of rabbit bucks. *Domest. Anim. Endocrinol.* **2018**, *64*, 84–92. [CrossRef]

16. *Statistical Analysis Systems*, version 9.1; SAS Institute Inc.: Cary, NC, USA, 2004.

17. Panyam, J.; Labhasetwar, V. Biodegradable nanoparticles for drug and gene delivery to cells and tissue. *Adv. Drug Deliv. Rev.* **2012**, *64*, 61–71. [CrossRef]

18. Danaei, M.; Dehghankhold, M.; Ataei, S.; Davarani, F.H.; Javanmard, R.; Dokhani, A.; Khorasani, S.; Mozafari, M.R. Impact of Particle Size and Polydispersity Index on the Clinical Applications of Lipidic Nanocarrier Systems. *Pharm.* **2018**, *10*, 57. [CrossRef]

19. Gan, Q.; Wang, T.; Cochrane, C.; McCarron, P. Modulation of surface charge, particle size and morphological properties of chitosan–TPP nanoparticles intended for gene delivery. *Colloids Surfaces B: Biointerfaces* **2005**, *44*, 65–73. [CrossRef]

20. Kumari, R.; Gupta, S.; Singh, A.R.; Ferosekhan, S.; Kothari, D.C.; Pal, A.K.; Jadhao, S.B. Chitosan Nanoencapsulated Exogenous Trypsin Biomimics Zymogen-Like Enzyme in Fish Gastrointestinal Tract. *PLoS ONE* **2013**, *8*, e74743. [CrossRef]

21. Agarwal, M.; Agarwal, M.K.; Shrivastav, N.; Pandey, S.; Das, R.; Gaur, P. Preparation of Chitosan Nanoparticles and their In-vitro Characterization. *Int. J. Life-Sciences Sci. Res.* **2018**, *4*, 1713–1720. [CrossRef]

22. Honary, S.; Zahir, F. Effect of zeta potential on the properties of nano-drug delivery systems-a review (Part 1). *Trop. J. Pharm. Res.* **2013**, *12*, 255–264.

23. Barbari, G.R.; Dorkoosh, F.A.; Amini, M.; Sharifzadeh, M.; Atyabi, F.; Balalaie, S.; Tehrani, N.R.; Tehrani, M.R. A novel nanoemulsion-based method to produce ultrasmall, water-dispersible nanoparticles from chitosan, surface modified with cell-penetrating peptide for oral delivery of proteins and peptides. *Int. J. Nanomed.* **2017**, *12*. [CrossRef]

24. Loutfy, S.A.; Alam El-Din, H.M.; Elberry, M.H.; Allam, N.G.; Hasanin, M.T.M.; Abdellah, A.M. Synthesis, characterization and cytotoxic evaluation of chitosan nanoparticles: In vitro liver cancer model. *Adv. Nat. Sci. Nanosci. Nanotechnol.* **2016**, *7*, 035008. [CrossRef]

25. Badawy, M.E.I.; El-Nouby, M.A.M.; Marei, A.E.-S.M. Development of a Solid-Phase Extraction (SPE) Cartridge Based on Chitosan-Metal Oxide Nanoparticles (Ch-MO NPs) for Extraction of Pesticides from Water and Determination by HPLC. *Int. J. Anal. Chem.* **2018**, *2018*. [CrossRef]

26. Amidi, M.; Romeijn, S.G.; Borchard, G.; Junginger, H.E.; Hennink, W.E.; Jiskoot, W. Preparation and characterization of protein-loaded N-trimethyl chitosan nanoparticles as nasal delivery system. *J. Control. Release* **2006**, *111*, 107–116. [CrossRef]

27. Lockman, P.R.; Mumper, R.J.; Khan, M.A.; Allen, D.D. Nanoparticle Technology for Drug Delivery Across the Blood-Brain Barrier. *Drug Dev. Ind. Pharm.* **2002**, *28*, 1–13. [CrossRef]

28. Mora-Huertas, C.E.; Fessi, H.; Elaissari, A. Influence of process and formulation parameters on the formation of submicron particles by solvent displacement and emulsification–diffusion methods: Critical comparison. *Adv. Colloid Interface Sci.* **2011**, *163*, 90–122. [CrossRef] [PubMed]

29. Cone, R.A. Barrier properties of mucus. *Adv. Drug Deliv. Rev.* **2009**, *61*, 75–85. [CrossRef]

30. Herrler, A.; Von Rango, U.; Beier, H.M. Embryo-maternal signalling: How the embryo starts talking to its mother to accomplish implantation. *Reprod. Biomed. Online* **2003**, *6*, 244–256. [CrossRef]

31. Nam, D.H.; Lee, S.H.; Kim, H.S.; Lee, G.S.; Jeong, Y.W.; Kim, S.; Kim, J.H.; Kang, S.K.; Lee, B.C.; Hwang, W.S. The role of gonadotropin-releasing hormone (GnRH) and its receptor in development of porcine preimplantation embryos derived from in vitro fertilization. *Theriogenology* **2005**, *63*, 190–201. [CrossRef]

32. Keyes, P.L.; Wiltbank, M.C. Endocrine Regulation of the Corpus Luteum. *Annu. Rev. Physiol.* **1988**, *50*, 465–482. [CrossRef]

33. Khan-Dawood, F.S.; Dawood, M. Chorionic gonadotropin receptors and immunoreactive chorionic gonadotropin in implantation of the rabbit blastocyst. *Am. J. Obstet. Gynecol.* **1984**, *148*, 359–365. [CrossRef]
34. Inskeep, E.K. Time-Dependent Embryotoxicity of the Endogenous Luteolysin Prostaglandin F2a in Ruminants. In *Endocrine Toxicology*, 3rd ed.; Informa UK Limited: London, UK, 2016; pp. 254–269.

MDPI
St. Alban-Anlage 66
4052 Basel
Switzerland
Tel. +41 61 683 77 34
Fax +41 61 302 89 18
www.mdpi.com

Animals Editorial Office
E-mail: animals@mdpi.com
www.mdpi.com/journal/animals